DATA HANDLING IN SCIENCE AND TECHNOLOGY–VOLUME 23

Nature-inspired Methods in Chemometrics: Genetic Algorithms and Artificial Neural Networks

DATA HANDLING IN SCIENCE AND TECHNOLOGY

Advisory Editors: B.G.M. Vandeginste and S.C. Rutan

Other volumes in this series:

Volume 1 Microprocessor Programming and Applications for Scientists and Engineers, by R.R. Smardzewski
Volume 2 Chemometrics: A Textbook, by D.L. Massart, B.G.M. Vandeginste, S.N. Deming, Y. Michotte, and L. Kaufman
Volume 3 Experimental Design: A Chemometric Approach, by S.N. Deming and S.L. Morgan
Volume 4 Advanced Scientific Computing in BASIC with Applications in Chemistry, Biology and Pharmacology, by P. Valkó and S. Vajda
Volume 5 PCs for Chemists, edited by J. Zupan
Volume 6 Scientific Computing and Automation (Europe) 1990, *Proceedings of the Scientific Computing and Automation (Europe) Conference, 12–15 June 1990, Maastricht, The Netherlands,* edited by E.J. Karjalainen
Volume 7 Receptor Modeling for Air Quality Management, edited by P.K. Hopke
Volume 8 Design and Optimization in Organic Synthesis, by R. Carlson
Volume 9 Multivariate Pattern Recognition in Chemometrics, illustrated by case studies, edited by R.G. Brereton
Volume 10 Sampling of Heterogeneous and Dynamic Material Systems: Theories of Heterogeneity, Sampling and Homogenizing, by P.M. Gy
Volume 11 Experimental Design: A Chemometric Approach (Second, Revised and Expanded Edition) by S.N. Deming and S.L. Morgan
Volume 12 Methods for Experimental Design: Principles and Applications for Physicists and Chemists, by J.L. Goupy
Volume 13 Intelligent Software for Chemical Analysis, edited by L.M.C. Buydens and P.J. Schoenmakers
Volume 14 The Data Analysis Handbook, by I.E. Frank and R. Todeschini
Volume 15 Adaption of Simulated Annealing to Chemical Optimization Problems, edited by J. Kalivas
Volume 16 Multivariate Analysis of Data in Sensory Science, edited by T. Næs and E. Risvik
Volume 17 Data Analysis for Hyphenated Techniques, by E.J. Karjalainen and U.P. Karjalainen
Volume 18 Signal Treatment and Signal Analysis in NMR, edited by D.N. Rutledge
Volume 19 Robustness of Analytical Chemical Methods and Pharmaceutical Technological Products, edited by M.W.B. Hendriks, J.H. de Boer, and A.K. Smilde
Volume 20A Handbook of Chemometrics and Qualimetrics: Part A, by D.L. Massart, B.G.M. Vandeginste, L.M.C. Buydens, S. de Jong, P.J. Lewi, and J. Smeyers-Verbeke
Volume 20B Handbook of Chemometrics and Qualimetrics: Part B, by B.G.M. Vandeginste, D.L. Massart, L.M.C. Buydens, S. de Jong, P.J. Lewi, and J. Smeyers-Verbeke
Volume 21 Data Analysis and Signal Processing in Chromatography, by A. Felinger
Volume 22 Wavelets in Chemistry, edited by B. Walczak
Volume 23 Nature-inspired Methods in Chemometrics: Genetic Algorithms and Artificial Neural Networks, edited by R. Leardi

DATA HANDLING IN SCIENCE AND TECHNOLOGY–VOLUME 23
Advisory Editors: B.G.M. Vandeginste and S.C. Rutan

Nature-inspired Methods in Chemometrics: Genetic Algorithms and Artificial Neural Networks

edited by

R. Leardi

*Department of Pharmaceutical and Food Chemistry and Technology,
University of Genova, Genova, Italy*

2003

ELSEVIER

Amsterdam–Boston–Heidelberg–London–New York–Oxford
Paris–San Diego–San Francisco–Singapore–Sydney–Tokyo

ELSEVIER B.V.
Sara Burgerhartstraat 25
P.O. Box 211, 1000 AE Amsterdam, The Netherlands

First edition 2003

Library of Congress Cataloging in Publication Data
A catalog record from the Library of Congress has been applied for.

British Library Cataloguing in Publication Data
A catalogue record from the British Library has been applied for.

ISBN: 0-444-51350-7
ISSN: 0922-3487 (Series)

∞ The paper used in this publication meets the requirements of ANSI/NISO Z39.48-1992 (Permanence of Paper).
Printed in Hungary.

To Marta and Paolo

PREFACE

In recent years Genetic Algorithms (GA) and Artificial Neural Networks (ANN) have progressively increased their importance among the techniques routinely used in chemometrics. This is mainly due to the increases in computing power, which makes it possible to perform calculations on a PC that previously required a very powerful mainframe. In addition, these methods are quite appealing since they are inspired by biological phenomena: the evolution of a species in the case of GA and how our brain learns in the case of ANN.

A third possible explaination of the increased usage of these two methods is that they are very much ''en vogue''. A consequence of this is that many papers are published in which a GA or ANN is used instead of much easier and simpler techniques because the authors think that the paper is more attractive if GA or ANN is among the keywords. This decision is generally independent of the fact that the application of these complex techniques was indeed required by the complexity of their problem and they produced a significantly better result than would have been obtained by simpler techniques.

The proper application of GA and ANN methodologies requires some expertise. They can be quite ''dangerous'' when used by people who are attracted by their theory and who use the first algorithm they can buy or download from the net. As a result, many papers containing gross errors can be found, especially concerning the correct validation of results.

This book contains a number of contributions from chemometricians for whom GA and/or ANN are their main research field. It is divided into two sections (GA and ANN) and each section starts with a tutorial chapter in which the theoretical basis of the technique will be thoroughly (but simply) described. These background chapters are followed by chapters describing the application of the methodology to real problems covering a wide range of interests. In the application chapters, special emphasis is given to the advantages of using GA or ANN to that specific problem, as compared to classical techniques, and also to the risks associated with misuse.

Therefore, the book could be recommended to all people who are using or are interested in GA and ANN. Beginners can focus their attention mainly on the tutorial part, while the most advanced readers should be more interested in how these techniques have been used to solve complex problems. The breadth of coverage of these two methodologies also means that the book can serve as a reference for students.

All those whose work has contributed to preparing this book are acknowledged and their efforts greatly appreciated, especially Brian Luke for correcting the English of some of the chapters.

R. Leardi
Genoa, July 2003

CONTENTS

PREFACE vii

LIST OF CONTRIBUTORS xvii

PART I: GENETIC ALGORITHMS 1

CHAPTER 1 3
GENETIC ALGORITHMS AND BEYOND
(Brian T. Luke)

1 Introduction 3
2 Biological systems and the simple genetic algorithm (SGA) 5
3 Why do GAs work? 6
4 Creating a genetic algorithm 7
4.1 Determining a fitness function 7
4.2 The genetic vector 8
4.3 Creating an initial population 13
4.4 Selection schemes 14
4.5 Mating operators 16
4.6 Mutation operators 23
4.7 Maturation operators 25
4.8 Processing offspring 26
4.9 Termination metrics 27
5 Exploration versus exploitation 28
5.1 The genetic vector 29
5.2 The initial population 30
5.3 Selection schemes 31
5.4 Mating operators 33
5.5 Mutation operators 34
5.6 Maturation operators 34
5.7 Processing offspring 34
5.8 Balancing exploration and exploitation 36
6 Other population-based methods 40
6.1 Parallel GA 41
6.2 Adaptive parallel GA 41
6.3 Meta-GA 42
6.4 Messy GA 42
6.5 Delta coding GA 43
6.6 Tabu search and Gibbs sampling 43
6.7 Evolutionary programming 44
6.8 Evolution strategies 44

6.9 Ant colony optimization 45
6.10 Particle swarm optimization 46
7 Conclusions 48

CHAPTER 2 55
HYBRID GENETIC ALGORITHMS
(D. Brynn Hibbert)
1 Introduction 55
2 The approach to hybridization 55
2.1 Levels of interaction 56
2.2 A simple classification 57
3 Why hybridize? 57
4 Detailed examples 59
4.1 Genetic algorithm with local optimizer 59
4.2 Genetic algorithm–artificial neural network hybrid optimizing quantitative structure–activity relationships 62
4.3 Non-linear partial least squares regression with optimization of the inner relation function by a genetic algorithm 63
4.4 The use of a clustering algorithm in a genetic algorithm 64
5 Conclusion 66

CHAPTER 3 69
ROBUST SOFT SENSOR DEVELOPMENT USING GENETIC PROGRAMMING
(Arthur K. Kordon, Guido F. Smits, Alex N. Kalos, and Elsa M. Jordaan)
1 Introduction 69
2 Soft sensors in industry 71
2.1 Assumptions for soft sensors development 72
2.2 Economic benefits from soft sensors 73
2.3 Soft sensor application areas 74
2.4 Soft sensor vendors 75
3 Requirements for robust soft sensors 76
3.1 Lessons from industrial applications 76
3.2 Design requirements for robust soft sensors 77
4 Selected approaches for effective soft sensors development 80
4.1 Stacked analytical neural networks 80
4.2 Support vector machines 85
5 Genetic programming in soft sensors development 90
5.1 The nature of genetic programming 90
5.2 Solving problems with genetic programming 96
5.3 Advantages of genetic programming in soft sensors development and implementation 98

6 Integrated methodology 99
6.1 Variable selection by analytical neural networks 100
6.2 Data condensation by support vector machines 101
6.3 Inferential model generation by genetic programming 102
6.4 On-line implementation and model self-assessment 102
7 Soft sensor for emission estimation: a case study 103
8 Conclusions 105

CHAPTER 4 109
GENETIC ALGORITHMS IN MOLECULAR MODELLING: A REVIEW
(Alessandro Maiocchi)
1 Introduction 109
2 Molecular modelling and genetic algorithms 110
2.1 How to represent molecular structures and their conformations 111
3 Small and medium-sized molecule conformational search 114
4 Constrained conformational space searches 119
4.1 NMR-derived distance constraints 120
4.2 Pharmacophore-derived constraints 121
4.3 Constrained conformational search by chemical feature superposition 122
5 The protein-ligand docking problem 124
5.1 The scoring functions 126
5.2 Protein–ligand docking with genetic algorithms 127
6 Protein structure prediction with genetic algorithms 131
7 Conclusions 134

CHAPTER 5 141
MOBYDIGS: SOFTWARE FOR REGRESSION AND CLASSIFICATION MODELS BY GENETIC ALGORITHMS
(Roberto Todeschini, Viviana Consonni, Andrea Mauri and Manuela Pavan)
1 Introduction 141
2 Population definition 143
3 Tabu list 143
4 Random variables 144
5 Parent selection 145
6 Crossover/mutation trade-off 145
7 Selection pressure and crossover/mutation trade-off influence 148
8 RQK fitness functions 151
9 Evolution of the populations 154
10 Model distance 155
11 The software MobyDigs 158
11.1 The data setup 158

11.2 GA setup 159
11.3 Population evolution view 161
11.4 Modify a single population evolution 162
11.5 Modify multiple population evolution 163
11.6 Analysis of the final models 164
11.7 Variable frequency analysis 165
11.8 Saving results 166

CHAPTER 6 169
GENETIC ALGORITHM-PLS AS A TOOL FOR WAVELENGTH SELECTION IN SPECTRAL DATA SETS
(Riccardo Leardi)
1 Introduction 169
2 The problem of variable selection 170
3 GA applied to variable selection 172
3.1 Initiation of population 172
3.2 Reproduction and mutation 173
3.3 Insertion of new chromosomes 173
3.4 Control of replicates 174
3.5 Influence of the different parameters 174
3.6 Check of subsets 175
3.7 Hybridisation with stepwise selection 176
4 Evolution of the genetic algorithm 176
4.1 The application of randomisation tests 176
4.2 The optimisation of a GA run 177
4.3 Why a single run is not enough 177
4.4 How to take into account the autocorrelation among the spectral variables 178
5 Pretreatment and scaling 181
6 Maximum number of variables 182
7 Examples 183
7.1 Data set Soy 183
7.2 Data set Additives 190
8 Conclusions 194

PART II: ARTIFICIAL NEURAL NETWORKS 197

CHAPTER 7 199
BASICS OF ARTIFICIAL NEURAL NETWORKS
(Jure Zupan)
1 Introduction 199
2 Basic concepts 200

	2.1 Neuron	200
	2.2 Network of neurons	202
3	Error backpropagation ANNs	204
4	Kohonen ANNs	206
	4.1 Basic design	206
	4.2 Self-organized maps (SOMs)	210
5	Counterpropagation ANNs	213
6	Radial basis function (RBF) networks	216
7	Learning by ANNs	220
8	Applications	223
	8.1 Classification	223
	8.2 Mapping	224
	8.3 Modeling	225
9	Conclusions	226

CHAPTER 8 231
ARTIFICIAL NEURAL NETWORKS IN MOLECULAR STRUCTURES—PROPERTY STUDIES
(Marjana Novic and Marjan Vracko)

1	Introduction	231
2	Molecular descriptors	231
3	Counter propagation neural network	233
	3.1 Architecture of a counter propagation neural network	233
	3.2 Learning in the Kohonen and output layers	235
	3.3 Counter propagation neural network as a tool in QSAR	236
4	Application in toxicology and drug design	237
	4.1 A study of aquatic toxicity for the fathead minnow	237
	4.2 A study of aquatic toxicity toward *Tetrahymena pyriformis* on a set of 225 phenols	239
	4.3 Example of QSAR modeling with receptor dependent descriptors	242
5	Conclusions	252

CHAPTER 9 257
NEURAL NETWORKS FOR THE CALIBRATION OF VOLTAMMETRIC DATA
(Conrad Bessant and Edward Richards)

1	Introduction	257
2	Electroanalytical data	257
	2.1 Amperometry	258
	2.2 Pulsed amperometric detection	259
	2.3 Voltammetry	259
	2.4 Dual pulse staircase voltammetry	259
	2.5 Representation of voltammetric data	261

3 Application of artificial neural networks to voltammetric data 261
  3.1 Basic approach 262
  3.2 Example of ANN calibration of voltammograms 263
  3.3 Summary and conclusions 269
4 Genetic algorithms for optimisation of feed forward neural networks 269
  4.1 Genes and chromosomes 269
  4.2 Choosing parents for the next generation 270
  4.3 Results of ANN optimisation by GA 272
  4.4 Comparison of optimisation methods 277
5 Conclusions 278

CHAPTER 10 281
NEURAL NETWORKS AND GENETIC ALGORITHMS APPLICATIONS IN NUCLEAR MAGNETIC RESONANCE (NMR) SPECTROSCOPY
(Reinhard Meusinger and Uwe Himmelreich)
1 Introduction 281
2 NMR spectroscopy 283
3 Neural networks applications 285
  3.1 Classification 286
  3.2 Prediction of properties 290
4 Genetic algorithms 303
  4.1 Data processing 304
  4.2 Structure determination 305
  4.3 Structure prediction 308
  4.4 Classification 308
  4.5 Feature reduction 309
5 Biomedical NMR spectroscopy 309
6 Conclusion 315

CHAPTER 11 323
A QSAR MODEL FOR PREDICTING THE ACUTE TOXICITY OF PESTICIDES TO GAMMARIDS
(James Devillers)
1 Introduction 323
2 Materials and methods 324
  2.1 Toxicity data 324
  2.2 Molecular descriptors 324
  2.3 Statistical analyses 329
3 Results and discussion 330
  3.1 PLS model 330
  3.2 ANN model 332
4 Conclusions 338

CONCLUSION 341

CHAPTER 12 343
APPLYING GENETIC ALGORITHMS AND NEURAL NETWORKS TO CHEMOMETRIC PROBLEMS
(Brian T. Luke)

1 Introduction 343
2 Structure of the genetic algorithm 345
3 Results for the genetic algorithms 350
4 Structure of the neural network 362
5 Results for the neural network 365
6 Conclusions 373

INDEX 377

LIST OF CONTRIBUTORS

Conrad Bessant Cranfield Centre for Analytical Science, Institute of BioScience and Technology, Cranfield University, Silsoe, Bedfordshire MK45 4DT, UK
E-mail: c.bessant@cranfield.ac.uk

Viviana Consonni Milano Chemometrics and QSAR Research Group, Dept. of Environmental Sciences, P.za della Scienza, I-20126 Milano, Italy

James Devillers CTIS, 3 Chemin de la Gravière, 69140 Rillieux La Pape, France
E-mail: j.devillers@ctis.fr

D. Brynn Hibbert School of Chemical Sciences, University of New South Wales, Sydney, NSW 2052, Australia
E-mail: b.hibbert@unsw.edu.au

Uwe Himmelreich Institute of Organic Chemistry, Technical University Darmstadt, Petersenstrasse 22, D-64287 Darmstadt, Germany

Elsa Jordaan The Dow Chemical Company, 61 N Bachelor Button, Lake Jackson, TX 77566, USA

Alex N. Kalos The Dow Chemical Company, 61 N Bachelor Button, Lake Jackson, TX 77566, USA

Arthur K. Kordon The Dow Chemical Company, 61 N Bachelor Button, Lake Jackson, TX 77566, USA
E-mail: akordon@dow.com

Riccardo Leardi Department of Pharmaceutical and Food Chemistry and Technology, University of Genoa, via Brigata Salerno (ponte) – I-16147 Genova, Italy
E-mail: riclea@dictfa.unige.it

Brian T. Luke SAIC-Frederick, Inc., Advanced Biomedical Computing Center, NCI Frederick, P.O. Box B, Frederick, MD 21702, USA
E-mail: lukeb@ncifcrf.gov

Alessandro Maiocchi Bracco Imaging S.p.A., Milano Research Center, via E. Folli 50, 20134 Milano, Italy
E-mail: Alessandro.Maiocchi@bracco.com

Andrea Mauri Milano Chemometrics and QSAR Research Group, Dept. of Environmental Sciences, P.za della Scienza, I-20126 Milano, Italy

Reinhard Meusinger Institute of Organic Chemistry, Technical University Darmstadt, Petersenstrasse 22, D-64287 Darmstadt, Germany
E-mail: meusi@oc.chemie.tu-darmstadt.de

Marjana Novic Laboratory of Chemometrics, National Institute of Chemistry, Ljubljana, Slovenia

Manuela Pavan Milano Chemometrics and QSAR Research Group, Dept. of Environmental Sciences, P.za della Scienza, I-20126 Milano, Italy

Edward Richards Cranfield Centre for Analytical Science, Institute of BioScience and Technology, Cranfield University, Silsoe, Bedfordshire MK45 4DT, UK

Guido F. Smits The Dow Chemical Company, 61 N Bachelor Button, Lake Jackson, TX 77566, USA

Roberto Todeschini Milano Chemometrics and QSAR Research Group, Dept. of Environmental Sciences, P.za della Scienza, I-20126 Milano, Italy
E-mail: roberto.todeschini@unimib.it

Marjan Vracko Laboratory of Chemometrics, National Institute of Chemistry, Ljubljana, Slovenia
E-mail: marjan.vracko@ki.si

Jure Zupan Laboratory of Chemometrics, National Institute of Chemistry, Ljubljana, Slovenia
E-mail: jure.zupan@ki.si

PART I

GENETIC ALGORITHMS

CHAPTER 1

Genetic algorithms and beyond

Brian T. Luke

SAIC-Frederick Inc., Advanced Biomedical Computing Center, NCI Frederick, P.O. Box B, Frederick, MD 21702, USA

1. Introduction

Genetic Algorithms (GAs) in the broadest sense are model techniques used by simple biological systems. These biological systems use reproduction to produce offspring that can better survive in the current environment. Similarly, GAs use Darwin's 'survival of the fittest' strategy and reproduction operators to improve the quality of solutions to a particular problem. The origin of this mathematical technique is not known. Bledsoe (1961) presented many of the basic concepts in 1961. Bagley (1967) was the first to use the term 'Genetic Algorithm' and the methodology was given a firm mathematical foundation by Holland (1975).

Though the remaining chapters in the first section of this book deal with the application of GAs to chemometric problems, it must be emphasized that GAs do not really solve these problems. Instead, a GA is a search strategy through a multi-dimensional parameter space. A given GA can be used for feature selection and/or the optimal adjustment of parameters, but these features or parameters must be paszsed to a function that evaluates how well they solve the problem. This function can be as simple as a partial least-square (PLS) fit, or as involved as a clustering or decision tree process.

In contrast, Neural Networks (NNs), which are described and used in the second half of this book, produce a response from an input vector. This response can represent a functional evaluation of the input parameters, or can be used to classify the object possessing these parameters. Studies showing that a GA is superior/inferior to a NN for a particular problem are not really comparing these two methodologies. In fact, the GA simply selects features or assigns values to parameters that are passed to an evaluation function, and all that the GA tries to do is select the optimal set of features or parameters for this function. The comparison is really between an optimal, or near-optimal, function (such as PLS) and a NN.

With this in mind, GAs and NNs can be used together to solve particular problems. As described in the second half of this book, a NN first divides all of the data into two sets—the training set and the test set. The former is used to train the NN while the latter is used to

Data Handling in Science and Technology, Volume 23
ISSN: 0922-3487

DOI: 10.1016/S0922-3487(03)23001-X

measure the trained network's effectiveness. Choosing the optimal partitioning of the data can greatly improve the results and a GA can be used to search through the space of all possible partitions. In addition, a GA can be used to select the optimum set of features so that the number of input nodes can be held to a reasonable number (Narayanan and Lucas, 1993). Finally, for feed-forward NNs, the backpropagation algorithm is used to find an optimal set of weights connecting the layers by a local optimization procedure. Since the resulting set of weights are not necessarily the global optimum, Boltzmann and Cauchy machines use a variant of Simulated Annealing to update the weights during training to find the global optimum. Conversely, a GA can be used to select an optimum set of initial weights for the Backpropagation algorithm, or it can be used to search for the optimum set of weights directly.

The next section presents some terminology that will be used in the rest of this chapter and correlates these terms with biological processes. It also includes the framework for a Simple Genetic Algorithm (SGA). Section 3 discusses why GAs are able to locate good feature sets and/or parameter values without the mathematical rigor presented in Holland (1975) and describes why this method may not always find the globally optimum values. This section also presents the two opposing factors that influence the search; exploitation and exploration. Section 4 presents many of the operators and processes that can be used to convert this search heuristic into an actual algorithm. These include determining the fitness function; selecting the form of the vector that contains each putative solution; building an initial population; various selection schemes; possible mating, mutation and maturation operators; processing the offspring and terminating the search.

Section 5 expands upon the information of Section 4 by describing how many of the operations and procedures promote either exploitation or exploration, and presents several new methods that emphasize one or the other. Section 6 describes other variants of the GA and shows how the operators presented in Sections 4 and 5 can be used to form a connection between GAs and other population-based search methods such as Evolutionary Programming (EP), Evolution Strategies (ES), Ant Colony Optimization (ACO) and Particle Swarm Optimization (PSO). This is followed by a general discussion of the information presented in this chapter (Section 7).

One final point to present here is magnitude of the search space for even simple problems. One of the simplest applications of GAs is the feature selection problem. If each object contains N features, the goal is to find the best set of features that describe or classify the objects. A simple representation of a putative solution is a bit string of length N where a 0 in the ith position means that this feature is not selected and a 1 in this position means that it is. With this notation, it is easy to see that this problem is equivalent to the (0,1)-knapsack problem.

The number of possible solutions for a given value of N is 2^N and for even a modest value of N, say 50, the number of solutions is very large (1.126×10^{15}). At present there is no known algorithm that finds the optimum solution to this problem and runs in a time that is proportional to a polynomial in N. This is not to say that such an algorithm does not exist, it is just that one has not been found. Therefore, this unknown algorithm runs in a Non-deterministic Polynomial amount of time and the problem is categorized as *NP-hard*. In addition, if someone presents a bit string of length N and stated that this is the optimal set of features for this problem, there currently is no known algorithm that can validate this claim and run in a time that is a polynomial in N. Therefore, this problem is actually known to be *NP-complete*. The purpose of a GA is to find the optimum, or a near-optimum, set of features by evaluating a very small fraction of the possible solutions in the overall search space.

2. Biological systems and the simple genetic algorithm

As stated in Section 1, a GA is a search methodology that mimics the basic techniques used by biological systems to thrive in a particular environment. In an SGA, this 'adaptation' is solely based on hereditary improvement of a population of organisms.

In a simple biological system, the ability of an organism to thrive in a given environment is directly related to its ability to utilize the nutrients and possibly fight off any harmful effects of the environment. For example, early life on Earth may have had to survive in a very different environment from the one present today. These differences may include the oxidative/reductive ability of the atmosphere, the intensity of ultraviolet radiation at the surface, and the surface temperature. Early life needed to possess enzymes that could metabolize the nutrients present and fight off any harmful effects of the environment. Simple organisms today need to do the same things, but a different, or modified, set of enzymes is necessary. A GA simulates the evolution of a simple organism to adapt to a given environment.

The ability of a simple organism to produce a given enzyme is contained in a code that is stored in its *chromosome*. This chromosome contains a string of deoxyribose nucleic acids (DNA), and in the general case each position in this string, or *locus*, can contain one of four nucleic acids (A, C, G, or T). These four acids constitute the *genetic alphabet* of the chromosome. Ignoring the concept of reading frames, a region or substring of the chromosome that contains the code for an enzyme is called the *gene* and the actual substring, and therefore the enzyme it codes for, is called the *allele*. The string of amino acids that constitute the chromosome represents the organism's *genotype*, while the set of enzymes it produces, and therefore the characteristics of the organism such as its ability to thrive in the current environment, is known as its *phenotype*.

In the current discussion of GAs, the chromosome is called the *genetic vector*. In the feature selection example presented in Section 1, the genetic vector is a bit string. Therefore, the genetic alphabet contains only two binary elements, 0 and 1. Each locus can be thought of as an individual gene and each allele is either Yes (1) or No (0). The bit string is the genotype of this organism or putative solution, and the ability of the set of selected features to solve the problem is the phenotype. The extent to which the putative solution solves the problem is also known as its *fitness*.

In the sexual reproduction of offspring, two parents are selected and the chromosome of the offspring is constructed by combining sections of the parents' chromosomes. There is a small probability that a mutation can occur in the offspring before its genotype is established. This genotype produces a phenotype for the offspring. Darwin's 'survival of the fittest' generally determines which parents are used for mating and whether or not the offspring is viable enough to become a parent.

Using this model as a guide, the steps in a SGA are as follows.

1. Create an initial population of genetic vectors and calculate their fitness.
2. Choose two members of this population based on their fitness to become parents.
3. Use a *mating operator* to construct a new genetic vector from the parents.
4. Use a *mutation operator* to probabilistically change the genetic vector.
5. Calculate the fitness of this offspring and have it replace the weakest member in the population.
6. Return to Step 2 until a sufficient number of offspring has been produced.

Details of each of these steps, as well as other factors needed to create a specific algorithm, will be presented in Section 4.

3. Why do GAs work?

The SGA presented in Section 2 is able to find good solutions to a problem by examining only a very small number of the total set of possible solutions. This occurs through a process called *focusing* (Luke, 1996).

During the mating, sections of the genetic vector from each parent will be copied to the offspring as long as the mating and mutation operators do not seriously disrupt these sections. Since the parents are chosen based on their fitness, there is a good chance that the offspring will also be reasonably fit, which means that they will not be replaced by future offspring too quickly. As the simulation proceeds, low order (i.e. short) patterns will form in all of the offspring. These patterns, or identical values, will appear at certain loci in all genetic vectors of the population. If each genetic vector contains 10 loci, an example of this pattern, or *schema* (Holland, 1975), is as follows.

$$(-, -, *, *, -, -, *, -, -, *)$$

In this example, a dash represents a variable locus where different genetic vectors can have different values from the genetic alphabet, and an asterisk represents a fixed locus where all genetic vectors have the same value. If the mutation operator does not change any of these values, all future offspring will have the same schema. The effect of this is to focus the search onto a six-dimensional space represented by the six variable loci instead of exploring the full 10-dimensional space of the problem.

As the simulation proceeds, the schemas grow to include more loci and eventually all genetic vectors in the population will be identical. Therefore, a GA finds a good solution relatively quickly by seeding the population with sections of fit genetic vectors and removing unfit genetic vectors.

The act of placing segments of the genetic vector from fit parents into the offspring is called *exploitation*, while creating offspring whose genetic vector differs substantially from fit members is called *exploration*. Exploitation promotes the formation of schema, thereby focusing the search in a of the full search space. If the mating operator does not allow segments to be preserved from parent to offspring, the offspring will be substantially different from its parents and exploration of other areas of search space will be performed.

Similarly, the mutation operator promotes exploration and this retards the formation of schema and the focusing of the search.

Therefore, the GA with the fastest convergence to a good solution would have a 1- or 2-point crossover as the mating operator (see Sections 4 and 5; Hasancebi and Erbatur, 2000) and no mutation operator. Though this algorithm will be relatively fast, there is a good chance that it will not converge on the globally optimum solution. The basic reason for this is that the procedure performs a local search around the most-fit solutions in the current population. If these fit members are close to a good, but not the best solution, the problem is called *deceptive* and the best solution will not be found. This means that an algorithm with a good balance between exploration and exploitation will have a harder time forming schema, but may be able to find a better solution.

4. Creating a genetic algorithm

This section fills in all of the details so that the basic methodology presented in Section 2 can be incorporated into an actual algorithm. It does not contain a complete list of all possible forms of each operator, but is intended to give the reader an understanding of the flexibility available within this methodology. Several of the options presented here will be revisited in Section 5 which deals with promoting exploitation versus exploration.

4.1. Determining a fitness function

As stated in Section 3, the SGA uses a selection procedure (see below) that chooses parents in an amount proportional to their fitness. This means that if fitness-based selection methods are used, better solutions must have a larger fitness and each member (genetic vector) in the population must have a non-negative fitness. This also means that the SGA is basically a *maximization* algorithm since it selects the features in a way to try and maximize the fitness of the offspring and population as a whole.

If the object is to maximize the score of a solution $s(X)$, where X represents the genetic vector with elements x_i ($X = \{x_i\}$), and this score can become negative, the fitness $f(X)$ can be set to non-negative values by selecting a 'minimum allowed value', $C_{\min}$, and subtracting this from all fitness values that are above this minimum.

$$\begin{aligned} f(X) &= s(X) + C_{\min} \quad \text{if } s(X) + C_{\min} > 0 \\ &= 0 \quad \text{otherwise} \end{aligned}$$

Conversely, the objective function may be a cost function $c(X)$, and the GA is used to select features or parameters that minimize this function. If the cost can become negative, a fitness value can be determined by subtracting the cost from a maximum allowed value $C_{\max}$.

$$\begin{aligned} f(X) &= C_{\max} - c(X) \quad \text{if } c(X) < C_{\max} \\ &= 0 \quad \text{otherwise} \end{aligned}$$

Rosenberg (1967) referred to the cost of a putative solution as its 'anti-fitness' and suggested using the inverse of the cost if this value is positive everywhere.

4.2. The genetic vector

All applications of GAs, and all population-based methods described in this chapter, assume that a putative solution to the problem can be stored in a small number of vectors. These *genetic vectors* represent the chromosomes of the organism and consist of a series of genes. Three different coding schemes will be described here; *gene-based*, *node-based*, and *delta coding*.

In gene-based coding there is a one-to-one correspondence between the gene number and a particular feature of the problem to be solved. For example, the first gene may correspond to a particular spectral intensity or a particular descriptor used to describe a set of molecules. If the goal is to select various spectral features to determine the presence or absence of a particular compound, a genetic vector of the form (0,0,1,0,1,0, ...) would mean that the third and fifth feature will be used while the first, second, fourth and sixth will not.

Node-based coding represents a route or schedule. For example, if a chemical laboratory has to run various experiments on the same piece of equipment, the genetic vector (2,7,5, ...) means that Experiment 2 will be run first, followed by Experiments 7 and 5, and so on. Conversely, this genetic vector may represent the order of features to use in a decision tree to separate individual compounds or classes of compounds from each other. In this case, a second static or varying gene-based genetic vector may be needed to store the threshold values for each feature.

Delta coding is used less because it can only be applied to certain types of problems. This is a gene-based coding scheme with the exception that the values of each gene are taken relative to a template genetic vector. If the goal is to optimize the conditions of a chemical reaction, the template genetic vector would contain reasonable values for the reaction conditions (concentrations, temperature, light frequency and intensity, etc.). Each putative solution would have a genetic vector that contains positive or negative changes to these base values.

More than one type of coding scheme can be used for a particular problem. If the object is to choose the best set of three descriptors from a set of 10 that can be used to describe the biological activity of a molecule, the gene-based coding (0,1,0,0,0,1,0,1,0,0) and node-based genetic vector (2,6,8) represent the same putative solution. It should be stressed that the latter representation has a sixfold degeneracy since the order of selected descriptors is not important. This degeneracy will affect the search, and so changing the coding scheme may affect the quality of the resulting algorithm.

The form of the genetic alphabet is dependent upon the problem and coding scheme. In a feature selection problem, a gene-based coding scheme can be used and the genetic alphabet is just 0 or 1 (depending upon whether the feature is not or is selected, respectively). Conversely if a node-based coding scheme is used, the genetic alphabet is just the set of integers from 1 to N, where N is the number of tasks to order. If the goal is to find an optimized set of parameters, real coding can be used. The only problem with using real numbers is that it is virtually impossible to build up schema unless some procedure is used to restrict the set to a finite number of values. A maturation operator (see below) can be used for this purpose. Conversely, an integer or binary coding can be used, but this reduces the precision of the parameter values.

In many studies binary coding is used since it is easier to build up schema (Goldberg, 1989). If the value of a given parameter varies from U_{min} to U_{max} and it is encoded in a bit string of length l, the precision (step size) of this mapping is

$$p_i = (U_{max} - U_{min})/(2^l - 1)$$

For example, if an 8-bit string is used to represent a gene and the value of the parameter varies between 0.0 and 500.0, the precision of the allele is 1.96. The parameter's value is formed by first converting the bit string into an integer (A) in the range [0,255] and then converting it to a floating-point number.

$$P = 1.96A$$

If the bit string representing this parameter is (00101101), the resulting integer is 45 and the parameter value is 88.20 ± 0.98. One problem with using binary coding is the effect of the mutation operator (see below). If this operator simply flips a bit and it happens to be the high-order (left-most) bit, the new bit string would be (10101101). Its integer value would be increased to 173, and the corresponding value of the parameter would be 339.08 ± 0.98. This means that it would be very hard to control the magnitude of the change caused by a single mutation.

This large change caused by flipping the high-order bit is called the Hamming cliff, and can be reduced using Gray coding. Gray coding is a non-unique one-to-one mapping between bit strings such that if the resulting integers differ by one, their Gray coded bit strings only differ by a single bit flip (i.e. their Hamming distance is 1). One example of Gray coding uses the following procedure.

$$\begin{aligned} g(1) &= b(1) && \text{for } i = 2 \text{ to } k \\ g(i) &= b(i) && \text{if } b(i-1) = 0 \\ g(i) &= \text{COMPLEMENT } (b(i)) && \text{if } b(i-1) = 1 \end{aligned}$$

In this notation, $b(i)$ is the ith element of the normal bit string and $g(i)$ is the corresponding value in the Gray coded string, $b(1)$ and $g(1)$ represent the high-order (left-most) bit, and k is the length of the bit string. With these rules, the 4-bit binary and Gray coding arrays for the numbers 0–15 are given in Table 1.

A Gray coded number can be converted to the standard binary form using similar rules.

$$\begin{aligned} b(1) &= g(1) && \text{for } i = 2 \text{ to } k \\ b(i) &= g(i) && \text{if } b(i-1) = 0 \\ b(i) &= \text{COMPLEMENT } (g(i)) && \text{if } b(i-1) = 1 \end{aligned}$$

Hollstein (1971) found that Gray coding produced better results because of the reduction of changes in the transcribed values caused by a single mutation. Though Gray coding

Table 1
Standard and Gray coding for the integers 0–15

Number	Standard	Gray
0	0000	0000
1	0001	0001
2	0010	0011
3	0011	0010
4	0100	0110
5	0101	0111
6	0110	0101
7	0111	0100
8	1000	1100
9	1001	1101
10	1010	1111
11	1011	1110
12	1100	1010
13	1101	1011
14	1110	1001
15	1111	1000

removes the Hamming cliff, only two of the possible bit flips result in an adjacent integer. The other $(k - 2)$ flips produce a larger change in the resulting integer. This means that if a 8-bit coding is used, only 25% of the possible flips change the resulting number by 1, and the remaining 75% of possible flips produce larger changes. For example, if the integer value of a parameter is 92, its Gray coded string is (01110010). If the sixth or eighth bit is flipped, the resulting integer changes by only one, and it changes by only three if the seventh bit is flipped. Conversely, if the first or second bit is flipped, the resulting integers are 163 and 35, respectively.

It is important to realize that a change in the coding not only affects the mutation operator, it can change the entire search landscape. For example, we assume that a parameter value of 92 (really 180.32 ± 0.98 in the example above) produces a good solution, but the best solution has a parameter value of 163 (319.48 ± 0.98). If regular binary coding is used and enough of a schema formed in the vicinity of 92, for example (-101-10-), the value of 163 will never be found since the binary code for 163 (10100011) differs from the binary code for 92 (01011100) in all eight loci. Conversely, if Gray coding is used and schema form at the same loci it would be (-111-01-). Both 92 and 163 would still be viable since the Gray coding for 92 (01110010) and 163 (11110010) differ only at the first locus. In other words, the conversion from normal to Gray coding merged two minima that were maximally separated in normal binary coding into a single good region of search space. There will be other cases, like 87 and 119, where the Hamming distance (number of different bits) is only one when normal coding is used and the distance is larger for Gray coding. This basically means that different coding schemes produce different distances between minima and this will affect the search for the best overall solution.

The last point to consider is how many copies of the chromosome are present. In certain viruses, only a single copy of the chromosome is present and they are known as *haploids*.

In most cellular organisms, there are two copies of each chromosome, *diploids*, and the alleles for a gene can be the same (*homozygous*) or different (*heterozygous*). Since much of the early work in this field attempted to mimic real biological systems, a diploid structure was used, though most of the current work using GAs assume a haploid structure.

The basic process of offspring production is different with each structure. In diploid systems, a single copy of each chromosome is given to each of two possible offspring, or one copy of the chromosome is randomly chosen from each parent if a single offspring is created. If information is exchanged between the two chromosomes, it can happen in several ways. For example, before a chromosome is selected for an offspring, sections can be switched (see below) between the chromosomes of each parent. Once the chromosome is selected from each parent and placed into the offspring, information can be switched between these two offspring chromosomes. In haploid studies, the single chromosome from each parent must be copied into a temporary organism that will have a diploid structure. Sections of these two chromosomes can be swapped before this temporary organism splits into two haploid offspring. This swapping of chromosome sections is not absolutely necessary in diploid organisms since it is very possible that the offspring will be different from either parent. In the haploid system, this swapping is required or else the offspring will be identical to the parents.

Diploid or polyploid (more than two copies of the chromosome) genetic structures are sometimes used to maintain diversity in the gene pool. Various studies have shown that this procedure is superior for problems that change with time, like the Dynamic (0,1)-Knapsack Problem (DKP) (Goldberg and Smith, 1987; Smith and Goldberg, 1992). If two or more copies of the chromosome are present in each organism, the problem becomes how to decode them to obtain a single value for a parameter if they are different.

Hollstein (1971) suggested that both chromosomes be used to generate a binary string that contains the genetic information. Enlarging the genetic alphabet does this. Each locus in the chromosome contains a 0, 1, or 2, where both 1 and 2 encode a 1 but 2 dominates 0 and 0 dominates 1. Therefore, the decoding is

	0	**1**	**2**
0	0	0	1
1	0	1	1
2	1	1	1

If one of the chromosomes contains (2110201...) and the other contains (1012021...), this organism is effectively represented by (1011111...) since the (2,1) combination at the first locus yields a 1, the (1,0) combination at the second locus yields a 0, and so on. Note that there are twice as many 1s as 0s and this would cause a drift to solutions that contain a large number of 1s, like the DKP examined in Smith and Goldberg (1992). If the genetic alphabet contains 0, 1, 2, 3 and even codes as 0 and odds code as 1, different dominance rules produce different results with the probability of obtaining a 0 equal to that for obtaining a 1. For example, if 0 dominates 1, 2 dominates 1 and 3 dominates all, the decoding is

	0	**1**	**2**	**3**
0	0	0	0	1
1	0	1	0	1
2	0	0	0	1
3	1	1	1	1

Therefore, chromosomes of (2031102...) and (1203213...) produce an equivalent binary chromosome of (0011001...).

If a higher number dominates a lower number except a homozygote codes as 0, the decoding is

	0	**1**	**2**	**3**
0	0	1	0	1
1	1	0	0	1
2	0	0	0	1
3	1	1	1	0

This means that chromosomes of (2031102...) and (1203213...) produce a binary chromosome of (0011011...).

Finally, if the absolute difference is odd it becomes a 1, otherwise it is 0, the decoding is

	0	**1**	**2**	**3**
0	0	1	0	1
1	1	0	1	0
2	0	1	0	1
3	1	0	1	0

An organism with chromosomes of (2031102...) and (1203213...) would therefore be represented by an equivalent binary chromosome of (1010111...).

Osmera and co-workers (1997) used a simple XOR to generate a bit vector from two chromosomal bit vectors. The decoding in this case is

	0	**1**
0	0	1
1	1	0

Chromosomes of (1011001...) and (1101010...) would therefore represent an equivalent binary chromosome of (0110011...). These decoding schemes do not represent all possible ones, but simply serve as an example of possible schemes. In addition, two different pairs of chromosomes can produce the same binary string upon decoding. This means that two parents with the same fitness can produce offspring with very different fitness values by simply using one chromosome from each parent.

All of these methods use both chromosomes to generate an expressed bit vector and do not deal with gene dominance. A simple procedure to include dominance is to use a masking bit-vector. This means that each solution contains two copies of the genetic vector and a masking vector. An offspring would contain one genetic vector from each parent and a masking vector. This vector can be randomly chosen from one parent, or it can be a combination of pieces from each parent. If the genetic vector contains 10 genes and the masking vector is (1001010100), the first, fourth, sixth, and eighth gene would come from one parent and the remaining genes from the other.

Conversely, each gene can contain an extra masking bit, and if they agree one chromosome's gene is expressed, while the other's is expressed if they disagree. Finally, each gene can contain a strength factor, which is related to the fitness of the chromosome over several generations. The stronger gene would then be expressed. This is slightly different than a biased uniform expression, where the fitness of the parent determines the probability that its gene will be expressed.

Kim and co-workers (1996) proposed a 'winner takes all' strategy. This method expresses both chromosomes and the most-fit solution is kept. In the mating process, the stronger chromosomes from each parent are combined in one offspring, and the weaker in the other. Yang (2002) used binary coding and really only kept a single chromosome. On evaluation, both this chromosome and its primal-dual were evaluated, and the chromosome with the higher fitness was kept. The primal-dual is simply the compliment of the chromosome (switch 1s and 0s). If the primal-dual scored better, it replaced the chromosome.

Calabretta and co-workers (1996) proposed a scheme where each gene contained two extra bits. The first bit of each gene was used with XOR to determine which gene could be dominant. If the second bits were different, the dominant gene is expressed. If the second bits were the same, both genes were decoded and the average value was used.

4.3. Creating an initial population

Once the form of the genetic vector is determined, an initial population needs to be created that will serve a parents in the first generation. The first question that needs to be addressed is how many putative solutions should be present in the population. If n_p is the number of parameters that are binary coded with a bit length of n_b, an intuitive order for the size of the population is (Goldberg et al., 1992)

$$n_\mathrm{pop} \approx \mathrm{O}(n_\mathrm{p}2^{n_\mathrm{b}})$$

If the goal is to determine the optimum value of 20 parameters, and each value is coded as a bit string of length 8, the number of putative solutions should be on the order of 5000. This is a rather large population size, but ensures that most of the search space is sampled. This means that exploitation can be used instead of exploration since all good regions of search space should be sampled by at least one member of the population.

In practice, smaller population sizes are often used and exploration needs to be promoted to some degree by the offspring generating operators described below.

This allows regions of search space that are not sampled with the initial population to be sampled at a later time.

The next step is to generate n_{pop} initial solutions. In most applications, this is done by randomly generating genetic vectors of length L. Section 5.2 describes methods to improve the diversity of this initial population. The average fitness of the initial population can be improved by using results of faster search methods as the initial parents, or the random solutions can be improved by local optimizations prior to the first generation. Unfortunately, improving the fitness of the initial population can often work against finding the best solution at the end of the search (Keser and Stupp, 1998). Unless a judicious scaling of the fitness values is used (see Section 4.4) a small number of fit local minima can dominate the parent selection process and the resulting search. This means that the search may become trapped in a region that does not contain the optimum result. This reasoning is very problem dependent, and running the simulation using various initial populations is preferred.

4.4. Selection schemes

A selection scheme is often used when choosing which parents to mate, and/or choosing which offspring to place in the next generation's parent population (see below). Since Darwin's 'survival of the fittest' strategy is used to exploit good solutions, a selection scheme based upon the fitness is generally used in both cases.

In practice, the selection of parents occurs in two ways. In the first, the two parents are simply chosen from the full population using a fitness-based method, while in the second the selection method is used to choose solutions from the population, with replacement, and these solutions are placed in a *mating pool*. After the mating pool is created, the two parents are randomly selected from this pool. Since the two methods are mathematically the same and the latter can only be used with a *generational algorithm* (see below), the former method will be assumed here.

The most common selection method is known as the *roulette wheel* or *proportional selection* method. The basic model is to place all solutions on a roulette wheel with the area of the wheel proportional to their fitness. In practice, the simplest procedure is to generate a random number between zero and the sum of the fitness values, and then start summing the fitness values until the sum is greater than or equal to this random number. The last solution in this summation becomes one of the parents. Mathematically, if r is a random number in [0,1], μ is the size of the population, and f_i is the fitness of the ith member of the population, the random value R is determined from

$$R = r \sum_{i=1,\mu} f_i$$

The kth member if the population is chosen such that

$$\sum_{i=1,k} f_i \geq R$$

It is clear from the above expressions that fitness must be used instead of a cost, and that each solution's fitness should be non-negative (as described above). Another way to state this is that the probability of choosing an animal, $p(i)$, is given by

$$p(i) = f_i \Big/ \sum_{i=1,\mu} f_i$$

For a small population size, the roulette wheel may become dominated by only one or two solutions. This can be removed by scaling the fitness before its region on the roulette wheel is established. One way to do this is linear scaling

$$f'_i = af_i + b$$

where the coefficients a and b vary with the population. One option is to require that

$$f'_{\max} = C_{\mathrm{mult}} f_{\mathrm{avg}}$$

$$f'_{\min} = C_{\min}$$

where C_{mult} is generally in the range 1.2–2.0, $C_{\min}$ is 0.0 or some other small, positive number, and f_{avg} is the average fitness of the population (Goldberg, 1989). This normalization yields

$$a = (C_{\mathrm{mult}} f_{\mathrm{avg}} - C_{\min})/(f_{\max} - f_{\min})$$

$$b = C_{\min} - af_{\min}$$

Another possible normalization condition is to have

$$f'_{\mathrm{avg}} = f_{\mathrm{avg}}$$

$$f'_{\max} = C_{\max} f_{\mathrm{avg}}$$

Again, $C_{\max}$ is in the range 1.2–2.0 for small (50–100) populations (Goldberg, 1989). This gives

$$a = f_{\mathrm{avg}}(C_{\max} - 1)/(f_{\max} - f_{\mathrm{avg}})$$

$$b = f_{\mathrm{avg}}(f_{\max} - C_{\max} f_{\mathrm{avg}})/(f_{\max} - f_{\mathrm{avg}})$$

A second fitness-based method that is often used is called K-Tournament Selection. With this procedure, K members of the population are randomly selected and the member with the highest fitness is selected. When $K = 1$, this is just a random selection, while if $K = \mu$ (the population size), only the most-fit member is selected. This procedure has the advantage that no scaling of the fitness is necessary and a cost can be used instead of fitness. In addition, non-generational algorithms (see below) constantly substitute new offspring for other members of the population and the roulette wheel method requires a new value for the sum of the fitness values and scaling constants with each substitution.

Instead of a fitness-based selection, a selection based upon the rank of each member when ordered from highest-to-lowest fitness or lowest-to-highest cost can be used. Brown and Sumichrast (in press) proposed a rank-based fitness that is used in a roulette wheel selection. If μ is again the population size and z is the rank number of the solution in *ascending* fitness, the rank-based fitness is

$$\mathrm{rankfit}(z) = 2z/[\mu(\mu + 1)]$$

Other rank based selection procedures can be used. For example, after ranking in order of *descending* fitness, the number of the selected member can be generated from the expression (Whitley, 1989)

$$I = \mu[\alpha - (\alpha^2 - 4(\alpha - 1)r)^{1/2}]/[2(\alpha - 1)]$$

where r is a random number in [0,1] and α a user-defined value that is greater than 1.0. As the value of α increases, the index I is more skewed towards 1.

Much simpler selection methods have also been used. For example, van Kampen and Buydens (1997) randomly selected the parents from the top 25% of the population when ranked by fitness.

4.5. Mating operators

All mating operators take genetic information from two parents and can create one or more offspring. There are a number of possible mating operators and only some of the most often used ones will be described here. Three points need to be discussed before the operators are presented. The first is that if the genetic vectors use node-based coding, the operators need to be very different than for gene-based or delta coding. This is because each task can only appear once in the genetic vector of the offspring and the mating operator must preserve this.

The second point is that the actual procedure for creating offspring is slightly different for GAs that use a haploid genetic vector (one copy) than those that use diploid vectors (two copies that do not need to be the same). For diploid organisms, the process of reproduction starts with placing one copy of the chromosome into a germ cell. Germ cells from each parent are then combined to produce the offspring. Combining genes from two chromosomes to produce two (potentially) different chromosomes can happen at two different times. The first is within the parent before the chromosomes are transferred to the germ cells. This means that the chromosome stored in a germ cell from an organism does not have to be identical to a chromosome in that parent. The second opportunity is after the two germ cells are combined in the offspring. This will produce two chromosomes where part of the genetic information on each chromosome came from one parent or the other, and if it came from one parent at a particular locus on one chromosome, the same locus on the other chromosome must have come from the second parent.

As stated above, most applications of GAs use a haploid structure for the organism, and the reproduction process does not correspond to a normal biological system. Here, the chromosome from each parent is placed into a 'temporary diploid organism.' It is in this temporary organism where genetic material from the chromosomes can be exchanged. This temporary organism then splits to form a complimentary pair of organisms.

If a diploid structure is used, and the genotype and phenotype of the offspring are formed from a combination of the elements from both genetic vectors at each locus, switching the alleles between the chromosomes after the germ cells have been combined will not change the resulting offspring. Therefore, a switching of the genetic information should only occur prior to the formation of the germ cells. If a masking vector is used to determine which genetic vector contains the dominant allele, mixing the genetic information can occur both before the germ cells are created and after they are combined in the offspring. This model also has a mixed haploid/diploid structure in that the masking vectors from the two parents can be combined to make complimentary pairs with each going to a different offspring.

The third point is that there can be a finite probability of applying the mating operator, P_{mate}. If a random number in [0.0,1.0] is less than P_{mate} the mating operator will be applied. Otherwise, the offspring will get unaltered genetic vectors from the parents.

In the examples shown below, a haploid model is assumed. The application of these methods to diploid organisms simply requires changing the labels from 'Parent A' and 'Parent B' to 'Chromosome A' and 'Chromosome B', respectively. Similarly, the labels 'Offspring 1' and 'Offspring 2' are changed to 'New Chromosome A' and 'New Chromosome B'.

For gene-based and delta coding schemes, the biologically inspired *crossover operator* is most heavily used. The simplest application of this operator is called a 1-point crossover. An example of a 1-point crossover is shown below.

```
a1 a2 a3 a4 a5 a6 a7 a8      Parent A
           |
         cut point
           |
b1 b2 b3 b4 b5 b6 b7 b8      Parent B

a1 a2 a3 b4 b5 b6 b7 b8      Offspring 1

b1 b2 b3 a4 a5 a6 a7 a8      Offspring 2
```

Here, the two genetic vectors are lined up and a cut point is randomly chosen. The two offspring are generated by taking pieces of the genetic vector before and after the cut point from different parents.

This crossover operator can be extended to a k-point crossover where $k \geq 1$. In the limit that each element of the genetic vector can come from one parent or the other, the mating operator is called a *uniform crossover* (Syswerda, 1989). If there is an equal probability of an element coming from either parent, it is called an *unbiased uniform crossover*. Conversely, it is possible to promote selection from the parent with the higher fitness. In this case, it is called a *biased uniform crossover*.

The difference between these crossover operators is shown in Fig. 1. Here the two parents contain genetic vectors that differ at only three loci, (..., a, ... , b, ... , c, ...) in one parent and (..., A, ... , B, ... , C, ...) in the other. These unique genes then correspond to the diagonal corners of a rectangular prism (shown as filled squares). A 1-point crossover between a and c in the first parent would only be able to generate the four unique offspring shown as open circles, while a 2-point crossover with one cut point between a and b and the other between b and c in the first parent would only produce the two unique offspring shown as closed circles. Only a uniform crossover would allow the possible production of all six unique offspring.

This reduction in the possible offspring increases with the dimensionality of the search space. For example, if the two parents differ in N genes, there are $2^N - 2$ possible unique offspring that can be generated by a uniform crossover. If a 1-point crossover is used, only

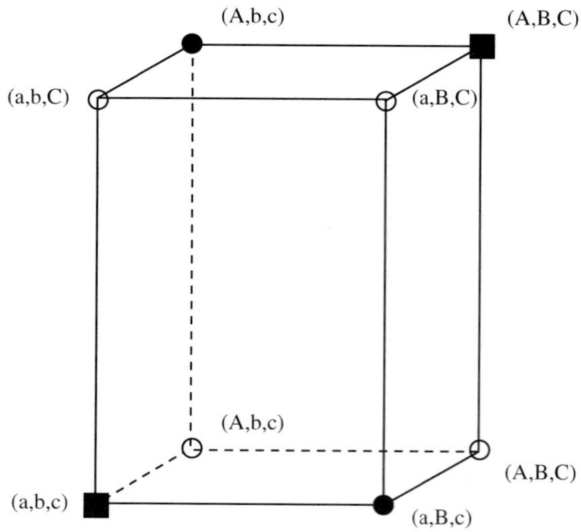

Fig. 1. Possible offspring that are generated by crossover operators. The black corners represent the parents, the red corners are offspring generated with a 1-point crossover, and the green are produced with a 2-point crossover.

$2(N-1)$ unique offspring could be generated, while a 2-point crossover can only generate $(N-1)(N-2)$ of them.

Another factor that needs to be considered when using a k-point crossover is that the probability of transferring two elements to an offspring from the same parent is a function of their relative positions. For example, if the genetic vector contains N elements, the probability that adjacent elements from the same parent will be present in one of the offspring is $(N-2)/(N-1)$ if a 1-point crossover is used. If there are j elements between them, the probability becomes $(N-j-2)/(N-1)$, which means that the first and last element in the genetic vector of an offspring can never come from the same parent if a 1-point crossover (or any k-point crossover where k is odd) is used. Conversely, the first and last elements of the genetic vector of an offspring have to come from the same parent if a k-point crossover is used when k is even. Only an unbiased uniform crossover can randomize the source of any genetic element in an offspring.

Another way to state this is that for small k, a k-point crossover promotes formation of schema since low-order patterns will be passed from the parent to the offspring. Conversely, as k increases, the chances of breaking a low-order pattern increases and this retards the formation of schema. Therefore, small values of k promote exploitation while larger values of k (such as a uniform crossover) promote exploration.

For non-binary genetic vectors the preceding arguments still hold, but in this case, there are mating operators that allow offspring to sample points within the N-dimensional rectangle instead of just the solutions represented by the corners (see Fig. 1). The first is called *intermediate recombination*. Each element of the offspring's genetic vector can be obtained from either of the following expressions

$$c_i = a_i + u(b_i - a_i)$$

or

$$c_i = a_i + u_i(b_i - a_i)$$

The first equation uses a single random number u between 0.0 and 1.0 to determine all elements of the offspring's genetic vector, while in the second a different random number is used for each element. The effect of the first is to place the offspring somewhere on the line segment connecting the two parents, while the second places the offspring somewhere in the N-dimensional rectangle. The compliment of this offspring is generated by substituting $(1 - u)$ for u in the equations above.

The offspring can be allowed to occupy a point in search space that is outside of this hyper-rectangle by using the *interval crossover* operator. In this case, the elements of the offspring's genetic vector can be determined from either

$$c_i = a_i + u(b_i - a_i + 2\delta) - \delta$$

or

$$c_i = a_i + u_i(b_i - a_i + 2\delta_i) - \delta_i$$

In practice, δ (or δ_i) is usually small, but this does not have to be the case. The major effect of this operator is to maintain the full dimensionality of the search space. In other words, the resulting value of c_i does not have to be a_i or b_i, even if they are equal. Therefore, any schema that may have been built up in the parent population will not necessarily be transferred to their offspring. The effect of this will be to slow down the convergence of the population and, depending upon the size of δ, it may also allow the population to escape from sub-optimal solutions. Again the compliment of this offspring can be generated by substituting $(1 - u)$ for u in the equations above.

For node-based coding problems, the mating operator must be very different. Since each element in the genetic alphabet can only be present once in the genetic vector, the mating operator must have a process that removes duplicates. In the examples listed below, it is assumed that 10 tasks, labeled A through J, need to be done in the order represented by a genetic vector, and the two parents selected for mating are

(A, B, C, D, E, F, G, H, I, J) Parent 1

(G, E, B, H, J, C, I, F, A, D) Parent 2

In a *partially matched crossover* (PMX) two cut points are randomly chosen and the elements between the cut points in the parents are paired. For example,

```
(A, B, C, D, E, F, G, H, I, J)   Parent 1
         |        |
(G, E, B, H, J, C, I, F, A, D)   Parent 2
```

pairs D and H, E and J, and F and C. Whenever one member of a pair is found in a parent, the other member is put in its location. Therefore, the two possible offspring are

(A, B, F, H, J, C, G, D, I, E) Offspring 1

(G, J, B, D, E, F, I, C, A, H) Offspring 2

In an *ordered crossover* (OX) the regions between the two cut points determine which entries in the other parent to remove (as denoted by dashes)

(A, B, –, | D, E, F, | G, –, I, –)

(G, –, B, | H, J, C, | I, –, A, –)

The missing elements are then returned to the region between the cut points by starting from the first position to the right of the second cut point and cyclically placing the existing elements.

(D, E, F, | –, –, –, | G, I, A, B)

(H, J, C, | –, –, –, | I, A, G, B)

The segment between the cut points are then moved to the other genetic vector.

(D, E, F, | H, J, C, | G, I, A, B) Offspring 1

(H, J, C, | D, E, F, | I, A, G, B) Offspring 2

This process can also start with the element to the left of the first cut point

(I, A, B, | –, –, –, | D, E, F, G)

(A, G, B, | –, –, –, | H, J, C, I)

(I, A, B, | H, J, C, | D, E, F, G) Offspring 1

(A, G, B, | D, E, F, | H, J, C, I) Offspring 2

The *cycle crossover* (CX) has the interesting property that the offspring must have an element in the same position of the genetic vector as one of it parents. The process starts by choosing an element from one of the parents and placing it in the same position of the offspring. If the first element (task) in Parent 1 is chosen, we have

(A, B, C, D, E, F, G, H, I, J) Parent 1

(G, E, B, H, J, C, I, F, A, D) Parent 2

(A, –, –, –, –, –, –, –, –, –) Partial Offspring

Since Task A cannot be used from Parent 2 (it can only appear once), the ninth element in the offspring must be Task I.

(A, –, –, –, –, –, –, –, I, –) Partial Offspring

Again, Task I cannot be taken from Parent 2, so the seventh element must be Task G.

(A, –, –, –, –, –, G, –, I, –) Partial Offspring

This ends the *cycle* portion since Parent 2 has Task G in the first position of the genetic vector. The remaining positions are filled in with the tasks from Parent 2.

(A, E, B, H, J, C, G, F, I, D) Offspring 1

The compliment can be generated by taking the first element from Parent 2 and repeating the process. The complimentary offspring is

(G, B, C, D, E, F, I, H, A, J)   Offspring 2

where the first, seventh and ninth elements are taken from Parent 2 and the rest from Parent 1.

A slight variation of this is the *position based crossover* (PBX) where only some of the elements in the offspring have the same position as one of the parents (Syswerda, 1990). A selected number of random positions along the genetic vector are chosen and the elements in those positions are directly transferred to the same positions in the offspring. These elements are removed from the second parent and the missing elements in the offspring are taken from this parent read from left-to-right. If the second, sixth and eighth positions are randomly selected, the process is as follows.

```
( A, B, C, D, E, F, G, H, I, J)        Parent 1
     |           |     |
(−, B, −, −, −, F, −, H, −, −)    Partial Offspring

(G, E, −, −, J, C, I, −, A, D)     Reduced Parent 2

(G, B, E, J, C, F, I, H, A, D)     Offspring 2
```

and

```
(G, E, B, H, J, C, I, F, A, D)     Parent 2
     |           |     |
(−, E, −, −, −, C, −, F, −, −)     Partial Offspring

(A, B, −, D, −, −, G, H, I, J)     Reduced Parent 1

(A, E, B, D, G, C, H, F, I, J)     Offspring 2
```

Conversely, the positions chosen from the second parent to produce the second offspring could be all of those not chosen in the production of the first (Davis, 1985, 1991). The second offspring would be the following.

```
(G, E, B, H, J, C, I, F, A, D)       Parent 2
 |     |  |  |     |     |  |
(G, −, B, H, J, −, I, −A, D)         Partial Offspring

(−, −, C, −, E, F, −, −, −, −, )     Reduced Parent 1

(G, C, B, H, J, E, I, F, A, D)       Offspring 2
```

Ahuja and co-workers (2000) presented the *swap path crossover* (SPX) operator. In the SPX, they started at a random point in the genetic vector, proceeding cyclically, and swapped elements so that each takes the others value. For example, if we use a five task case and start at the first element, this procedure generates the following.

(A, B, C, D, E) Parent 1
|
(E, B, C, A, D) Parent 2

Switch A and E in both parents

(E, B, C, D, A)
|
(A, B, C, E, D)

Jump to position 4 since positions 2 and 3 are the same, so switch D and E in both.

(D, B, C, E, A)
|
(A, B, C, D, E)

Switch A and E in both.

(D, B, C, A, E) Offspring 1

(E, B, C, D, A) Offspring 2

In an *edge recombination crossover* (ERX) only a single offspring is created from each mating pair. Here, both parents are examined to determine the cyclic connections to each task and the task with the smallest number of edges is chosen and removed from the list. Ties are broken by a random choice. To simplify the example, only five tasks will be considered.

(A, B, C, D, E) Parent 1 A : B, E, C, D

(E, B, C, A, D) Parent 2 B : A, C, E

C : B, D, A

D : C, E, A

E : D, A, B

Randomly choose between B, C, D and E; say D.

(D, –, –, –, –) A : B, E, C

B : A, C, E

C : B, A

E : A, B

Randomly choose between C and E; say C.

$$(\mathrm{D,C,-,-,-})\quad \mathrm{A: B,E}$$
$$\mathrm{B: A,E}$$
$$\mathrm{E: A,B}$$

Randomly choose between A, B and E; say A

$$(\mathrm{D,C,A,-,-})\quad \mathrm{B: E}$$
$$\mathrm{E: B}$$

Randomly choose between B and E; say E

(D, C, A, E, B) Offspring

A single-point crossover can also be used where the sequence before the cut is taken from one parent, the other parent is appended, and duplicates in the other parent are removed.

(A, B, C, D, E, F, G, H, I, J) Parent 1
|
(G, E, B, H, J, C, I, F, A, D) Parent 2

(A, B, C, D, E, F|G, E, B, H, J, C, I, F, A, D) (G, E, B, H, J, C|A, B, C, D, E, F, G, H, I, J)

(A, B, C, D, E, F, G, H, J, I) Offspring (G, E, B, H, J, C, A, D, F, I)

A 2-point crossover can also be used, where the region between the cut points are kept from one parent and the other parent is read from left-to-right to fill it in.

(A, B, C, D, E, F, G, H, I, J) Parent 1
| |
(G, E, B, H, J, C, I, F, A, D) Parent 2

(–, –, –, D, E, F, –, –, –, –) (–, –, –, H, J, C, –, –, –, –)

(G, B, H, D, E, F, J, C, I, A) Offspring (A, B, D, H, J, C, E, F, G, I)

This is known as a *linear order crossover* (LOX) (Falkenauer and Bouffouix, 1991).

4.6. Mutation operators

As with the mating operator, the mutation operator depends upon the coding method and, for gene-based coding, on the genetic alphabet. If each genetic vector contains a bit string (gene-based coding with a binary genetic alphabet), the simplest mutation operator flips a bit in a binary-coded GA. The basic assumption in GAs is that this mutation is small, but especially with standard binary coding (Hamming cliff) and with Gray coding (see above) a single bit flip can make a relatively large change in the value.

Two different applications of a mutation probability are used in practice. The first applies to the entire genetic vector and the second to each element in this vector. In the first case, if a randomly generated number R in [0,1] is less than P_{mut}, a single bit is randomly chosen and it is flipped. In the second case, a random number is generated for each element of the vector and in cases where this number is less than P'_{mut}, the bit is flipped. Obviously for the mutation to be relatively infrequent, P'_{mut} must be less than P_{mut}.

De Falco and co-workers (2002) presented two different mutation operators for binary genetic vectors; the *frame-shift operator* and the *translocation operator*. In the frame-shift operator, a block of the genetic vector is shifted one position. This can occur by deletion first or insertion first. In deletion first, the bit before the block is deleted, the block is shifted one position to the left, and a random bit is placed after the block. If *b* represents the random bit, an example of this would be

```
(0, 1, 1, 1, 0, 0, 1, 0, 1, 1, 0, 0, 0, 1)   Copied Parent
             |             |  block
(0, 1, 1, 1, –, 0, 1, 0, 1, 1, 0, 0, 0, 1)   Deletion

(0, 1, 1, 1, 0, 1, 0, 1, –, 1, 0, 0, 0, 1)   Shift

(0, 1, 1, 1, 0, 1, 0, 1, b, 1, 0, 0, 0, 1)   Offspring
```

In insertion first, a random bit is placed before the block. This shifts the block one place to the right and the first bit after the block is deleted.

```
(0, 1, 1, 1, 0, 0, 1, 0, 1, 1, 0, 0, 0, 1)   Copied Parent
             |             |  block
(0, 1, 1, 1, 0, b, 0, 1, 0, 1, 0, 0, 0, 1)   Offspring
```

In translocation, two blocks of the same size are selected and are swapped. The only caveat is that both blocks must completely reside within different genes. If there are three genes and each are coded with five bits, we have

```
|  Gene 1  |  Gene 2  |  Gene 3  |

(0, 1, 1, 1, 0, 0, 1, 0, 1, 1, 0, 0, 0, 1, 0)   Copied Parent
  |   |                           |   |  blocks
(0, 0, 1, 1, 0, 0, 1, 0, 1, 1, 0, 0, 1, 1, 0)   Offspring
```

The length of the block is randomly selected to lie between a minimum and maximum value. In practice, this minimum value can be zero which gives a finite probability of no mutation.

If real-valued coding is used, an element of the genetic vector, c_i, can be mutated by the following expression.

$$c_i = c_i + (2R - 1)\delta_i$$

where δ_i is the maximum allowed mutation for this element. δ_i can be constant throughout the search, it can monotonically increase or decrease as the simulation proceeds, or it can be changed based upon the statistics of the current population. This latter procedure has been used in several studies to maintain diversity in the population (see Section 5).

For node-based coding, a mutation operator has to change the order of the elements in the genetic vector. Examples of node-based mutation operators include *switching*, *relocation*, and *inversion*. The switching operator randomly chooses two loci in the genetic vector and switches the elements in these locations.

```
(G, J, B, D, E, F, I, C, A, H)   Offspring
       |        |
(G, J, F, D, E, B, I, C, A, H)   Mutated Offspring
```

The relocation operator randomly selects a position between two elements and an element and moves the element to this position. The other elements can be (cyclically) shifted to the left or the right. The following examples illustrate the effects of this operator.

```
(G, J, B, D, E, F, I, C, A, H)   Offspring
  element |        | position
(G, J, D, E, F, B, I, C, A, H)   Mutation   (H, G, J, D, E, F, B, I, C, A)
         Left Shift                                   Right Shift

(G, J, B, D, E, F, I, C, A, H)   Offspring
  position |        | element
(J, B, I, D, E, F, C, A, H, G)   Mutation   (G, J, B, I, D, E, F, C, A, H)
         Left Shift                                   Right Shift
```

In certain problems, a particular element must be located in the first position of the genetic vector (like the Traveling Salesman Problem where all routes must start from a given city) and only certain shifts will be allowed.

The inversion operator selects two cut points and inverts the order of the elements between the points. If this is done in a cyclic fashion, two mutations are possible.

```
(G, J, B, D, E, F, I, C, A, H)   Offspring
          |     | cut points
(G, J, F, E, D, B, I, C, A, H)   Mutation   (C, I, B, D, E, F, J, G, H, A)
      inner-inversion                              outer-inversion
```

In general, crossovers that retain low-order patterns are exploitation operators while mutation is an exploration operator.

4.7. Maturation operators

The mating and mutation operators described so far represent the biological cycle for simple organisms. In more advanced species, the environment also plays a role in their eventual fitness and ability to produce further offspring. This process is termed a *maturation operator* and either lets the offspring adapt to their local environment or change to better augment or assist the population.

If the adaptation to the local environment employs an optimization or hill-climbing process to maximize its fitness, the overall algorithm is known as a Memetic Algorithm (Moscato, 1989). In the literature, this is also known as a Genetic Local Search Algorithm or a Hybrid Genetic Algorithm. For example, studies by Bosworth and co-workers (Bosworth et al., 1972; Foo and Bosworth, 1972; Zeigler et al., 1973) used a

Fletcher-Reeves optimization, though they called this a mutation operator, and Park and Froment (1998) used a Levenberg-Marquardt optimization procedure.

If the adaptation is meant to improve the collective behavior of the population, it usually takes the form of a process to enhance the population's diversity. In Konig and Dandekar (1999) a *pioneer search* was used every tenth generation. This search required every offspring to be different from any member of the current population if the length of the genetic vector was relatively small, while for larger genetic vectors they must differ from all other members in four or more positions. A similar *uniqueness operator* was used in (Luke, web) to find the maximum common substructure between a query structure and compounds in a chemical database.

4.8. Processing offspring

Using the standard notation, the population size is μ and this should remain the same size throughout the simulation. Two parents are used to generate one or more offspring, and the question is what to do with the offspring.

4.8.1. Non-generational algorithms

In a non-generational algorithm, also known as an algorithm with a zero generation gap, any offspring that is kept replaces a member of the population and can be immediately used to generate offspring. Care must be used if a mating pool and/or various selection procedures are employed (see Section 4.4).

The original GA had two parents generate a single offspring, and this offspring replaced the weakest member of the population. This increases the rate at which schema are formed and therefore increases the rate of convergence of the population to a good solution. On the other hand, all genetic information present in that weakest member was removed from the *gene pool* before it had a chance to be passed to an offspring, unless that solution was one of the parents.

To preserve possibly good genetic information, the offspring can be compared to one or both of the parents and the best one or two solutions are placed in the population. This comparison can either be deterministic or probabilistic. In a deterministic process the offspring's fitness is compared to the fitness of one or both parents and the solution with the lowest fitness is discarded. In a probabilistic process, the offspring and one or both of its parents are used with one of the selection procedures described in Section 4.4. Another option is to replace the most similar member of the population.

Cavicchio (1970) used 'pre-selection' where a good offspring replaces one of its parents in the hope of maintaining population diversity. Ahuja and co-workers (2000) used a combination of replacement methods. If the offspring has a higher fitness than both parents, it replaced the most similar parent. Otherwise it replaced the weaker parent.

4.8.2. Generational algorithms

In a generational algorithm, all offspring are placed into a new population. Production of offspring continues until λ offspring are present in the new population. At this point, several options are available for generating a parent population of size μ.

One option is to create the parent population from the new population. This is denoted as a (μ, λ) process and requires that λ is at least as large as μ. These μ members of the new population can either be selected deterministically or probabilistically, with or without replacement. A variation of this is to transfer one or more (van Kampen and Buydens, 1997) of the most-fit solutions from the old parent population to the next generations' parent population and select the remaining members from the offspring population. This variation is known as the *elitist strategy*.

Another option is to combine the parent and offspring populations and select μ members from this combined pool; again either deterministically of probabilistically. This is called a $(\mu + \lambda)$ process, and if a deterministic selection is used, the μ solutions with the highest fitness are selected. If a probabilistic selection is used (see Section 4.4), the elitist strategy can again be enforced. One way to implicitly incorporate the elitist strategy into a probabilistic selection is to use a modified tournament selection. Here, each solution in the combined population is compared to ρ other randomly selected solutions and then scored. This score represents the number of solutions that have fitness less than the one considered. When finished, each solution in the combined population will have a score between 0 and ρ. By ranking the solutions from highest to lowest, the μ solutions with the highest score are used as parents in the next generation.

A third option is to combine the populations and select only a few (η) of the best solutions. This again can be based on their fitness or on their scores from the modified tournament selection. These η solutions represent *focus points* in search space, and in a cyclic fashion each focus point is used to select the solution from those remaining in the combined population that is most similar to it. This continues until a total of μ solutions are selected.

Each of the generational procedures can be augmented to maintain more of the genetic diversity in the parent population of the next generation. This is done by forcing the complimentary pair of offspring to stay together. In other words, if a particular offspring is chosen to be placed in the next generation's parent population, its compliment must also be placed in this population. This would mean that during offspring generation, both offspring are placed in the new population. In addition, each chromosome should carry an extra bit that is set to 0 if it does not have a compliment (which would be true only for the randomly generated initial population) and to 1 if it does. Since the initial, random population is not expected to contain very good solutions, the parent population will consist of only complimentary pairs of offspring after a relatively small number of generations. In addition, if any mating produced a reasonably fit offspring, all of the genetic information from both parents will be transferred to the next generation.

4.9. Termination metrics

To this point, all aspects of a GA have been described. After a fitness function and the form of the genetic vector are determined, an initial population of putative solutions can be generated. Two parents are selected and generated offspring using a probabilistic application of mating, mutation, and possibly maturation operators. A selection procedure is then used to incorporate selected offspring into the population (non-generational) or build a new parent population for the next generation. The only aspect left to consider is when to stop the GA.

Several simple procedures can be used to terminate the GA if the value of the optimum solution is not known. Obviously if the optimum fitness is known, the simulation stops as soon as an optimum offspring is produced, and this method can be used to test various reproduction operators, selection schemes, population sizes, and forms of the genetic vector. The simplest termination metric for a generational algorithm is to stop the search after a user-defined number of generations. For non-generational algorithm, a maximum number of matings can be used instead.

Other termination metrics use properties of the population from one generation to the next (generational algorithms) or after a given number of matings (non-generational algorithms). For example if the most-fit solution has not changed in a given number of generations/matings, or if the change in the average fitness of the population between one generation and the next is below a threshold, the search stops. Finally, the convergence of the population towards a single solution can be used as a termination metric. For example, the search can stop if the size of the schema is equal to or greater than a given fraction of the number of loci, or if the distance between the most-fit and least-fit genetic vectors (Hamming distance for binary coding and Euclidean distance for real or integer genetic alphabets) is less than a threshold value.

Care should be used if a termination metric based on the population is employed and there is a non-zero mutation probability. An example presented in Chapter 13 shows that even if the entire parent population converges to a single solution, a non-zero mutation probability can allow one or more offspring to search a new, potentially better region of search space. In this example, the population left this converged state and found a better solution.

5. Exploration versus exploitation

As with many search procedures, the construction of a specific search algorithm using the GA methodology requires deciding between making every effort to find the globally optimum set of features and/or parameter values and the need to have the algorithm produce a result within a reasonable amount of computer time. Any decision that promotes the formation and preservation of schema promotes exploitation. This means that low-order patterns from fit individuals are transferred to multiple offspring so that these patterns quickly appear in all members of the population. As stated in Section 3, this focuses the search onto a sub-dimension of the full search space and aids in rapidly finding good, but sometimes sub-optimum, solutions. Conversely, any procedure that produces an offspring with different values at multiple loci from either parent promotes exploration into other regions of search space not sampled by the current population. This will assist in finding the global optimum, but can greatly increase the running time of the program. This section examines many of the choices presented in Section 4 to determine whether they promote exploration or exploitation.

5.1. The genetic vector

As stated above, diploid organisms inherently have greater diversity than haploid organisms. This is because two organisms can have large differences in one or both chromosomes and still represent identical solutions. When they are used to create offspring, these differences can emerge and the offspring can examine different regions of search space. This mechanism can be very useful in the search for the optimal solution, but it is not heavily used in practice. Therefore, the remainder of this section deals with haploid organisms.

For gene-based or delta coding, the size of the search space is given by the product of the genetic alphabet's length (N) at each locus in the genetic vector of length L. For example, if the genetic vector is a bit vector ($N = 2$) of length L, the size of the search space is $N^L = 2^L$. For node-based coding that represents an order of tasks or decisions, the size of the search space is $L!$.

Exploration can be considered a local search around a good solution. Since GAs work by combining the genetic vectors of two solutions, exploration assumes that the distance between these parents is relatively small. This occurs when the parents have many schema and/or the schemas are relatively large. As the distance between the parents increases, the mating promotes exploration to a greater extent since exploration requires that the offspring are relatively different from either parent. As the size of the search space increases, the members of the initial population will, on average, be further apart. This means that larger search spaces start with exploration and it takes longer for schema to be constructed so that exploitation is promoted.

The size of the search space increases with increasing N and L for gene-based and delta coding schemes, and increasing L (number of tasks or decisions) in node-based coding. Since the user cannot change the size of the search space for node-based coding problems, it will not be discussed further. Similarly, feature selection problems have solutions with a genetic vector that is a binary vector whose length is fixed to the number of available features. Conversely, if the problem is to optimize parameters, the user has control over the size of the search space for gene-based and delta coding.

If binary coding is used (either regular or Gray coding), there is a trade-off between the size of the search space and the accuracy of the decoded value. For example, if each value is coded as a 4-bit gene and there are K genes, $L = 4K$ and the size of the search space is 2^{4K}. If one wishes to increase the accuracy of each value, an 8-bit gene can be used, but the size of the search space will be squared. Therefore, increasing the accuracy of the result increases the average distance between members of the initial population. This means that the GA will take longer to create enough schemas to promote exploration.

A similar argument holds for integer coding. If each locus/gene in the genetic vector has an allele that can vary between 1 and 1000, the genetic alphabet can be restricted to all multiples of five. If there are L loci/genes, the size of the search space is 200^L. If the genetic alphabet contains all even numbers, the accuracy of the result increases but the size of the search space increases to 500^L. The most accurate solution has a genetic alphabet that includes all integers, but the search space is a maximum (1000^L). For even small numbers of parameters (say $L = 10$) the increase in the size of the search space is dramatic (approximately 10^{23}, 10^{26}, and 10^{30}, respectively).

The problem is compounded even further if real coding is used. Here the genetic alphabet contains an infinite number of entries and therefore the search space is uncountably infinite (∞^L). This means that schema can never be formed unless a mutation operator is used, and only a limited exploitation occurs. The search space can be made finite if binary or integer coding is used, but this introduces approximations into the answer. The author has found that if an optimization procedure is used as the maturation operator, the finite number of minima greatly reduces the search space and schemas are formed, focusing the search (Luke, 1999).

5.2. The initial population

As stated above, as the size of the search space increases, the GA starts out in a purely exploratory phase if the initial population is randomly generated, since the average distance between members of this population is quite large. This can be circumvented by using some other rapid search procedure to build an initial population with high-fitness members. Unfortunately, Keser and Stupp (1998) found that seeding the initial population with known members of high fitness caused a premature convergence of the algorithm to a sub-optimal solution. Again, this is because the algorithm will quickly enter an exploitation phase and the global optimum will not be found unless it is reasonably close to one or more of the most-fit members of this population.

The probability of having the global optimum close to a member of the initial population is increased if the size of the initial population is large enough. But as the size of the population increases, more computer time is needed for each generation and the formation of schema is slowed. Therefore, unless the length of the genetic vector and genetic alphabet are both small, large population sizes are generally avoided.

With smaller population sizes, the GA must start out in an exploratory phase and then shift to an exploitation phase (see below). This will only yield good results if the initial population spans the entire search space, or if the mutation operator starts out having dramatic effects on the phenotype of the offspring (see below). Several researchers have developed schemes to build an initial population with as much diversity as possible.

For example, Bandyopadhyay and co-workers (1998) separated the population into two classes (M and F). The initial population contained equal numbers of Ms and Fs. They used a binary coding and the M population was generated first by randomly assigning a binary value to each locus of the genetic vector. For each member of the F population, the value at each locus was determined by randomly selecting a member of the M group and using the compliment of its value. Since a different M was chosen for each locus, the Ms and Fs did not form complimentary pairs, but any bias present at a locus in the M group was probably offset by an opposite bias in the F group. Also, since all matings have to be between an M and an F, the probability that they have a large Hamming distance is increased. This large distance increases the exploratory characteristic of the mating.

Guo and Zhao (2002) used a different method for generating a diverse initial population of binary-coded genetic vectors. Each time a new solution is randomly generated, it is compared with all previously generated solutions. If the Hamming distance with a previous solution is less than a threshold, it is discarded and another one is generated.

5.3. Selection schemes

Any procedure that selects parents or offspring using a fitness-based metric (see Section 4.4) promotes exploitation. This is because a more-fit member of the parent population will be selected more often and has more of its genetic material present in the offspring. In addition, fit parents have a greater chance of generating fit offspring, especially for relatively small population sizes. If these offspring immediately replace members of the current population, as is true in a non-generational algorithm, local patterns from the fit parents will quickly grow in the population and generate schema. In generational algorithms, schema can quickly form in the offspring population by choosing more fit parents with a greater regularity and this will be transmitted to the next generation's parent population. As stated in Section 4.4, scaling the fitness values before selection can reduce the probability that the most-fit members are chosen as parents and can retard the rate of schema formation.

This argument can also be related to the Hardy–Weinberg law. This law states that for an infinitely large population that takes part in random mating without selection pressure or mutations, the allele frequencies at each locus will not change from generation to generation (Gillespie, 1998). This means that if the population contains 40% 0s and 60% 1s in a given position of the binary genetic vector, the parents are randomly chosen and the population size is sufficiently large, the fraction of 0s and 1s at this location will not change. As the population size decreases, some parents will be chosen more often than others and the probabilities of finding a 0 or 1 at this position will change until all members of the population have the same allele at this position (i.e. a schema forms). This is called *genetic drift* for random parent selections, but the same process occurs for non-random parent selections independent of the size of the population, assuming that the allele at this position has an effect on the fitness of the organism. Therefore, a fitness-based selection procedure can be thought of as a procedure to reduce the effective size of the population and promote this drift, and reduce the diversity of the population (promote schema formation).

In the Population Genetics literature there are two equivalent terms that describe the mating of parents with similar features, *assortive mating* and *assortative mating*. Choosing parents with dissimilar features is known as *dissortative mating*. Conversely, Hollstein (1971) used *positive assortive mating* and *negative assortive mating* to label these two extremes. Since the selection is based on observable features, the measure of similarity is based on the phenotype of the parents. This was extended so that the parents can be chosen by either *genotypic assortative mating* or *phenotypic assortative mating* (Strickberger, 1985; Gardner et al., 1991; De et al., 1998). To properly distinguish between the four possible similarity-based selection procedures, they will be denoted *positive phenotypic assortative mating*, *negative phenotypic assortative mating*, *positive genotypic assortative mating*, and *negative genotypic assortative mating*. For each of these methods, it is assumed that one of the parents is chosen based upon its fitness.

In positive phenotypic assortative mating, the second parent is the one with the closest fitness value. To promote negative phenotypic assortative mating, the second parent should be randomly selected, but then rejected if is fitness is within a threshold value of the first parent. Note that this is different than maximizing the difference in fitness values since this latter case would require that the most-fit or least-fit parent be the second parent in all cases. Also note that either mating does not necessarily promote exploitation beyond the fact that the first parent is chosen based on fitness, because the comparison uses the phenotype. It is possible, especially in the early generations, that two parents have virtually the same fitness, but very different genetic vectors. Similarly, small changes in the genetic vector can produce large changes in the resulting fitness; so negative phenotypic assortative mating may select two parents with very similar genetic vectors. Positive phenotypic assortative mating is expected to have a greater exploitative effect and increase the drift of the population because the first parent is chosen based on fitness and the second parent is deterministically chosen. Therefore, a very fit parent pair will be chosen often and have their genetic information passed on to many more offspring than if the selection was random. Negative phenotypic assortative mating will still promote exploitation since the first parent is chosen based on fitness, but since the second parent is randomly chosen, the drift and formation of schema will be reduced.

In positive genotypic assortative mating, the second parent is chosen based on the smallest difference in the genetic vector relative to the first parent. For binary genetic vectors, the difference between two genetic vectors is simply their Hamming distance, which is simply a count of the number of loci that have different alleles. For genetic vectors that use integer or real coding, this difference can be the sum of the absolute differences between the alleles across all loci (their Manhattan distance), their Euclidean distance, or any other distance metric. *Inbreeding* is similar to positive genotypic assortative mating, but is not the same. If an inbreeding operator is used (Hollstein, 1971) the simulation needs to start with a single parent pair that represents the origin of a particular *family*. They are then used to produce other members of this family and another parent pair is used to create another family. This is repeated until the initial population is constructed. The inbreeding operator then requires that mating can only occur between members of the same family. Because each family is relatively small, the drift of the family towards a homogeneous population is accelerated, though each family can drift towards a different final genetic vector. This is one example of a process known as *niching*.

Negative genotypic assortative mating forces the second parent to have a genetic vector that is sufficiently different from the first. Eshelman (1991a) called this *incest prevention* and simply required that the distance between the parents' genetic vectors be above a threshold value. Craighurst and Martin (1995) proposed a different definition of incest prevention where two parents cannot mate if they share a common ancestor within a given number of generations. These two methods of incest prevention can produce very different results since two genetic vectors can be very close without having a recent common ancestor because of the fitness-based selection of parents for the next generation. In addition, if the required generation gap to a common ancestor is large enough, two parents may not be allowed to mate with the Craighurst and Martin criterion even though the difference in their genetic vectors is substantial. Only the Eshelman procedure qualifies as negative genotypic assortative mating, but both methods try to require a significant difference between the parents. This difference promotes exploration, and retards the drift of the population and schema formation.

Hollstein (1971) also presented the *line breeding operator*, which increases the exploitation around the best solution. With this method, the solution with the highest fitness was always used as one of the parents and was mated with each member of the parent population. The offspring of these matings become the subsequent parents. Konig and Dandekar (1999) explored increasing the rate of schema formation by seeding each new population with good solutions from fit parents. At the beginning of each generation, the most-fit solution and a second solution chosen by a roulette wheel procedure were mated using a series of 1-point crossovers. The cut point traveled down the genetic vector to produce a group of offspring. The fitness of each offspring was determined and the two most-fit members were placed into the new population. This process was repeated five more times with both parents chosen using the roulette wheel procedure. This seeded the new population with 12 solutions that tried to maximize the exploitation of the current population.

5.4. Mating operators

As stated in Section 4.5, a mating probability, P_{mate}, can be used with this operator. As this probability decreases, the offspring will become more populated with unaltered (except from mutation and/or maturation) genetic vectors from the parents. This increases the presence of fit members in the population and promotes the formation of schema. In addition, it was shown that crossover operators having a small number of cut points promote the formation of schema while those that have a larger number promote exploration. This is because operators with a large number of cut points are more likely to break local patterns in the parents and slow the process of schema formation. This argument was supported in the study of Hasancebi and Erbatur (2000). They analyzed the performance of GA searches on various test problems, and came to the conclusion that as the number of crossover points increases from single-point to uniform crossover, the algorithm changes from exploitation to exploration, though in many problems a 2-point crossover outperformed a 1-point crossover. Therefore, they suggested that if the global optimum is close to a fit member of the population, the number of crossover points should be small (say two).

Rosenberg (1967) used an Offspring Generation Function (OGF) where each possible crossover point contained a likelihood number X_i between 0 and 7 that was stored in the chromosome. The probability of cutting at a point is $P_i = X_i / \Sigma_i (X_i)$. If a fit offspring is produced, the likelihood number at that point can be increases in the offspring and the likelihood at other points can be reduced. If keeping a pair of adjacent genes together in an offspring produces high fitness, the likelihood will eventually be reduced to zero and a schema will be quickly generated. This procedure is relatively unique in that memory of good and bad crossover points can be retained over many generations.

Conversely, the diversity in a new population can also be promoted by generating offspring using an *orthogonal crossover operator* (OCX) (Yu et al., 2000). If a mating produces t offspring and the length of the genetic vector is m, they used a $t \times m$ table where each vector is orthogonal to the rest. If this operator is used, t offspring are generated by looking at each vector and choosing the gene from one parent if the table element was 0, and from the other if it was not.

5.5. Mutation operators

As stated in Section 4.6, the mutation operator promotes exploration since it can break schema and reduce the similarity between an offspring and its parents. Many authors have suggested that the probability of mutation, P_{mut}, changes during the simulation. For example, if this probability decreased with each new generation, the search would convert from an exploration to an exploitation process. Conversely, other authors have suggested that the mutation probability be increased at certain points to reduce the rate of early schema formation and convergence on a sub-optimal solution. For example, Keser and Stupp (1998) proposed that if the Hamming distance between a new binary-coded offspring before mutation and a member of the new population is below a threshold, the mutation rate is increased. This would increase the diversity in the new population and reduce the rate of schema formation. A similar procedure is to use the incest prevention method of Eshelman (1991a). Here two parents are only mated if their Hamming distance is above a threshold and an increase in the mutation probability is not required.

5.6. Maturation operators

Since the maturation operator can be any operator that helps an individual offspring or the population as a whole, it can promote either exploration or exploitation. If this operator performs a local or global optimization, it can move offspring to the same region in search space. This increases the agreement between the offspring and promotes exploitation in the next generation. As stated above, this maturation operator is recommended if real coding is used in the genetic vector.

If the maturation operator promotes diversity in the offspring population, such as continuous mutation until an offspring is sufficiently different from the others or using the uniqueness operator, it will increase or maintain the distance between members of the next generation's parent population and promote exploration.

5.7. Processing offspring

Different procedures for choosing offspring to become future parents can also promote exploitation or exploration. The procedure used in the SGA is designed to promote rapid focusing of the population and therefore promote exploitation. In this algorithm, the parents are chosen based upon their fitness and the offspring immediately replaces the weakest member of the population. This non-generational algorithm increases the presence of local patterns from fit individuals with every mating. By replacing the weaker parent instead of the weakest member, the growth of these local patterns is slowed and genetic information in weaker members of the parent population can be used before this member is removed.

For generational algorithms, the procedure that increases the focusing of the next generation's parent population the most is a probabilistic $(\mu + \lambda)$ selection with replacement. In this procedure, the μ members of the parent populations are combined with the λ members of the offspring population and a fitness-based selection procedure is used. If a solution is chosen, it is not removed from the combined population so that fit members have a good chance of being chosen more than once. A slightly less focusing method is a deterministic $(\mu + \lambda)$ selection. Here the populations are combined and the most-fit μ members become the next generation's parents.

An algorithm that maintains the genetic diversity from generation to generation would use each parent only once to generate a complimentary pair of offspring and a (μ, λ) selection procedure (with $\lambda = \mu$). This means that the offspring will have the same genetic information as the parents (excluding changes caused by mutation), but the mating operator produces genetic vectors with different combinations of genes. With this method and others that use a probabilistic selection, it may be advantageous to use an elitist strategy to ensure that the fitness of the most-fit member does not decrease from one generation to the next. This can be done by simply copying the most-fit member and its compliment from the parent to the offspring population. The remaining $\mu - 2$ parents are used to produce complimentary pairs of offspring and the new population becomes the next generation's parent population.

Yu and co-workers (2000) suggested other methods to maintain diversity in the population. The first is to use a Boltzmann acceptance in non-generational algorithms. If the offspring's fitness is greater than the weaker parent it automatically replaces this parent, while if the weaker parent has a fitness that is Δf greater than the offspring, the probability that it is replaced is given by $e^{-\Delta f/T}$ where T is an 'effective temperature.' In this algorithm, the effective temperature starts at a large value. If a random number between 0.0 and 1.0 is less than this Boltzmann probability, the offspring replaces the parent; otherwise the population is unchanged. This replacement method makes the Boltzmann GA very similar to an ensemble Simulated Annealing (Kirkpatrick et al., 1983), though the latter would produce an offspring by mutating a single parent.

These authors (Yu et al., 2000) also presented a method for maintaining population diversity in a generational algorithm. This procedure uses a deterministic $(\mu + \lambda)$ selection and then applies an *effective crowding operator*. This operator ranks the best μ members from the combined populations from highest to lowest fitness and then compares adjacent members. If the difference in their genetic vectors is less than a threshold value, the least-fit member of the pair is replaced by a randomly generated solution. In Yu et al. (2000) this operator was only applied if the difference between the highest fitness and average fitness of the population was less than a threshold value, and acted to retard premature convergence to a sub-optimal population.

Kim and co-workers (2002) also presented two methods of processing the offspring, one that promotes exploitation and one that promotes exploration. In the first, they used a separate population to store the μ best unique solutions found so far. While a fitness-based, probabilistic $(\mu + \lambda)$ selection method was used most of the time, every n generations a *generation-apart elitism* was used where these μ solutions became the parent population for the next generation. This procedure basically ran a probabilistic and a deterministic $(\mu + \lambda)$ selection method in parallel, where the latter only selected unique solutions. The probabilistic selection was interrupted every n generations and the deterministic selection was used instead. This raises the average fitness of the population and forces the procedure to continue using only the most-fit solutions found to date in the simulation. Therefore, this procedure also uses information from past generations that may have been lost during the probabilistic selection, and as stated above, this deterministic selection increases the focusing of the population into one or more regions of search space that contain good solutions.

Their second method was designed to increase the diversity of the population. Every m generations, h of the members for the next generation's parent population were randomly selected from the offspring population without regard to their fitness. The remaining $\mu - h$ members were selected by a fitness-based method. By randomly selecting some of the offspring for the next generation, it is possible to break schema that were present in the previous generation and therefore increase the diversity.

Eshelman (1991b) also proposed a method of forcing diversity onto a population when it may have become trapped around a sub-optimal solution. He proposed a *cataclysmic mutation* when matings were not able to generate better offspring. If a deterministic $(\mu + \lambda)$ selection does not change the parent population from one generation to the next, or the best η members of the population do not change in a deterministic or probabilistic selection, the best solution is used to generate the remainder of the population using only a mutation operator. This procedure forces diversity into the population while retaining good patterns from the best genetic vector in many of the other genetic vectors.

5.8. Balancing exploration and exploitation

As stated above, decreasing the probabilities of mating, P_{mate}, and mutation, P_{mut}, increases the presence of fit members in the parent population of later generations and increases the exploitation of these solutions. Conversely, increasing these probabilities and using crossover operators with more cut points promotes exploration. Bagley (1967) suggested, but did not try, to store the crossover and mutation probabilities within the genetic vector so that they can be modified during the simulation. Cavicchio (1970) proposed that the crossover and mutation probabilities change with the average fitness of the population.

The 'sex-based' mating algorithm of Bandyopadhyay and co-workers (1998) uses sex, selection, and several mating operators to balance exploitation and exploration. For example, an elitist strategy is used where the most-fit M and F are placed into the new population. Then, a roulette wheel selection procedure is used to pick a selected number of parents to place in a mating pool, which means that the number of Ms and Fs will probably not be the same. The next step is to use all members of the mating pool to generate offspring. This is done by randomly choosing an M and F without replacement, and applying mating and mutation operators to generate offspring. If the mating pool becomes emptied of one sex, the most-fit member of this sex in the mating pool is used as one parent for all remaining members of the other sex. The sex coding is such that a mating can produce two Ms, two Fs, or an M and an F. Therefore, at some point, the mating pool may only contain members of one sex. In this case, an offspring is generated from each parent using only a mutation operator. Therefore, the combination of sex, selection, and offspring generation methods produces an algorithm that can mate fit parents with potentially large (exploration), small (exploitation), or intermediate (mixed exploration and exploitation) Hamming distances; mate a dominant (most fit) animal of one sex with the remaining members of the mating pool (partial to full exploitation), which is the *line breeding operator* proposed in Hollstein (1971), or can produce offspring using only the mutation operator (exploration).

Guo and Zhao (2002) rank-ordered the solutions based on fitness and from top down only chose points that have a decoded, Euclidean distance greater than a threshold from all previously selected points. Only these points are used in mating and the parents and offspring are used to *cast out* several new offspring. In the initial part of the simulation, this casting distance is large, and it becomes smaller (local search around selected points) in later generations. This means that a mutation-like operator is used to generate new solutions from these solutions. This operator generates very different solutions at the start of the simulation (global search or exploration), and only slightly changes the solutions (local search or exploitation) later. By requiring the distance between the parents to be greater than a threshold, different regions of the search space are simultaneously explored.

Hasancebi and Erbatur (2000) proposed two new mating operators; *mixed crossover* and *direct design variable exchange crossover*. The first technique is simply a prescription for using various crossover operators. For their test problems, a 2-point crossover promoted exploration the most, so it was used last. For example, if the simulation ran for 100 generations, they would use a 3-point crossover (exploration) for the first 20 generations, a 1-point crossover (partial exploitation) for the next 40, and a 2-point crossover (full exploitation) for the last 40 generations. The second technique copies the bit strings from each parent to each offspring. The substrings are switched between offspring with a probability P_{cross}, and this probability decreases with each generation. In essence this is a biased uniform crossover between integer-coded strings, and the bias increases with each generation.

Yoon and Moon (2002) examined four strategies for choosing between multiple crossover operators in a single simulation. The first strategy counts the number of times each operator generated an offspring that was selected as member of the parent population in the next generation. From this, a probability that a particular operator generates a good offspring can be determined, and this probability is used to select the operators in the next generation. The second strategy can be thought of as the inverse of the first. After a generation is complete and the frequency of using each operator is determined, the highest frequency is assigned to the operator with the lowest and so on, so that the operator that generated the most offspring has the smallest probability of being used in the next generation. The third strategy simply tries to use each operator to the same extent in that the operator that has generated the fewest number of offspring to date is used next. The fourth strategy simply uses an unbiased selection to pick the next crossover operator. For a small number of operators, it is also possible to use them all for each pair of parents and simply keep the best offspring (with or without its compliment).

Simões and Costa (2001a,b) proposed using a genetic operator called transformation instead of crossover to solve dynamic problems. Transformation consists of transferring small pieces of the chromosome between organisms. If the incorporation fails (restriction), the additional genetic material is destroyed, as what may happen to viral DNA. Conversely, the incorporation can succeed (recombination) and the added genetic material replaces some of the cell's genetic information.

This algorithm, known as a Transformation-based Genetic Algorithm (TGA) proceeds as follows.

1. Generate an initial population and initial gene segment pool. This pool contains chromosomal fragments that can be transferred to members of the population. A binary genetic alphabet is used, so each gene segment is a small bit-string. In addition, the chromosome is assumed to be circular (see below).
2. In each generation, a roulette wheel selection procedure is used to choose a member.
   (a) A gene segment is randomly chosen from the segment pool and replaces a random segment in the chromosome (a circular chromosome keeps the length constant).
   (b) This transformed individual is placed in a new population.
   (c) At the end of the generation, the new population replaces the old and the gene segment pool is updated.
      (i) 70% (for example) of the segments are taken from segments of the old population and the remaining 30% are randomly generated.

Simões and Costa (2002a) continued the preceding study and used the same tests. In Simões and Costa (2002b) they examined the parameters and adjusted them to optimize the quality of the algorithm. In particular, in the original study the mutation rate was 0.1%, the transformation rate was 70%, the replacement rate was 70%, and the segment lengths were random. In Simões and Costa (2002b), empirical tests showed that values of 0, 90, and 40%, and a fixed segment length of 5 performed best.

They compared the original TGA and this Enhanced TGA (ETGA) with two other methods; the Triggered Hypermutation GA (THGA also called HMGA) and the Random Immigrants GA (RIGA). The THGA (Cobb, 1990) states that if the algorithm becomes stable the mutation rate is increased, while if it is changing (time-averaged best performance) the mutation rate decreases. The RIGA (Grefenstette, 1992) uses a fitness based selection procedure to choose a fraction of the next generation's parents while the remainder, as determined by the replacement rate, is composed of randomly generated genetic vectors.

The TMGA/HMGA method generally did the best in the tests, though the ETGA sometimes did better. The ETGA always did better than TGA and RIGA, but this is expected since the algorithm was optimized for these particular problems.

In the form presented above, the TGA/ETGA method behaves like a mutation operator that is applied to a contiguous substring of the genetic vector. Obtaining the substring from a fit parent from the previous generation would not be of much help since this substring can be placed anywhere in the genetic vector of the current parent. This is probably the reason why a minority of the substrings came from previous parents in the ETGA (40% from previous parents, 60% randomly generated). This procedure therefore maintains an exploratory nature. The exploitation of good results would be enhanced if each substring contained an index that gave the locus number of the first bit if the substring came from a previous parent. Randomly generated substrings would have an index of − 1, for example, and they would be placed anywhere in the parent string. If the substrings mostly came from previous parents, this transformation operator could be used more than once, meaning that an offspring could contain genetic information from three or more 'parents'.

This modified TGA method can also be applied to non-binary-coded genetic vectors. In this case, the substrings would contain the integer or real values for one or more genes and the same indexing scheme would be used.

Sorensen and Jacobsen (web) presented an interesting implementation of a GA that contains several unique features. Each animal is diploid with real value coding, but only one of the chromosomes is used to determine the solution and its fitness. There is also a spatial distribution of animals amongst grids on a torus. Each grid can hold a maximum number of animals, and each grid has a 'field of view' which corresponds to that grid and the eight adjacent grids. The standard deviation of the fitness for animals in a grid region and its neighbors control the storing and retrieving of genetic information between the active and inactive chromosomes for all animals in that grid. If the deviation is high, information is probabilistically stored in the inactive chromosome. If the diversity is low, information is probabilistically retrieved; otherwise, no inter-chromosome communication occurs. In addition, each animal is allowed to probabilistically migrate from one grid region to another, based upon the average fitness of these grids. If the desired field is already fully occupied or there are no individuals in the field of view, the motivated movement is chosen at random. They used a Gaussian mutation that decreased with each generation, $e^{-1/(1+\sqrt{t})}$, where t is the generation number. The overall process is

1. initialize world population (chromosomes and location)
2. for each generation
   (a) calculate fitness
   (b) move individual
   (c) calculate deviation
   (d) store/recall material
   (e) produce offspring
   (f) mutate individual

Offspring are produced for each animal by randomly selecting a partner in its field of view and performing a 1-point crossover of the active chromosome only. Both offspring are mutated and their fitness values are compared to that of the originating parent. The one with the highest fitness stays in the population.

Herrera and co-workers (1997) presented multiple crossover operators that promoted exploration or exploitation and one that was an intermediate operator. Assuming that each parameter i has a minimum and maximum value, $a_{i,\min}$ and $a_{i,\max}$, respectively, and the mating is between two parents with phenotypes of $a_{i,1}$ and $a_{i,2}$, where $a_{i,1} < a_{i,2}$. The range of values for this parameter is divided into three sections: F $[a_{i,\min}, a_{i,1})$, M $[a_{i,1}, a_{i,2}]$, and S $(a_{i,2}, a_{i,\max}]$. Creating an offspring in region M (M-crossover) is exploitation, while regions F (F-crossover) and S (S-crossover) are explorations. They define a fourth region L (L-crossover) where the offspring is contained in $[a_{i,s}, a_{i,m}]$ with $a_{i,s} < a_{i,1}$ and $a_{i,m} > a_{i,2}$ which is called relaxed exploitation.

They presented four different forms for each of these four crossover operations. Their study used real-valued coding and each operator generated a single offspring, though M- and L-crossovers can generate a complimentary pair by replacing λ with $(1 - \lambda)$. In their studies they use the operators to create 1 or 2 offspring each and the 2–4 best offspring replace their parents and other members of the population (their parents are always replaced). A possible problem with this is that each operator is used to generate the entire offspring, so that it is completely exploratory, exploitatory, or relaxed exploitatory. It would be interesting to see whether better results are obtained if different operators are used for different genes.

Other methods include restart mechanisms, adaptive mutation, and addition of memory. The restart mechanism was used in Grefenstette and Ramsey (1992) while increasing the mutation during a change in the problem, called hypermutation, was tried in Grefenstette (1992) and Cobb and Grefenstette (1993). The memory can be implicitly incorporated by using redundant representations (diploidy, tetraploidy) (Ng and Wong, 1995; Hadad and Eick, 1997), or explicitly included by adding an extra-memory. This extra-memory can be retrieved or updated depending upon the conditions of the problem (Trojanowsky et al., 1997).

Finally, the incest prevention mechanism of Eshelman (1991a) can be used to promote exploration at the start of the search. If the minimum difference between parents is relatively large, the offspring will be sufficiently different to promote exploration. As this required difference decreases in later generations, the similarity of the parents and therefore the offspring increases and this focuses the search into a particular region of search space. Similarly, the incest prevention method of Craighurst and Martin (1995) requires that the parents have a sufficient generational distance to a common ancestor and can therefore be thought of as requiring each parent to come from different families. If the population then converges on a particular region of search space, it must do so by relatively independent paths.

6. Other population-based methods

The options presented in Sections 4 and 5 allow the heuristic of a GA to be represented in a large number of algorithms. For example, the standard crossover (mating) operator is only valid for binary-coded genetic vectors. For integer and real coded gene-based genetic vectors, these operators only choose points at the corners of the hyper-rectangle represented by the parents (Fig. 1). When these codings are used, an intermediate recombination or interval crossover operator should be used.

There has also been a great deal of work done on balancing or selectively promoting exploitation and exploration. This is accomplished by maintaining diversity in the population and allowing the algorithm to focus in several areas of search space simultaneously (niching). An example of the latter is the use of focus points when selecting a parent population for the next generation. For example, the combined $(\mu + \lambda)$ parent and offspring populations can be rank orders and the best η members can be selected such that their distance from all previously selected points is greater than a threshold value. The remaining members of the parent population are chosen based upon their similarity to these focus points. This generates a population that is comprised of sub-populations that are located in different regions of search space. It is similar to the formation of families proposed by Hollstein (1971) and is also considered a niching algorithm. If Hollstein's inbreeding operator is used for parent selection, mating only occurs within a family. This allows the algorithm to exploit the genetic information contained in fit parents while searching multiple regions of search space. The cataclysmic mutation scheme of Eshelman (1991b) can also be used to construct families if it is extended to use multiple parents with a given threshold separation instead of simply using just the most-fit solution. It must be emphasized that this process generates multiple offspring from a single parent by only using a mutation operator.

Finally, the OGF of Rosenberg (1967) and generation-apart elitism of Kim and co-workers (2002) allow information collected over many generations to be used by the current population. These ideas are used in other population-based methods, some of which are described here.

6.1. Parallel GA

A parallel GA is an extreme example of niching where separate GAs are run (Muhlenbein et al., 1991). These can be sequentially or simultaneously run on different compute nodes of a cluster. Each GA will focus onto its own region of search space. Each population will be 'shocked' every *n* generations by having the most-fit solution from another population replace its weakest member. One implementation of this is to arrange the populations in a ring and copy the most-fit solution to the adjacent population in a clockwise fashion. When this procedure was used to search for the most stable conformation of small polypeptides (Luke, 1999) it was found that a similar solution was continually copied to the same population. Better results were obtained when a hopping mechanism was used. In the first copying the nearest clockwise neighbor received the solution, while in the second copying the second-nearest clockwise population received the solution.

Since it is assumed that each population will converge on a different solution, the introduction of a different, fit solution to this population will cause this new solution to be used as a parent fairly regularly. This will decrease the similarity between the parents being mated and will cause the algorithm to explore new regions of search space each time it is used.

Care must be taken to ensure that the population is not too small. In this case, this new solution can dominate the parent selections and after a sufficient number of copying, all populations will be focused around the same, potentially sub-optimal, solution (Dymek, 1992). Conversely, if each population is too large, introducing a single solution that represents a different region of search space may be too small of a perturbation to affect the search since its probability of being selected as a parent is decreased.

6.2. Adaptive parallel GA

The Adaptive Parallel GA (Liepins and Baluja, 1991) is a slight variation of the parallel GA. Since each population is assumed to focus on a different region of search space, copying the best solution from each to a central population should create a diverse population of fit solutions. Therefore, this algorithm uses these fit solutions in a separate GA. In other words, every *n* generations the best solution from each population is copied to a central population. This central population can be expanded by choosing more than one solution from each population or by adding randomly generated solutions. This central population evolves using a GA and for the first several generations this simulation explores new regions of search space because of the dissimilarity of its fit members. After a given number of generations, the most-fit solution may be reasonably different from some or all members of its initial population, and it is copied to all of the other populations. These separate populations then evolve independently and the process continues.

6.3. Meta-GA

Weinberg (1970) proposed, but did not test, a method of optimizing the parameters controlling a GA. His proposal has two GAs, one that optimizes the crossover and mutation parameters and one that does the GA. The outer-most (meta-level) GA has a population that contains the control parameters. Each then runs a GA that examines the problem. The quality of the results for each simulation determines the fitness of that member of the meta-level GA.

Care must be taken to ensure that the meta-level GA does not converge on a single set of control parameters too quickly. It has been argued that a given set of parameters may quickly improve the quality of the population (highest and/or average fitness) but will cause the population to converge on sub-optimal solutions (Wolpert and Macready, 1995). Other parameter sets that cause the quality to improve more slowly can actually yield better final results since they may spend more time exploring the search space before exploitation dominates.

Mansfield (1990) applied this technique without using the meta-level GA. In this application, each population of a parallel GA has its own set of parameters. If the performance of a population that is donating a copy of its best solution is better than the population receiving this copy, some or all of the control parameters can also be transferred.

6.4. Messy GA

A Messy GA differs from all others considered in this chapter in that the genetic vector may be incomplete (Goldberg et al., 1989; Goldberg and Kerzic, 1990). Each genetic vector is doubly indexed; the first gives the locus of the gene and the second gives its allele. Any missing genes are taken from a template solution that is generated at the start of the simulation. Therefore, each solution represents a modification of this template. This non-generational algorithm is also different from the others in that offspring are generated from either one or two parents and the parent or parents are destroyed after offspring generation.

The two major operators for offspring generation are SPLIT and JOIN. As the names suggest, SPLIT cuts the genetic vector at a random point between genes to create two offspring, while JOIN joins the genetic vectors of two parents to create a single offspring. If this offspring contains two alleles for the same gene, the left-most allele in the genetic vector is expressed. Studies showed that better results were obtained when the SPLIT probability was larger than the JOIN. This increased the number of members in the population, so a periodic pruning was necessary. This pruning simply removed the least-fit members from the population.

6.5. Delta coding GA

Mathias and Whitley (1995) proposed that a Delta Coding GA be run after a 'regular' GA converged on a solution. This best solution is known as the *interim solution* and as with the messy GA it acts as a template. In contrast to the messy GA, each member of a randomly generated population contains all the genes and their values are simply changes to the template's values. In the first iteration, the magnitude of the changes has a maximum value for each gene that guarantees it can span the full search space. This ensures that the algorithm starts with a purely exploratory scheme. These delta genetic vectors are used in a Simple GA without mutation (i.e. crossover only) and the simulation continues until the population converges. The final set of delta values is applied to the interim solution and it becomes the new template if it has a higher fitness. A new population of delta genetic vectors is generated and the process repeats.

If the template solution changes from one iteration to the next, the maximum allowed change for each gene is reduced when building the next initial population. Conversely, if the template stays the same (meaning the best solution has a fitness that is lower than for the unaltered template) the maximum allowed change is increased. All alleles are assumed cyclic which means that if the value of a gene exceeds its maximum value, the amount it exceeds this maximum is added to the minimum value. If the template does not change for a given number of iterations, the search stops and this template represents the final result.

6.6. Tabu search and Gibbs sampling

In many ways, the Delta Coding GA is similar to two other search procedures, Tabu Search and Gibbs Sampling, in that both of these methods generate a population of offspring relative to a template structure. In contrast to the Delta Coding GA, these methods produce offspring that only sample a sub-space of the full problem. For example, if the genetic vector consists of 10 real coded genes, offspring are generated that vary by only one or a small set of these genes. This can be a local search or the offspring can sample the full range of values for these selected genes.

In Tabu Search (Glover, 1986, 1990), the sampled region is placed at the top of a tabu list. Each randomly chosen region to check is compared with those stored in the tabu list. If it is on the list, it is either excluded from further search or can only yield a new result if the new fitness is the best to date (aspiration criterion). Each time a new region is sampled it is added to the top of the list and the region at the bottom of the list is removed, meaning that it can be sampled again. A deterministic selection is used which means that the highest fitness offspring becomes the new template even if its fitness is less than that of the old template. This means that the best-to-date solution must be stored and this is the reported solution at the end of the search.

Gibbs Sampling (Gelfand and Smith, 1990; Gelfand et al., 1990) does not use a tabu list, meaning that a region of search space can be re-sampled at any time. In addition, this method uses a roulette wheel selection procedure to probabilistically choose the next template from the old template (parent) and offspring. Therefore, this is a probabilistic $(1 + \lambda)$ selection. If f_{max} is the maximum fitness of this set, the roulette wheel areas are given by

$$e^{-(f_{\max}-f_i)/T}$$

where f_i is the fitness of the ith genetic vector and T is an effective temperature. The probability that a given genetic vector is chosen is the above value divided by the sum of these values over all genetic vectors. As with Simulated Annealing, the temperature starts at a high value, increasing the probability that a less-fit solution will be selected, and is slowly decreased. In contrast to Simulated Annealing, this method takes multiple steps at once and can span the full range of values for a sub-dimension, while Simulated Annealing only takes a single step that randomly changes all genes by a relatively small amount. Both methods have the ability to 'walk away' from the global minimum and therefore best-to-date solutions need to be stored.

6.7. Evolutionary programming

Evolutionary Programming (EP) (Fogel et al., 1966, 1991; Fogel, 1993) produces an offspring from a single parent using a mutation operator. Each of the μ parents produces an offspring and the offspring is placed into a new population. A probabilistic selection scheme is then used to choose parents for the next generation from the combined parent and offspring populations. Therefore, this is a generational $(\mu + \lambda)$ algorithm with probabilistic selection.

As with the SGA, this method can be augmented to include a maturation operator, the elitism strategy, or deterministic selection of the next generation's parents from the combined populations. The major difference between EP and GA is that the mutation operator acts on all loci. This means that schema cannot form and the search always maintains the full dimensionality of the problem. The mutation operator can start by making large changes (exploration) and can be reduced at each generation to perform local searches in good regions (exploitation). This also means that the algorithm is able to search multiple good regions simultaneously (niching) if they exist in the search space. Maintaining the full dimensionality of the search space is most likely the reason why an EP-based feature selection program was able to outperform a GA-based algorithm in QSAR/QSPR generation (Luke, 1994). If a required feature is dropped from the population in a given generation, an EP-based method can re-sample it because of its reliance on mutation while a GA-based method cannot without incorporating some of the procedures outlined above to enhance diversity.

6.8. Evolution strategies

Along with GA and EP, ES is a third, independently developed method of using biologically inspired methods to optimize parameters (Rechenberg, 1973; Schwefel, 1977). This method started using a population size of one and therefore only used a mutation operator to generate offspring, but has since been expanded to use multi-member populations and a mating operator (Back et al., 1995). The main difference between ES and GA or EP is that it was designed to include the offspring generation parameters in the genetic vector. These parameters evolve through mutation and mating along with the rest of the information and are used to generate offspring in the next generation. New parents are chosen by probabilistic selection from either the offspring population, (μ, λ), or the combined parent and offspring populations, $(\mu + \lambda)$.

6.9. Ant colony optimization

The transformation operator proposed by Simões and Costa (2001b) (see Section 5.8) allowed an offspring to be created by combining pieces of genetic vectors from a pool generated by the previous generation's parents with the genetic vector of a current parent. If the population is focusing onto a favorable region of search space, many of the solutions will have similar values for some or all of the genes. Therefore, many of the genetic pieces will contain this information and it will be passed to the next generation's offspring. If a better region of search space is found, the old information will be removed from later gene pools and replaced with the better values.

This use of previous knowledge is similar to the process used by ants and is the basis of the ACO procedure (Colorni et al., 1991). As ants travel a route from a starting point to a food source and then to a destination, they deposit pheromone. Subsequent ants will generally choose paths with more pheromone and after many trials will converge on an optimal path. Therefore, this method was designed to handle node-based problems, but it can also be applied to gene-based problems, where each gene has a finite genetic alphabet. To ensure that the search does not get stuck in a sub-optimal solution, a pheromone evaporation rate is used. This means that if a path or optional value has not been used recently, the strength of this pheromone trail will be small.

$\tau_{i,J}(t)$ is the un-normalized probability (strength of the pheromone trail) of either choosing task J after task i, where J is part of a list of tasks that have not been completed yet, or the probability of choosing a given value J for a particular parameter i in generation t. Initially, all un-normalized probabilities are set to the same small, non-zero value.

$$\tau_{i,J}(0) = c$$

This means that the initial order of tasks or parameter values is randomly generated for each genetic vector (ant). If a transition between particular tasks or a given value for a parameter is used in genetic vector k, the pheromone level is increased by an amount proportional to its fitness, $f(k)$, which is always positive. This increase can take the form

$$\Delta\tau_{i,j}(t) = \Delta\tau_{i,j}(t) + f(k)^{\beta}$$

Each possible change in the pheromone level is set to zero at the start of each generation and β is a constant in the range (0.0,1.0]. When all genetic vectors have been evaluated and all changes in the pheromone levels calculated, the new pheromone levels for each sub-path or parameter value are determined from the expression

$$\tau_{i,j}(t) = (1 - \rho)\tau_{i,j}(t - 1) + \rho\Delta\tau_{i,j}(t)$$

where ρ represents the evaporation rate.

For node-based problems, the next generation starts by randomly choosing the first task. Conversely, a separate $\tau_{0,J}(t)$ can be used where this represents the un-normalized probability that a particular task is tried first, and J is the set of all tasks since none has been removed from consideration. The next task is determined from either of the following expressions.

$$
\begin{aligned}
j &= \text{MAX}\{\tau_{i,J}(t)\} \quad && \text{if } r \leq r_0 \\
&= J && \text{otherwise}
\end{aligned}
$$

where the probability of choosing a particular task from the set of available tasks is given by

$$P_{i,j}(t) = [\tau_{i,j}(t)]^{\alpha} \Big/ \sum_{u \in J} [\tau_{i,u}(t)]^{\alpha}$$

In these expressions, r is a random number in [0.0,1.0], r_0 is a threshold value, and α is a non-negative constant. If r is less than or equal to r_0, a deterministic selection is used since the top equation simply instructs the algorithm to choose the next task with the largest pheromone level. If r is greater than r_0, the bottom equation simply states that a roulette wheel selection of the next task should be made. For gene-based coding problems, the above expressions can be used to determine the value for each gene in an offspring.

This procedure requires four parameters to be set: ρ controls the evaporation of pheromone; r_0 controls the splitting between deterministic selection (exploitation) and probabilistic selection (exploration); β controls the extent to which the fitness of an offspring increases the amount of pheromone deposited; and α controls how much the pheromone level affects the probability that a particular task or parameter value will be chosen.

Different values of these parameters can be tried in different searches, or a meta-level GA can be used (Botee and Bonabeau, 1998). The meta-GA simply controls the values of these parameters and for each offspring (set of parameter values) an ACO is run. The quality of the final result determines the fitness for these parameters.

6.10. Particle swarm optimization

Both the transformation operator in GAs and ACO use information from many solutions that was acquired over many generations to produce good offspring. This is similar to a flock of birds searching for food where each bird (particle, member of the population, or genetic vector) remembers the most favorable region it has encountered and, through communication with other members of the population, the most favorable region encountered by any member. This model is the basis for PSO (Kennedy and Eberhart, 1995; Kennedy, 1997).

For problems that use real coded, gene-based genetic vectors, the alleles in the ith vector at a particular time or generation, $X_i(t)$, determines its position in the search space of the problem. In PSO, each particle also has a velocity vector $V_i(t)$. If unit time steps are used, the position (genetic vector) at the next time interval (generation) is given by

$$X_i(t) = X_i(t-1) + V_i(t-1)$$

The velocity vector is influenced by both the best solution found by this particle, $B_i(t)$, and the best solution found by any member of the population up to this time, $B_p(t)$. The velocity vector at generation t is given by

$$V_i(t) = V_i(t-1) + RA[B_i(t) - X_i(t)] + R'B[B_p(t) - X_i(t)]$$

where R and R' are random vectors with elements in [0,1], and A and B are constant vectors that control the maximum allowed change in the velocity vector at each position. If $B = 0$, it becomes a 'cognition-only' model where each particle is only influenced by the best solution it has encountered so far. Conversely, if $A = 0$, the algorithm becomes a 'social-only' model where only the best solution found by any particle has an affect on the motion of each particle. The authors obtained best results when $A = B = 2$.

The algorithm starts by randomly generating each particle's position vector, $X_i(0)$, and velocity vector, $V_i(0)$. Its fitness is calculated and, since these must be the best values found to date, they are stored and $B_i(0) = X_i(0)$. For each generation, the position and velocity vectors are added to generate the next position vectors, $X_i(t)$. If the fitness of this solution is better than the stored fitness it replaces this value and this position becomes the best-to-date position for this particle. If the fitness is less, $B_i(t) = B_i(t-1)$. The $B_i(t)$ vector with the highest fitness becomes $B_p(t)$, and the expression above is used to update the velocity vectors.

This algorithm can be described as a generational GA, or generational EP, where each offspring is generated from a single parent. The mutation operator is not random, but instead depends upon its current value, and the distances between the current position in search space and the positions of the best solution found by this particle and the best solution found overall.

All of the discussion of this algorithm assumes that the problem requires real coded, gene-based genetic vectors. If integer coding is needed instead, the same algorithm can be used by allowing the genetic vector to have real values and then just round the values to the nearest integer before evaluating the fitness function. Binary coding is the same as integer coding if the binary gene must be converted into an integer before the fitness can be determined. Conversely, if this is used on a feature selection problem, a real genetic vector can still be used and the decision of whether or not a particular feature is used can be determined based on the sign of the allele. If the value in a particular position is negative that feature is not used, otherwise that feature is used.

This algorithm can also me modified to include the concept of families. In this case, an index needs to be included in the genetic vector which stores the family number. In addition, the best solution found by each particle, $B_i(t)$, is replaced by the best solution found by any member of its family, $B_f(t)$, in the expressions above. The simulation starts by randomly generating the genetic vectors for a relatively small number of particles that act as seeds for each family. The other members of each family are created by applying a small mutation vector to each gene in the seed's genetic vector. The velocities of all particles are still generated randomly and the velocity, or mutation, vector contains terms proportional to the distance from the family's best solution to date and the global best solution to date.

The final point is to ensure that the value of each element in the genetic (position) vector stays within the range of values allowed for this parameter. The easiest way to do this is to force each parameter to be cyclic. This means that if a particular element j in the ith genetic vector, $x_{j,i}$, is greater than the maximum allowed value, $x_{j,\max}$, by an amount $\Delta x_{j,i}$, this value is added to the minimum allowed value.

$$\text{if } x_{j,i} = x_{j,\max} + \Delta x_{j,i} \text{ then } x_{j,i} = x_{j,\min} + \Delta x_{j,i}$$

Similarly,

$$\text{if } x_{j,i} = x_{j,\min} - \Delta x_{j,i} \text{ then } x_{j,i} = x_{j,\max} - \Delta x_{j,i}$$

Another way to handle this is to assume that the particles are contained in a hyper-rectangle with reflective walls. In this case

$$\text{if } x_{j,i} = x_{j,\max} + \Delta x_{j,i} \text{ then } x_{j,i} = x_{j,\max} - C\Delta x_{j,i}$$

$$\text{if } x_{j,i} = x_{j,\min} - \Delta x_{j,i} \text{ then } x_{j,i} = x_{j,\min} + C\Delta x_{j,i}$$

where C is the reflectively of the walls and could be a constant between zero and one, or a random number in [0,1] to account for imperfections in the walls.

7. Conclusions

Many of the options for generating an algorithm that are presented in this chapter are listed in outline form in Table 2. In addition, a flowchart connecting many of the methods is presented in Fig. 2.

The simplest connection is to construct a Simple GA (SGA) with the mating probability, P_{mate}, set to zero. This converts the SGA into an EP algorithm. This algorithm can also be gradually formed from an SGA if parents are chosen based upon their genotypic similarity and an interval crossover mating operator is used. This selection method chooses parents who are close to each other in search space and the limiting case is when they occupy the same location. In this case, the interval crossover operator is reduced to a mutation operator that acts on all alleles. If either of these methods use a Boltzmann acceptance criterion to have an offspring replace its parent in a non-generational algorithm, an Ensemble Simulated Annealing algorithm is produced.

As discussed in Section 5.8, the Transformation-based GA can be modified so that information from multiple parents can be incorporated into the genetic vector of an offspring. Section 6 then describes how this model can be modified to produce the Ant Colony and PSO procedures. In PSO, the mutation operator that is applied to all alleles is simply treated as a velocity vector that updates the position of each particle.

Finally, the Delta-Coding GA uses a single solution as a template and the genetic vector for each member of the population can be considered a mutation, or step vector. This algorithm is slightly different than a Messy GA in that in the former the template can change from generation to generation and the mutation vector modifies all alleles, while in the latter the genetic vectors may be incomplete and they replace the alleles in the static template vector.

Table 2
Representative options that can be used in various steps of a GA

Options

- A. Genetic Vector
  - 1. Coding scheme
    - a. Gene based
    - b. Node based
    - c. Delta coding
  - 2. Genetic alphabet
    - a. Binary
      - i. Standard coding
      - ii. Gray coding
    - b. Integer
    - c. Real
  - 3. Copies
    - a. One (haploid)
    - b. Two (diploid)
    - c. Three or more (polyploid)
- B. Initial population
  - 1. Random
  - 2. High fitness
  - 3. Diverse
  - 4. Family groupings
- C. Parent selection
  - 1. Fitness based
    - a. Scaled
    - b. Unscaled
  - 2. Rank based
  - 3. Similarity based
  - 4. Diversity based
  - 5. Dominant member
  - 6. Random
  - 7. Complete
- D. Mating operator
  - 1. Value of P_{mate}
  - 2. Gene-based and delta coding
    - a. k-Point crossover
    - b. Uniform crossover
    - c. Intermediate recombination
    - d. Interval crossover
  - 3. Node based
    - a. Partially matched crossover
    - b. Ordered crossover
    - c. Cycle crossover
    - d. Position based crossover
    - e. Swap path crossover
    - f. Edge recombination crossover
    - g. Linear order crossover
  - 4. Transformation operator
- E. Mutation operator
  - 1. Value of P_{mut}
  - 2. Binary
    - a. Bit flip
    - b. Frame-shift operator
    - c. Translation operator
  - 3. Non-binary
    - a. Random change
  - 4. Node-based coding
    - a. Switching
    - b. Relocation
    - c. Inversion
- F. Maturation operator
  - 1. Hill climbing (memetic algorithm)
  - 2. Diversity requirement
- G. Processing offspring
  - 1. Non-generational
    - a. Replace weakest member
    - b. Replace parent
      - i. Deterministic
      - ii. Boltzmann probability
    - c. Replace most similar member
  - 2. Generational
    - a. Deterministic
      - i. (μ, λ)
      - ii. $(\mu + \lambda)$
    - b. Probabilistic
      - i. (μ, λ)
      - ii. $(\mu + \lambda)$
      - iii. Elitist strategy

If the Delta-Coding GA is modified such that the genetic vectors in each member only produce a change in the value of one or a few genes, a list is kept to ensure that the same sub-space is not searched too frequently, and the most-fit offspring becomes the new template solution, a Tabu Search is produced. Conversely, if a Boltzmann acceptance is used with a combination of the offspring and template, a Gibbs Sampling algorithm is generated.

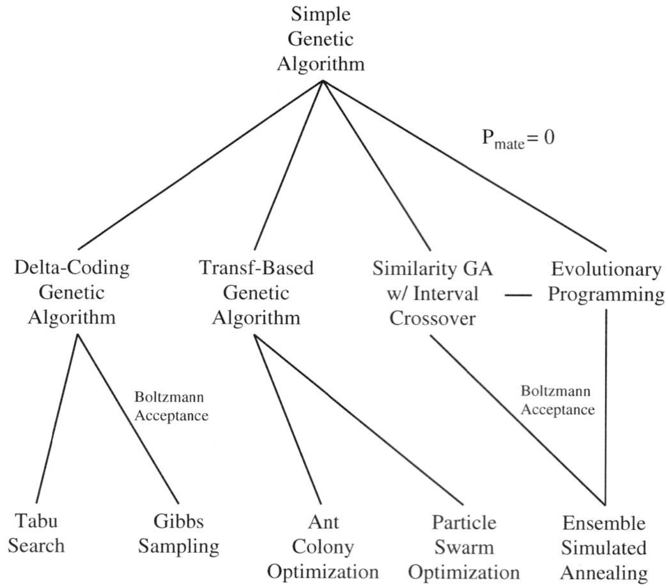

Fig. 2. Flowchart connecting various population-based search strategies.

This chapter presents many options that can be used within the general framework of GAs and shows how small modifications to this framework can produce many other, independently developed, search strategies. Therefore, a GA is really a search heuristic and not a specific algorithm. This means that the framework can be adjusted to properly handle a specific problem, but also means that many options may have to be tried to generate an optimal algorithm.

A simple interpretation of the 'No Free Lunch Theorems' (Wolpert and Macready, 1995) states that because there are an uncountably infinite number of combinatorial problems, any specific algorithm that solves a class of them (and therefore at most a countably infinite number of problems) has a zero probability of solving a randomly selected problem. The extension of this is that the problem has to dictate the solution method, and the flexibility inherent in GAs and other population-based search methods allows specific algorithms to be constructed that are tailored to specific problems. In addition, the specific algorithm can be built to generate a reasonably good answer quickly or a better answer in a longer period of time.

Acknowledgments

This work was funded in whole or in part with federal funds from the US National Cancer Institute, National Institutes of Health, under contract no. NO1-CO-12400. The content of this publication does not necessarily reflect the views or policies of the Department of Health and Human Services, nor does any mention of trade names, commercial products or organizations imply endorsement by the US Government.

References

Ahuja, R.K., Orlin, J.B., Tiwari, A., 2000. A greedy algorithm for the quadratic assignment problem. Computers Oper. Res. 27, 917–934.

Back, T., Rudolph, G., Schwefel, H.P., 1995. Private communication.

Bagley, J.D., 1967. The behavior of adaptive systems which employ genetic and correlation algorithms. Doctoral dissertation. University of Michigan, Dissertation Abstracts International, 28, 5106B (University Microfilms No. 68-7556).

Bandyopadhyay, S., Pal, S.K., Maulik, U., 1998. Incorporating chromosome differentiation in genetic algorithms. Informat. Sci. 104, 293–319.

Bledsoe, W.W., 1961. The use of biological concepts in the analytical study of systems. Paper presented at ORSA-TIMS National Meeting, San Francisco.

Bosworth, J., Foo, N., Zeigler, B.P., 1972. Comparison of Genetic Algorithms with Conjugate Gradient Methods (CR-2093), National Aeronautics and Space Administration, Washington, DC.

Botee, H.M., Bonabeau, E., 1998. Evolving ant colony optimization. Adv. Complex Syst. 1, 149–159.

Brown, E.C., Sumichrast, R.T., 2003. Impact of the replacement heuristic in a grouping genetic algorithm. Computers Oper. Res. 30, 1575–1593.

Calabretta, R., Galbiati, R., Nolfi, S., Parisi, D., 1996. Two is better than one: a diploid genotype for neural networks. Neural Process. Lett. 4, 149–155.

Cavicchio, D.J., 1970. Adaptive search using simulated evolution. Unpublished Doctoral Dissertation. University of Michigan, Ann Arbor.

Cobb, H., 1990. An investigation into the use of hypermutation as an adaptive operator in genetic algorithms having continuous, time-dependent nonstationary environments. Technical Report AIC-90-001.

Cobb, H., Grefenstette, J.J., 1993. Genetic algorithms for tracking changing environments. Proceedings of the Fifth International Conference on Genetic Algorithms, Morgan Kaufmann, Los Altos, CA, pp. 523–530.

Colorni, A., Dorigo, M., Maniezzo, V., 1991. Distributed optimization by ant colonies. In: Varela, F., Bourgine, P., (Eds.), Proceedings of the First European Conference on Artificial Life, MIT Press, Cambridge, MA, pp. 134–142.

Craighurst, R., Martin, W., 1995. Enhancing GA performance through crossover prohibitions based on ancestry. Proceedings of the Sixth International Conference on Genetic Algorithms, pp. 115–122.

Davis, L., 1985. Job shop scheduling with genetic algorithms. Proceedings of an International Conference on Genetic Algorithms and their Application, pp. 162–164.

Davis, L., 1991. Order-based genetic algorithm and the graph coloring problem. Handbook of Genetic Algorithms, Van Nostrand Reinhold, New York, Chapter 6.

De, S., Pal, S.K., Ghosh, A., 1998. Genotypic and phenotypic assortative mating in genetic algorithms. J. Inform. Sci. 105, 209–226.

De Falco, I., Della Cioppa, A., Tarantino, E., 2002. Mutation-based genetic algorithm: performance evaluation. Appl. Soft Comput. 1, 285–299.

Dymek Captain, A., 1992. An examination of hypercube implementations of genetic algorithms. Thesis. Air Force Institute of Technology AFIT/GCS/ENG/92M-02.

Eshelman, L.J., 1991a. Preventing premature convergence in the genetic algorithms by preventing incest. In: Belew, R.K., Booker, L.B., (Eds.), Proceedings of the Fourth International Conference on Genetic Algorithms, San Diego, July 1991, pp. 115–122.

Eshelman, L., 1991b. The CHC adaptive search algorithm. How to have safe search when engaging in nontraditional genetic recombination. In: Rawlings, G., (Ed.), Foundations of Genetic Algorithms, Morgan Kaufmann, San Mateo, CA, pp. 265–283.

Falkenauer, E., Bouffouix, S., 1991. A Genetic Algorithm for Job-Shop. Proceedings of the IEEE International Conference on Robotics and Automation, Sacramento, vol. 1, pp. 824–829.

Fogel, D.B., 1993. Applying evolutionary programming to selected traveling salesman problems. Cybernet. Syst. (USA) 24, 27–36.

Fogel, L.J., Owens, A.J., Walsh, M.J., 1966. Artificial Intelligence Through Simulated Evolutions, Wiley, New York.

Fogel, D.B., Fogel, L.J., Porto, V.W., 1991. Evolutionary methods for training neural networks. IEEE Conference on Neural Networks for Ocean Engineering, 91CH3064-3, pp. 317–327.

Foo, N.Y., Bosworth, J.L., 1972. Algebraic, Geometric, and Stochastic Aspects of Genetic Operators (CR-2099), National Aeronautics and Space Administration, Washington, DC.

Gardner, E.J., Simmons, M.J., Snustad, D.P., 1991. Principles of Genetics, Wiley, New York.

Gelfand, A.E., Smith, A.F.M., 1990. Sampling-based approaches to calculating marginal densities. J. Am. Statist. Assoc. 85, 398–409.

Gelfand, A.E., Hils, S.E., Racine-Poon, A., Smith, A.F.M., 1990. Illustration of Bayesian inference in normal data models using Gibbs sampling. J. Am. Statist. Assoc. 85, 972–985.

Gillespie, J.H., 1998. Population Genetics: A Concise Guide, Johns Hopkins University Press, Baltimore.

Glover, F., 1986. Future paths for integer programming and links to artificial intelligence. Computers Oper. Res. 5, 533–549.

Glover, F., 1990. Tabu search: a tutorial. Interfaces 20, 74–94.

Goldberg, D.E., 1989. Genetic Algorithms in Search, Optimization, and Machine Learning, Addison-Wesley, Berkeley, CA.

Goldberg, D.E., Kerzic, T., 1990. mGA 1.0: a common LISP implementation of a messy genetic algorithm. NASA-CR-187260, Cooperative Agreement NCC 9-16, Research Activity AI.12, N91-13084.

Goldberg, D.E., Smith, R.E., 1987. Nonstationary function optimization using genetic algorithms with dominance and diploidy. In: Grefenstette, J.J., (Ed.), Proceedings of the Second International Conference on Genetic Algorithms, Laurence Erlbaum Associates, Hillsdale, NJ, pp. 59–68.

Goldberg, D.E., Korb, B., Deb, K., 1989. Messy genetic algorithm: motivation, analysis, and first results. Complex Syst. 3, 493–530.

Goldberg, D.E., Deb, K., Clark, J.H., 1992. Genetic algorithms, noise, and the sizing of populations. Complex Syst. 6, 333–362.

Grefenstette, J.J., 1992. Genetic algorithms for changing environments. In: Maenner, R., Manderick, B., (Eds.), Parallel Problem Solving from Nature 2, North-Holland, Amsterdam, pp. 137–144.

Grefenstette, J.J., Ramsey, C.L., 1992. An approach to anytime learning. In: Sleeman, D., Edwards, P., (Eds.), Proceedings of the Ninth International Conference on Machine Learning, Morgan Kaufmann, Los Altos, CA, pp. 180–195.

Guo, L.X., Zhao, M.Y., 2002. A parallel search genetic algorithm based on multiple peak values and multiple rules. J. Mater. Process. Technol. 129, 539–544.

Hadad, B., Eick, C., 1997. Supporting polyploidy in genetic algorithms using dominance vectors. In: Angeline, P., Reynolds, R.G., McDonnell, J.R., Eberhart, R., (Eds.), Proceedings of the Sixth International Conference on Evolutionary Programming, Volume 1213 of LNCS, Springer.

Hasancebi, O., Erbatur, F., 2000. Evaluation of crossover techniques in genetic algorithm based optimum structural design. Computers Struct. 78, 435–448.

Herrera, F., Lozano, M., Verdegay, J.L., 1997. Fuzzy connectives based crossover operators to model genetic algorithms population diversity. Fuzzy Sets Syst. 92, 21–30.

Holland, J., 1975. Adaptation in Natural and Artificial Systems, University of Michigan Press, Ann Arbor, MI.

Hollstein, R.B., 1971. Artificial genetic adaptation in computer control systems. Doctoral dissertation, University of Michigan, Dissertation Abstracts International, 32, 1510B. (University Microfilms No. 71-23, 773).

van Kampen, A.H.C., Buydens, L.M.C., 1997. The ineffectiveness of recombination in a genetic algorithm for the structure elucidation of a heptapeptide in torsion angle space. A comparison to simulated annealing. Chemom. Intel. Lab. Syst. 36, 141–152.

Kennedy, J., 1997. The particle swarm: social adaptation of knowledge. Proceedings of the 1997 International Conference on Evolutionary Computation, Indianapolis, IN, IEEE Service Center, Piscataway, NJ, pp. 303–308.

Kennedy, J., Eberhart, R.C., 1995. Particle swarm optimization. Proceedings of the IEEE International Conference on Neural Networks, Perth, Australia, IEEE Service Center, Piscataway, NJ, IV, pp. 1942–1948.

Keser, M., Stupp, S.I., 1998. A genetic algorithm for conformational search of organic molecules: implications for materials chemistry. Computers Chem. 22, 345–351.

Kim, Y., Kim, J.K., Lee, S.-S., Cho, C.-H., Lee-Kwang, H., 1996. Winner take all strategy for a diploid genetic algorithm. The First Asian Conference on Simulated Evolution and Learning.

Kim, J.O., Shin, D.-J., Park, J.-N., Singh, C., 2002. Atavistic genetic algorithm for economic dispatch with valve point effect. Electric Power Syst. Res. 62, 201–207.

Kirkpatrick, S., Gelatt, C.D., Vecchi, M.P., 1983. Optimization by simulated annealing. Science 220, 671–680.

Konig, R., Dandekar, T., 1999. Improving genetic algorithms for protein folding simulations by systematic crossover. BioSystems 50, 17–25.

Liepins, G.E., Baluja, S., 1991. apGA: an adaptive parallel genetic algorithm. Fourth International Conference on Genetic Algorithms, San Diego, CONF-910784-2.

Luke, B.T., 1994. Evolutionary programming applied to the development of quantitative structure–activity relationships and quantitative structure–property relationships. J. Chem. Informat. Computer Sci. 34, 1279–1287.

Luke, B.T., 1996. An overview of genetic methods. In: Devillers, J., (Ed.), Genetic Algorithms in Molecular Modeling, Academic Press, London, pp. 35–66.

Luke, B.T., 1999. In: Truhlar, D.G., Howe, W.J., Hopfinger, A.J., Blaney, J., Dammkoehler, R.A., (Eds.), Applications of Distributed Computing to Conformational Searches, The IMS Volumes in Mathematics and Its Applications, vol. 108. Springer, New York, pp. 191–206.

Luke, B.T., web, Substructure searching using genetic methods. http://members.aol.com/btluke/mcdga2.htm

Mansfield Squadron Leader, R.A., 1990. Genetic Algorithms. Dissertation. School of Electrical, Electronic and Systems Engineering, College of Cardiff, University of Wales, Crown Copyright.

Mathias, K.E., Whitley, L.D., 1995. Private communication.

Moscato, P., 1989. On evolution, search, optimization, genetic algorithms and martial arts: towards memetic algorithms. Tech. Rep. Caltech Concurrent Computation Program, Report 826, California Institute of Technology, Pasadena, CA, USA.

Muhlenbein, H., Schpmisch, M., Born, J., 1991. The parallel genetic algorithm as a function optimizer. Parallel Comput. 17, 619–632.

Narayanan, L., Lucas, S.B., 1993. A genetic algorithm to improve a neural network to predict a patient's response to Warfarin. Methods of Information in Medicine 32, 55–58.

Ng, K.P., Wong, W.C., 1995. A new diploid scheme and dominance change mechanism for non-stationary function optimization, Proceedings of the Sixth International Conference on Genetic Algorithms, Morgan Kaufmann, Los Altos, CA, pp. 159–166.

Osmera, P., Kvasnicka,V., Pospichal, J., 1997. Genetic algorithms with diploid chromosomes. Mendel '97, PC-DIR Brno, ISBN 80-214-0084-7, pp. 111–116.

Park, T.-Y., Froment, G.F., 1998. A hybrid genetic algorithm for the estimation of parameters in detailed kinetic models. Computers Chem. Engng 22, S103–S110.

Rechenberg, I., 1973. Evolutionsstrategie: Optimierung technischer Systeme nach Prinsipien der Biologischen Evolution, Frommann-Holzboog Verlag, Stuttgart.

Rosenberg, R.S., 1967. Simulation of Genetic Populations with Biochemical Properties. Doctoral dissertation. University of Michigan, Dissertation Abstracts International, 28, 2732B. (University Microfilms No. 67-17, 836).

Schwefel, H.-P., 1977. Numerische Optimierung von Computer-Modellen Mittels der Evolutionstrategie, Interdisciplinary Systems Research, vol. 26, Birkhauser, Basel.

Simões, A., Costa, E., 2001a. Using biological inspiration to deal with dynamic environments. Proceedings of the Seventh International Conference on Soft Computing (MENDEL'01), Brno, Czech Republic, 6–8 June.

Simões, A., Costa, E., 2001b. On biologically inspired genetic operators: transformation in the standard genetic algorithm. Proceedings of the Genetic and Evolutionary Computing Conference (GECCO'2001), San Francisco, USA, July.

Simões, A., Costa, E., 2002a. Using GAs to deal with dynamic environments: A comparative study of several approaches based on promoting diversity. In: Langdon, W.B., et al. (Eds.), Proceedings of the Genetic and Evolutionary Computation Conference (GECCO'02), Morgan Kaufmann, New York, New York, 9–13 July.

Simões, A., Costa, E., 2002b. Parametric study to enhance the genetic algorithm's performance when using transformation. In: Langdon, W.B., et al. (Eds.), Proceedings of the Genetic and Evolutionary Computation Conference (GECCO'02), Morgan Kaufmann, New York, New York, 9–13 July.

Smith, R.E., Goldberg, D.E., 1992. Diploidy and dominance in artificial genetic search. Complex Syst. 6, 251–285.

Sorensen, H., Jacobsen, J.F., web, Maintaining diversity through triggerable inheritance. http://www.daimi.au.dk/~manaic/ToEC/triggerEA.ps

Strickberger, M.W., 1985. Genetics, Prentice-Hall of India, New Delhi.

Syswerda, G., 1989. Uniform crossover in genetic algorithms. Proceedings of the Third International Conference on Genetic Algorithms, Morgan Kaufman, New York, pp. 2–9.

Syswerda, G., 1990. Schedule optimization using genetic algorithms. In: Davis, L., (Ed.), Handbook of Genetic Algorithms, Van Nostrand Reinhold, New York, Chapter 21.

Trojanowsky, K., Michalewicz, Z., Xiao, J., 1997. Adding memory to the evolutionary planner/navigator. IEEE International Conference on Evolutionary Computation, pp. 483–487.

Weinberg, R., 1970. Computer simulation of a living cell. Doctoral dissertation. University of Michigan, Dissertation Abstracts International, 31, 5312B. (University Microfilms No. 71-4766).

Whitley, D., 1989. The GENITOR algorithm and selection pressure: why rank-based allocation of reproductive trials is best. In: Schaffer, J.D., (Ed.), Proceedings of the Third International Conference on Genetic Algorithms, Morgan Kaufmann, Los Altos, CA, pp. 116–121.

Wolpert, D.H., Macready, W.G., 1995. No Free Lunch Theorems for Search, Technical Report TR-95-02-010, The Santa Fe Institute, Santa Fe, NM, USA.

Yang, S., 2002. Genetic algorithms based on primal-dual chromosomes for royal road functions. In: Grmela, A., Mastorakis, N.E., (Eds.), Advances in Intelligent Systems, Fuzzy Systems, Evolutionary Computation, WSEAS Press, Athens, pp. 174–179, ISBN 960-8052-49-1.

Yoon, H.-S., Moon, B.-R., 2002. An empirical study on the synergy of multiple crossover operators. IEEE Trans. Evolut. Comput. 6, 212–223.

Yu, H., Fang, H., Yao, P., Yuan, Y., 2000. A combined genetic algorithm/simulated annealing algorithm for large scale system energy integration. Computers Chem. Engng 24, 2023–2035.

Zeigler, B.P., Bosworth, J.L., Bethke, A.D., 1973. Noisy Function Optimization by Genetic Algorithms, Technical Report No. 143, 143. Department of Computer and Communication Sciences, University of Michigan, Ann Arbor.

CHAPTER 2

Hybrid genetic algorithms

D. Brynn Hibbert

School of Chemical Sciences, University of New South Wales, Sydney NSW 2052, Australia

1. Introduction

The power of evolutionary methods is evidenced by the wide adoption of methods following the early papers of Holland and others (detailed in Chapter 1). The ability to process enormous parameter spaces and cope with multiple local optima are the hallmarks of these methods. It must also be said that the seduction of methods that are inspired by Nature has also been responsible for the high level of interest. However, it has been recognized that there are limitations, both in terms of the systems that can be efficiently tackled and the ability of the methods to find the optimum. Different types of optimizers have different strengths. It may be of no surprise, therefore, that the possibility of combining optimizing strategies to give a quicker, better result was mooted early in the piece.

In this chapter different methods of combining a genetic algorithm with another optimizer will be discussed, giving examples from the chemistry-related literature. Where appropriate examples from medicine, which has a great interest in genetic algorithms and their hybrids, and computer science in which much of the basic theory has been developed will be given.

2. The approach to hybridization

Evolutionary methods lend themselves to hybridization because of their flexible construction and also their strengths tend to complement those of other methods. In terms of the problem to be optimized the set up of a genetic algorithm requires only the specification of the decoding of the chromosome to give the fitness function. Thus a genetic algorithm can operate with another optimizer in a number of ways. It can be used to optimize the optimizer, provide a method of generating search conditions, or be totally integrated with other methods. One way of discussing hybrid genetic algorithms is to look at the sibling method, and the degree to which the hybridization is achieved by interaction of the methods.

Data Handling in Science and Technology, Volume 23
ISSN: 0922-3487

DOI: 10.1016/S0922-3487(03)23002-1

2.1. Levels of interaction

At one end of the scale, genetic algorithms can be used in conjunction with another method (or methods). If the problem is sufficiently large, then a genetic algorithm can find itself being used before or after another method, but without any interaction between them. Peña-Reyes (Pena-Reyes and Sipper, 2000) calls this an 'uncoupled' system, citing Medsker (1994). Examples in medical diagnosis are found where a genetic algorithm solves one sub-problem while an expert system tackles another sub-problem. In analytical chemistry, an automatic system with (limited) intelligence has been described for the interpretation of 2D NMR, involving two expert system sub-modules with a genetic algorithm in a third sub-module (Wehrens et al., 1993). This is hybridization by virtue of being used for the same super-problem, but there is no synergy arising from this, and hence will not be considered further.

The most tenuous level of interaction is when the methods only share data through external files or memory resident structures. One method takes the output data of another as the input to its method. The genetic algorithm can come first or second in this process. Examples may be found in medical diagnosis (Pena-Reyes and Sipper, 2000). As genetic algorithms produce a number of solutions, as represented by the final population, an heuristic method in an expert system can be used to assess the population in wider terms than the fitness function and present a more 'intelligent' solution. Shengjie and Schaeffer (1999) have used a LISP-based expert system with genetic algorithm to optimize debinding cycles in powder injection molding.

The level of coupling is greater when the methods are integrated to the extent that one optimizes the parameters of the other.

In an early review (Lucasius and Kateman, 1994) the distinction was made between serial hybrids, those that employed a genetic algorithm before or after another optimizer, or chains of genetic algorithms, and parallel hybrids. The latter covers problems that may be partitioned into smaller sub-problems, each of which is amenable to a genetic algorithm treatment. For example, a large protein structure problem need not be tackled with one genetic algorithm having a long chromosome containing all molecular parameters, but may be split into a series of fragments that can be treated, at least initially, as separate optimizations. An alternative to problem partitioning, population partitioning has operators that may be independently searched. This is amenable to parallel computing in which each processor can solve an independent system. Migration between sub-populations can then be used to communicate information. The incorporation of domain-specific search heuristics (so-called greedy heuristics) can be used to direct the search to achieve quick and dirty results. This approach is often found in commercial problems in which solutions are required quickly and an absolute optimum is not particularly desired, nor often conceivable. An example of this may be found in scheduling of vehicle movements described by Kwan et al. (2001).

Finally, there are hybrids of hybrids. An example of such a multi-hybrid is that between an artificial neural network and a hybrid genetic algorithm given by Han for the optimization of the production of CVD silica films. Experimental data are used to produce a neural network model of the process which is then optimized by a steepest descent–genetic algorithm hybrid.

Table 1 shows published hybrids in terms of the interactions.

2.2. *A simple classification*

It seems that all the published genetic algorithm hybrids can be classed in three configurations (Fig. 1, Table 2). The most prevalent is the genetic algorithm that provides input to a second optimizer. The results may or may not be cycled back into the genetic algorithm in an iterative manner. The genetic algorithm's great ability to search a parameter space makes it ideal as a formulator of input data for a more directed optimizer.

The other two have one method embedded in the other; either a genetic algorithm embedded within an optimizer with the task of optimizing some facet of the method, for example the number of nodes in an artificial neural network, or vice versa in which an optimizer undertakes some service for the genetic algorithm.

3. Why hybridize?

Before considering some mathematical or computational reasons for hybridization, it is interesting to consider the arguments of Davis (1991) in one of the early 'bibles' of genetic algorithm research. He points out that single crossover, binary coded genetic algorithms are often not particularly the best for specific real world problems. The strength of a genetic algorithm—that it is robust across a number of problems—is seen as a drawback to a user who has one single problem to solve. Davis writes "People with real world problems do not want to pay for multiple solutions to be produced at great expense and then compared" (Davis, 1991, p. 55). His approach is to start from the current algorithm and then hybridize this with a genetic algorithm suitably adapted. Using the current algorithm (he assumes that a real problem will already have some sort of solution), the grafting on of a genetic algorithm must improve the optimization and should also outperform a more generic genetic algorithm. Taking the horticultural analogy, the hybrid will be more vigorous than its parents (the conventional algorithm and a standalone genetic algorithm). Since this time a great number of possible combinations have been published, some developed by the Davis method, some for more academic reasons.

Davis has answered the question "why hybridize?". The only reason to hybridize a genetic algorithm is to obtain better results. Hybridizing is almost certainly to do this, certainly with respect to a simple genetic algorithm, or the current method (if it exists). The concern that may arise is the time taken to develop and validate a more complex hybrid, if the improvement is only minor.

Table 1
Hybrid genetic algorithms classified by the second method and the role of the genetic algorithm

Optimization method	Role of the genetic algorithm	References
k-Nearest neighbor	Provides weights to attributes of kNN	Raymer et al. (1997) and Anand et al. (1999)
Partial least squares	Feature selection	Leardi and Lupiáñez Gonzalez (1998)
Discrete canonical variate analysis	Optimizes DVCA loadings and scores	Kemsley (2001)
Clustering algorithms	Refines population	Hanagandi and Nikolaou (1998)
Simplex	Provides starting guesses for Simplex search	Hartnett et al. (1995), Han and May (1996), Cela and Martinez (1999), and Lee et al. (1999)
'Simplex'	Reproduction algorithm enriches population with variants of most fit member.	Shaffer and Small (1996a,b)
Steepest descent methods (e.g. Gauss–Newton, Pseudo-Newton, Newton–Raphson, Powell)	Provides starting guess for steepest descent optimization (Steepest descent method may also provide the value of the fitness function for the genetic algorithm)	Hibbert (1993), de Weijer et al. (1994), Del Carpio et al. (1995), Cho et al. (1996), Del Carpio (1996a,b), Han and May (1996), Ikeda et al. (1997), Handschuh et al. (1998), Heidari and Ranjithan (1998), Kim and May (1999), Balland et al. (2000, 2002), Vivo-Truyols et al. (2001a,b), and Yang et al. (2002)
Levenberg–Marquadt	Provides starting point for optimization	Park and Froment (1998)
Other hill climbing	Interacts with 'alopex' allocation algorithm	Xue et al. (2000)
Artificial neural networks	Optimizes problem using ANN-generated fitness function	Devillers (1996), Han and May (1996), So and Karplus (1996), Kim and May (1999), Liu (1999), Shimizu (1999), Parbhane et al. (2000), Zuo and Wu (2000), Mohaghegh et al. (2001)
Artificial neural networks	Trains or optimizes parameters of ANN	Gao et al. (1999), Dokur and Olmez (2001), and Nandi et al. (2001)
Fuzzy methods	Optimizes fuzzy neural net system	Ouchi and Tazaki (1998), Wang and Jing (2000), Chen et al. (2001)
Expert systems	Evolves network structure. Provides solutions for input into ES Optimizer using Pareto optimality Input to heuristic method	Haas et al. (1998), Shengjie and Schaeffer (1999), Mitra et al. (2000), and Kwan et al. (2001)
Finite element	Performs optimization with fitness from FE	Wakao et al. (1998)

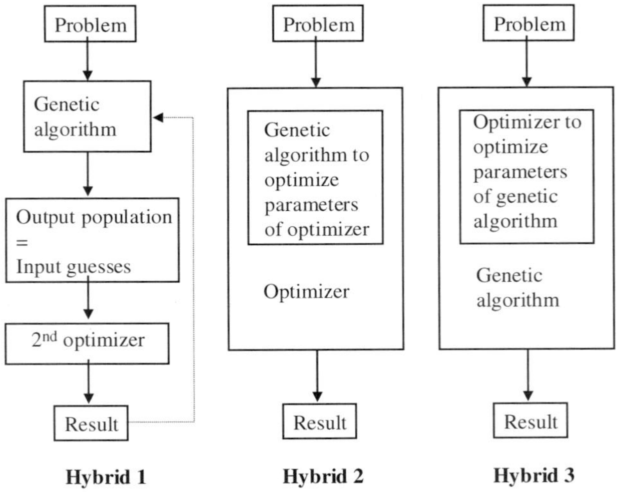

Fig. 1. Three configurations of genetic algorithm hybrids.

4. Detailed examples

4.1. Genetic algorithm with local optimizer

The hybrid described by Hibbert (1993) is a typical example of using a steepest descent optimizer on a population generated by a genetic algorithm. There are a number of alternative modes of use that could be considered by a potential user.

The problem was to determine kinetic rate constants by fitting experimental data to an integrated rate equation. This optimization is typically solved by an iterative steepest

Table 2
Genetic algorithms classified by configuration (see Fig. 1)

Hybrid 1	Hybrid 2	Hybrid 3
Genetic algorithm as precursor to a second optimizer with or without iteration.	Genetic algorithm, embedded in an optimizer, that configures the parameters of the optimizer	Genetic algorithm with an optimizer determining an aspect of the genetic algorithm
Hibbert (1993), de Weijer et al. (1994), Hartnett et al. (1995), Del Carpio (1996b), Devillers (1996), Shaffer and Small (1996a,b), So and Karplus (1996), Gunn (1997), Wakao et al. (1997, 1998), Handschuh et al. (1998), Park and Froment (1998), Zacharias et al. (1998), Liu (1999), Yamaguchi (1999), Balland et al. (2000, 2002), Kemsley (2001), Mohaghegh et al. (2001), Nandi et al. (2001), and Vivo-Truyols et al. (2001a,b)	Raymer et al. (1997), Yoshida and Funatsu (1997), Leardi and Lupiáñez González (1998), Anand et al. (1999), and Gao et al. (1999)	Hanagandi and Nikolaou (1998)

ascent method, such as a pseudo-Newton, Newton–Raphson or Gauss–Newton, for example. The equation for which the rate constants $k_1, \ldots, k_4$ need to be determined is

$$y_t = \left(k_1 + \frac{2k_3k_4}{k_4 - k'} + \frac{k_1k_2}{k_2 - k'}\right)\frac{\exp(-k't) - 1}{k'} + \frac{2k_3[\exp(-k_4t) - 1]}{k_4 - k'} + \frac{k_1[\exp(-k_2t) - 1]}{k_2 - k'} \quad (1)$$

where y_t is a measured concentration, and $k' = k_1 + k_3$. The fitness function is the reciprocal of the sum of squares of the residuals, $F(k_1, \ldots, k_4) = \left(\sum (y_t - \hat{y}_t)^2\right)^{-1}$, where $\hat{y}_t$ is the estimated value at time t. The response surface (F as a function of the parameters, k) shows a long valley of about the same F in the k_1, k_3 plane in which these parameters can compensate each other. Thus there are local minima at high k_1, low k_3 and high k_3, low k_1. In terms of the chemistry, Eq. (1) is the solution of the rate equations of parallel mechanisms described by k_1 and k_2, and k_3 and k_4. Depending on the initial guesses a steepest ascent optimizer discovers the nearest local optimum. The role of the genetic algorithm in the hybrid is to provide a suitable range of initial guesses that can properly represent possible solutions.

The genetic algorithm used eight bits for each of the four parameters, a population of 20, stochastic remainder selection with a probability of 0.9 of mating by a single point crossover, and a 0.01 mutation rate. The paper by Hibbert (1993) details comparisons between this simple genetic algorithm, a real number coded genetic algorithm, and a genetic algorithm with incest prevention. In the hybrid, a binary coded genetic algorithm with incest prevention (Eshelman and Schaffer, 1991) was employed.

Three coupling strategies may be explored (Fig. 2).

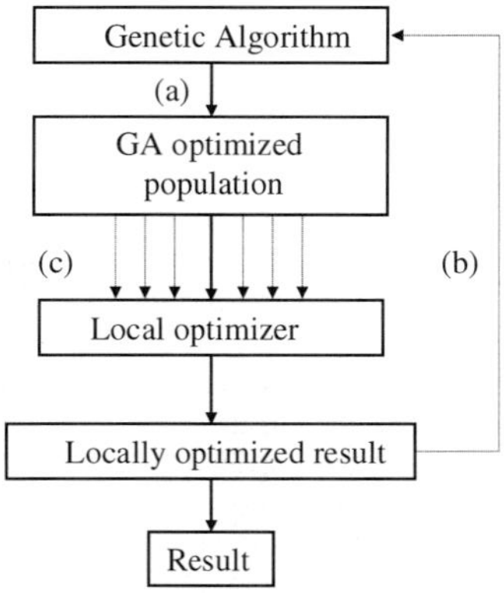

Fig. 2. Schematic of ways through a hybrid genetic algorithm in which a genetic algorithm provides starting points for a local optimizer (e.g. steepest ascent, Simplex, ANN). The main route is with solid arrows and corresponds to the first hybrid described in the text. Dotted lines, labeled (c), are for hybrid 3 where a number of solutions are fed through to the local optimizer. Dotted line (b) indicates hybrid 2 in which the converged result from the local optimizer is used to make a starting population for the genetic algorithm.

(a) The genetic algorithm is run using the best of a generation as the starting point of the steepest descent algorithm. This is done at the end of any generation that improves the function.
(b) As in hybrid (a), but after the steepest descent step, the optimized parameters are put back into the genetic algorithm as one of the new population. The genetic algorithm is then run until the function improves, when it again provides a start for the steepest descent optimizer.
(c) The genetic algorithm is cycled to completion, then the steepest descent optimizer is run on every member of the final population. This gives good information about the response surface. To be useful, the genetic algorithm needs to have an incest-preventing strategy to make sure the population retains sufficient diversity (Eshelman and Schaffer, 1991).

The second hybrid must give a better optimum than the first, and it may be shown (Fig. 3) that in terms of function evaluations, it is worth allowing the genetic algorithm to find a good starting point. Fig. 3 also shows the vagaries of genetic algorithms. It required at least 93 generations of the genetic algorithm to find a starting point that converged on the optimum (value 5400). It may be noted that after 53 generations the genetic algorithm had found a starting point from which the steepest descent optimizer rapidly converged to a reasonable fitness function in a small number of function evaluations. Thereafter it took longer to find better solutions. However, it was found that allowing the genetic algorithm to run for about 200 generations was always sufficient to ensure convergence to the optimum within only a few iterations of the steepest descent optimizer. In the trade off between genetic algorithm and steepest descent optimizer it is seen that it is certainly worth allowing the genetic algorithm to find good solutions (compare the 93 generation and 211 generation results in Fig. 3).

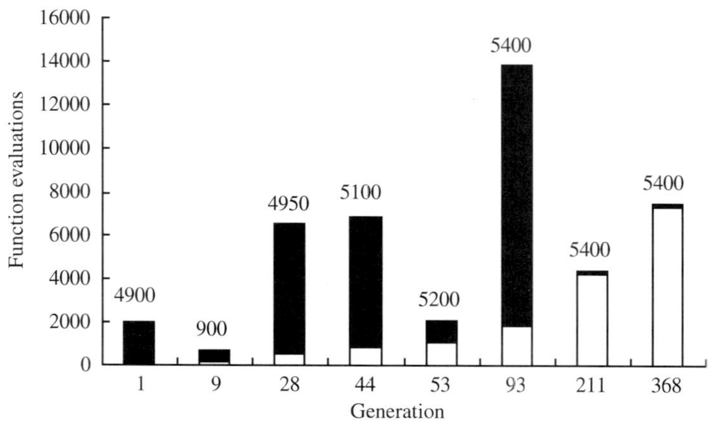

Fig. 3. Number of function evaluations during a hybrid genetic algorithm (lower, white bar) followed by a steepest descent optimizer (upper hatched bar), against generation. Figures on each bar give the final value of the fitness function.

The population of the third hybrid tracks the valleys of the response surface well and provides good starting values for the steepest descent optimizer. In fact the final optimum from this hybrid was obtained from a population member that was not the fittest.

4.2. Genetic algorithm–artificial neural network hybrid optimizing quantitative structure–activity relationships

So and Karplus (1996) report a hybrid method that combines a genetic algorithm and artificial neural network to determine quantitative structure–activity relationships. The method described in their paper is discussed here as an excellent example of a clearly explained methodology, unfortunately unusual in the literature on this subject. The method was applied to a well-known set (the Selwood data set) of 31 antifilarial antimycin analogues, with 53 physiochemical descriptors, such as the partial atomic charges, van der Waals volume, and melting point. The input to the neural network are three descriptors chosen from the pool of 53 by the genetic algorithm, and the outputs are the activity values of the analogues.

The artificial neural network was a 3-3-1 (Fig. 4), i.e. with three hidden nodes in a layer between input (three descriptors) and one output (drug activity). Although not germane to the nature of the hybrid, the neural network used a steepest descent back-propagation algorithm to train the weights, and a pseudo-second derivative method was also in use.

The genetic algorithm provided the three descriptors for input into the artificial neural network. A population of 300 sets of descriptors was created at random with constraints that no two sets could be identical, and descriptors in a given individual should be different. These constraints were maintained throughout the optimization.

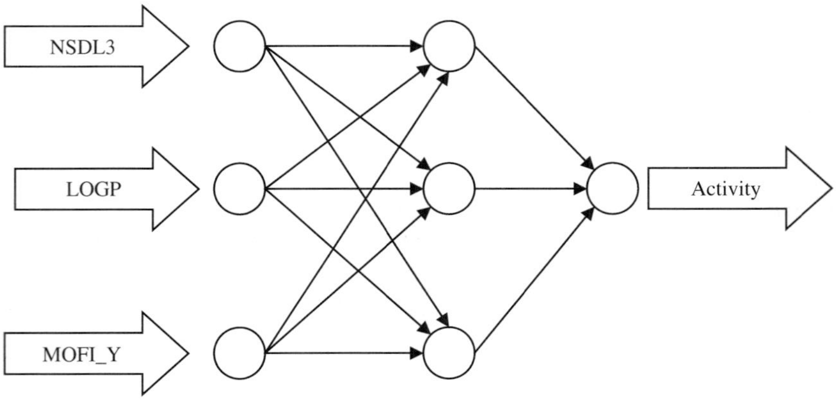

Fig. 4. 3-3-1 Artificial neural network used to determine the activity of a drug from three input descriptors. The examples shown are the best found, NSDL3, nucleophilic superdelocalizability for atom 3; LOGP, calculated log partition coefficient for octanol/water; MOFI_Y, principal moment of inertia in the *y*-direction.

The fitness function was the reciprocal of the residual root mean square error: $F = (\text{RmsE})^{-1}$, determined from the training set.

$$\text{RmsE} = \sqrt{\frac{\sum_{j=1}^{j=N}(\text{activity}_{\text{calc},j} - \text{activity}_{\text{obs},j})^2}{N}} \tag{2}$$

Because of concerns that the fitness function may create models that train the known data well but fail to find the best value of unknown sets, an alternative function was studied. Here three of the data set having high, low, and middle activities were removed from the training set and used as an independent test set. The fitness function was the RmsE of these three, predicted from a model trained with the remaining 28 analogues. The optimum was still found, and the models showed better predictive power. In a similar exercise, the cross-validated correlation coefficient was used as the fitness function, $F = 1 + R_{\text{calc,obs}}$. As the correlation coefficient may take values between -1 and $+1$, F lies between 0 and 2. The cross-validation method was to leave one analogue in the data set out, train the method using the remaining 30 analogues, and then predict the activity of the missing one. This is repeated in turn for all members and the correlation coefficient calculated between the vector of calculated values and their observed activities. Although a powerful method of validation, a complete optimization must be performed N (here 31) times which is costly in terms of computer time. The authors conclude that the three analogues used as a test set were entirely sufficient and much more efficient in computer time.

Reproduction was by the stochastic remainder method which ensures that individuals having greater than the average fitness are reproduced at least once with the best member of each generation going through to the next without change. Mating between individuals was based on a choice weighted by fitness, with the offspring having two descriptors from one parent and one from the other. Mutation is applied, it appears, to every child, with one descriptor being randomly changed.

The data set can be exhaustively studied with 23,426 combinations of three descriptors chosen from a possible 53. Therefore the efficiency of any algorithm may be unambiguously determined—it either finds the best descriptors or it does not. The hybrid genetic algorithm–artificial neural network found the optimum combination in only 10 generations. Moreover, the best 10 sets of descriptors were found by the 14th generation.

4.3. Non-linear partial least squares regression with optimization of the inner relation function by a genetic algorithm

An interesting hybrid that uses the power of a genetic algorithm has been reported by Yoshida and Funatsu (1997). In quadratic partial least squares (QPLS) regression the model equations are:

$$\mathbf{X} = \sum_{i=1}^{i=A} \mathbf{t}_i \mathbf{p}_i' + \mathbf{E} \tag{3}$$

$$\mathbf{y} = \sum_{i=1}^{i=A} \mathbf{u}_i q_i' + \mathbf{f} \tag{4}$$

$$\mathbf{u} = g(\mathbf{t}) + \mathbf{h} \tag{5}$$

$$\mathbf{t} = \mathbf{X}\mathbf{w} \tag{6}$$

The function $g(\mathbf{t})$ is a quadratic in $\mathbf{t}$. The model is conventionally solved by calculation of the weight vector $\mathbf{w}$ by linear PLS, optimization of the quadratic coefficients, then updating $\mathbf{w}$ by a linearization of the inner relation function. This process is iterated to convergence. The authors identify two problems with this method; the initial guess from linear PLS may not be appropriate, and the optimization of the quadratic inner relation function may not always converge or converge quickly. The solution proposed uses a genetic algorithm to determine the latent variable $\mathbf{t}$ by optimizing $\mathbf{w}$ followed by a conventional least squares solution of the quadratic coefficients of g. $\mathbf{w}$ is chosen because of the reduced dimension of $\mathbf{Xw}$ compared with $\mathbf{t}$. The genetic algorithm coded each w as a 10-bit string with 90 members in the population. One point crossover with a probability of 50% was employed and a relatively high mutation rate of 2%. Return of the best individual was also found to improve the result. The fitness function was the residual square error, so it may be assumed that the genetic algorithm was run as a minimization algorithm.

The example given in the paper is the optimization of the auto ignition temperature of 85 organic compounds predicted from six physicochemical parameters. Inherent non-linearity means that QPLS is indicated, and the authors show that the conventional optimization of the inner relation function does not lead to an adequate solution. The use of the genetic algorithm, however, produces good results within 40–50 generations. The improvement obtained by using a genetic algorithm appears to arise from the scope of the parameter space searched allowing a good starting point for the quadratic optimization to be found.

4.4. The use of a clustering algorithm in a genetic algorithm

A unique instance of a hybrid genetic algorithm is the use of a clustering algorithm to ensure diversity of populations. Hanagandi and Nikolaou (1998) have reported such an algorithm for the optimization of pump configuration in a chemical engineering problem. The motivation behind this work was the observation that a published crowding algorithm (De Jong, 1975) did not dissolve the clusters in the problem at hand. This is one of the few examples of a hybrid genetic algorithm in which the second method is used within the genetic algorithm. The philosophy behind such an approach is to observe that in nature, when a number of individuals inhabit the same space, crowding is likely to reduce the fitness of all and cause a reduction in the population even if the niche is highly suitable. This was the motivation of the work of De Jong. The authors of the paper cited (Hanagandi and Nikolaou, 1998) show that an approach by Torn (1977, 1978) could be used to perform the task of causing the dissolution of clusters while keeping a suitable representative in the population. The algorithm is as follows (taken from Hanagandi and Nikolaou (1998))

1. Choose uniformly random points
2. Use a local search algorithm for a few steps
3. Find clusters using a cluster analysis technique
4. Take a sample point from each cluster
5. Go to step 2

This is embedded within a genetic algorithm formalism as shown in the schematic of Fig. 5. This hybrid was applied to a number of classical bench mark problems. Typically the population size was 30, the mutation rate was 0.01, crossover probability 1.0, with a chromosome length of 30 per variable. While a genetic algorithm with De Jong's crowding algorithm performed little better than a simple genetic algorithm, the present method converged on solutions much more rapidly. The method was further applied to a problem in engineering, that of configuring pipes to maximize flows within constraints of total mass transport, etc. Interestingly the hybrid genetic algorithm found a new better, more simple solution that was, unfortunately, impossible.

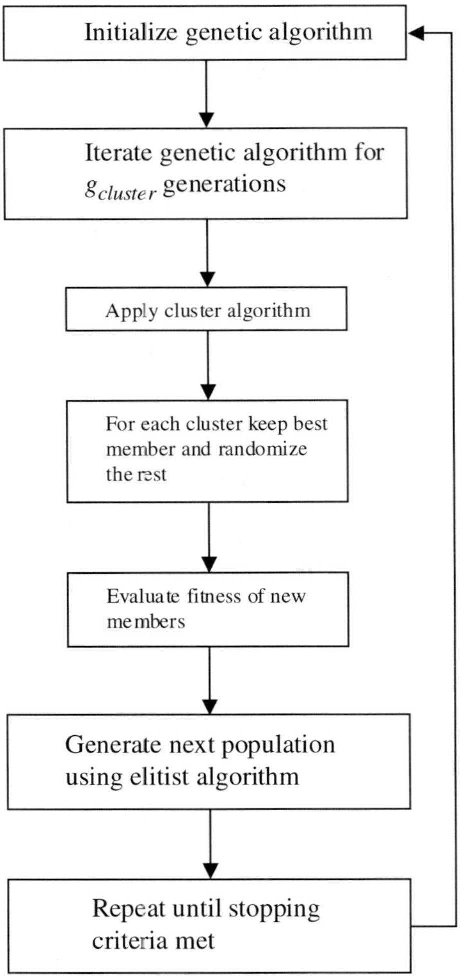

Fig. 5. Schematic of a genetic algorithm with cluster algorithm embedded to maintain population diversity.

5. Conclusion

Optimization algorithms, be they evolutionary-based or not, will always be augmented by hybridization. The beauty of genetic algorithms is that they can be integrated with other algorithms with evidently superior results. The majority of published methods have a genetic algorithm as a precursor, feeding in starting points to a local search method. There are many ways of implementing this relationship, with iterations or parallel processing of members of the population. Although not every publication explores this point, the consensus appears to be that the extra effort of coding and implementation of the genetic algorithm is to significantly improve the quality of the result. Many of the methods date to the mid-1990s and it is not clear to the author that many of these methods are now in common currency.

Some methods embed a genetic algorithm within an optimizer to tweak the optimizer's parameters. Artificial neural networks have been the targets for many of these hybrids. The number of nodes and weights can be determined by the genetic algorithm. Another example is the use of a genetic algorithm to determine the weights in a *k*-nearest neighbor classification. Finally a genetic algorithm has been used to optimize the inner relation function of a partial least squares regression.

In one case, a cluster algorithm was used to prune the population of a genetic algorithm. Other examples are found in which some level of further searching is incorporated as part of the genetic algorithm.

References

Anand, S.S., Smith, A.E., Hamilton, P.W., Anand, J.S., Hughes, J.G., Bartels, P.H., 1999. An evaluation of intelligent prognostic systems for colorectal cancer. Artif. Intell. Med. 15, 193–214.

Balland, L., Estel, L., Cosmao, J.M., Mouhab, N., 2000. A genetic algorithm with decimal coding for the estimation of kinetic and energetic parameters. Chemom. Intell. Lab. Syst. 50, 121–135.

Balland, L., Mouhab, N., Cosmao, J.M., Estel, L., 2002. Kinetic parameter estimation of solvent-free reactions: application to esterification of acetic anhydride by methanol. Chem. Engng Process. 41, 395–402.

Cela, R., Martinez, J.A., 1999. Off-line optimization in HPLC separations. Quim. Anal. (Barcelona) 18, 29–40.

Chen, W.C., Chang, N.-B., Shieh, W.K., 2001. Advanced hybrid fuzzy-neural controller for industrial wastewater treatment. J. Environ. Engng (Reston, VA) 127, 1048–1059.

Cho, K.-H., Hyun, N.G., Choi, J.B., 1996. Determination of the optimal parameters for meson spectra analysis using the hybrid genetic algorithm and Newton method. J. Korean Phys. Soc. 29, 420–427.

Davis, L., 1991. Handbook of Genetic Algorithms, Van Nostrand Reinhold, New York.

De Jong, K.A., 1975. An Analysis of the Behavior of a Class of Genetic Adaptive Systems, University of Michigan, Ann Arbor.

Del Carpio, C.A., 1996a. A parallel genetic algorithm for polypeptide three dimensional structure prediction. A transputer implementation. J. Chem. Inf. Comput. Sci. 36, 258–269.

Del Carpio, C.A., 1996b. A parallel hybrid GA for peptide conformational space analysis. Pept. Chem. 34, 293–296.

Del Carpio, C.A., Sasaki, S.-i., Baranyi, L., Okada, H., 1995. A parallel hybrid GA for peptide 3-D structure prediction. Genome Inf. Ser. 6, 130–131.

Devillers, J., 1996. Designing molecules with specific properties from intercommunicating hybrid systems. J. Chem. Inf. Comput. Sci. 36, 1061–1066.

Dokur, Z., Olmez, T., 2001. ECG beat classification by a novel hybrid neural network. Computer Meth. Programs Biomed. 66, 167–181.

Eshelman, L.J., Schaffer, J.D., 1991. Preventing premature convergence in genetic algorithms by preventing incest. Fourth International Conference on Genetic Algorithms, 115–122.

Gao, F., Li, M., Wang, F., Wang, B., Yue, P., 1999. Genetic algorithms and evolutionary programming hybrid strategy for structure and weight learning for multilayer feedforward neural networks. Ind. Engng Chem. Res. 38, 4330–4336.

Gunn, J.R., 1997. Sampling protein conformations using segment libraries and a genetic algorithm. J. Chem. Phys. 106, 4270–4281.

Haas, O.C., Burnham, K.J., Mills, J.A., 1998. Optimization of beam orientation in radiotherapy using planar geometry. Phys. Med. Biol. 43, 2179–2193.

Han, S.-S., May, G.S., 1996. Recipe synthesis for PECVD SiO_2 films using neural networks and genetic algorithms. Proc. Electron. Compon. Technol. Conf. 46, 855–860.

Hanagandi, V., Nikolaou, M., 1998. A hybrid approach to global optimization using a clustering algorithm in a genetic search framework. Comput. Chem. Engng 22, 1913–1925.

Handschuh, S., Wagener, M., Gasteiger, J., 1998. Superposition of three-dimensional chemical structures allowing for conformational flexibility by a hybrid method. J. Chem. Inf. Comput. Sci. 38, 220–232.

Hartnett, M.K., Bos, M., van der Linden, W.E., Diamond, D., 1995. Determination of stability constants using genetic algorithms. Anal. Chim. Acta 316, 347–362.

Heidari, M., Ranjithan, S.R., 1998. A hybrid optimization approach to the estimation of distributed parameters in two-dimensional confined aquifers. J. Am. Water Resour. Assoc. 34, 909–920.

Hibbert, D.B., 1993. A hybrid genetic algorithm for the estimation of kinetic parameters. Chemom. Intell. Lab. Syst. 19, 319–329.

Ikeda, N., Takayanagi, K., Takeuchi, A., Nara, Y., Miyahara, H., 1997. Arrhythmia curve interpretation using a dynamic system model of the myocardial pacemaker. Meth. Informat. Med. 36, 286–289.

Kemsley, E.K., 2001. A hybrid classification method: discrete canonical variate analysis using a genetic algorithm. Chemom. Intell. Lab. Syst. 55, 39–51.

Kim, T.S., May, G.S., 1999. Optimization of via formation in photosensitive dielectric layers using neural networks and genetic algorithms. IEEE Trans. Electron. Packag. Manufact. 22, 128–136.

Kwan, R.S., Kwan, A.S., Wren, A., 2001. Evolutionary driver scheduling with relief chains. Evolut. Comput. 9, 445–460.

Leardi, R., Lupiáñez González, A., 1998. Genetic algorithm applied to feature selection in PLS regression: how and when to use them. Chemom. Intell. Lab. Syst. 41, 195–207.

Lee, B., Yen, J., Yang, L., Liao, J.C., 1999. Incorporating qualitative knowledge in enzyme kinetic models using fuzzy logic. Biotechnol. Bioengng 62, 722–729.

Liu, H.-L., 1999. A hybrid AI optimization method applied to industrial processes. Chemom. Intell. Lab. Syst. 45, 101–104.

Lucasius, C.B., Kateman, G., 1994. Understanding and using genetic algorithms Part 2. Representation, configuration and hybridization. Chemom. Intell. Lab. Syst. 25, 99–145.

Medsker, L.R., 1994. Hybrid Neural Network and Expert System, Kluwer, Boston.

Mitra, P., Mitra, S., Pal, S.K., 2000. Staging of cervical cancer with soft computing. IEEE Trans. Biomed. Engng 47, 934–940.

Mohaghegh, S., Platon, V., Ameri, S., 2001. Intelligent systems application in candidate selection and treatment of gas storage wells. J. Petrol. Sci. Engng 31, 125–133.

Nandi, S., Ghosh, S., Tambe, S.S., Kulkarni, B.D., 2001. Artificial neural-network-assisted stochastic process optimization strategies. AIChE J. 47, 126–141.

Ouchi, Y., Tazaki, E., 1998. Medical diagnostic system using Fuzzy Coloured Petri Nets under uncertainty. Medinfo 9 (Pt 1), 675–679.

Parbhane, R.V., Unniraman, S., Tambe, S.S., Nagaraja, V., Kulkarni, B.D., 2000. Optimum DNA curvature using a hybrid approach involving an artificial neural network and genetic algorithm. J. Biomol. Struct. Dyn. 17, 665–672.

Park, T.-Y., Froment, G.F., 1998. A hybrid genetic algorithm for the estimation of parameters in detailed kinetic models. Comput. Chem. Engng 22, S103–S110.

Pena-Reyes, C.A., Sipper, M., 2000. Evolutionary computation in medicine: an overview. Artif. Intell. Med. 19, 1–23.

Raymer, M.L., Sanschagrin, P.C., Punch, W.F., Venkataraman, S., Goodman, E.D., Kuhn, L.A., 1997. Predicting conserved water-mediated and polar ligand interactions in protein using a *K*-nearest-neighbors genetic algorithm. J. Mol. Biol. 265, 445–464.

Shaffer, R.E., Small, G.W., 1996a. Comparison of optimization algorithms for piecewise linear discriminant analysis: application to Fourier transform infrared remote sensing measurements. Anal. Chim. Acta 331, 157–175.

Shaffer, R.E., Small, G.W., 1996b. Genetic algorithms for the optimization of piecewise linear discriminants. Chemom. Intell. Lab. Syst. 35, 87–104.

Shengjie, Y., Schaeffer, L., 1999. Optimization of thermal debinding process for PIM by hybrid expert system with genetic algorithms. Braz. J. Mater. Sci. Engng 2, 29–40.

Shimizu, Y., 1999. Multi-objective optimization for site location problems through hybrid genetic algorithm with neural networks. J. Chem. Engng Jpn 32, 51–58.

So, S.-S., Karplus, M., 1996. Evolutionary optimization in quantitative structure–activity relationship: an application of genetic neural networks. J. Med. Chem. 39, 1521–1530.

Torn, A.A., 1977. Cluster analysis using seed points and density-determined hyper-spheres as an aid to global optimization. IEEE Trans. Syst. Man Cybernet. 7, 610.

Torn, A.A., 1978. A search-clustering approach to global optimization. In: Dixon, L.C.E., Szego, G.P., (Eds.), Towards Global Optimization, North-Holland, Amsterdam.

Vivo-Truyols, G., Torres-Lapasio, J.R., Garcia-Alvarez-Coque, M.C., 2001a. A hybrid genetic algorithm with local search: I. Discrete variables: optimisation of complementary mobile phases. Chemom. Intell. Lab. Syst. 59, 89–106.

Vivo-Truyols, G., Torres-Lapasio, J.R., Garrido-Frenich, A., Garcia-Alvarez-Coque, M.C., 2001b. A hybrid genetic algorithm with local search II. Continuous variables: multibatch peak deconvolution. Chemom. Intell. Lab. Syst. 59, 107–120.

Wakao, S., Onuki, T., Ogawa, F., 1997. A new design approach to the shape and topology optimization of magnetic shields. J. Appl. Phys. 81, 4699–4701.

Wakao, S., Onuki, T., Tatematsu, K., Iraha, T., 1998. Optimization of coils for detecting initial rotor position in permanent magnet synchronous motor. J. Appl. Phys. 83, 6365–6367.

Wang, F.-S., Jing, C.-H., 2000. Application of hybrid differential evolution to fuzzy dynamic optimization of a batch fermentation. J. Chin. Inst. Chem. Engr. 31, 443–453.

Wehrens, R., Lucasius, C., Buydens, L., Kateman, G., 1993. HIPS, a hybrid self-adapting expert system for nuclear magnetic resonance spectrum interpretation using genetic algorithms. Anal. Chim. Acta 277, 313–324.

de Weijer, A.P., Lucasius, C., Buydens, L., Kateman, G., Heuvel, H.M., Mannee, H., 1994. Curve fitting using natural computation. Anal. Chem. 66, 23–31.

Xue, D., Li, S., Yuan, Y., Yao, P., 2000. Synthesis of waste interception and allocation networks using genetic-alopex algorithm. Comput. Chem. Engng 24, 1455–1460.

Yamaguchi, A., 1999. Genetic algorithm for SU(N) gauge theory on a lattice. Nucl. Phys. B, Proc. Suppl. 73, 847–849.

Yang, M., Zhang, X., Li, X., Wu, X., 2002. A hybrid genetic algorithm for the fitting of models to electrochemical impedance data. J. Electroanal. Chem. 519, 1–8.

Yoshida, H., Funatsu, K., 1997. Optimization of the inner relation function of QPLS using genetic algorithm. J. Chem. Inf. Comput. Sci. 37, 1115–1121.

Zacharias, C.R., Lemes, M.R., Dal Pino, A. Jr., 1998. Combining genetic algorithm and simulated annealing: a molecular geometry optimization study. Theochem 430, 29–39.

Zuo, K., Wu, W.T., 2000. Semi-realtime optimization and control of a fed-batch fermentation system. Comput. Chem. Engng 24, 1105–1109.

CHAPTER 3

Robust soft sensor development using genetic programming

Arthur K. Kordon[a], Guido F. Smits[b], Alex N. Kalos[a], Elsa M. Jordaan[b]

[a]*The Dow Chemical Company, Freeport, TX 77566, USA*
[b]*Dow Benelux NV, Terneuzen, The Netherlands*

1. Introduction

One of the unexpected results from the research activities of nature-inspired chemometric methods is their fast acceptance in industry. The paradox is that some of these approaches have been successfully applied to resolve real world problems even before their theoretical development has reached its maturity level. The business potential of evolutionary computing in the area of engineering design was rapidly recognized in the early 1990s by companies like GE, Rolls Royce, and British Aerospace (Parmee, 2001). In a short period of several years, many industries like aerospace, power, chemical, etc., transferred their research interest for evolutionary computing into various practical solutions. Contrary to its very nature, evolutionary computing entered industry in a revolutionary way. Different factors contributed to this phenomenon such as the ease of understanding of the approach and finding potential business benefits, the fast growing computational power that allows effective implementation even on cheap hardware, and the availability of software. Among the various application areas of evolutionary computing (a good survey can be found in Banzhaf et al., 1998) and neural networks (Haykin, 1998), the most mature and widespread industrial implementation is the inferential or soft sensor. It has been implemented in thousands of applications in various industries by several well-established vendors. The economic benefit from these implementations is estimated in hundreds of millions of dollars and vendors' revenues are in the range of tens of millions of dollars (Aspentech, 2003; Pavilion, 2003). The development and support of soft sensors in industry is a small industry by itself. What is the nature of soft sensors? Why are they so popular in industry? What are the unique features that make the contribution of nature-inspired chemometrics methods like neural networks and evolutionary computing so important? These are some of the questions that will be addressed in this chapter.

Data Handling in Science and Technology, Volume 23
ISSN: 0922–3487

DOI: 10.1016/S0922-3487(03)23003-3

Some critical parameters in chemical processes are not measured on-line (composition, molecular distribution, density, viscosity, etc.) and their values are captured either by lab samples or off-line analysis. However, for process monitoring and quality supervision, the response time of these relatively low frequency (several hours, or even days) measurements is very slow and may cause loss of production due to poor quality control. When critical parameters are not available on-line in situations with potential for alarm 'showers', the negative impact could be significant and eventually could lead to shutdowns. One of the approaches to address this issue is through development and installation of expensive hardware on-line analyzers. Another solution is by using soft or inferential sensors that infer the critical parameters from other easy-to-measure variables like temperatures, pressures, and flows.

The idea of using direct measured variables to deduce process quality parameters was first discussed in the control community in late 1960s–early 1970s (Rotatch and Hadjiski, 1966; Weber and Brosilow, 1972; Joseph and Brosilow, 1978). However, the proposed estimators required linear open loop models and a priori knowledge of the disturbances. Both assumptions pose significant limitations for real applications and the proposed solutions have not been implemented in industry.

The first wave of industrial soft sensors appeared in different manufacturing areas in the early 1990s (Di Massimo et al., 1991; Piovoso and Owens, 1991; Tham et al., 1991). The breakthrough technology was neural networks because of their ability to capture nonlinear relationships and with their adequate framework for industrial use. The detailed description of neural networks is given in the second part of the book; however, we will emphasize the following features, which are relevant to practical application of inferential sensors (Haykin, 1998):

- neural networks are universal approximators;
- no a priori knowledge on the process and the disturbances is required for model building;
- the model development process is based on machine learning approaches (especially for neural networks based on the back-propagation algorithm);
- the key requirement for model development is a representative data set for model training, testing, and validation.

Several software vendors like Pavilion Technologies, Neuralware, Aspentech, Gensym Corporation, etc., offer off-line development and on-line run-time packages based on different neural network architectures. They claim thousands of soft sensors applications in the chemical, petro-chemical, pharmaceutical, power generation, and other industries. The common methodology of building neural net soft sensors and the practical issues of their implementation have been discussed in detail in Qin (1996). Despite their successes, most commercially available neural net packages are still based on classical back-propagation algorithm. As a result, those commercial neural networks generally exhibit poor generalization capability outside the range of training data (Haykin, 1998). This can result in poor performance of the model and unreliable prediction in new operating conditions. Another drawback is that such packages usually yield neural net structures with unnecessary high complexity. Selection of the neural net structure is still an ad hoc process and very often leads to inefficient and complex solutions. This 'fat' dimensionality significantly reduces the robustness of empirical models. Of special importance is the selection of only those inputs that have a major influence on the output. In order to achieve proper input selection, a sensitivity analysis is needed on the influence of each input on the output. This type of analysis is very difficult to perform by the conventional back-propagation-based neural nets. As a result of this inefficient structure and reduced robustness, there is a necessity of frequent re-training of the empirical model. The final effect of all of these problems is an increased maintenance cost and gradually decreased performance and credibility (Lennox et al., 2001).

The need for robustness toward process variability, the ability to handle industrial data (e.g. missing data, measurement noise, operator intervention on data, etc.) and ease of model maintenance are key issues for mass-scale application of reliable inferential sensors. Several machine-learning approaches have the potential to contribute to the solution of this important problem. Stacked analytical neural networks (internally developed in The Dow Chemical Company) allow very fast model development of parsimonious black-box models with confidence limits. Genetic Programming (GP) can generate explicit functional solutions that are very convenient for direct on-line implementation in the existing process information and control systems (Koza, 1992). Recently, Support Vector Machines (SVM) give tremendous opportunities for building empirical models with very good generalization capability (Vapnik, 1998). At the same time, each approach has its own weaknesses which reduces the implementation space and makes it difficult to design the robust soft sensor based on separate computational intelligence techniques.

An alternative, more integrated approach for a 'second generation' soft sensor development is described in this chapter. It combines a nonlinear sensitivity and time-delay analysis based on Stacked Analytical Neural Nets with outlier detection and condensed data selection driven by the SVM. The derived soft sensor is generated by GP as an analytical function. The integrated methodology amplifies the advantages of the individual techniques, significantly reduces the development time, and delivers robust soft sensors with low maintenance cost.

The chapter is organized in the following manner. Section 2 covers the current state of the art of soft sensors, including a survey of key application areas and vendors. The requirements for robust soft sensors are defined in Section 3, followed by a description of selected approaches for effective soft sensors development, such as stacked analytic neural networks and SVM. The nature of GP, its difference from genetic algorithms, and its advantages for inferential sensors development are described in Section 5. Finally, the integrated methodology is described and illustrated with an industrial application for volatile materials emission estimation.

2. Soft sensors in industry

Soft sensors infer important process variables (called outputs) from available hardware sensors (called inputs). Usually the outputs are measured infrequently by lab analysis, material property tests, expensive gas chromatograph analysis, etc. Very often the output measurement is performed off-line and then introduced into the on-line process monitoring and control system. It is assumed that soft sensors' inputs are available on-line either from cheap hardware sensors or from other soft sensors.

Different inference mechanisms can be used for soft sensor development. If there is a clear understanding of the physics and chemistry of the process, the inferred value can be derived from a fundamental model. Another option is to estimate the parameters of the fundamental model via Kalman Filter or Extended Kalman Filter. There are cases where the input/output relationship is linear and can be represented either by linear regression or by multivariate model (Russel et al., 2000). The most general representation of the soft sensor, however, is as a nonlinear empirical model. The generic view of the soft sensor is represented in Fig. 1, where $\mathbf{x}$ is a vector of input variables, y_k is a sequence of output off-line measurements, and $\hat{y}$ is the on-line output prediction, based on the nonlinear relationship $f(\mathbf{x})$.

2.1. Assumptions for soft sensors development

As any model, the inferential sensor is derived under certain assumptions that define its validity limits. The first assumption for soft sensor design is that the developed input/output relationship could be nonlinear. This assumption broadens the implementation areas, especially in the chemical industry. However, it imposes challenges, typical for nonlinear empirical models such as unpredictable extrapolation, lack of model confidence limits, and multiplicity of model solutions. Most of these issues are the object of intensive research from the statistical and machine learning communities (Hastie et al., 2001) and of efforts to design effective soft sensor supervision to detect areas of poor performance (Martin and Morris, 1999; Kordon and Smits, 2001; Jordaan, 2002). The second assumption is that the derived empirical model will guarantee reliable performance with acceptable accuracy of prediction inside the range of input data used for model development (also called training data.) On the basis of this assumption are the excellent interpolation qualities of machine learning approaches (Mitchel, 1997) in general and the universal approximator property of neural networks (Haykin, 1998) in particular. This assumption is critical for industrial applications of soft sensors because it gives the necessary theoretical basis for reliable predictions and to support the investment for model development. The third assumption is the expectations of performance deterioration in

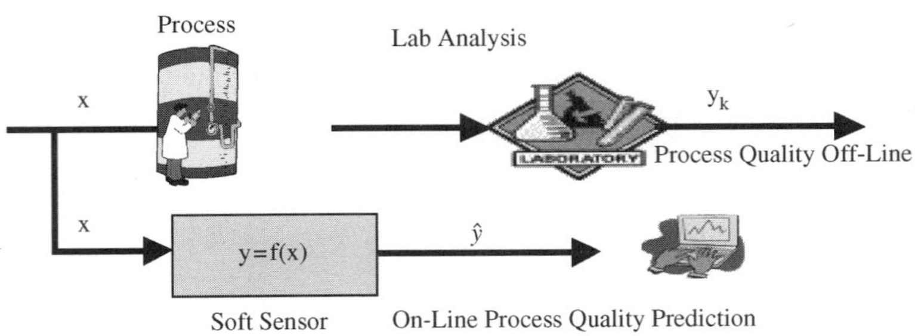

Fig. 1. Generic structure of a soft sensor.

new operating conditions (i.e. outside the range of training data), which is a logical consequence from the previous two assumptions. As a result, model re-training or even complete re-design is recommended. Since process and operating condition changes are more a rule than an exception; increased robustness in these circumstances becomes the central issue of soft sensors design. The fourth assumption is that in order to develop and maintain inferential sensors we need measurements based on hardware sensors, i.e. soft sensors do not entirely eliminate hardware sensors. The economic benefit is realized not by strict replacement, but by (a) intelligent use of cheap hardware sensors, or (b) by more efficient use of expensive hardware sensors, especially when the routine lab samples would need to be taken in a dangerous chemical environment.

The fifth assumption is that the successful empirical model building requires different types of data, such as historical process data, lab samples, data generated by Design Of Experiments (DOE), and data from modeling sources.

- *Historical process data that is data-rich and information-poor.* Usually this type of data represents model inputs and some of them are highly correlated (for example, several temperatures representing the temperature profile in a chemical reactor.) The key issue is how to condense the data set with informational-rich data only.
- *Lab test data represent the low-frequency output data.* Very often, there is a substantial difference in data collection frequency between process and lab data (several hours even days for lab data versus several minutes for process data). One of the issues is the exact time alignment between taking the sample, doing the lab analysis, and entering the results in the data historian.
- *DOE-generated data is the best source for soft sensor development.* Each data point contains unique information and the inputs are statistically orthogonal. This approach for data generation is generally used in data collection campaigns for the development of emission estimation soft sensors.
- Data generated by different models (fundamental or empirical) or simple relationships like material balances can also be used in soft sensor development and maintenance. However, source model validation is a pre-condition for using this type of data.

2.2. Economic benefits from soft sensors

The defined assumptions formulate the requirements and the realistic expectations for industrial implementation of inferential sensors. However, the key issue that can turn the technical capabilities of this modeling approach into a successful application is the potential for value creation. The sources for economic benefits from inferential sensors are as follows:

- soft sensors allow tighter control of the most critical parameters for final product quality and as a result the product consistency is significantly improved;
- on-line estimates of critical parameters reduce process upsets through early detection of problems;
- inferential sensors improve working conditions by reducing or eliminating lab measurements in a dangerous environment;

- very often soft sensors are at the economic optimum. Their development and maintenance cost is lower in comparison to the alternative solutions of expensive hardware sensors or more expensive fundamental models;
- one side effect of the implementation of inferential sensors is the optimization of the use of expensive hardware i.e. they reduce capital investments;
- soft sensors can be used not only for parameter estimation but also for running 'What-If' scenarios in production planning.

These economic benefits have been realized rapidly by the industry and from the early 1990s a spectacular record of successful applications is reported by the vendors and in the literature. We will illustrate the large range of applications with several typical cases.

2.3. Soft sensor application areas

The poster child of soft sensors applications is environmental emission monitoring (Qin et al., 1997; Pavilion Technology, 2003). Traditionally emission monitoring is performed by expensive analytical instruments with costs between \$100 000 and \$200 000 and maintenance costs of at least \$15 000 per year (Dong and McAvoy, 1995). The inferential sensor alternative, implemented as a 'classical' neural network, is much cheaper and with accuracy acceptable by the federal, state, and local regulations (Eghneim, 1996) NO*x* emissions in burners, heaters, incinerators, etc. are inferred by associated process variables, mostly temperatures, pressures, and flows. According to Pavilion Technologies, the leading vendor in soft sensors for emission monitoring, more than 250 Predictive Emission Monitoring (PEMs) have been installed in the United States since the mid-1990s (Pavilion Technology, 2003) A case study of using neural net soft sensor to predict O_2 contents in a boiler that can significantly improve combustion efficiency is described in Al-Duwaish et al. (2002).

Another area of successful inferential sensors implementation is biomass estimation in different continuous and fed-batch bioprocesses (Di Massimo et al., 1991; Tham et al., 1991; Willis et al., 1992). Estimating the biomass is of critical importance for successful control of fermenters, especially during the growth phase of organisms. Usually the biomass concentrations are determined off-line by lab analysis every 2–4 h. However, this low measurement frequency can lead to poor control and on-line estimates are needed. In Di Massimo et al. (1991) a soft sensor for biomass estimation based on two process variables—fermenter dilution rate and carbon dioxide evolution rate (CER)—successfully estimated biomass in continuous *mycelial* fermentation. The neural net model included six inputs that incorporated process dynamics for three consecutive sampling periods, two hidden layers with four neurons and one output, the biomass estimate. Another successful implementation of a biomass soft sensor for a penicillin fed-batch process is described in Willis et al. (1992). The topology of the neural net in this case is (2-3-1), where the two inputs are the oxygen uptake rate (OUR) and the batch time, and the output is penicillin biomass estimates. Of similar nature to biomass soft sensors are inferential sensors for on-line estimation of microbial activity in activated sludge systems (Zhao and McAvoy, 1996; Sotomayor et al., 2002). By using an Extended Kalman Filter, estimates of both OUR and the oxygen transfer function (OTF) have been achieved, based on dissolved oxygen (DO) and respiration rate. Another interesting application is a soft sensor for phosphorus level estimation in municipal wastewater (Jansson et al., 2002).

One of the first popular applications of soft sensors was estimation of product composition in distillation columns (Piovoso and Owens, 1991; Hadjiski et al., 1992; Willis et al., 1992). An example of using a neural network-based soft sensor for the estimates of the catalytic reformer octane number and gasoline splitter product quality is described in Brambilla and Trivella (1996). According to Martin (1997), several soft sensors were implemented on a crude tower to infer kerosene flash point, distillate flash point, atmospheric gas oil% boiled at 450°F, and atmospheric reduced crude% boiled at 500°F. In another application at Albemarle's alpha olefins plant in Belguim, a large empirical model with 320 inputs has been used for product quality predictions (Martin, 1997).

The most widespread implementation of soft sensors in the chemical industry is for prediction of polymer quality (Tham et al., 1991; Zhang et al., 1997; Rallo et al., 2002). Several polymer quality parameters such as melt index, average molecular weight, polymerization rate, and conversion are inferred from reactor temperature, jacket inlet and outlet temperatures, and the coolant flow rate through the jacket. According to Zhang et al. (1997) it is also possible to estimate on-line the amount of reactor impurities during the initial stage of polymerization. Of special interest is the nonlinear controller developed by Pavilion Technology, called Process Perfecter, that optimizes the transition between different polymer products (Pavilion Technology, 2003).

There are many other interesting applications such as inferential sensors for cupola iron-melting furnaces (Abdelrahman and Subramanian, 1998), digester quality estimation in batch pulping process (Rao et al., 1993), and particle size monitoring (Del Villar et al., 1996). The presented application areas, however, highlighted the main directions for successful industrial implementation of inferential sensors.

2.4. Soft sensor vendors

A substantial factor for full-scale industrial application of soft sensors is the quality of the software and the services of the specialized vendors. The dominant vendor has been Pavilion Technologies with more than 1700 installations worldwide, of which 100 are nonlinear inferential control systems based on Process Perfecter (Pavilion Technology, 2003). Pavilion offers several products like Process Insights for general soft sensor development, Software CEM for specialized emission estimation sensors, and Power Perfecter for inferential based optimization of energy generation assets.

Several other vendors include specialized software for inferential sensors development and on-line implementation as part of their modeling or control systems. Aspen Technology offers linear or nonlinear inferential sensors in its Aspen IQ package (Aspen Technology, 2003). The broad suite of tools for developing inferential sensors include Partial Least Squares (PLS), fuzzy PLS, neural networks, and hybrid neural networks. Gensym Corporation offers NeurOn-Line, a specialized product for development and on-line deployment of neural network-based soft sensors (Gensym Corporation, 2003). One of the advantages of this tool is that the designed models can be easily integrated into the G2 real-time expert system environment, which is the industrial standard for intelligent process monitoring systems. Several key control systems vendors offer inferential sensor capabilities as well. Fisher Rosemount Systems has developed the DeltaV Neural functional block, which is capable of soft sensors development and run-time implementation within the DeltaV controller (DeltaV, 2003). Honeywell offers a specialized product Profit SensorPro within the ProfitPlus control system (Honeywell, 2003). It includes a combination of PCA and linear and nonlinear PLS techniques. Siemens has used the software product PRESTO (Property Estimator Toolkit) for development of inferential sensors based on linear identification, neural networks, PLS, and fuzzy logic (Siemens, 2003).

In summary, inferential sensors fill the growing need in industry for sophisticated nonlinear estimators of process quality parameters. For several years, a number of well-established vendors have implemented thousands of soft sensors almost in any industry. The benefit from improved quality and reduced process upsets is estimated in the hundreds of millions of dollars. And the potential market is much bigger.

3. Requirements for robust soft sensors

Along with successes, the process of mass-scale implementation of classical neural network-based soft sensors in the mid 1990s also involved a lot of nontechnical components that played a negative role: like the push from top management to “replace all hardware sensors with soft sensors”. Several vendors oversold the technology beyond the capabilities and limitations of classical neural networks. Many managers embraced the slogan to “Transfer Data into Gold” and initiated a lot of mass-scale applications without the necessary understanding, skill set, and data quality (Gensym, 2003; Pavilion, 2003). Unfortunately, a few years later, the support of many of these initiatives gradually evaporated. The average economic results from soft sensors use have been below expectations because of the growing maintenance costs. It became evident that the technical support of soft sensors is much more difficult even in comparison to complicated hardware sensors. Gradually, the initial ‘irrational exuberance’ about inferential sensors was replaced by more realistic expectations. A lot of bitter lessons have been learned during this transition. The key lessons can be summarized in Section 3.1.

3.1. Lessons from industrial applications

Lesson 1: Soft sensors are not a silver bullet. It became clear that soft sensors cannot deliver ‘voodoo’ solutions as was initially promised by the vendors. Due to lack of experience, this unpleasant fact was recognized in later phases of project development. In some cases soft sensors were inappropriate solutions and even it was technically infeasible to develop empirical models with the available data. This created a lot of disappointment at the top management levels that had started the implementation campaigns. Another source of ‘confusion’ to management was the realization that soft sensors could not replace hardware sensors entirely because these were needed for model development and validation.

Lesson 2: Data quality is of critical importance. Since inferential sensors were one of the first mass-scale applications of empirical models, their developers were initially unprepared for the challenges brought on by the nasty reality of real process data. In most of the cases the data was not generated by DOE. The information content of the available undesigned data was not explored effectively, and as a result, the developed models were inefficient.

Lesson 3: Process knowledge is as important as the data. The initial selling point that soft sensors were entirely data-driven approach and there was no need for process knowledge was eventually corrected by the industrial reality. Of special importance is process knowledge in chemical processes that are influenced by different disturbances with long-term effects such as product impurities and degrading catalyst performance.

Lesson 4: Changing operating conditions require frequent model re-training. A common disease of the first wave of soft sensors was incomplete data sets for model development. Very often, the data did not include all possible operating conditions due to the limited storage capacity of data historians or simply lack of available data. Due to the poor extrapolation capabilities of neural networks, any change of operating conditions requires a total model re-design with all of the following consequences: new data collection, neural net re-training, new structure selection, model validation, and installment of a new run-time model.

Lesson 5: Long-term maintenance becomes the Achilles heal of soft sensors implementation. Since changing operating conditions is more a rule than an exception, frequent re-training became a regular procedure in the soft sensor life cycle. It added a significant maintenance cost and requirement for highly qualified (PhD level) specialists for model technical support. In addition, the specialized software for neural networks implementation also required support, upgrades, and training. It is very difficult and inefficient for a manufacturing organization to afford such type of efforts and resources long term.

Lesson 6: Engineers hate black boxes. Process engineers and operators are the ultimate users of the inferential sensor and its fate depends on their acceptance of this type of empirical model. It is a common observation that engineers have a negative predisposition toward black boxes and look suspiciously at their estimates. They challenge the model with their physical insight and capture even the smallest disagreement.

The final effect of these implementation issues resulted in evaporated support and reduced credibility of inferential sensors. Some internal surveys in the chemical industry show that very few initiated soft sensor projects survived longer than 2–3 years (Massop and Hommersom, 1998) Obviously, the first wave of 'classical' neural network type soft sensors has exhausted their implementation potential. In order to resolve the accumulated issues and to build the basis for large number of industrial applications, new design goals based on new modeling approaches are needed.

3.2. Design requirements for robust soft sensors

There is a very simple and clear criterion for mass scale acceptance of soft sensors in industry—they must have the same level of reliability, ease of use, and maintenance efforts as hardware sensors. This is the natural way to integrate the new technology within the existing work processes and support infrastructures in manufacturing. The defined criterion can be separated into different requirements for design of a 'second' wave of robust soft sensors with low sensitivity to process changes, performance self-assessment capabilities, and reduced maintenance cost. The key design requirements are given in following sections.

3.2.1. Model complexity control

Inferential sensors are often prone to underfitting or overfitting the training or learning data (Gribock et al., 2000). Underfitting occurs when the model fails to capture all variability in the data. Overfitting is exactly the opposite: The model also fits the noise present in the data. In both cases, the inferential sensor will have a sub-optimal ability to predict in the long run new, unseen data points. This may seriously affect its reliability as well as its lifespan.

The root cause of underfitting and overfitting can be explained in terms of the complexity of the model. A model's inherent complexity determines how much variability in the data it can account for. If too much complexity is used, the variability in the data due to noise is modeled as well. Thus, overfitting occurs. If the complexity is too low and the model fails to account for the true variability, it is said to underfit (Cherkassky and Mullier, 1998).

It seems straightforward that the solution to the problem is to choose the model from a set of functions that have the appropriate complexity. However, it is seldom known what the true complexity of the functions should be (Cherkassky and Mullier, 1998; Vapnik, 1998). In classical modeling a set of functions is defined a priori without knowing whether this set of functions actually possesses the right level of complexity. Therefore, many methods try to overcome this problem by adjusting the complexity recursively, until an appropriate level of complexity is found (e.g. by selecting the number of neurons in the hidden layer of neural networks). There are, however, no guarantees that the complexity is in fact optimal. Finally, even if the complexity level is known, there is often no way of controlling the complexity during modeling.

Recently, Statistical Learning Theory and SVM give the necessary theoretical basis for direct selection and control of model complexity, based on an assessed quality of generalization.

3.2.2. Robustness

Inferential sensors often have to operate with data that contain noise and several outliers, and cope with the possibility that all kinds of changes may occur in the plant. The noise present in industrial data sets is often such that it is neither constant nor normally distributed. Furthermore, the presence of outliers is often not known beforehand and they are often not easy to identify.

In the development of a soft sensor, one has to take the noise into account as well as keep in mind that the noise is rarely constant over the whole input space. With respect to outliers, there are two ways of dealing with outliers; namely, remove them, or use techniques that are insensitive to them (Cherkassky and Mullier, 1998). As many outlier detection algorithms do not work well with high-dimensional data sets, the presence of some outliers in the data cannot be ruled out (Aggarwal and Yu, 2001). It is very important that decisions about the process are not made based on the information conveyed by outliers (Pell, 2000). The inferential sensor therefore has to be made insensitive to outliers. The term *robustness* is often used in industry to describe the inferential sensor's sensitivity to perturbations in the variables, parameters or learning data (Gribock et al., 2000; Pell, 2000).

One characteristic aspect of any inferential sensor is that it should not only make accurate predictions, but also be robust to moderate changes in the data (Gribock et al., 2000). Therefore, the learning machine that generates the empirical model is required to resolve the subtle trade-off between accuracy and robustness.

3.2.3. Good generalization capabilities

Although many measurements may be taken in a process, a process (even in a pilot plant) cannot be run over the whole range of possible process conditions, in order to obtain information over the whole input space. It is too expensive and very time consuming. The result is that the training data often cover only a small part of the input space. Therefore, when a process is in operation in a plant it may venture into operating regions that were unknown at the time of modeling. As was discussed already, most empirical models, such as neural networks do not extrapolate well. In many inferential sensor applications, efforts are made to restrict the inferential sensor's predictions to the known input space (Qin et al., 1997). However, it is often expected by process engineers that the model is able to make good predictions from unseen data within a reasonable distance from the known input space. And for unseen data that are too far away, therefore, a 'graceful degradation' of the model is preferable. This means that the model does not become instable and exhibit erratic behavior. Therefore, the requirement is that the inferential sensor is able to predict unseen data in regions of low data density in the known input space, as well as regions that are outside the known input space.

3.2.4. Incorporating prior knowledge

As processes get more complicated and more knowledge is being accumulated during their operation in different regimes, more information about the physical laws, constraints and conditions is becoming available. It is to be expected that when prior knowledge is included into the learning process, the resulting model will have better generalization abilities. The inferential sensor built on such a model may be more intelligent and should be more reliable (Keeler and Ferguson, 1996). It is therefore required that the new generation of inferential sensors should not only be built on empirical data, but also try to incorporate any other form of information that is available.

3.2.5. Adaptivity

Due to the dynamic behavior of many of the processes involved in industry, soft sensors need to be retrained regularly. This leads to high maintenance costs and limited use of inferential sensors in industrial processes (Keeler and Ferguson, 1996). Therefore, the inferential sensor has to become adaptive to the changing conditions, in order to increase its lifespan and reduce costs (Xu et al., 1999).

However, in order to adapt an inferential sensor to new information or conditions, the soft sensor first needs to 'know' when something novel has occurred. It is not a question of recognizing obvious changes in the process such as new equipment or procedures, but rather one of detecting subtle changes, e.g. seasonal behavior or the slow degradation of a catalyst (Xu et al., 1999). The final requirement is that the inferential sensor can perform novelty detection and implement a procedure to adapt the model to the changed conditions.

3.2.6. Self-diagnostic capabilities

In industry, soft sensors are often used for monitoring the performance of a process (Qin et al., 1997). Therefore, it is necessary that the inferential sensor supply the process engineer with information about the accuracy of its predictions. Such a soft sensor thus monitors its own performance (Gribock et al., 2000). This requires that more intelligence be built into the inferential sensor, so that a process engineer is warned early enough that the current model is no longer applicable to the current process conditions, thus reducing the manufacturing of off-specification products and preventing a false sense of trust. The requirement for the inferential sensor is that it should have self-diagnostic capabilities in order to evaluate its own reliability.

3.2.7. Easy on-line implementation and maintenance

The way of on-line implementation is of key importance for reducing soft sensor's cost and guarantee long-term support from process engineers and operators. From that perspective the preferable solution is to avoid additional specialized software and to implement the empirical model directly into the process monitoring or control system. Ideally, the final user must not differentiate the inferential sensors from the other sensors in the system.

4. Selected approaches for effective soft sensors development

It is very difficult to satisfy the defined requirements for industrial soft sensors by a specific empirical technique only, especially by the 'classical' neural networks. However, several new machine-learning approaches can effectively resolve some specific issues and become the building blocks of an integrated methodology for robust soft sensor development. Of special interest are the following three approaches—Stacked analytical neural nets (SANN), SVM, and GP. The first two will be described in this section and a more detailed explanation of GP will be given in Section 5.

4.1. Stacked analytical neural networks

Stacked analytical neural networks are based on a collection of individual, feedforward, single layer neural networks where the weights of the input to hidden layer have been initialized according to a fixed distribution such that all hidden nodes are active. The weights of the hidden to output layer can then be calculated directly using least squares (i.e. no iterative learning procedure such as back propagation is needed). The advantages of this method are: it is fast and each neural network has a well-defined, single, global optimum. Time delays between inputs can be handled through convolution functions.

In addition, the use of a collection of networks gives more robust models that include confidence limits based on the standard deviation of stacked neural nets.

4.1.1. Principles of analytical neural networks

Let us start by looking at the structure of both a linear system and a feed-forward neural network (see Figs. 2 and 3).

The variables $X_1, X_2, \ldots, X_n$ are the inputs for the two structures and Y is the output. In Fig. 2, the coefficients of the linear regression are $a_0, a_1, \ldots, a_n$. The special box marked bias is a convenient way of constructing a constant input required for a_0. In Fig. 3, the coefficients a_{10} to a_{43} and b_0 to b_1 are the neural network weights. F_h is a nonlinear transformation function like the sigmoid function. We would like to be able to gradually move from linear to nonlinear type models. To do this we will modify the neural network structure in such a way that setting the number of hidden nodes equal to zero reduces the model to a linear regression type model. This can be done easily by adding additional weights (c_1 to c_3) between the output and the input layer (see Fig. 4).

If we examine this last neural network structure more carefully, we notice that, assuming we know the weights from the input to the hidden layer a_{ij}, the remaining part can be reduced to a standard linear regression problem. In case F_0 is a nonlinear function, the output Y can be replaced by $F_0^{-1}(Y)$.

The weights b_i and c_j can be found directly by solving the over determined system:

$$F_0^{-1}(Y) = [\,1 \quad X \quad Z\,]\begin{bmatrix} b_0 \\ c_i \\ b_j \end{bmatrix} \tag{1}$$

Y, X, and Z are matrices where the number of rows is equal to the number of patterns in the data set available for training. The number of columns of X is equal to the number of inputs n_i. The number of columns of Z is equal to the number of hidden nodes n_h. The unknown

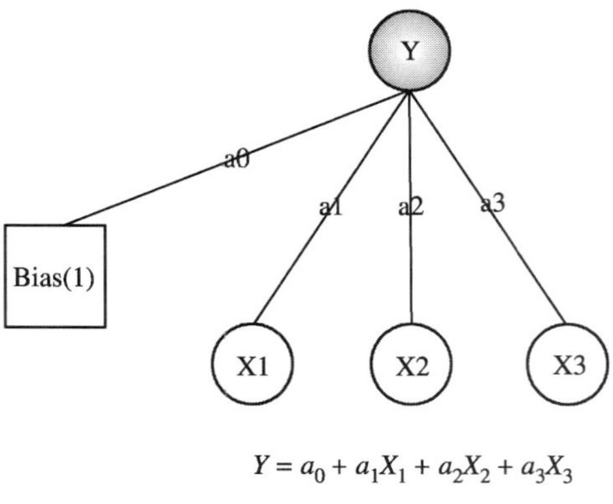

$Y = a_0 + a_1X_1 + a_2X_2 + a_3X_3$

Fig. 2. Graphical representation of a linear regression.

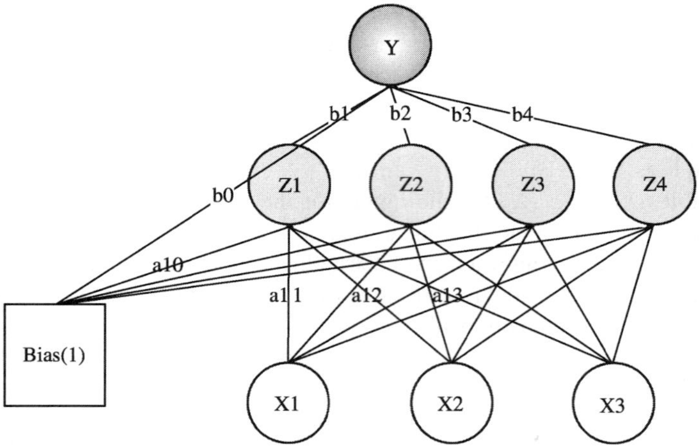

$$Z_1 = F_h(a_{10} + a_{11} X_1 + a_{12} X_2 + a_{13} X_3)$$

$$Z_2 = F_h(a_{20} + a_{21} X_1 + a_{22} X_2 + a_{23} X_3)$$

$$Z_3 = F_h(a_{30} + a_{31} X_1 + a_{32} X_2 + a_{33} X_3)$$

$$Z_4 = F_h(a_{40} + a_{41} X_1 + a_{42} X_2 + a_{43} X_3)$$

$$Y = F_0(b_0 + b_1 Z_1 + b_2 Z_2 + b_3 Z_3 + b_4 Z_4)$$

Fig. 3. Graphical representation of a feed forward neural network.

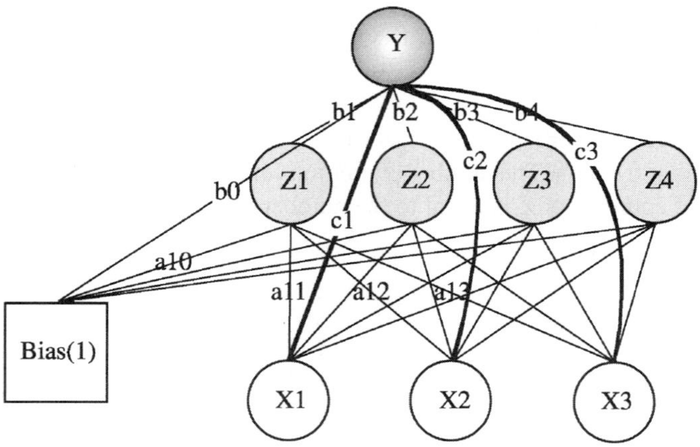

$$Z_1 = F_h(a_{10} + a_{11} X_1 + a_{12} X_2 + a_{13} X_3)$$

$$Z_2 = F_h(a_{20} + a_{21} X_1 + a_{22} X_2 + a_{23} X_3)$$

$$Z_3 = F_h(a_{30} + a_{31} X_1 + a_{32} X_2 + a_{33} X_3)$$

$$Z_4 = F_h(a_{40} + a_{41} X_1 + a_{42} X_2 + a_{43} X_3)$$

$$Y = F_0(b_0 + b_1 Z_1 + b_2 Z_2 + b_3 Z_3 + b_4 Z_4 + c_1 X_1 + c_2 X_2 + c_3 X_3)$$

Fig. 4. Graphical representation of a hybrid structure between a linear regression and a feed forward neural network.

coefficients $[b_0, c_i, b_j]$ can be computed by doing a least squares fit that minimizes the sum of squares of $F_0^{-1}(Y)$ from the model. This problem has a unique analytical solution as long as the input vectors X_i and Z_j are all linearly independent.

A key issue of analytical neural networks is how to initialize the weights from the inputs to the hidden node. One possible solution is to initialize these weights in such a way that the activities of the hidden nodes are within the active region of the corresponding nonlinear transformation functions. For the case of a sigmoidal function, these active regions are defined by the so-called 'temperature' which is a parameter controlling the steepness of the sigmoid. According to Smits (1997), there is an empirical relationship between the number of inputs of the neural network n_i and the normalized temperature of the sigmoid function T_n, defined by:

$$T_n = \eta \frac{\log(2 + \sqrt{3})}{\sqrt{n_i} - 0.5} \tag{2}$$

The factor η can be used to control the effective nonlinearity and is usually set to one.

Because the weights from the input-to-hidden layer are sampled from a normal distribution, the analytical neural networks are different each time they are rebuilt. This is not a problem as long as one realizes that each of the interpolative predictors is just one of the numerous realizations from an infinite number of possible options. The fact that we can generate many different interpolative models with a similar (good) performance actually is a feature that can be exploited in the context of mixtures of models.

4.1.2. Key features of stacked analytical neural networks related to soft sensors

Stacked analytical neural networks contribute to the soft sensor development process by allowing a fast feasibility test of the hypothesis that a nonlinear empirical model can be built, by developing stacked predictors with self-assessment capabilities, by performing an extensive nonlinear sensitivity analysis and input variable selection, and by dealing with time-delayed data.

One of the key advantages of analytical neural networks is that it is no longer possible to get stuck in local minima. The new algorithm is no longer iterative. Since the time limiting step is the solution of a linear system of algebraic equations, there is a clear bound on the required computing time. As a result, the development time is significantly faster than with back propagation-based neural networks. This gives the opportunity to very quickly test the hypothesis that it is possible to build a nonlinear model with the available data.

Another feature of SANN of benefit to inferential sensors is their ability to generate a confidence interval of the predicted values. At the basis of this property is that many different models can be generated and combined. Alternative methods to generate different models with similar performance can be obtained by varying the data set used to generate the model (e.g. by using cross-validation or bootstrap samples), by selecting different inputs to the model in the case of redundancy between the candidate inputs or by varying the model structure and/or complexity. Next to the possibility of generating a combination model with better performance compared to the individual models, this collection of models can also be used to estimate the standard error of a predicted value SE_y. This standard error can then be used to calculate a confidence interval for a predicted value $\hat{y}$ according to:

$$\hat{y} \pm t\left(n-1, 1-\frac{\alpha}{2}\right)\mathrm{SE}_y \tag{3}$$

where t is the value of the t-distribution corresponding to $n-1$ degrees of freedom and $\alpha\%$ confidence.

The advantages of this method to calculate confidence limits is that it is model-based, with no restrictions on the model type and it gives a local estimate of the confidence for a given prediction. The only possible disadvantage is that it is potentially computationally intensive because the method requires building the model N times. This is why having a fast type of neural network is critical.

Sensitivity analysis of process inputs relative to the output is the next key feature of analytical neural networks, which helps to reduce the dimensionality of soft sensors. In order to establish a procedure which can be used together with combination models using neural networks, let us first examine the statistic used to rank the significance of the candidate variables. The partial F-statistic with 1 and ν degrees of freedom, for testing the hypothesis that a given parameter $b_j = 0$ versus the hypothesis that $b_j \neq 0$ is equal to the t-statistic with ν degrees of freedom, is obtained via $t = b_j/\mathrm{SE}(b_j)$ where the standard error $\mathrm{SE}(b_j)$ is the square root of the appropriate diagonal term of $(\mathbf{X}'\mathbf{X})^{-1}s^2$ and s^2 is the mean squared error. In case of a linear equation the parameter value $b_j = \partial Y/\partial X_j$ is a direct measure of the 'influence'. It is also unique and global, i.e. it is valid within the entire input space. The corresponding derivative for a nonlinear model, like a neural network, is only a local measure of the 'influence' of that specific variable and depends on the specific values the other variables have. A good statistical measure of 'influence' of a specific variable in the nonlinear case could be the sum of the absolute values of the partial derivatives, over all patterns in the training set, weighted by the standard error:

$$\mathrm{SI}_j = \frac{\dfrac{1}{Np}\displaystyle\sum_{p=1}^{Np}\left|\left[\frac{\partial Y}{\partial X_j}\right]_p\right|}{\sqrt{(\mathbf{X}'\mathbf{X})_{jj}^{-1}}} \tag{4}$$

Another feature of analytical neural networks that contribute to dimensionality reduction is the effective handling of time delays. The classical approach to handle time series by neural nets is to add additional inputs for the previous time steps. Unfortunately, this technique increases the dimensionality of the neural net significantly. For example, if one has a problem with five inputs and one wants to use the current input plus the inputs from five previous time-steps as inputs to the network, then one needs a network with 30 inputs as opposed to the original five. This increase in the dimensionality of the input vectors has a big impact on the number of required data points for proper model identification. Therefore, it would be desirable to include information from previous time-steps without increasing the dimensionality of the input to the network. This can be achieved by performing a convolution on the input using an appropriately shaped function. This delay function is represented by Eq. (5).

$$f_k(t) = \beta\left(\frac{t-d}{k}\right)^n \exp\left[-n\left(\frac{t-d}{k}\right)\right] \tag{5}$$

with

β = normalization parameter
d = shift parameter
k = peak placement parameter
n = width parameter

The parameter k in Eq. (5) controls the placing of the peak of the delay curve; it also influences the width of the curve, because the starting point of the curve is always the same. Changing n can also influence the width of the curve. These two parameters that shape the convolution function (the peak k and the width n of the delay curve) are estimated through iterative runs of neural nets with inputs, convoluted by $f_k(t)$, for a range of expected time delays. In principle, using convolution functions to capture time delays are independent of the modeling approach. However, since this optimization algorithm requires building hundreds, even thousands of candidate models, the fast stacked analytic neural network is the proper implementation solution.

4.2. Support vector machines

4.2.1. Principles of SVM

The next approach related to inferential soft sensors is the SVM that has its foundation in Statistical Learning Theory and is particularly useful for learning with small sample sizes (Vapnik, 1998). Statistical Learning Theory formalizes the role of model complexity and gives statistical guarantees for the validity of the inferred model (Scholkopf and Smola, 2002). The theory provides a formal method to select the optimal model capacity from a set of nested subsets of functions $S_1 \subset S_2 \subset \cdots \subset S_m \subset \cdots$ where the elements of this structure are ordered according to their VC-dimension (which is a direct measure of their complexity) such that $C_1 \le C_2 \le \cdots \le C_m \le \cdots$. A very simple example is a set of polynomial functions with increasing power. Statistical Learning Theory provides an explicit procedure to estimate the difference between the unknown true risk (e.g. the prediction error) and the known empirical risk (e.g. the sum of squared errors) as a function of sample size n and the properties of the set of approximating functions. These bounds on the generalization properties of a specific set of models provide a quantitative characterization of the trade-off between the complexity of the approximating functions and the quality of fitting the training data.

At the basis of Statistical Learning Theory is the Structural Minimization Principle (SRM) that minimizes the empirical risk for the functions in the subset S_k for which the guaranteed risk (i.e. the combination of empirical risk and generalization ability), is minimal. The SRM defines a trade-off between the quality of the approximation of the given learning data and the generalization ability (defined in Vapnik (1998) as VC confidence interval) of the learning machine. If the learning machine uses a too high complexity, the learning ability may be good, but the generalization ability is not. The learning machine will overfit the data to the right of the optimal complexity with VC dimension of h^* in Fig. 5.

On the other hand, when the learning machine uses too little complexity, it may have good generalization ability, but not good learning ability. This underfitting of the learning machine corresponds with the region to the left of the optimal complexity. The optimal complexity of the learning machine is the set of approximating functions with lowest VC-dimension and lowest training error.

There are two approaches of implementing the SRM inductive principle in learning machines:

1. keep the VC confidence interval fixed and minimize the empirical risk;
2. keep the empirical risk fixed and minimize the VC confidence interval.

Neural networks algorithms implement the first approach, since the number of hidden nodes is defined a priori and therefore the complexity of the structure is kept fixed.

The second approach is implemented by the SVM method where the empirical risk is either chosen to be zero or set to an a priori level (the value of the ε-insensitive zone) and the complexity of the structure is optimized. In general, the SVM maps the input data into a higher dimensional feature space. The mapping can be done nonlinearly and the transformation function is chosen a priori. Usually it is a kernel that satisfies Mercer's conditions (Vapnik, 1998) and the selection is problem-specific. Typical cases for kernels are the Radial Basis Function (RBF) and the polynomial kernel. In the feature space the SVM finally constructs an optimal approximating function that is linear in its parameters. In the classification case, the function is called a decision function or hyperplane, and the optimal function is called an optimal separating hyperplane. In the regression case, it is called an approximating function or in statistical terms a hypothesis. The general scheme of the SVM is given in Fig. 6.

The key notion of SVM is the support vector. In classification, the support vectors are the vectors that lie on the margin (i.e. the largest distance between two closest vectors to either side of a hyperplane) and they represent the input data that are the most difficult to

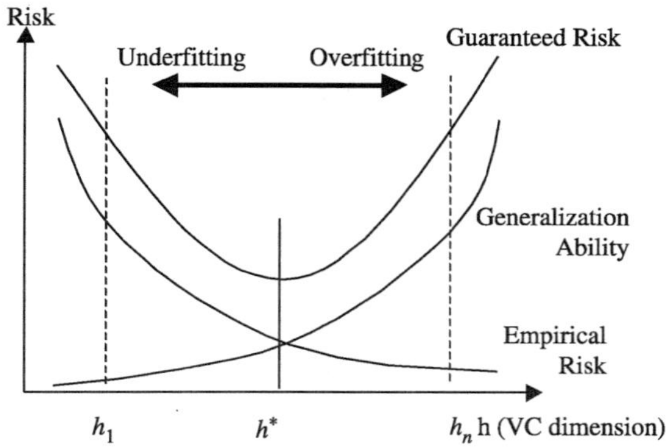

Fig. 5. The Structural Risk Minimization Principle.

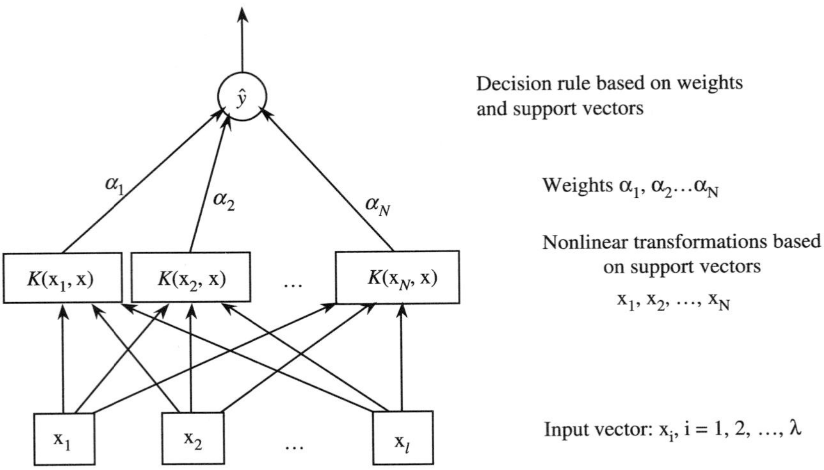

Fig. 6. The general scheme of SVM.

classify. From mathematical point of view, the support vectors correspond to the positive Lagrangian multipliers from the solution of a Quadratic Programming (QP) optimization problem. Only these vectors and their weights are then used to define the decision rule or model. Therefore this learning machine is called the SVM.

For the purpose of soft sensor development, however, we are using SVM for regression (Vapnik, 1998; Jordaan, 2002). In this case, a specific loss function with an insensitive zone (called ε-insensitive zone) is optimized. The objective is to fit a tube with radius ε to the data (see Fig. 7).

We assume that any approximation error smaller than ε is due to noise and accept it, i.e. the method is said to be insensitive to errors inside this zone (tube.) The data points

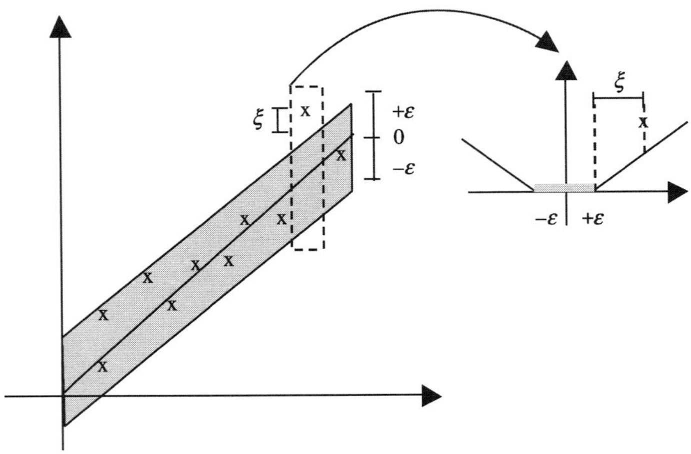

Fig. 7. The ε-insensitive zone for support vector regression.

(vectors) that have approximation errors outside, but are close to the insensitive zone, are the most difficult to predict. Usually the optimization algorithm picks them as support vectors. Therefore, the support vectors in the regression case are those vectors that lie immediately outside the insensitive zone and as such contain the most important information from the data.

4.2.2. Key features of SVM related to soft sensors

The SVM is a very robust method and has a unique contribution to soft sensor development providing automatic outlier and novelty detection. The fact that the SVM model is a sparse representation of the learning data allows the extraction of a condensed data set based on the support vectors. Finally, by using certain types of kernels, the extrapolation capabilities of the model can be increased dramatically, especially by incorporating prior information (Smits and Jordaan, 2002).

The fact that the outliers are part of the support vector set makes it potentially suitable to use as a model-based outlier detection method. At the basis of SVM outlier detection is the inspection of the values of the Lagrange multipliers and in particular those that hit the upper bound of the constraint, or with a large corresponding slack variable. In both cases this is an indication for unusual data points and can be considered as potential outliers. However, it is difficult to identify that a data point is a possible outlier based only on a single model. Several models of varying complexity should be constructed. For each model, a data point is identified as a suspected outlier, if it has a Lagrange multiplier value close to the upper boundary and its corresponding slack variable is large. Next, we determine the number of times that a data point is suspected to be an outlier and plot the frequency of suspicion at increasing rates. Usually, there is a sharp increase in the frequency at the tail end of the detection rate in the case of outlier presence (Jordaan, 2002). This effect is illustrated in Fig. 8 where in Fig. 8(a) are shown which data points are suspected to be outliers per iteration. The frequency rates for outlier detection are illustrated in Fig. 8(b) in increasing order and the predictions of the corresponding SVMs as well as the detected outlier is shown in Fig. 9.

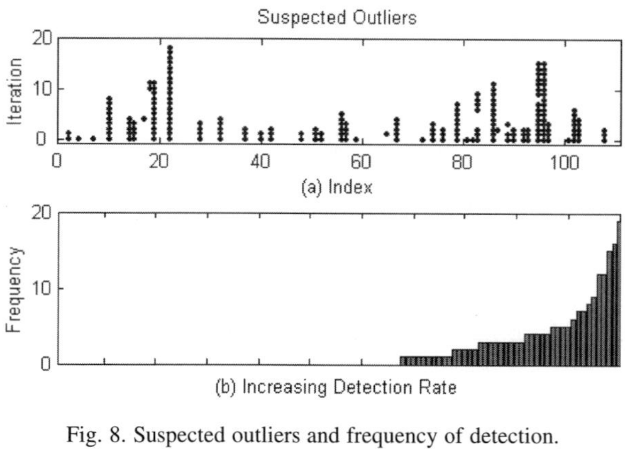

Fig. 8. Suspected outliers and frequency of detection.

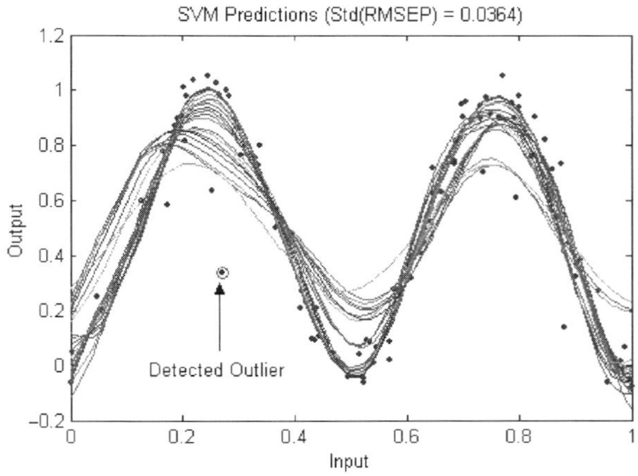

Fig. 9. Predictions of various SVM models and detected outlier.

Data condensation is an inverted data selection to that used for outlier detection. In this case, we make use of the fact that those vectors that contain information are identified as support vectors. Therefore, in order to detect data points that are redundant, one has to inspect the nonsupport vector data points, i.e. the data points with zero weights. Again, one should not conclude from a single model that a data point is in fact redundant. Therefore, several models with varying complexity need to be constructed to determine whether the data point is consistently not needed as a support vector.

The improved generalization capabilities of SVM models based on mixed kernels are of great importance to robust soft sensor performance (Smits and Jordaan, 2002). The best effect is achieved by combining the advantages of global and local kernels. It is observed that a global kernel (like a polynomial kernel) shows better extrapolation abilities at lower orders, but requires higher orders for good interpolation. On the other hand, a local kernel (like the RBF kernel) has good interpolation abilities, but fails to provide longer-range extrapolation.

There are several ways of mixing kernels. What is important though, is that the resulting kernel must be an admissible kernel (Scholkopf and Smola, 2002). One way to guarantee that the mixed kernel is admissible, is to use a convex combination of the two kernels K_{poly} and K_{rbf}, for example

$$K_{\mathrm{mix}} = \rho K_{\mathrm{poly}} + (1 - \rho)K_{\mathrm{rbf}} \tag{6}$$

where the optimal mixing coefficient ρ has to be determined. The results from an investigation of different data sets show that only a 'pinch' of a RBF kernel ($1 - \rho = 0.01 \div 0.05$) needs to be added to the polynomial kernel to obtain a combination of good interpolation and extrapolation (Smits and Jordaan, 2002). This effect is illustrated on an industrial data set, used for soft sensor development in the chemical industry. The input variable, a ratio of two temperatures, was selected by sensitivity analysis from stacked analytical neural networks. The relative input range for training (learning) data is between 0 and 1 and the range of the test data is between -1.2 and 1.5, i.e. the extrapolation ability

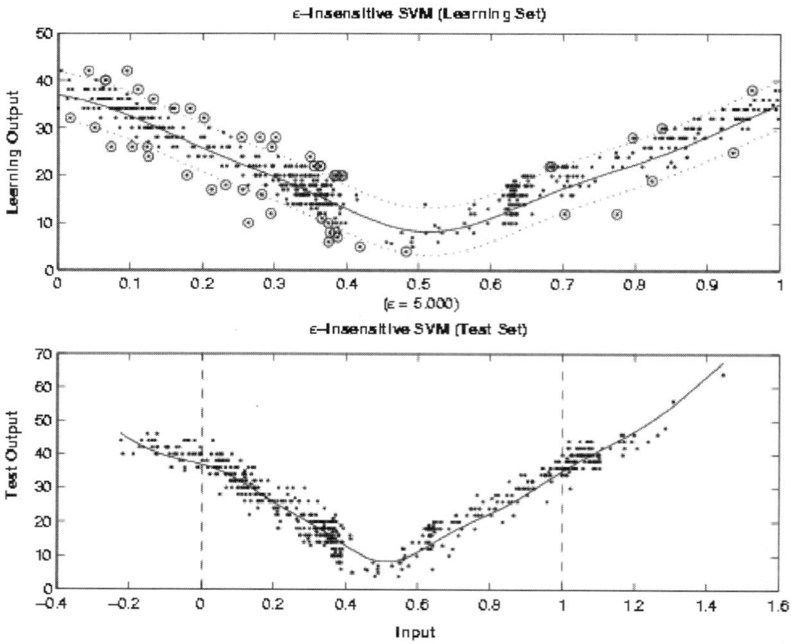

Fig. 10. SVM model based on a mixture of polynomial kernel of second degree, an RBF kernel of width 0.15, and a mixing coefficient of 0.98.

of 50% outside the training range is explored. While the polynomial kernel-based and the RBF kernel-based SVM models fail to extrapolate outside the training range, the mixed kernel, shown in Fig. 10 is able to both interpolate the sharp turning point of the training data as well as extrapolate outside the known input space. The top graph in Fig. 10 shows the prediction of the training set as well as the ε-insensitive tube and support vectors (encircled points).

The ability to both interpolate and extrapolate well, now opens the door for making use of prior information. If, for example, the asymptotic behavior of a process is known from fundamental models, this information can be incorporated in the empirical model.

5. Genetic programming in soft sensors development

5.1. The nature of genetic programming

GP is an evolutionary computing approach to generate soft structures, such as computer programs, algebraic equations, electronic circuit schemes, etc. GP and genetic algorithms have a similar nature. Since the reader has already been introduced to the nature of genetic algorithms, the emphasis in this chapter will be focused on the differences between both the approaches and the unique features of GP.

There are three key differences between GP and the classical genetic algorithm. The first difference is in the solution representation. While GA uses strings to represent solutions, the forms evolved by GP are tree-structures. The second difference relates to the length of the representation. While standard GA is based on a fixed-length representation, GP trees can vary in length and size. The third difference between standard GA and GP is based on the type of alphabets they use. While standard GA uses a binary alphabet to form a bit string, GP uses an alphabet of varying size and content depending on the solution domain.

With genetic algorithms, the emphasis is mostly on the optimization of some objective function in a fixed parameter space (Goldberg, 1989). The coding consists of binary strings of a fixed length. In the case of GP, the mechanism of evolving a population of potential solutions to a population of improved solutions is quite similar to what happens in the simple genetic algorithm. The major distinction is that the objective of GP is to evolve computer programs (i.e. perform automatic programming) rather than evolving chromosomes of a fixed length. This puts GP at a much higher conceptual level, because suddenly we are no longer working with a fixed structure, but with a dynamic structure equivalent to a computer program. The whole concept of a building block, which has a precise definition with a fixed length binary chromosome, is still present intuitively, but becomes a lot less precise in its definition. GP essentially follows a similar procedure as for genetic algorithms: there is a population of individual programs that is initialized, evaluated, and allowed to reproduce as a function of fitness. Again, crossover and mutation operators are used to generate offspring that replaces some or all of the parents.

Because of the need to apply crossover to dynamic, varying length structures, very often the representation is in the form of a tree-structure with terminals (input variables) and functions as potential nodes. The original GP implementation of Koza was in LISP, a computer language where tree-structures are a natural form of representation of code. Let us illustrate this with an example (Koza, 1992) of a small computer program represented in C.

```
int foo (int time)
{
 int temp1, temp2;
 if(time > 10)
   temp1 = 3;
 else
   temp1 = 4;
 temp2 = temp1 + 1 + 2;
 return(temp2);
}
```

In LISP the same program would look like:

```
(+1 2(IF(>TIME 10)3 4))
```

which can also be represented as a tree-structure with functions (+, if, >) and terminals (1, 2, 3, 4, time, 10) (Fig. 11):

There is no difference in the function of these three programs and they produce exactly the same result. The tree representation, however, has the right balance between a structure that has enough rigor to allow reproduction operations and one that is flexible enough to allow arbitrary programs to evolve.

One of the key differences between GA and GP is in the mechanism of the genetic operation of crossover. When we select a hierarchical tree structure as a representation, crossover can be used to interchange randomly chosen branches of the parent trees. This operation can occur without disrupting the syntax for the child trees. We will illustrate the crossover between two algebraic expressions in Fig. 12 (Negnevitsky, 2002):

The first expression (or parent 1) has the following LISP representation, shown as a tree structure in Fig. 12:

$$(/(-\mathrm{sqr}(+(^*aa)(-ab))a)(^*ab)),$$

which is equivalent to

$$\frac{\sqrt{a^2+(a-b)}-a}{ab}$$

The LISP expression for parent 2 is:

$$(+(-(\mathrm{sqr}(-(^*bb)a))b)(\mathrm{sqr}(/ab))),$$

which is equivalent to:

$$(\sqrt{b^2-a}-b)+\sqrt{\frac{a}{b}}$$

In GP, any point in the tree structure can be chosen as a crossover point. In this particular case, the function (∗) is selected to be the crossover point for parent 1. The selection for the crossover point for parent 2 is the function sqr. The crossover sections in both parents are shown in grey color in Fig. 12. On the bottom part of Fig. 12 are shown the two offspring, generated by the crossover operation. Replacing the crossover section of the first parent with the crossover material from the second parent creates offspring 1. In the same way, the second offspring is created by inserting the crossover section from the first parent into the place of the crossover section of the second parent.

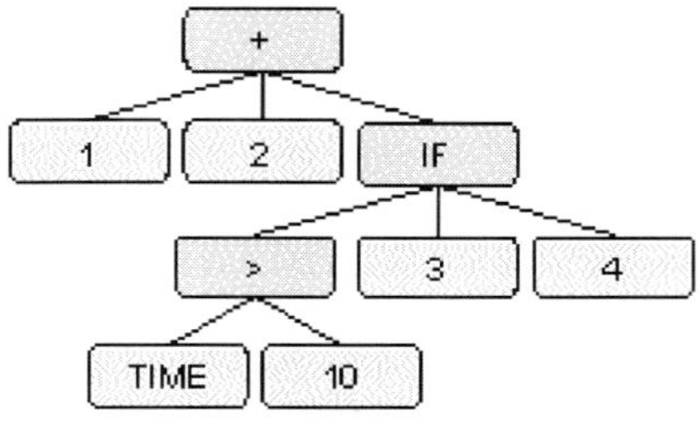

Fig. 11. Tree structure representation of programs and functions.

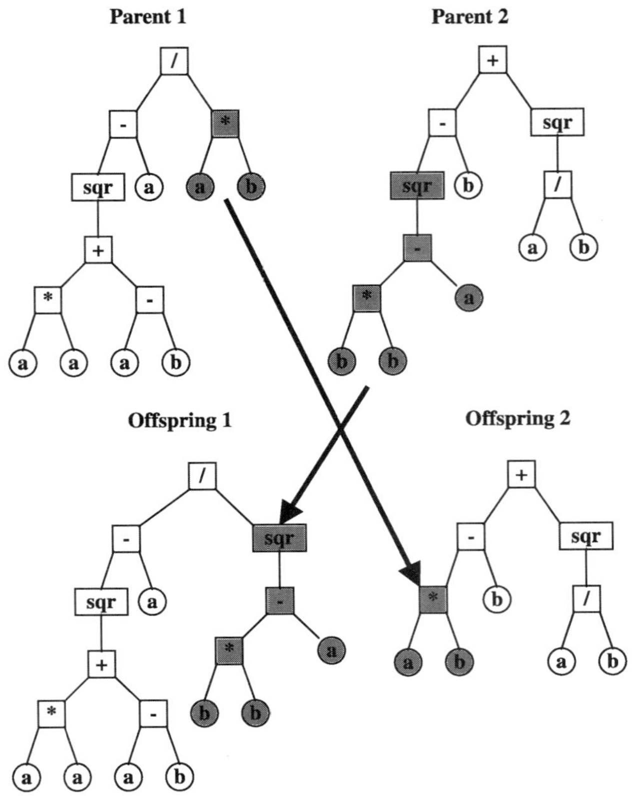

Fig. 12. Crossover operation between tree-structures with different parents.

The mutation operator can randomly change any function, terminal, or sub-tree by a new one. An example of a mutation operator on the two parents expressions is shown in Fig. 13. The randomly changed function (for parent 1) and terminal (parent 2) are shown with arrows.

Because the structures are dynamic, one of the problems that can occur is an excessive growth of the size of the trees over a number of generations. Very often, these structures possess large portions of redundant or junk code (also called introns). Quite a bit of research is focused on gaining an understanding of the function of these introns, since there is some evidence that the presence of introns is beneficial for the development of high-quality solutions (Banzhaf et al., 1998).

The primitives of GP are functions and terminals. Terminals provide a value to the structure while functions process a value already in the structure. Together, functions and terminals are referred to as nodes. The terminal set includes all inputs and constants used by the GP algorithm and a terminal is at the end of every branch in a tree structure. The function set includes the available functions, operators, and statements that are available as building blocks for the GP algorithm. The function set is problem-specific and can include problem domain-specific relationships like Fourier transforms (for time series problems,

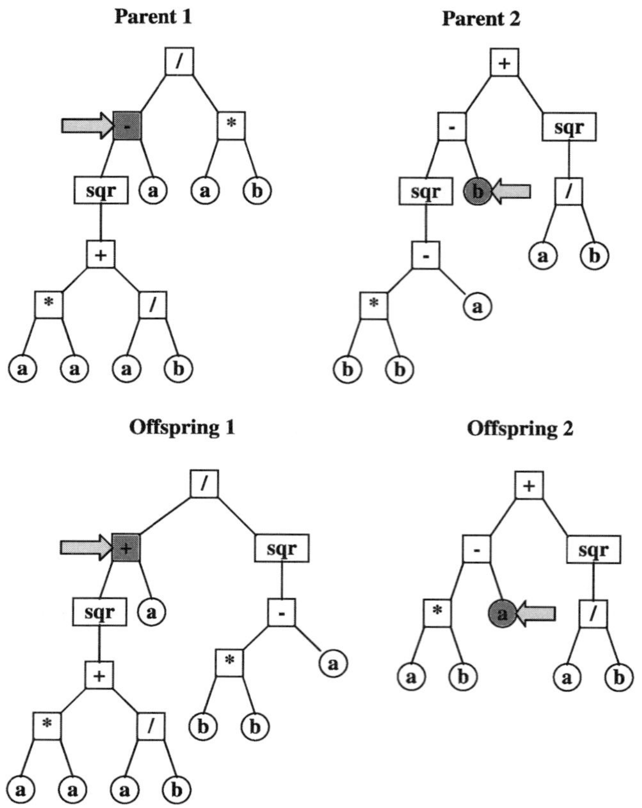

Fig. 13. Examples of mutation operation on tree-structures.

for example), or the Arrhenius law (for chemical kinetics problems). An example of a generic functional set, including the basic algebraic and transcendental functions, is given in Eq. (7):

$$F = \{+, -, *, \div, \ln, \exp, \text{sqrt}, \text{power}, \cos, \sin\} \tag{7}$$

The basic GP evolution algorithm is very similar to a GA and has the following key steps (Banzhaf et al., 1998):

Step 1: GP parameters selection

Includes definitions of the terminal and functional sets, the fitness function characterization, and selection of population size, maximum individual size, crossover probability, selection method, and maximum number of generations.

Step 2: Initialization with random population of structures (functions)
Step 3: Simulated evolution of the selected population

This includes fitness evaluation of the individual structures, selection of the winners from the competition and performing genetic operations (copying, crossover, mutation, etc.) on them, replacing the losers with the winners, and so on, until some termination criterion is fulfilled.

Step 4: Selection of the 'best and the brightest' structures (functions) as the output from the algorithm.

Usually the GP algorithm is very computationally intensive, especially in cases with large search spaces and without complexity penalties for the generated structures. The lion share of the computational burden is taken by the fitness evaluation of each individual member of the population.

In a standard GP algorithm, the average fitness of the population increases from generation to generation. In modeling, we are frequently looking not only for the best performing equation in terms how well the measured output data is reproduced, but we also require this solution to be as parsimonious as possible. Since evolution is driven by survival of the fittest, the easiest way to achieve parsimony as well as small reproduction error is to include a penalty for the complexity of the equation into the fitness measure. Although the true complexity of the function is difficult to assess e.g. in terms of a VC-dimension (Vapnik, 1998), simple measures that can be used are either the total number of nodes or the number of levels in a function. The effect of a parsimony pressure penalty on the effective fitness is illustrated in Fig. 14, represented as number pp in the range [0…1], with zero being equivalent to no parsimony pressure applied. Increasing equation size decreases the effective fitness of the equation.

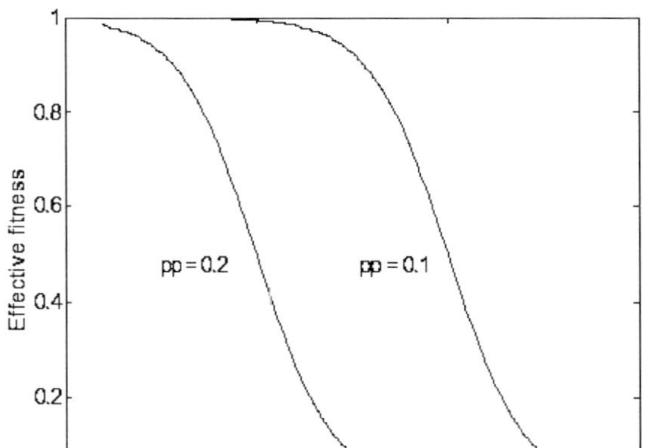

Fig. 14. Effect of parsimony pressure on the effective fitness of an equation.

5.2. Solving problems with genetic programming

As any data-driven approach, GP depends very strongly on the quality and consistency of the data. A necessary pre-condition for applying GP to real problems is that the data has been successfully pre-processed in advance. Of special importance is removing the outliers and increasing the information contents of the data by removing insignificant variables and duplicate records. That is why it is necessary to use GP in collaboration with other methods, such as neural networks, SVM, and principal component analysis.

The specific steps for applying GP will be illustrated with a simple problem, deriving Kepler's third law. The starting point is some data about the nine planets in the Solar system, as shown in Table 1.

All variables are relative to Earth's measures. We are looking for a potential relationship between two inputs × 1 (planet's diameter) and × 2 (the length of planet's semi-major axis of the ellipse, denoted as radius) and one output y (planet's period of revolution). In this particular case each data point includes unique information and there are no obvious outliers. We know in advance that there is no relationship between the planet's diameter and the period of revolution. The purpose of including this parameter is to test if GP will automatically ignore this input.

The preparatory step before running the GP algorithm includes determination of the following parameters.

- Terminal set (inputs used for model generation): × 1 (diameter) and × 2 (radius)
- Functional set (selected functions used for genetic operations): standard arithmetic operations +, −, *, and /, and the following mathematical functions: sqr, ln, exp, and power
- Fitness function: the absolute value of the correlation coefficient between actual and calculated output (the period, in this case)
- Genetic operators parameters:
  - Probability for random vs. Guided crossover: 0.5
  - Probability for mutation of terminals: 0.3
  - Probability for mutation of functions: 0.3

Table 1
Data set for deriving Kepler's third law

Planet	Diameter	Radius	Period
Mercury	0.384	0.387	0.24
Venus	0.972	0.723	0.62
Earth	1.000	1.000	1.00
Mars	0.533	1.524	1.88
Jupiter	9.775	5.203	11.86
Saturn	9.469	9.569	29.46
Uranus	3.692	19.309	84.01
Neptune	3.492	30.284	164.79
Pluto	0.447	39.781	247.69

– Simulated evolution control parameters:
  ○ Number of generations: 30
  ○ Number of simulated evolutions: 20
  ○ Population size: 100

Usually the genetic operators parameters are fixed for all practical applications. They are derived after many simulations and represent a good balance between the two key genetic operators crossover and mutation. In order to address the stochastic nature of GP, it is necessary to repeat the simulated evolution several times (usually 20 runs are recommended). The other two parameters that control the simulated evolution—population size and the number of generations—are problem-size dependent. The advantages of using bigger populations are that they increase genetic diversity, explore more areas of the search space, and improve convergence. However, they increase significantly the calculation time. According to Koza (1992), a population size between 50 and 10,000 can be used for model development of almost any complexity. The number of generations depends on the convergence of the simulated evolution. Usually it is obtained experimentally with a starting range of 30–100 generations and is gradually increased until a consistent convergence is achieved. Deriving the third Kepler's law based on two inputs and nine data samples is a relatively simple problem that could be solved within 30 generations and a population size of 100 functions.

The next step for solving the problem is implementing the GP algorithm. There are many software products on different platforms that can be used (see Banzhaf et al. (1998) for survey and list of web sites). In this particular case an implementation in MATLAB is used. During the different runs it is possible to keep track on the quality of simulated evolution by looking at the average fitness of the entire population and the fitness of the best equation, population fitness distribution, error distribution, etc. Of special importance is to have information about the overall complexity of the equations in the population. An example of the changing complexity during one of the GP runs for Kepler's third law derivation is shown in Fig. 15.

The plot illustrates how one of the measures of complexity—the total number of nodes in all equations—changes during the 30 generations of one simulated evolution. It is

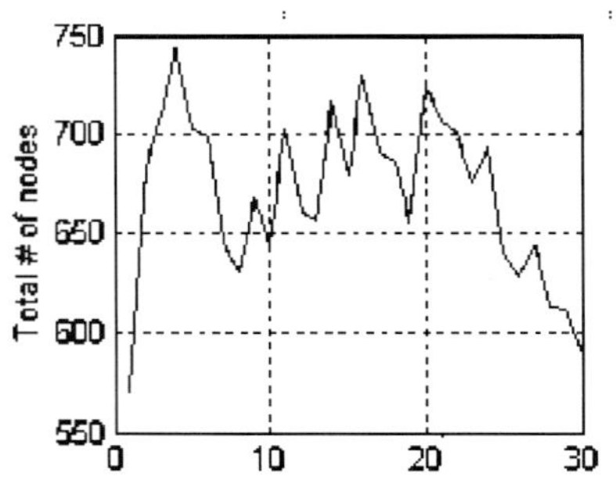

Fig. 15. Average complexity of the population of equations (total number of nodes on the Y-axis vs. number of generations on the X-axis) during Kepler's third law GP derivation.

obvious that in this particular case, the average complexity of seven nodes (a population of 100 with $\sim$700 total nodes) has been reduced to six nodes per equation at the end of the run.

The most nontrivial step in applying GP for solving problems is the model selection. As a result of different GP runs we have several potential solutions that have approximately equivalent fitness. Very often there is a whole series of variations of the same equation. Here are some of the equations for the Kepler's third law at the end of the simulated evolution lasted 30 generations:

$x2$ *sqr($x2$)
sqr($x2$^$^{2.98}$)
$x2$^$^{1.509}$
etc.

All of these equations include only the planet's radius and turn out to be equivalent to Kepler's third law after simplification. It is noted that the law defines the following relationship between the planet's radius and period of revolution:

$$\text{Period} = c\ \text{Radius}^{\wedge 3/2},$$

where c is a constant that depends on the particular unit system.

One additional result from the simulated evolution is that the other parameter—input × 1, planet's diameter has been automatically rejected during the evolutionary process. This demonstrates the unique capability of GP to perform a nonlinear variable selection based on the probability of choosing different variables and their nonlinear transforms during the simulated evolution.

On the negative side, GP also generates ineffective solutions with high complexity and inefficient blocks, called introns (Banzhaf et al., 1998). An example of such type of solution is shown in Fig. 16, where the intron is captured in a rectangle.

5.3. Advantages of genetic programming in soft sensors development and implementation

GP is the breakthrough technology from the point of view of implementation of soft sensors. Of special importance to soft sensor development are the following unique features of GP:

- no a priori modeling assumptions are required
- derivative-free optimization is possible
- few design parameters are needed
- natural selection of the most important process inputs
- parsimonious analytical functions as a final result.

The last feature has double benefit. On one hand, a simple inferential sensor often has better generalization capability, increased robustness, and needs less frequent re-training. On the other hand, process engineers and developers prefer to use nonblack box empirical models and are thus much more open to implement inferential sensors based on functional

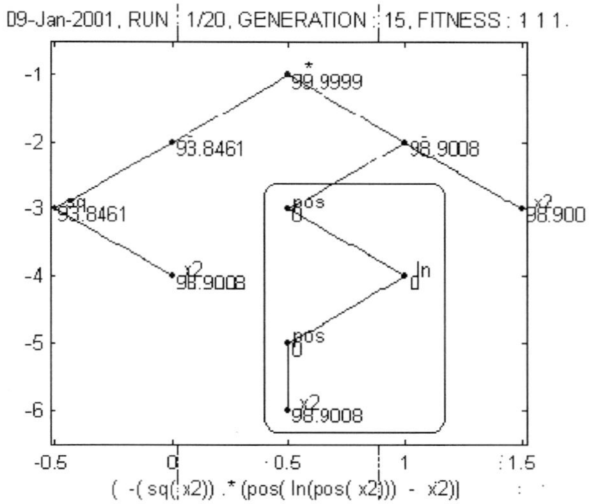

Fig. 16. Example of a solution containing a so-called intron, a part of the equation that does not contribute to its overall fitness.

relationships. An additional advantage is the low implementation cost of such type of soft sensors: they can be imbedded directly into the existing Distributed Control Systems (DCS) avoiding the need of additional specialized software packages, typical for most commercially available neural net-based inferential sensors. The robustness of GP-generated empirical models is a key factor for selecting them over classical neural nets for industrial applications. One of the significant factors is the ability to examine the behavior of the model outside the training range. With a functional solution, examination outside the training range can be done in an easy and direct way, whereas this is more difficult for the case of a black-box model. Another factor in favor of GP-generated empirical models, for example, is the ability to impose external constraints in the modeling process and to limit the extrapolation level of the final model.

At the same time, there are still significant challenges in implementing industrial-scale soft sensors generated by GP alone: function generation with noisy industrial data, dealing with time delays, and sensitivity analysis of large data sets, to name a few. Of special importance is the main drawback of GP—the slow speed of model generation due to the inherent high computational requirements of this method. For industrial-scale applications the calculation time is on the order of hours and days, even with the current high-end workstations.

6. Integrated methodology

The objective of the proposed integrated methodology is to deliver successful industrial soft sensors with reduced development time, better generalization capability, and minimal implementation and maintenance cost. The main blocks of the methodology are shown in Fig. 17.

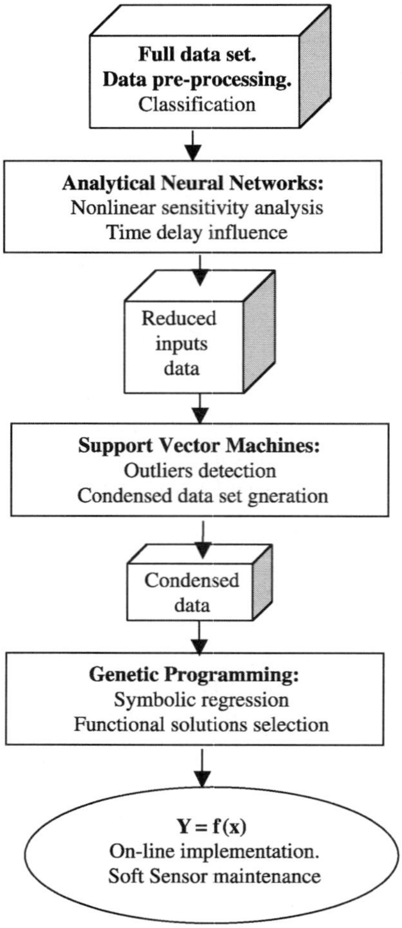

Fig. 17. Main blocks of an integrated methodology for inferential sensor development.

The purpose of the Data pre-processing and classification block, that starts the development process with the full data set, is to supply the inferential model development process with clean data that cover the broadest possible operating range. The key modeling blocks for inferential sensor development (variable selection by analytical neural networks, data condensation by SVM, and the model generation by GP) as well as the issue of on-line performance self-assessment (of significant importance for on-line implementation and maintenance) will be discussed in the following sections.

6.1. Variable selection by analytical neural networks

The main purpose of this block is to reduce the number of inputs to those with the highest sensitivity toward the output. Another objective is to test, via simulation, the hypothesis whether some form of nonlinear relationship between the selected inputs and outputs exists. This is a critical point in the entire methodology, because if a neural net model cannot be built, the inferential sensor development process stops here. The conclusion in this case would be that if a universal approximator, like a neural net, cannot capture a nonlinear relationship, there would be no basis for variable dependence and no need to look for other methods. The sensitivity analysis is based on stacked analytical neural nets. A big advantage of this type of neural nets is the reduced development time. Within a few hours, the most sensitive inputs are selected, the performance of the best neural net models is explored, and the data for the computationally intensive symbolic regression step (GP-function generation) is prepared. Typically, 30 stacked neural nets are used to improve generalization and to determine the neural net model agreement error. This step begins with the most complex structure of all possible inputs. During the sensitivity analysis, decreasing the number of inputs gradually reduces the initial complex structure. The sensitivity of each structure is the average of the calculated derivatives on every one of the stacked neural nets. The procedure performs automatic elimination of the least significant inputs and generates a matrix of input sensitivity vs. input elimination.

Another important task performed by analytical neural networks is to deal with time delays by optimizing the parameters of the convolution function (5). As a result of this block of the integrated methodology, the size of the full data set is reduced to the number of the most sensitive inputs.

6.2. Data condensation by support vector machines

The purpose of the next block, based on SVM, is to further reduce the size of the data set to only those data points that represent the substantial information about the nonlinear model. Outlier detection is the first task in this process. For outlier detection, we make use of the fact that the data points containing important information are identified by the SVM method as support vectors. When the weight of a data point is nonzero, it is a support vector. The value of a support vector's weight factor indicates to what extent the corresponding constraint is violated. Nonzero weight factors hitting the upper and lower boundaries indicate that their constraints are very difficult to satisfy at the optimal solution. Such data points are often so unusual with respect to the rest of the samples, that they might be considered as outliers. An outlier detection tool, using the SVM method, typically constructs several models of varying complexity. Data points with a high frequency of weight values on the boundaries are assumed to be outliers.

One of the main advantages of using SVM as a modeling method is that the user has direct control over the complexity of the model (i.e. the number of support vectors). The complexity can be controlled either implicitly or explicitly. The implicit method controls the number of support vectors by controlling the acceptable noise level. To explicitly control the number of support vectors, one can either control the ratio of support vectors or the percentage of nonsupport vectors. In both cases, a condensed data set that reflects the appropriate level of complexity is extracted for effective symbolic regression.

An additional option in this main block is to deliver an inferential sensor based solely on SVM, given the good extrapolation features of mixed global and local kernels. If a soft sensor generated by GP does not have acceptable performance outside the range of training data, the SVM-based inferential model is a viable on-line solution.

6.3. *Inferential model generation by genetic programming*

The next block of the integrated methodology for inferential sensor development uses the GP approach to search for potential analytical relationships in a condensed data set of the most sensitive inputs. The search space is significantly reduced by the previous steps and the effectiveness of GP is considerably improved. The set of possible functions that can be generated in GP is the set of all possible functions that can be composed from the list of available terminals $T = \{X_1, X_2, ..., X_n\}$ and the set of available functions $F = \{F_1, F_2, ..., F_m\}$.

Various parameter settings control the type and complexity of equations that are generated. The most important parameters are the list of available functions as well as the list of available inputs. Another parameter that is quite important in controlling the average complexity of the equations being generated is the probability for function selection (default value equals 0.6). This parameter controls whether to grow a specific branch of a tree by selecting a function, or terminating the branch by selecting a terminal (a number or a variable) as the next node. The larger this probability value is, the higher the complexity of the functions being generated.

The final result of a GP-based simulated evolution is a list of several analytical functions and sub-equations that satisfy the best solution according to a defined objective function. The analytical function selection for the final inferential sensor on-line model is still more of an art than a well-defined procedure. Very often the most parsimonious solution is not acceptable due to specific manufacturing requirements. It is preferable to deliver several potential functions with different levels of complexity and let the final user make the decision. The generalization capabilities of each soft sensor are verified for all possible data sets. Of special importance is the performance outside the training range. It is also possible to design a model agreement-type confidence indicator based on stacked symbolic predictors.

6.4. *On-line implementation and model self-assessment*

The final block of the integrated methodology for robust soft sensors development relates to their on-line implementation and maintenance. Each one of the discussed approaches generates models that can be implemented on-line. The preferred type of model, however, is symbolic regression, generated by GP. As was discussed earlier, in addition to their nonblack-box nature, analytical functions can be directly implemented on any software platform without specialized packages. Their maintenance is also much easier and usually is limited to periodic re-adjustment of function parameters.

One key advantage of the proposed integrated methodology is the model self-assessment capability by a confidence indicator of on-line soft sensors. The purpose of a confidence indicator is to estimate the validity of the model in the absence of physical measurements. Depending on the implementation, two different types of performance indicators are used: a 'model agreement indicator' and a 'parameters-within-range indicator'. If a single estimation is based on several 'parallel' or 'stacked' models, then a 'model agreement indicator' can provide a quantitative assessment of the reliability of that model's output. The model agreement indicator assumes that if the standard deviation between the models is above a threshold value (usually 3σ of the standard deviation), the average model prediction is unreliable.

The within-the-range indicator is more complicated. The purpose of this indicator is to evaluate if the inputs are within the training range of the model. The necessity of the within-the-range indicator is due to the limited number of product types and operating conditions usually available for training. Since it is generally not economically feasible to conduct full-blown DOE with a running industrial process, it is unrealistic to expect to collect a data set that covers all possible operating conditions and to design the 'perfect' soft sensor. With a within-the-range indicator, however, the estimates based on known operating conditions are assumed as reliable. The out-of-range data are explicitly identified and can be used for the next re-design. In this way, only the trustworthy predictions are presented to the operator. Another advantage of the within-the-range indicator is that it can serve to prompt for re-training and re-design. If the out-of-range data are more than some acceptable threshold, a new re-design procedure is recommended. For the inputs within-the-range indicator it is assumed that the maximal confidence of 1 is in the middle of the range, and that the confidence decreases linearly to 0 outside the range. The overall within-the-range indicator of the soft sensor is a product of the within-the-range indicators of all inputs, i.e. if any of the inputs is outside the training range, the predictions are deemed to be unreliable. For actual on-line implementation, the confidence indicators are assigned to tags in the processes information system, and are displayed on-line in real time along with the model outputs. This way, the operators can quickly assess the performance of the model, rather than blindly relying on the model output.

7. Soft sensor for emission estimation: a case study

Soft sensors for emission estimation are one of the most popular application areas and a viable alternative to hardware analyzers. Usually an intensive data collection campaign is required for empirical model development. However, during on-line operation the output measurement is not available and some form of soft sensor performance self-assessment is highly desirable. Since it is unrealistic to expect that all possible process variations will be captured during the data collection campaign, a soft sensor with increased robustness is required. Such type of soft sensors, based on the proposed integrated methodology, was developed and implemented in one of The Dow Chemical Company plants in Freeport, TX. The key results from implementation of the main blocks are as follows:

A representative data set from eight potential process input variables and the measured emission as output included 251 data points for training and 115 data points for testing. The test data is 140% outside the range of the training data, which by itself is a severe challenge for the extrapolation capability of the model. As a result of the nonlinear sensitivity analysis based on the SANN, the data set was reduced to five relevant inputs. The performance of such type of potential model with five inputs, 10 neurons in the hidden layer, and a model disagreement indicator based on the standard deviation of 30 stacked predictors is shown in Fig. 18. The possibility for nonlinear model building and the potential of the model agreement indicator for performance self-assessment are clearly demonstrated.

The extraordinary extrapolation capability of a potential empirical model based on SVMs is shown in Fig. 19. The model is based on a mixture of a second order polynomial global kernel and an RBF local kernel with width of 0.5 in a ratio of 0.95 between kernels. An additional benefit from this phase of the integrated methodology is that the model is based on 34 support vectors only.

As a result, the representative data set for deriving the final symbolic regression model is drastically reduced to only 8.44% of the original training data set. As shown in Fig. 20, the performance of the GP-generated model, based on the condensed data set, is comparable with the other two approaches.

The initial functional set for the GP includes: {addition, subtraction, multiplication, division, square, sign change, square root, natural logarithm, exponential, and power}. Function generation takes 20 runs with population size of 500, number of generations of 100, number of reproductions per generation of 4, probability for function as next node of 0.6, parsimony pressure of 0.05, and correlation coefficient as optimization criterion. Eight symbolic predictors with different number of inputs and nonlinear functions were selected in eight stacked models. The average value of the stacked models is used as the soft sensor prediction and the standard deviation is used as a model disagreement indicator. At the time of this writing, the soft sensor for emission estimation has been in operation for 20 months.

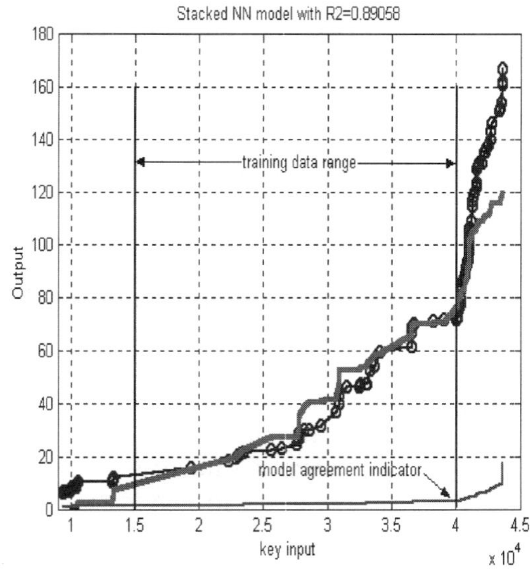

Fig. 18. Performance of a Stacked Analytical Neural Net model with model agreement indicator.

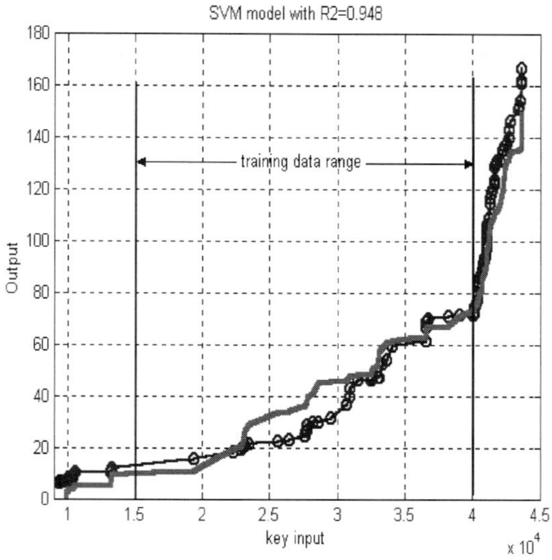

Fig. 19. Performance of an SVM model using a mixture of polynomial and RBF kernels.

8. Conclusions

Inferential or soft sensors are one of the success stories of implementing nature-inspired methods in chemometrics. However, the first wave of classical neural net-based soft sensors has exhausted its potential to satisfy the growing requirements for robust performance and reduced maintenance cost. A new integrated methodology based on

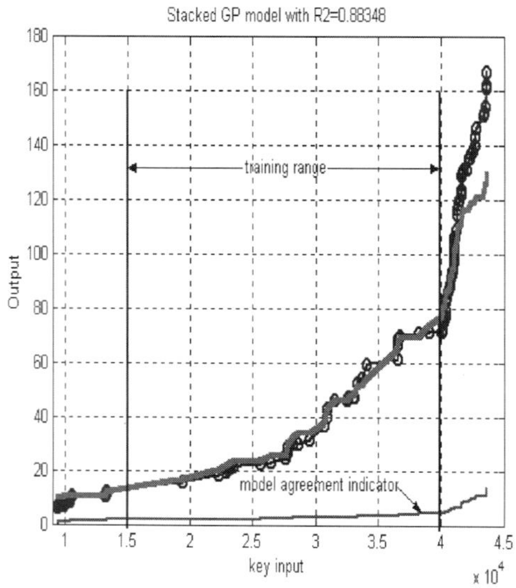

Fig. 20. Performance of a Stacked Symbolic Regression model with model agreement indicator.

different computational intelligence approaches (stacked analytical neural nets, GP, and SVM) can resolve many of the problems and satisfy the requirements of industry for soft sensors with increased robustness. A significant part of the methodology is based on the unique features of GP to deliver the final solution as a very simple analytical function. In this way, the methodology significantly increases the robustness of the models and generates simple models that can be implemented directly in the plant process information system. The models deployed on-line need less maintenance than classical neural nets.

References

Abdelrahman, M., Subramanian, S., 1998. Inferential sensors for copola iron-melting furnaces. Proceedings of American Control Conference 1998, Philadelphia, pp. 461–464.

Aggarwal, C., Yu, P., 2001. Outlier detection for high dimensional data. Proceedings of the SIGMOD Conference.

Al-Duwaish, H., Ghouti, I., Halawani, T., Mohandes, M., 2002. Use of artificial networks process analyzers: a case study, Proceedings of ESANN'2002, Bruges, Belgium, pp. 465–470.

Aspen Technology web site www.aspentech.com, 2003.

Banzhaf, W., Nordin, P., Keller, R., Francone, F., 1998. Genetic programming: an introduction, Morgan Kaufmann, San Francisco.

Bhartia, S., Whitely, J., 2001. Development of inferential measurements using neural networks. ISA Trans. 40, 307–323.

Brambilla, A., Trivella, F., 1996. Estimate product quality with ANNs. Hydrocarbon Process. 7 (9), 61–66.

Cherkassky, V., Mullier, F., 1998. Learning from data, concepts, theory, and methods, Wiley, New York.

Del Villar, R., Thibault, J., Del Villar, R., 1996. Development of a soft sensor for particle size monitoring. Miner Engng. 9, 55–72.

DeltaV web site www.easydelta.com, 2003.

Di Massimo, C., Willis, M., Montague, G., Tham, M., Morris, A., 1991. Bioprocess model building using artificial neural networks. Bioprocess Engng. 7, 77–82.

Dong, D., McAvoy, T., 1995. Emission monitoring using multivariate soft sensors. Proceedings of the American Control Conference 1995, Seattle, pp. 761–764.

Eghneim, G., 1996. On predictive emissions monitoring from a regulatory perspective. J. Air Waste Manag. Assoc. 46, 1086–1092.

Gensym Corporation web site www.gensym.com, 2003.

Goldberg, D., 1989. Genetic Algorithms, Addison-Wesley, Reading, MA.

Gribock, A., Hines, J., Uhrig, R., 2000. Use of kernel based techniques for sensor validation in nuclear power plants. Proceeding of the International Topical Meeting on Nuclear Plant Instrumentation, Controls, and Human–Machine Interface Technologies, Washington, DC, November.

Hadjiski, M., Elenkov, G., Hadjiski, L., Mohadjer, A., 1992. Neural network-based steady-state observation and reference control of distillation columns. Proceedings of the Third IFAC DYCORD'92 Symposium, College Park, MD, pp. 393–398.

Hastie, T., Tibishirani, R., Friedman, R., 2001. The Elements of Statistical Learning, Springer, New York.

Haykin, S., 1998. Neural Networks: A Comprehensive Foundation, Prentice-Hall, New York.

Honeywell web site www.iac.honeywell.com, 2003.

Jansson, A., Rottorp, J., Rahmberg, M., 2002. Development of a soft sensor for phosphorus in municipal wastewater. J. Chemom. 16, 542–547.

Jordaan, E., 2002. Development of Robust Inferential Sensors: Industrial Application of Support Vector Machines for Regression. PhD Thesis. Technical University Eindhoven.

Joseph, B., Brosilow, B., 1978. Inferential control of processes. I: steady state analysis and design. AIChE J. 24, 485–496.

Kalos, A., Kordon, A., Smits, G., Werkmeister, S., 2003. Hybrid model development methodology for industrial soft sensors. Proceedings of the American Control Conference 2003, Denver, CO, pp. 5417–5422.

Kecman, V., 2001. Learning and Soft Computing, Support Vector Machines, Neural Networks, and Fuzzy Logic Models, MIT Press, London.

Keeler, J., Ferguson, R., 1996. Commercial Applications of Soft Sensors®: TheVirtual On-line Analyzer® and the Software CEM®. Proceedings of the International Forum for Process Analytical Chemistry, Orlando, FL.

Kordon, A., Smits, G., 2001. Soft sensor development using genetic programming, Proceedings of GECCO'2001, San Francisco, pp. 1346–1351.

Kordon, A., Smits, G., Jordaan, E., Rightor, E., 2002. Robust soft sensors based on integration of genetic programming, analytical neural networks, and support vector machines, Proceedings of WCCI 2002, IEEE Press, Honolulu, HW, pp. 896–901.

Koza, J., 1992. Genetic Programming: On the Programming of Computers by Means of Natural Selection, MIT Press, Cambridge, MA.

Lennox, B., Montague, G., Frith, A., Gent, C., Bevan, V., 2001. Industrial applications of neural networks—an investigation. J. Process Control 11, 497–507.

Martin, G., 1997. Consider soft sensors. Chem. Engng. Prog. 47 (7), 66–70.

Martin, E., Morris, A., 1999. Artificial Neural Networks and Multivariate Statistics, Statistics and Neural Networks, Oxford University Press, NY.

Massop, B., Hommersom, G., 1998. Personal communication.

Mitchel, T., 1997. Machine Learning, McGraw-Hill, Boston, MA.

Morrison, S., 1996. The importance of automatic design techniques in implementing neural net intelligent sensors. Adv. Instrum. Control 51, 293–302.

Neelakantan, R., Guiver, J., 1998. Applying neural networks. Hydrocarbon Process 9.

Negnevitsky, M., 2002. Artificial Intelligence: A Guide to Intelligent Systems, Addison-Wesley, Harlow, UK.

Parmee, I., 2001. Evolutionary and Adaptive Computing in Engineering, Springer, London.

Pavilion Technology web site (http://www.pavtech.com), 2003.

Pell, R., 2000. Multiple outlier detection for multivariate calibration using robust statistical techniques. Chemometr. Intell. Lab. Sys. 52, 87–104.

Piovoso, M., Owens, A., 1991. Sensor data analysis using artificial neural networks. Proceedings of Chemical Process Control (CPC-IV), Padre Island, TX, pp. 101–118.

Ponton, J., Klemes, J., 1993. Alternatives to neural networks for inferential measurement. Computers Chem. Engng. 17, 991–1000.

Qin, S., 1996. Neural networks for intelligent sensors and control—practical issues and some solutions, Neural Systems for Control, Academic Press, New York.

Qin, S., Yue, H., Dunia, R., 1997. Self-validating inferential sensors with application to air emission monitoring. Ind. Engng. Chem. Res. 36, 1675–1685.

Rallo, R., Ferre-Gine, J., Arenas, A., Giralt, F., 2002. Neural virtual sensor for the inferential prediction of product quality from process variables. Computers Chem. Engng 26, 1735–1754.

Rao, M., Corbin, J., Wang, Q., 1993. Soft sensors for quality prediction in batch chemical pulping process. IEEE Symposium on Intelligent Control, Chicago, pp. 150–155.

Rotatch, V., Hadjiski, M., 1966. Optimal Static Inferential Systems, Proceedings of the Institute of Technical Cybernetics, vol. 3. Academy of Sciences of the USSR, pp. 199–205 (in Russian).

Russel, E., Chiang, L., Braatz, R., 2000. Data-Driven Methods for Fault Detection and Diagnosis in Chemical Processes, Springer Verlag, London.

Scholkopf, B., Smola, A., 2002. Learning with Kernels, MIT Press, Cambridge, MA.

Sharkey, A. (Ed.), 1999. Combining Artificial Neural Nets, Springer, London.

Siemens web site www.ad.siemens.de/sw-haus/apc, 2003.

Smits, G., 1997. Personal communication, 1997.

Smits, G., Jordaan, E., 2002. Using Mixtures of Polynomial and RBF Kernels for Support Vector Regression, Proceedings of WCCI'2002, IEEE Press, Honolulu, HW, pp. 2785–2790.

Sotomayor, O., Park, S., Garcia, C., 2002. Software sensor for on-line estimation of microbial activity in activated sludge system. ISA Trans. 41, 127–143.

Tham, M., Morris, A., Montague, G., Lant, P., 1991. Soft sensors for process estimation and inferential control. J. Process Control 1, 3–14.

Vapnik, V., 1998. Statistical Learning Theory, Wiley, New York.

Weber, R., Brosilow, C., 1972. The use of secondary measurements to improve control. AIChE J 18, 614–623.

Willis, M., Montague, G., Di Massimo, C., Tham, M., Morris, A., 1992. Solving process engineering problems using artificial neural networks. In: Mc Ghel, Grimble, M., Mowforth, P., (Eds.), Knowledge-Based Systems for Industrial Control, Peter Peregrinus, London, pp. 123–142.

Xu, X., Hines, J., Uhrig, R., 1999. Sensor validation and fault detection using neural networks. Proceedings of the Maintenance and Reliability Conference (MARCON 99), Gatlinburg, TN, pp. 106–112.

Zhang, J., Yang, X., Morris, A., Kiparassides, C., 1995. Neural network-based estimators for a batch polymerization reactor. Proceedings of DYCORD'95, Helsingor, Denmark, pp. 129–133.

Zhang, J., Martin, E., Morris, A., Kiparassides, C., 1997. Inferential estimation of polymer quality using stacked neural networks. Computers Chem. Engng. 21, 1025–1030.

CHAPTER 4

Genetic algorithms in molecular modelling: a review

Alessandro Maiocchi

Milano Research Center, Bracco Imaging s.p.a., Via. E. Folli 50, 20134, Milano, Italy

1. Introduction

Computer-assisted molecular modelling systems can be briefly described as the marriage of molecular graphics tools with computational chemistry methods. Such systems have proven themselves quite useful through the years and, as a matter of fact, nowadays all major chemical and pharmaceutical companies do employ a variety of software packages as standard research tools.

Molecular modelling systems can offer a wide variety of visualizations and representations of the three-dimensional geometry of molecular systems, thanks to increasingly faster and more sophisticated computer graphics. Furthermore, they allow manipulations and modifications of virtually all molecular systems, from simple organic molecules to metallo-organic and inorganic complexes, to large macromolecular structures, including proteins and nucleic acid models. Modelling tools, often in combination with X-ray or NMR-derived structural models, take advantage of theoretical methods like ab initio and semi-empirical quantum mechanics or molecular mechanics (also known as *force fields*) to predict energetically accessible geometrical states of a given molecular system, by evaluating its associated physical and chemical properties (e.g. thermodynamic stabilities, reactivities and electronic properties). The choice of which relevant conformations/conformers to consider, given the specific objectives of a modelling study, is often the major task to face. As an example, in computer-aided drug design (CADD) applications, the relevant conformations of a molecular system are those that closely mimic the so-called active conformation, which is the specific conformational state responsible of an observed biological response. By the same token, if the objective of a study is the calculation of the macroscopic property of a molecular system in water, then the relevant conformations are those that are more populated and represented by low-energy conformational states in that medium. It is worth noting at this stage that the objective of modelling does not affect the nature of the conformational states to be sampled, but only the strategy of the search for relevant conformations. The characterization of the conformational states of a molecular system falls into the category of optimization problems where the definition of the 'optimal' conformational state does

Data Handling in Science and Technology, Volume 23
ISSN: 0922-3487

DOI: 10.1016/S0922-3487(03)23004-5

depend upon the main objective of the modelling study at hand. Many sophisticated search methods are available today, and some of them are widely used to characterize relevant conformational states of the molecular system of choice, but none of them can assure the completeness of the search for optimal solutions—with the exception for trivial cases. The complexity of the response surface associated with a conformational search may be very high due to the size of the parameter space, the occurrence of discontinuity and the presence of many local minima.

In the last ten years genetic algorithms (GAs) have emerged as new and powerful search methods and have been put forward in many molecular modelling applications to tackle the problem of exhaustively exploring the conformational space of small to medium to large molecular systems.

This chapter will not give an in-depth view on the fundamentals of GAs, as they have already been presented and discussed in previous chapters. Thus, the next section will point out some basic elements needed to adapt the *genetic metaphor* to handle molecular structures and conformations. In the rest of the chapter attention will be focused on the implementation of GAs in some of the most important and widely used molecular modelling applications, such as small and medium-sized molecule conformational searches, simulations of protein–ligand docking, conformational searches under constrained conditions and protein folding investigations. We have attempted to review all those papers that have contributed in a relevant way to the assessment of the performance of GAs in the above application areas. However, since currently available publications constitute a large, continuously growing body of literature, we do apologize to those authors whose contributions have been inadvertently overlooked.

2. Molecular modelling and genetic algorithms

A genetic algorithm (Davis, 1991a; Goldberg, 1989a; Forrest, 1993) is a general purpose optimization scheme having some basic elements as depicted in Fig. 1. The main feature of a GA is that it operates on a population of individuals, which are configurations in the search space, representing a possible solution to a given problem. Within the molecular modelling framework, these individuals are conformations that must be encoded with a suitable string describing all the degrees of freedom of the problem. Each member of the population is evaluated using a so-called *fitness function* so as to select those individuals (or parents) that are enabled to produce *offspring* (or children).

In molecular modelling applications, the fitness function may be the internal energy of a molecule calculated by means of a more or less simplified force field or a quantum-mechanical Hamiltonian. The fitness function is commonly applied to the three-dimensional representation of the molecular system that, within the genetic metaphor, is assimilated to the *phenotype* associated with the *genotype* string encoding the parameters to be optimized. Obviously, a mapping between the phenotype and genotype spaces must be provided. The nature of the fitness function depends on both the dimension of the molecular system under investigation and the objective of the study. However, since most of the computing time must be spent in evaluating the fitness function, a compromise between the accuracy and the complexity of the fitness calculation must be found.

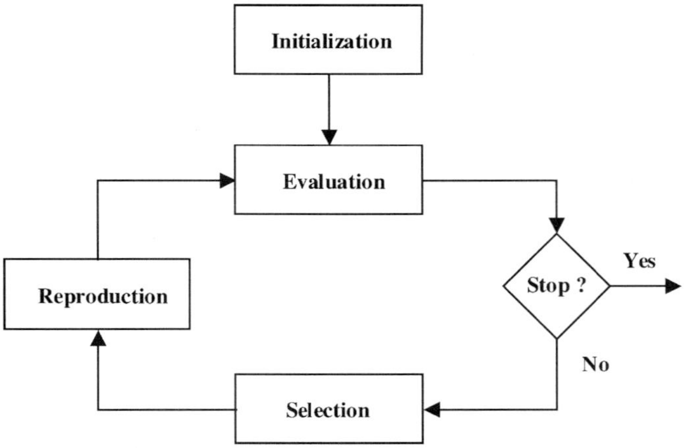

Fig. 1. Schematic representation of the basic elements of the evolution procedure in a standard GA.

The standard genetic operators used to create the new generation of individuals, or potential solutions, are *crossover* and *mutation*. With crossover, two parents exchange the information contained in their genotype while, with mutation, the genotype of a parent is randomly modified by altering one or more bits of its genetic string. The children are then inserted into the population, thus replacing the previous members, and are evaluated accordingly to the chosen fitness function. This sequence of steps, from parent selection to the population updating, is iteratively repeated until a predefined convergence statement is reached.

However, even though the general GA scheme outlined above is quite simple and more or less independent from the specific application, its implementation is a challenging task due to the large number of available possibilities entailed by the parameters, encoding rules, the nature of the fitness function or its transformation, the parents selection method, the kind of crossover, and the use of sharing operators and generation gaps. In addition, a number of parameters is to be established for the chosen procedures, all of which may affect the success rate of the designed GA.

Typically, several GA parameters, such as the population size, number of generations, and crossover and mutation rates, must be chosen. A key point in the design and tuning of a GA is the search for those elements and conditions that can assure a balance between optimizing the fitness function and maintaining population diversity, so that a variety of individuals may be selected at each iteration, thus avoiding a premature convergence to local minima. A number of operators such as forced mutation, or the use of individually evolving subpopulations that periodically exchange the best part of their genetic material, may increase population diversity.

2.1. How to represent molecular structures and their conformations

As has already been described, a prerequisite of every GA implementation is the choice of the encoding rules to describe all the degrees of freedom of a specific problem, while avoiding all redundancies. The application of the encoding rules will result in a genotype representation of the problem solutions, also called a chromosome. There are at least four genetic alphabets which can be used in a gene-based coding scheme: integer, binary, Gray-coded binary and real-value numbers. Among these, for historical reasons, the standard binary alphabet was the first gene-coding scheme applied in the conformational analysis problem, subsequently replaced by binary Gray coding and the real numbers. Table 1 illustrates the difference between the two binary schemes encoding an integer in the case of a three-bit string. With the Gray coding scheme, adjacent values are encoded by two binary strings differing only in the switch of one bit. However, it is not true that any variation of one bit in a Gray-coded binary string will result in an adjacent integer (e.g. $000 = 0$ and $100 = 7$).

In the conceptually simplest conformational problem, the internal energy of an isolated molecular system is analysed with respect to the variation of the dihedral angles defined for each available rotatable bond, as depicted in Fig. 2. In the absence of specific restraints, each dihedral angle can assume values in the -180 to $180°$ range, and the dimensionality of the conformational space to be searched is simply given by the number of rotatable bonds.

If the genotype string is built using real numbers, the dimensionality of the search space will not be modified, but each dihedral angle's gene may assume any value in the -180 to $180°$ range. Conversely, if a binary coding scheme is used to generate the genotype, the dimensionality of the search space will be augmented by a k factor, which is the number of bits used to encode each dihedral angle value. With a binary encoding scheme, the number of levels at which each dihedral angle will be explored is finite and equal to 2^k. Hence, it becomes clear that the precision of the proposed solutions can be modulated by the number of bits used to encode the dihedral angle values, as shown in Table 2.

It is also important to point out that the use of different coding schemes will affect the landscape of the internal energy of the molecular system with respect to the genotype search space.

Regardless of the choice of the genetic encoding schemes, the dimensionality of the genes may be increased by the nature of the conformational search problem. An example of this is the GA implementation of the conformational search of a small ligand enclosed in an active site of a biological macromolecule; this is known as the protein–ligand docking problem. In this case, the ligand is not an isolated system since it is at least

Table 1
Comparison between the standard binary and Gray coding schemes

Standard Binary	Decimal Values	Gray Code
000	0	000
001	1	001
010	2	011
011	3	010
100	4	110
101	5	111
110	6	101
111	7	100

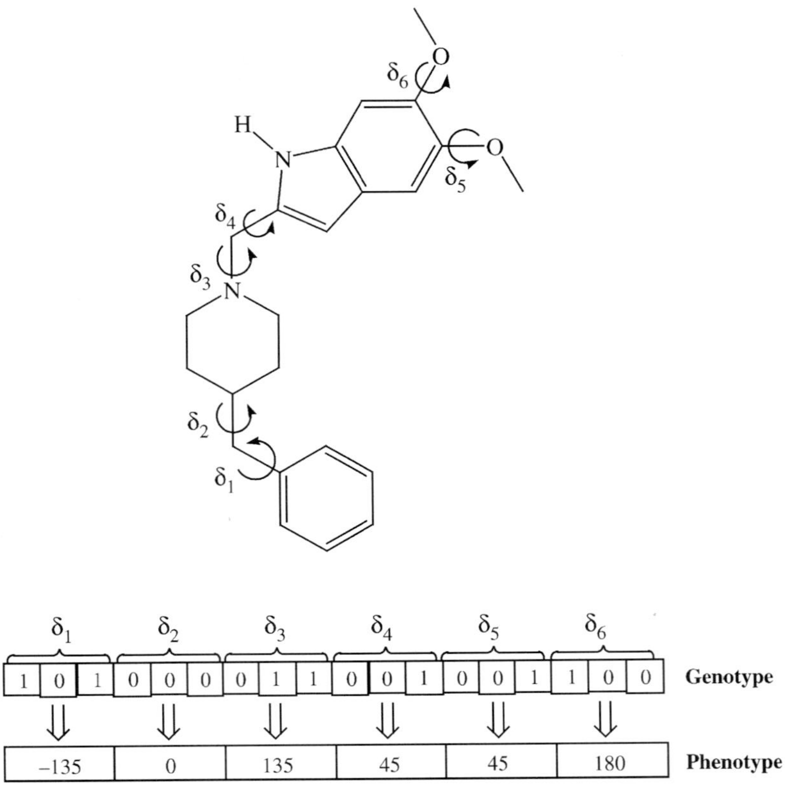

Fig. 2. Representation of the conformation of a molecule by means of the standard binary scheme to encode its dihedral angles δ_i. Each gene associated to the dihedral angle has a length of 3-bits.

partially surrounded by the interacting walls of the receptor cavity. Hence, the conformational states accessible to the ligand are also dependent on both its relative spatial position and orientation with respect to the three-dimensional structure of the active site. The genotype representation, needed to perform a conformational search of a ligand in a receptor cavity must be augmented to allocate genes containing information about the orientation of the ligand with three extra genes encoding the angles of rotation of the molecule (around the x, y and z axes) and the other three genes encoding the translation of the ligand (along the x, y and z directions) with respect to a standard orientation.

Table 2
The resolution of the solution space with the number of bits used for each gene

Bits	Levels	Resolution
2	4	90
3	8	45
4	16	22.5
5	32	11.25

The translational motion of a ligand in the receptor cavity is commonly bounded in a user-defined spatial region centred on the mass or geometric centre of the ligand docked into the receptor in a preliminary orientation. For this reason, when the binary coding scheme is used for the translational genes, it became essential to decode the binary gene, while accounting for appropriate shifting and scaling, so as to return back in the original spatial coordinates.

To decode the binary translational genes, a simple equation can be applied as follows:

$$v_{i,j} = v_i^{\min} - \left(v_i^{\max} - v_i^{\min}\right)\frac{d_j}{(2^k - 1)} \tag{1}$$

where v_{ij} are the values of the spatial coordinates in the bounded region to be decoded; $v_i^{\min}$ and $v_i^{\max}$ are the minimum and maximum permitted values of the spatial coordinates; d_j are the decimal values corresponding to the jth binary number; and k is the number of bits in the variable gene.

3. Small and medium-sized molecule conformational search

A basic problem in the prediction of a three-dimensional structure of a certain molecule is the characterization of its conformational states. The common approach is to search for low-energy conformations under the assumption that a molecule mainly populates conformational states close to the minimum global energy. Regardless of the nature of the computational theory used (e.g. molecular mechanics or quantum mechanics methods), the potential energy of a molecule can be represented as a hypersurface with ridges and wells with respect to its torsional space. The complexity of this hypersurface is related to the number of local energy minima that increase with the flexibility of the molecular system under study. For highly flexible molecules (15–20 rotatable bonds), an exhaustive, systematic search for the global minimum of the potential energy in the torsional space becomes impossible even given the performances of modern computational platforms.

Several methods have been proposed to tackle the problem of an effective exploration of the conformational states of small and medium-sized molecular systems and they are very well reviewed elsewhere (Leach, 1991).

Conformational search is a general molecular modelling procedure of great importance for both small and large molecular systems. In this section, we will focus our attention on GA applications in conformational searches of small and medium molecular systems, while in a next section, macromolecular applications, including protein folding, will be considered. In this section, we will also review those published works that discuss the effectiveness of GA-based conformational search procedures as compared to other widely used search methods. The reader should be aware that comparing such search methods is not a trivial task. To assess the relative performances of these search methods, such comparisons should be made only after each one has actually been tuned with respect to all the internal parameters, which, in turn, can affect their exploration/exploitation behaviour. Obviously, in a comparative study all the search methods must share the same potential energy function. Furthermore, all those methods that are based on algorithms having a stochastic component, such as Metropolis Monte Carlo (Metropolis et al., 1951), simulated annealing (Kirkpatrick et al., 1983), and GAs, should be compared with the results obtained from multiple-run protocols evaluating both the best and averaged calculated solutions. Another important point is the definition of the objective of the conformational search performed. In fact, the effectiveness of a search method may be assessed by examining several properties as follows: (i) the energy of the solutions retrieved; (ii) the conformational diversity of the explored solutions; (iii) the number of function evaluations or the CPU time required to find the minimum energy conformer; (iv) a combination of points (i)–(iii).

Probably one of the first thorough attempts to assess GA performance in conformational searches of small and medium-sized organic molecules was undertaken by Judson et al. (1993). In this study, a set of 72 molecules having up to 12 rotatable bonds was chosen from the Cambridge Structural Database (Allen et al., 1991). To perform the GA-based conformational search, the dihedral angles of the rotatable bonds (bonds in rings were kept fixed) were encoded in the chromosome by using a binary scheme with six bits, allowing a search resolution of about 5°. A typical generational GA with the *elitist* strategy turned on was used, together with a common implementation of mutation and crossover as genetic operators. A population size of ten times the number of dihedral angles, up to a maximum of 100, was used. The fitness of each conformation in the evolving populations was assessed using the MM2 force field (Allinger, 1977), as implemented in the MacroModel program (Mohamadi et al., 1990). At the end of the GA evolution phase, the best structures found were minimized by using a steepest descent gradient minimization routine. The results demonstrated that the GA was able to find solutions with an energy close to that of the relaxed crystallographic structures for almost all the molecules in the test set.

McGarrath and Judson (1993) published a conformational search on a cyclic hexaglycine, exploring 24 dihedral angles. They also studied the effect of niches on GA performance: the results showed that the use of intercommunicating subpopulations can promote a broader conformational space search as long as the subpopulations do not exchange their best individuals too frequently. The authors also suggested that, for medium-sized molecular systems, even infrequent local minimizations during the GA evolution (hybrid GA) of the conformations in the current subpopulations, may increase the search performance of the whole procedure. It is important to note that in the 'hybrid' GA procedure, local minimization steps only updated the energies of the solutions, but not their atomic coordinates.

Brodmeier and Pretsch (1994) attempted to assess the effect of the initial setting of GA parameters in the conformational search problem using two alkanes as test molecules: *n*-decane and 3-methyl-nonane. Their GA implementation encoded the dihedral angles as real numbers and used a fitness function that was an exponential transformation of the molecular strain energy calculated with a modified version of the MM2 force field. The one-point crossover using a randomly generated point was also applied. To reduce the occurrence of premature convergence of the whole search, they introduced the so-called *sharing* technique. This technique was designed to maintain a certain degree of diversity among the individuals of the current population. The proposed *sharing* technique required that each new configuration generated from a crossover or mutation be compared with all the existing configurations already placed in the new population so as to verify if the new configuration was distinct enough. The sum of the unsigned differences for all the dihedral angles between two conformations had to be larger than a predefined threshold value. If this condition was not verified, random increment values were added to each dihedral of the new conformation. By working with these basic operators, Brodmeier and Pretsch concluded that: (i) larger populations did not increase the effectiveness of the search procedure but, on the contrary, small populations (30 individuals) performed more efficiently; (ii) high mutation rates hindered the usefulness of the crossover, and the best performances were found with a mutation rate of 0.01; (iii) higher crossover rates gave better search performances even though the effect of this parameter was less critical than the mutation rate; (iv) the *sharing* operator actually increased the diversity of the final population of solutions; (v) using elitism a further improvement of the performance of the GA was achieved. For both of the tested molecular systems, the GA was able to find low-energy conformers.

Hermann and Suhai (1995) published the application of a GA method to the energy minimization of some small peptides using a real-number encoding scheme of the dihedral angles. The internal energy of the test molecules was evaluated using the AM1 Hamiltonian (Dewar et al., 1985) implemented in the MOPAC program (Stewart, 1990). For all the dipeptide and the tetrapeptide systems studied in this work, the GA was able to locate the global minimum-energy conformations.

Meza et al. (1996) compared their GA implementation and the direct search method, PDS (Dennis and Torczon, 1991). They found that the two methods were equally effective and efficient at minimizing the energy of a test set containing 19 organic molecules having up to 39 dihedral angles. In their conclusions, they suggested that multiple, independent GA runs could be a better strategy than a single run having the same total number of function evaluations.

Jin et al. (1999) compared the performances of three versions of their GA program GAP searching for the conformational space of the pentapeptide [Met]-enkephalin. The three GAP versions differed with respect to the initialization of the GA scheme, as well as in the crossover operators: uniform crossover; 'three-parent' crossover; and uniform crossover preceded by a 'population splitting scheme'. These alternative crossover operators were all implemented so as to enhance the diversity of the new conformations with respect to that of the old parents. An interesting result was that, although the crossover operators influenced the propagation of the useful information in the chromosome (in this case a binary encoding scheme was used), this aspect did not emerge from the sampling of the conformational space or from the evolution of both the average energy of the populations and the lowest energy conformation. A similar conclusion was also outlined in the previous work by Brodmeier and Pretsch (1994). Hence, Jin et al. argued that the crossover operator becomes redundant when highly fit individuals appear in the evolving population; all the other stochastic operators that should increase individual diversities, such as mutation or sharing operators, will have more likely generated a worst individual than really new low-energy conformations.

Most of the GA implementations proposed only explored the conformational subspace limited to rotamers. To account for the flexibility of cyclic fragments, Payne and Glen (1993) included flips of the so-called ring-free corners that involve rotations around two

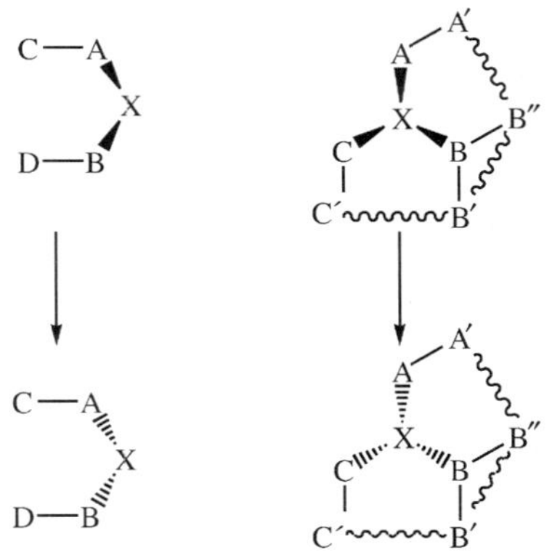

Fig. 3. In (a) are represented the two states of the free corner (A–X–B) in a cyclic moiety. All the atoms A–D and X are in the same cycle. In (b) are represented the two states obtained from the reflection of a pyramid involving an atom X with hybridization sp^3 belonging to at least two condensed cycles.

bonds from a ring (Fig. 3a). Mekenyan et al. (1999) extended the treatment of cyclic moieties by means of the so-called pyramid reflection involving an X site with an sp^3 hybridization belonging to at least two condensed rings together with its first neighbours A, B and C (Fig. 3b). The pyramid reflection consists of its mirrored reflection with respect to the base plane of the pyramid itself. Since the reflection of the pyramid inverts all the potential stereocentres, including A, B and C, with the exception of those depicted in Fig. 3b, the stereochemistry of the involved centres was controlled.

In the same work, Mekenyan et al. introduced a modification in the standard GA scheme aiming to perform an optimal sampling of the conformational space of a molecule by means of a fitness function related to a property of the whole population rather than just some individuals. Accordingly, a fitness function measuring the average dissimilarity among the conformations in each evolving population was used, as follows:

$$\mathrm{rms}(S) = \frac{1}{N(N-1)} \sum_{\substack{i,j \in S \\ i \neq j}} \mathrm{rms}_{ij} \tag{2}$$

where S is a set of N conformations (the scored population); rms_{ij} is the root-mean-square distance between the two conformations i and j; and $N(N-1)$ is the total number of conformation pairs.

Another relevant modification with respect to the standard GA implementation was introduced in the selection step of the procedure, whereby the evolving N_p size population was expanded by N_c new individuals generated by the mutation and crossover operators applied to the selected parents. Moreover, a new conformation obtained from the above genetic operators was not directly admitted in the N_c extended population fraction: it might be rejected if it closely resembled already available conformations in the $N_p + N_c$ population. When the extended population was filled, it was again further reduced to N_p under the condition that the selected subset of conformations maximized the rms(S)

function. This is a combinatorial problem that was exactly solved using the branch-and-bound algorithm (Lawler and Wood, 1966).

In view of the results obtained with both flexible and more rigid organic compounds (a test set of four molecules was used), Mekenyan et al. (1999) suggested that, in their genetic protocol, the reproducibility of the resulting populations, along with the conformational space coverage, mainly related to the ratio between parents and children generated in the reproduction step. Thus they observed that, by selecting a lower N_p/N_c ratio, the diversity of the final populations increased and the resulting evolutionary process was more dynamic and less reproducible.

Wehrens et al. (1998) have tackled the problem of defining a set of quality criteria in order to assess the performance of GAs, also in comparison with other methods, as far as regards optimization in general and the conformational search problem in particular. Instead of evaluating only the best solutions found by the algorithm, Wehrens et al. proposed the use of four quality criteria based on a multiple-run strategy as follows: the coverage of both the total search space and the solutions space, the last one being the relevant part of the former, and the reproducibility of the coverage of these spaces when the GA is run several times. The search space coverage was evaluated simply by dividing first the search space into several equally sized hypercubes and subsequently counting the number of hypercubes occupied by at least one conformation during a run of the GA scheme applied. The coverage of the solution space was calculated by grouping similar solutions into a cluster and then counting the clusters in the final populations of the pooled repeated runs, occupied by solutions having low fitness values. Hence, the reproducibility of the solution space was evaluated simply by counting the number of conformations obtained from different runs in each cluster. The reproducibility of the total search space coverage was evaluated by means of principal component analysis (PCA) (Flury, 1988) of all the runs. All the conformations belonging to the evolving populations of one run were used to generate a PCA space and subsequently, the conformations observed during the evolution of all the other replicated runs were projected into that PCA space. The degree of reproducibility was calculated from the residuals after projection using the ratio between the mean residual of all the projected replicated runs and the mean residual of the run used to generate the PCA space. This ratio approximates to 1 if the replicate runs cover the same total search space, otherwise it will assume larger values.

More recently, Wehrens et al. (1999) extended their preliminary work by also evaluating the effects that some common parameters of a GA procedure can produce in any of the quality criteria proposed. In particular, with a series of eight experiments (each one consisting of five replicated runs), organized according to a Plackett–Burman design (Plackett and Burman, 1946), it was possible to explore the effects of seven parameters, thus reaching the following conclusions: the search space coverage increased for higher mutation and crossover rates values, while the selection pressure had a detrimental effect on the exploration property of the procedure; the solution space coverage increased by setting higher population sizes and a higher number of generations, in conjunction with the use of higher crossover rates, while the mutation rates were reduced; the reproducibility of the search space coverage was reduced by means of a higher selection pressure and number of generations, while better reproducibility was achieved combining higher mutation rates with a two-point crossover and small populations. The definition of the solution space reproducibility in this work was slightly modified with respect to the previous one since, in the new definition, this quality parameter was related to the mean of the distances between the closest clusters with elements derived from other runs. It was found that the reproducibility of the solution space was diminished in larger populations owing to the increased number of clusters formed.

This fine-tuning of a GA scheme was applied to the dermorphine conformational search under several distance constraints obtained from NMR measurements, but it was suggested that the four quality criteria should also be used in the performance evaluation of other optimization algorithms, regardless of the specific applications. The value of the work of Wehrens et al. (1999) lies in their having outlined a general method of fine-tuning a GA scheme, although the reader should be aware that their results are not necessarily transferable to any other GA implementation that uses other operators or generation gaps. Morover, a complete fine-tuning protocol should involve several diverse molecular systems so as to avoid bias in setting the final parameters by the particular form of the potential energy surface of the molecular system being studied.

4. Constrained conformational space searches

An important area of molecular modelling techniques is the exploration of the conformational space of a molecule under a number of predefined, three-dimensional constraints. The nature of these constraints ultimately depends on available knowledge and the objective of the modelling study. The simplest constraint is the molecular distance between two atoms in a molecule or the intermolecular distance between atoms belonging to two interacting molecules. This kind of information may be properly derived through multi-dimensional NMR spectra. A second level of complexity is represented by the constraints derived from a three-dimensional pharmacophoric hypothesis. A three-dimensional pharmacophore may be defined as the essential spatial arrangement of atoms or functional groups necessary to produce a given biological response. Even in its simplest form, the pharmacophore definition requires that atom types and interatomic distances are specified, but it can be extended to include the orientation of functional groups or more general functional properties (e.g. hydrophobic/hydrophilic centres, hydrogen-bond donor/acceptor atoms, ring centres and lone pairs). If a pharmacophoric hypothesis is available, it can be used to search for new molecules that can adopt conformations satisfying the spatial relationships of the structural features identified in the pharmacophore. If a pharmacophore is not known, it can be derived from the overlay of molecular structures of biologically active molecules. Again, in the search for a pharmacophoric hypothesis, a conformation for each biologically active molecule must be selected under the constraint of matching some predefined structural features.

A conformational search may also be subjected to external constraints due to the steric and physicochemical properties of a binding site provided either by a pseudo-receptor model or an experimental protein structure (the protein–ligand docking problem).

Each one of the aforementioned constrained conditions in which a conformational search may be performed, was also approached by exploiting the GA paradigm. This section will briefly review the use of GAs in all these applications except for the docking problem, which will be discussed in more detail in Section 5.

4.1. NMR-derived distance constraints

The identification of molecular conformations satisfying NMR-derived constraints is a common problem in the molecular modelling framework. Besides, the availability of NMR-derived inter-proton distances allows the validation of the results obtained by previous conformational searches.

Payne and Glen (1993) published a fundamental study describing the use of GAs for constrained conformational searches. In order to generate conformations satisfying distance constraints derived from Nuclear Overhauser Effect (NOE) data, their GA implementation employed 8-bit binary strings to encode the dihedral angles of rotatable bonds. The scoring function was built using two contributions: the first was a distance constraints term calculated as the square root of the sum over all the distance constraints of the squared differences between the target distance and the corresponding distance in the conformation; the second was a penalty term for bad van der Waals contacts to avoid occurrences of atom clashes.

Roulette-wheel selection, one-point crossover and mutations were the genetic operators used to generate new solutions during the GA evolution.

A very similar GA was described by Sanderson et al. (1994) to determine possible conformations of a cyclic Arg-Gly-Asp peptide analogue that were consistent with NMR data. The dihedral angles of 22 rotatable bonds were encoded in the chromosome using 8-bit binary strings. In order to facilitate the exploration of the conformational space of the cyclic oligopeptide, the macrocycle was broken by removing a disulfide bond, but new constraints were added to the inter-proton distances derived from the NOE data to ensure a ring closure condition in the final solutions. The molecular conformations obtained from the GA procedure were further submitted to geometry minimization under the experimental inter-proton distance constraints.

Van Kampen and Buydens (1997) examined the effectiveness of GA with respect to the simulated annealing (SA) method (Kirkpatrick et al., 1983) by performing the conformational search of a linear eptapeptide under NMR-derived distance constraints. Their study mainly focused on assessing the effectiveness of the crossover operator on the GA optimization properties. The results of the analysis showed that, in the design of the GA, the crossover operator was not actually effective in promoting improved trial solutions. Consequently, SA outperformed GA—even though both methods generated conformations of comparable quality, SA converged three times faster.

A different GA application was proposed by van Kampen et al. (1996), who developed a procedure combining a distance geometry program with a GA. Molecular conformations consistent with NMR data are often generated using distance geometry, but the algorithm can generate low-quality structures, especially when experimental NOE data are incomplete or imprecise. The method devised by van Kampen et al. (1996), also called DGΩ (distance geometry-optimized metric matrix embedding by genetic algorithms), combined well-defined parts of individual structures generated by the DGII distance geometry program (Havel, 1991) with a GA that, in turn, identified new lower and upper distance bounds within the original experimental restraints. The aim was to restrict the sampling of the metrization algorithm to more promising regions of the conformational space. In the devised procedure, a complete set of modified restraints was encoded in a chromosome using real numbers within the corresponding original bounds. Each chromosome represented a trial structure, and replaced the original set of experimental restraints as input for the DGII algorithm. Then a DGII calculation was carried out for each chromosome, and a fitness function depending on the number of constraints and the entity of their violations was evaluated. The chromosomes were ranked and selected accordingly with the rank-based threshold selection (Lucasius and Kateman, 1994). The selected chromosomes were subjected to a uniform crossover operator and a modified version of the mutation operator in order to generate a new population of trial solutions. The whole evolutionary procedure was tested for the restrained conformational analysis of cyclosporin A having 58 NMR-derived distance constraints. The results confirmed that the proposed procedure was able to enhance both the convergence and the sampling properties of the conformational space with respect to the standard distance geometry approach used in DGII.

Several other examples of GA applications in the conformational analysis of biological macromolecules under NMR-derived constraints can be found in the literature (Beckers et al., 1997; Baylay et al., 1998) and are reviewed elsewhere (Sanctuary, 2000).

4.2. Pharmacophore-derived constraints

The first work elucidating the use of a GA for conformational searches under pharmacophore-derived constraints was published by Payne and Glen (1993), who reported the use of the GA approach to fit three *N*-methyl-D-aspartate antagonists to a putative NMDA pharmacophore.

Two constraints were imposed by the pharmacophore definition: (i) the distance between an amine nitrogen and a phosphonate sp^2 oxygen, and (ii) the distance between a carboxylic oxygen and the same phosphonate sp^2 oxygen (Fig. 4). The scoring functions

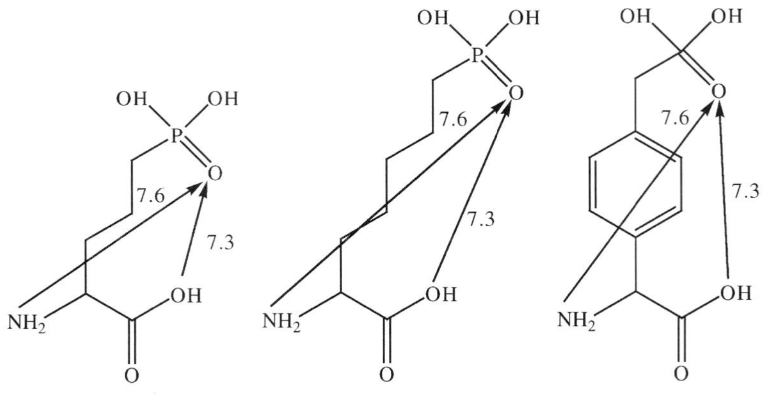

Fig. 4. The representation of the pharmacophore constraints for three NMDA antagonists. The distances are in Ångstroms.

were based solely on the distance constraints term and on the penalty term for bad van der Waals contacts.

Starting from randomly chosen conformations and orientations, reasonable conformations that closely matched the target distances were obtained for each antagonist. This preliminary work inspired the research of Clarke et al. (1992, 1994), who developed a general procedure for a flexible substructure-searching system for three-dimensional chemical structures. The substructure was represented by the functional features and their relative spatial arrangement as defined in the pharmacophore. Searching a database of three-dimensional chemical structures was envisaged by Clarke et al. as a three-stage procedure for exploring the conformational space of each molecule in order to match the pharmacophoric query. In the first stage, the molecule in the database is analysed to verify the existence of the pharmacophoric structural features. In the second stage, a geometric search is performed to determine potential hits based on the consistency between the bounded distance matrices of both the pharmacophore and the structure in the database. In the third stage, the molecules surviving the previous steps are submitted to the conformational search in order to verify whether the low-energy conformations are able to match the pharmacophore constraints. With the aim to improve the effectiveness of the third stage Clark et al. (1992, 1994) used a GA-based approach to perform the conformational search. The GA scheme they used was very similar to that proposed by Payne and Glen (1993) with a scoring function having only two weighted contributions, namely a distance penalty term and an energy penalty term. The former was defined as the sum of the unsigned differences between the solution distance and the closest bound of the target distance; the latter was based on a 6–12 Lennard–Jones potential using the TAFF force field parameters (Clark et al., 1989). In order to assess the effectiveness of the GA implementation, a database of 1538 three-dimensional structures was searched with eight pharmacophore queries and the results of the conformational search were compared against several alternative searching methods: distance geometry (Crippen, 1978), systematic search and directed tweak (Hurst, 1994). The comparison showed that the GA and directed tweak methods were more effective than the others, even though directed tweak was noticeably faster than GA, as confirmed by a later work based on the search of a database of approximately 10,000 structures (Jones et al., 1996).

4.3. Constrained conformational search by chemical feature superposition

When in a drug discovery program some basic knowledge about the key interactions between a potential drug and the target receptor is not available, a pharmacophore hypothesis is generated using the activity data of a series of compounds that are assumed to interact with the receptor in the same binding mode. Under this assumption, the compounds in the series are overlaid in order to identify common structural features that may be responsible for their observed biological activity. The alignment of the compounds can be done using both a rigid-body and conformationally flexible superposition. Moreover, the structural features to be superimposed may be either defined a priori by the user or searched with an automatic procedure designed to map the possible chemical features in each compound of the series.

Again, it was Payne and Glen (1993) who were the first to report how a GA was able to drive the conformationally flexible overlay of a series of compounds that are constrained to match predefined structural features. They elucidated the concept using the same set of NMDA antagonists already discussed above, but in this case only the two oxygen atoms and the amine nitrogen were employed to define the constraints of the further conformational search. In order to optimize both the conformation and orientation of each compound, the chromosome in their GA was adapted to concatenate the bit-strings coding the dihedral angles of rotatable bonds for each compound in the series. Each chromosome thus represented all the compounds in one conformation and was scored after a least-squares fitting of the compounds to the specified constraints. The fitness function used a distance penalty term calculated as the sum of the distances between the position of the selected features of all the molecules with respect to a reference position provided by one molecule of the set. In order to decrease the total volume of the combined overlaid molecules, a second term was also added to the fitness function to measure the atom-by-atom overlapping integral between atoms in different molecules.

This approach was also followed by Jones et al. (1995a, 1996), who devised the GASP program (genetic algorithm similarity program) for flexible molecular overlay and pharmacophore elucidation. Within the GASP program, molecules were represented by a chromosome containing binary strings to encode the conformational information (a binary Gray coding scheme is used) as well as integer strings to allow the intermolecular mappings between important structural features likely to be required for activity. This part of the chromosome constituted the major modification in comparison with the GA implementation of Payne and Glen (1993), since the algorithm in this form did not require any prior knowledge or assumption about the nature of the structural features involved in the pharmacophore. In GASP, each molecule in a data set was analysed to identify the number and nature of the available structural features. Then the molecule with the minimum number of recognized structural features within the whole set was selected to act as the base molecule. The molecules in their conformations were then overlaid to the base molecule using a least-squares fitting procedure to optimize the number of structural equivalences suggested by the intermolecular mapping. The fitness function used to rank the solutions was a weighted sum of contributions as follows: a first term was calculated as the mean (over all the overlaid molecules) of the differences between the internal steric energy of the input conformation and the conformation encoded in the chromosome (steric energy was evaluated using the 6–12 Lennard–Jones potential based on the TAFF force field parameters); the second term was the mean volume integral calculated from the sum of the pairwise common volumes between the base molecule and each one of the other molecules; the third term was a similarity score calculated by comparing the similarity between the base molecule and the other molecules with respect to the position, orientation and type of hydrogen-bond donor and hydrogen-bond acceptor, and the position and orientation of aromatic rings. For more details about the calculation of the similarity score see the referenced works (Jones et al., 1995a, 1996). In GASP, a steady-state-with-no-duplicate GA was used (Davis, 1991b), in conjunction with crossover and mutation as genetic operators. The crossover operator performed differently depending on the part of the chromosome that was recombined: a common one-point crossover is applied to the binary parts of the chromosome, while a two-point crossover is applied to the integer strings using the PMX (partially matched crossover) crossover operator (Goldberg 1989b). The GA-based procedure implemented in GASP to perform a flexible overlay of molecular structures for pharmacophore elucidation, has been proven to be successful with several data set (Jones et al., 1995a, 1996). A recent work (Patel et al., 2002) reported the comparison of GASP with two other pharmacophore elucidation programs, namely Catalyst/HipHop (Barnum et al., 1996) and DISCO (Martin et al., 1993). The three programs were compared on the basis of their ability to generate known pharmacophores deduced from the protein–ligand complexes extracted from the Protein Data Bank available at the Cambridge Crystallographic Data Centre. Five different protein families were included in the study: Thrombin, Cyclin Dependent Kinase 2, Dihydrofolate Reductase, HIV Reverse Transcriptase and Thermolysin. From the inspection of the available three-dimensional structures of the protein–ligand complexes, a target pharmacophore was defined for each protein family, as the set of pharmacophoric features common to all the ligands. The three programs were then tested on their ability to generate the target pharmacophore. The authors found that GASP and Catalyst/HipHop clearly outperformed DISCO. However, it was difficult to differentiate GASP from Catalyst/HipHop as the two programs provided almost equivalent performance even though the results were not consistent for all the data sets.

5. The protein–ligand docking problem

The protein–ligand docking problem refers to the prediction of the conformation of a ligand when it is non-covalently bound to a protein molecule. The binding process between a small molecule and a target protein is an essential step in many biological functions, so the understanding of the key interactions that determine a tight binding between a molecule and its target protein or enzyme, provides a more direct approach to designing compounds likely to function as new drugs. All the molecular modelling methodologies that make use of the three-dimensional structure of a target protein with the aim of searching for new compounds with improved affinity and selectivity, are commonly referred to as structure-based drug design methods (Kuntz, 1992). The growing number of protein–ligand structures solved by X-ray crystallography (Priestle and Paris, 1996) or by Nuclear Magnetic Resonance (NMR) spectroscopy (Markley 1989; Fesik 1991) has provided the basic condition for a substantial increase in the number of research projects embracing a structure-based drug design approach by pharmaceutical companies. If the experimental structure of a target protein is not available, homology-derived protein structure models (Blundell et al., 1987) or pseudo-receptor models (Vedani et al., 1995) can in principle be used. However, it should be emphasized that, regardless of its source, the quality of the three-dimensional protein structure will affect the results of any docking method currently available.

Broadly speaking, docking procedures essentially require two elements: a search strategy and a scoring function. An optimal search strategy should cover all the possible binding modes between a ligand and a receptor as well as finding the experimentally observed binding mode among them. In a protein–ligand docking problem, the dimensionality of the search space can be controlled on the basis of the permitted degree of flexibility of the ligand and the side chains of the receptor site. In the simplest molecular docking procedure, a rigid molecule in its conformation is docked into a rigid receptor, and the dimensionality of the search space corresponds to the six roto-translational degrees of freedom of the ligand. This approach cannot be considered as a general method of choice, since it requires the knowledge of the conformation adopted by the ligand after the binding with the receptor. In practice, the rigid-body docking procedure is restricted to the cases where the ligand has a low number of rotatable bonds. When the conformational flexibility of both the ligand and the receptor is taken into account, the dimensionality of the search space can be so high as to become practically unsearchable in a systematic way. To show more emphatically such complexity of the search space, it may be sufficient to calculate the number of configurations theoretically to be evaluated in the ligand docking procedure for a fairly simple system: a small molecule with three rotatable bonds and a rigid receptor. If any angle (torsional and positional) is rotated with increments of 10°, and the ligand modifies its position in a cubic region with three edges of 10 Å sampled every 0.5 Å, the number of non-redundant configurations available for the system would be approximately 1.6×10^{13}. This is a very large number and even supposing it may be possible to perform 100 configuration evaluations per second, it would take at least 5000 years to complete the search.

From the aforementioned example it becomes clear that only a limited portion of the search space can be explored for small-size protein–ligand docking problems, and this explains why the searching procedure must be efficient and computationally inexpensive.

Several computational methods have been applied to the protein–ligand problem to account for the different level of flexibility of ligands and proteins. These methods are extensively reviewed elsewhere (Lybrand, 1995; Muegge and Rarey, 2001; Taylor et al., 2002).

GAs have often been evaluated in comparison with other search methods commonly employed in the protein–ligand docking problem, such as molecular dynamics (MD) trajectories, Monte Carlo (MC) methods, fragment-based (FB) methods and distance geometry (DG) methods. The comparisons were generally carried out by examining the root mean squared distances (RMSD) of the atomic positions of the ligand in the docked conformation from the equivalent atoms in the crystallographic conformation. A well-predicted docked conformation should have an RMSD near to 1 Å or lower, providing that the main experimental protein–ligand interactions are closely reproduced. It is also useful to analyse the results of a docking procedure by looking at scatter plots of RMSD against the calculated energy in order to recognize the possible presence of solution clusters identifying specific binding modes of the ligand.

In order to validate a new docking method, several authors compared the results of their procedure with experimentally derived protein–ligand structures, and evaluated the effectiveness of their methods as compared with the results obtained with previous, well-established docking procedures. However, the validation of a new docking method requires that:

a) a large number of protein–ligand complexes are employed;
b) the protein families from which the selected protein–ligand systems are derived be as diverse as possible;
c) the protein–ligand structures in the test set have high crystallographic resolution ($\leq$ 2.5 Å);

d) stochastic search methods are evaluated through multiple runs, and their robustness against adjustable parameters be assessed.

GAs form a class of non-deterministic algorithms in an intermediate position between the systematic and stochastic methods. As in common practice, the implementation of the protein–ligand docking procedures with genetic algorithms will have an intrinsic stochastic behaviour, ideally, all the requirements listed above should be fulfilled if a more consistent validation and comparison of methods are to be achieved.

5.1. The scoring functions

Different ligands can interact with the receptor in many different conformations with varying affinities. The scoring functions are then used to rank the putative protein–ligand complexes generated by the search algorithm. In principle, a scoring function should enable the discrimination of the experimental binding modes among all the other explored alternatives, and is a compromise between the computational cost required to evaluate complex functions and the loss of accuracy derived from the use of some simplified functional forms. Several scoring functions have been proposed, most of which are based on the calculation of the interaction energy of the protein–ligand complex using only the non-bonded atomic pairwise interaction terms, such as the Coulombic electrostatic term, 6–12 Lennard–Jones potentials and 12-10 hydrogen bonding term as formulated in some common and well-established force fields, i.e. AMBER (Weiner et al., 1984, 1986; Pearlman et al., 1995), OPLS (Jorgensen and Tirado-Rives, 1988; Jorgensen et al., 1996) and CHARMM (Brooks et al., 1983). Non-bonded interaction terms in scoring functions have also been combined with solvation and entropy estimates.

Some scoring functions are built empirically by means of a multivariate regression analysis over a training set of protein–ligand complexes of known binding affinity. The LUDI program (Bohm, 1992, 1994) for de novo design and the FlexX program (Rarey et al., 1996) for protein–ligand docking, are examples of computational procedures that exploit simplified scoring functions to estimate the binding free energies of protein–ligand complexes. Broadly speaking, this class of scoring functions attempts to account for the binding free energy contributions due to solvation, conformational changes, and protein–ligand intermolecular interactions through a sum of terms associated with the number of rotatable bonds in the ligand, hydrogen bonds, ion-pair interactions, the hydrophobic and π-stacking interactions of aromatic groups along with lipophilic interactions. Other approaches have been suggested to develop knowledge-based scoring functions that use atom-pair potentials derived from statistical analysis of structural databases (Verkhivker et al., 1995; Wallqvist et al., 1995). Several other more heuristic score functions have been described, but a detailed discussion of their formulation is beyond the scope of the present work. For detailed information the reader is referred to the excellent reviews covering the topic (Oprea and Marshall, 1998; Muegge and Rarey 2001; Böhm and Stahl, 2002). The use of some simplified scoring functions greatly speeds up the docking procedure and becomes very helpful in the so-called 'virtual' screening methods, where libraries of virtual (hypothetical and not yet synthesized) compounds are screened for bioactivity against a predefined protein target.

When a scoring function contains potential energy contributions, a relevant reduction of the computational cost can be achieved using pre-computed grids of points encompassing the whole protein volume or a volumetric portion in the surroundings of the docking cavity. It is noteworthy that the most widely used docking software tools do not implement the same scoring function demonstrating that an optimal choice is not yet available.

5.2. Protein–ligand docking with genetic algorithms

Genetic algorithms have been applied to the protein–ligand docking problem for at least ten years and a significant number of approaches have been already attempted.

The first recognized molecular docking application of GAs was described by Dixon (1993) for the dihydrofolate reductase–methotrexate system. The designed procedure was based on matching the ligand atoms and the spheres generated into the active site following the method implemented inside the DOCK program (Kuntz, 1982). The author used a standard implementation of the genetic algorithm accounting for both the roto-translational and the torsional degrees of freedom of the ligand. Moreover, the chromosome encoding the position and the orientation of the ligand was extended to encode the pairs of ligand atoms and sphere centres that had to be matched.

This preliminary work was further extended by Oshiro et al. (1995), who also included the thymidylate synthase–phenolphtalein and HIV protease–thioketal haloperidol complexes in the test systems. The binary Gray coding scheme was used to encode the roto-translational and positional states of the docked conformations. A molecular mechanics interaction energy, based on the AMBER force field and accounting only for van der Waals and electrostatic Coulombic potentials, was employed as scoring function. In the final step, after the GA runs, the low-scoring solutions were subjected to rigid-body minimization using either a quasi-Newton or a simplex optimizer.

For all the three tested protein–ligand systems the RMSD (heavy atoms only) were lower than 1 Å. Judson et al. (1994, 1995) presented their GA implementation of a flexible molecular docking as an extension of the methods they had previously developed for the conformational analysis of small molecules (Judson et al., 1993). One of the most relevant variants they introduced to their GA implementation was a 'growing' algorithm in which only a small portion of the ligand was initially docked. This variant was introduced to increase the GA efficiency during the first evolving step and to reduce the number of conformations that failed to fit the receptor cavity. After a certain number of generations, the ligand was grown by adding another small portion of the whole molecule, and the resulting larger submolecule was allowed to search for new low-energy conformations. Another specific feature of this docking method was the definition of a 'pivot' atom in the ligand as a reference point to control the rotation and translation of the molecule. The pivot atom was to be contained in the first substructure positioned in the receptor cavity at the beginning of the search, and was to be either a hydrogen bond donor or an acceptor atom that was expected to be involved in the binding process or bound to such an atom. In this case as well, the chromosome was built by means of a binary Gray coding scheme for both the roto-translational and the torsional degrees of freedom of the ligand. The GA search strategy was designed using independent niches of conformations evolving with a generational model in elitist mode and exchanging the best individual every 20 generations. The conformations in each population were ranked according to their internal energy evaluated with the CHARMM force field. An evaluation of the solvation energy was also made possible by using a continuum solvation model envisaged by Hasel et al. (1988). To accelerate the whole calculation procedure, the internal energy of each conformation, with or without the solvation contribution, was evaluated only after a check on the number of 'bad' van der Waals contacts; if no bad contacts were revealed, the energy calculation was performed. The results obtained with ten ligands and the three proteins thermolysin, carboxypeptidase A and dihydrofolate reductase (eight ligands were thermolysin inhibitors) were considered to be encouraging, especially when the solvation contribution was turned on.

A similar independently developed approach was published in the same period by Clark and Ajay (1995), who created the DIVALI software to treat rigid and flexible ligand docking into a fixed receptor. The authors employed a typical generational GA with a binary Gray coding scheme, which evolved in elitist mode. An AMBER-type potential function without the hydrogen bond term was used as scoring function. To enforce the diversity in the translation component of the chromosome, eight portions of the permitted translational space were generated by fixing a most significant bit for each spatial directions. All eight sub-populations, each one defined by a triad of the most significant bits, were evolved simultaneously, and a 'masking' operator was devised to constrain the search to sub-populations. If a conformation belonging to one sub-population fell outside its octant of standard GA iterations, owing to a mutation of one or more of the significant bits, the operator masked the event and brought the conformation back into the original octant. The effectiveness of the masking operator inside DIVALI was demonstrated when the rigid-body docking of glucose and glycyl-L-tyrosine into a periplasmic protein and carboxypeptidase A, respectively, was unable to generate reliable orientations in the absence of this operator.

Jones et al. (1995b) described a GA-based docking procedure designed to dock a flexible ligand to a partially flexible protein. The main modification in comparison with the previous studies was the particular representation of a chromosome. This was divided into four independent parts, two of which were devoted to encode the values of the dihedral angles of the ligand and of the protein side chains, respectively (by means of the binary Gray code). The other two parts were filled with such information as required to map the hydrogen bond interactions between the ligand and the protein binding site. The devised procedure required, firstly, placing the ligand into the binding site, and secondly, defining the values of the dihedral angles (chromosome decoding) for both the ligand and protein. Moreover, a least-squares rigid-body fit was carried out with the aim of maximizing the number of hydrogen bonds mapped in the third and forth parts of the chromosome. As a consequence, the GA solutions would be biased toward the complexes with a higher number of inter-molecular hydrogen bonds. The scoring function was defined taking into account the number and strength of the protein–ligand hydrogen bonds formed, the van der Waals energy of the complex and the internal energy of the ligand conformation.

Their original ideas were later implemented, with some modifications, in the GOLD software tool (Jones et al., 1997). The chromosome was modified to treat also the ligands with less than three polar atoms potentially available for hydrogen bond formation. In this case, of course, rigid-body least-squares fitting is not possible. In addition the *nicheing* technique was also introduced to improve population diversity. Even the scoring function was improved adding the ligand internal energy contribution to the hydrogen bond and the protein–ligand van der Waals energy contributions. To date GOLD has been validated using a test set of 305 protein–ligand complexes (Nissink et al., 2002). It was found that the first-ranked solutions had reproduced the crystallographic data with an RMSD lower than 2.0 Å in almost 70% of the protein–ligand complexes.

Wang at al. (1999) proposed a rigid-body GA-based docking procedure aimed at predicting both peptide–protein and protein–protein complexes. The authors proposed a two-stage GA approach. In the first stage, the protein was searched for possible binding sites using a simplified steric energy function; in the second stage, the possible orientations of the ligand in the recognized binding sites were more finely adjusted, according to the interaction energy evaluated using a more thorough scoring function based on the AMBER force field parameters. A standard generational GA scheme was used with mutation and crossover as genetic operators. The chromosome of each solution encoded only the position and the orientation of the ligand (six genes), as the position of the target molecule was kept fixed. Eight complexes randomly chosen from the Brookhaven Protein Data Bank were used in the test set. The smallest RMSD solutions had an RMSD lower than 1.0 Å for seven ligand–protein complexes and a certain degree of correlation was found between the interaction energy and the RMSD of the solutions.

Quite recently, Budin et al. (2001) presented the FFLD program, a fragment-based strategy for docking flexible ligands into the active site of a rigid protein. The procedure started with a mapping of the active site of the protein using a library of rigid functional groups. The method proposed by Majeux et al. (1999, 2001) was used to search for the most favourable binding modes of the fragment library according to an accurate binding energy, including electrostatic solvation effects. Moreover, three fragments from the ligand were selected, and their similarity with the functional groups used to map the active site was evaluated. The ligand in a certain conformation was then positioned and oriented in the binding site by matching the position of the most similar triads of functional groups with the three selected ligand fragments. An algorithm minimizing the square of the distance between the geometric centres of both the functional groups and the ligand fragments was applied. The genetic algorithm was used to explore only the conformational space of the ligand, as each population of ligand conformations was docked into the binding site following the procedure described above. Since each ligand conformation can have more different locations in the binding site, only the location with the best score was retained. The scoring function used to rank the ligand binding modes contained three contributions as follows: a van der Waals intra-ligand energy based on a 6–12 Lennard-Jones potential using the CHARMM parameters set; a ligand–receptor polar interaction term depending on the number of hydrogen bonds and the number of unfavourable polar contacts between two donor or two acceptor atoms; and a van der Waals protein–ligand interaction term based on a modified 6–12 Lennard–Jones potential to increase its softness for lower protein–ligand atom distances. The FFLD program was applied to docking NAPAP and XK263 into thrombin and HIV-1, respectively, obtaining an averaged RMSD of <1.8 Å.

An apparent limitation in the current implementation of the procedure is that no less than three anchoring fragments are required for the ligand to be docked.

Morris et al. (1998) introduced a hybrid genetic procedure based on the so-called Lamarckian genetic algorithm (LGA) in AUTODOCK 3.0. Essentially, an LGA results from a combination of a common genetic algorithm, which acts as global optimizer, with a local search operator that minimizes the internal energy of the ligand–protein complex. With this approach, the ligand conformations docked into the protein site and obtained after energy minimization were used to update the fitness function values of the current population. The energy minimization step was performed in the phenotype space after the parameters in the chromosome had been decoded. In AUTODOCK 3.0, the use of Solis and Wets (1981) local search operator avoided the genotype decoding and the phenotype re-coding steps as it can work directly on the genotype space. The typical docking parameters are coded into the chromosome as real numbers. Only a user-defined, randomly selected, fraction of the offspring generated by crossover (two-points) and mutation operators was submitted to a local search procedure. The scoring function contained five energy contributions as follows: a 12–6 Lennard–Jones potential; a directional 12-10 hydrogen bonding term; a Coulombic electrostatic potential; a term proportional to the number of sp^3 bonds in the ligand to take into account the entropy loss due to the conformational restriction of the ligand in the protein site; and a desolvation term. The scoring function was calibrated against 30 protein–ligand complexes with known experimental inhibition constants. The whole docking procedure was tested on seven protein–ligand complexes obtaining an RMSD of < 1.15 Å against the crystallographic conformations of the ligand for all the lowest energy solutions.

Taylor and Burnett (2000) presented a similar approach when combining a genetic algorithm with local energy minimization. In the DARWIN program, a standard GA with mutation and crossover was implemented by means of a standard binary scheme to encode the conformations and positions of the ligand. The energy of the conformations generated by the genetic operators were further minimized using a gradient method, and the resulting conformers passed to a new population. The fitness of the conformations was calculated using the CHARMM force field, with a solvent contribution evaluated with a modified version of the DelPhi program (Nicholls and Honig, 1991), which calculates the electrostatic potential energy of a molecular system from finite difference solutions of the Poisson–Boltzmann equation. The performance of the implemented GA was evaluated against three protein–ligand complexes with known crystal structure. Although the procedure was able to find conformations close to the experimental ones, some other conformations with lower energy than the experimental binding mode were also retrieved. These 'false positive' solutions were attributed to limitations in the scoring function. However, several of these solutions were eliminated when the water molecules found in the crystal close to the active sites were retained.

As has been already mentioned, several requirements should be satisfied when the objective of a study is to compare different docking procedures, especially when these procedures are built using complex algorithms. Nevertheless, some authors have attempted to perform such comparisons and even though their conclusions might be questionable a brief review is given here.

Vieth et al. (1998) have attempted to compare the efficiency of molecular dynamics, Monte Carlo and genetic algorithms for docking five protein–ligand complexes. These procedures were compared by means of a modified CHARMM-based scoring function while keeping the protein rigid. The authors used two search spaces based on spheres with radii of 11 and 2.5 Å centred on the active site. The authors concluded that all the algorithms performed reasonably well, but the molecular-dynamics-simulated annealing provided the best efficiency in docking structures in the large space, whereas the genetic algorithm did better in the small search space.

Morris et al. (1998) in their work presenting the program AUTODOCK, compared the genetic algorithm and its Lamarckian version with the simulated annealing method. They found that the two GA implementations did better giving the lower averaged RMSD for a comparison among seven protein–ligand complexes.

In a previous work, Westhead et al. (1997) had compared four search methods for flexible docking inside the PRO_LEADS program. The four methods were simulated annealing, genetic algorithm, evolutionary programming and tabu search. The comparison was carried out over five complexes using the PLP scoring function proposed by Gehlhaar et al. (1995) for fast docking applications. From the results, the authors argued that the genetic algorithm was the most effective search algorithm in terms of the median energy of solutions, but tabu search was generally found to locate the assumed global minimum more reliably.

6. Protein structure prediction with genetic algorithms

Protein folding is a process by which a polypeptide chain made up of a linear sequence of amino acids adopts a well-defined three-dimensional native structure under certain physiological conditions. As the tertiary structure of a protein is responsible for its biological function, the availability of computational methods for solving the protein structure prediction problems may be of great value for modern molecular biology. Unfortunately, the computational protocols for describing protein folding are not as straightforward as one might think. From a theoretical point of view, the dynamics of the folding process can be described by the classical Newtonian equations of motion, and folding may be directly monitored by a molecular dynamics trajectory for an appropriately long time. However, even for single-domain proteins, the time-scales needed to reach their biologically active native conformations are typically on the order of 10–1000 ms, whereas current molecular dynamics simulation methods can generate trajectories over a time window of 10^{-5} ms. During the folding process, amino acid chains can adopt a very large number of conformations and, as observed by Levinthal (1968), there is a clear contradiction between the almost infinite number of possible states that the system can sample and the relatively short time-scale required for actual protein folding. It would not be feasible for any protein to try out all of its conformations on a practical time-scale. The apparent conclusion is that proteins do not fold by randomly sampling all possible conformations until the lowest free energy is encountered. Conversely, the folding process is under kinetic control, and the native state of the protein is the most accessible free energy minimum, which may be different from the global minimum.

On the other hand, Anfinsen's hypothesis (Anfinsen, 1973) suggests that the protein folding process is under thermodynamic control and consequently the native structure corresponds to the lowest free energy conformation, as demonstrated by the reversibility of the folding process of globular proteins under physiological conditions (Kim and Baldwin, 1990).

From a computational point of view, the two theories delineate completely different objectives: according to the kinetic hypothesis, the computational method employed would be able to map the shape of the energy hypersurface for predicting folding pathways; according to the thermodynamic hypothesis, the computational method employed would be able to sample the hypersurface minima searching for the global minimum. GAs belong to the category of optimizers, and their use to approach the folding problem is mainly consistent with the thermodynamics hypothesis.

Several procedures were developed to solve the folding problem by means of GAs. The principal factor affecting GA implementation is the type of protein representation used to perform the folding simulation. In principle, there are at least three different levels of complexity that have been used to represent a single-domain protein structure: lattice models, united-atom models and all-atom models.

Lattice models of single chains have been widely used in polymer physics to derive several universal properties (i.e. scaling of the size of the polymer with N, distribution of end-to-end distances, etc.) of real homopolymer chains (Orr, 1947). In spite of their intrinsic simplicity, lattice models were first introduced by Taketomi and Ueda (1975) in the context of protein folding studies, and have demonstrated an ability to capture the essence of some important protein folding components (Lau and Dill, 1990). In the simplest lattice model, N amino acids are represented by N backbone beads, representing the $C\alpha$ carbons of a protein backbone, and the side chains are not explicitly considered. The N beads occupy the vertices of a two- or three- dimensional regular lattice, depending on the simplicity of the representation of the space where the folding simulation is carried out.

In an early study, Unger and Moult (1993) proved the better performance of a GA with respect to the Metropolis Monte Carlo method for folding eight sequences varying from 20 to 64 amino acids into a simple two-dimensional lattice. The amino acids were solely of two types: hydrophilic and hydrophobic. The scoring function was very simple: -1 for each non-bonded hydrophobic–hydrophobic direct contacts. The chromosome were built encoding with a two-bits binary scheme, the angles along the chain being the bond angles and restricted to the values 0, 90, 180 and 270°.

Several other studies were published on the GA folding simulations of amino acid sequences or polymers in both two- and three- dimensional lattice models (Judson et al., 1992; Sun et al., 1998; König and Dandekar, 1999), and all of them demonstrated the effectiveness of the method also in comparison with simulated annealing, conjugate gradient minimization and random search.

The united-atom model increases the complexity of the protein structure description, which is moved from a lattice to the dihedral angle space. In this type of structure representation, the simplification involves the side chains, which are commonly replaced by one or two atoms at the centre of the original side chain. With the aim of speeding up the calculations, the dihedral angles of the side chains are selected from a library of preferred conformations.

Bowie and Eisenberg (1994) demonstrated the effectiveness of GAs by using an empirical scoring function containing contributions from the profile fit, hydrophobicity, accessible surface area, atomic overlap and the sphericalness of the structure. A library of peptide fragment conformations was used to build the starting population thus improving the local structure accuracy with respect to a random procedure. The chromosome was built with genes representing the set of dihedral angles of the structures and mutations simply modifying one angle. Moreover, mutation and crossover were likelier to occur at fragment junctions.

Dandekar and Argos (1994) successfully folded small helical proteins by using Cα backbone models of proteins in real space. The protein backbone was modelled by tacking different, φ and ψ dihedral values from a set of seven possible standard conformations, representative and frequently populated in known tertiary structures (Rooman et al., 1991). The conformations of all residues were collected together and encoded by a three-bit binary string for each residue to allocate the seven possible conformational states. The fitness function was built as a linear combination of terms which scored for van der Waals overlap, secondary structure formation, tertiary structure formation, distribution of hydrophobic residues, and hydrogen bonding. The potential function was further refined for non-helical proteins (Dandekar and Argos, 1996).

Sun (1993) described the protein with its full backbone, with one virtual atom substituting the side chain. The chromosome contained genes encoding the φ, ψ dihedral angles for the non-secondary structure residues. Binary numbers were used to encode a state on the Ramachandran map for each residue. A library of conformations derived from mono- and dipeptides were used to reduce the conformational search space. The crossover and mutation operators were allowed to act only on non-secondary structure residues, and the mutation operator was controlled by the library of peptide segment conformations. By using a simple fitness function summing hydrophobic contacts, hydrogen bonds and steric overlap, the two proteins melittin and apamin were successfully folded. More recently, the algorithm performed well on two test sets of ten small proteins (Sun 1995; Sun, et al., 1995).

One common limit of all the techniques described above that used the united-atom models is that they require the knowledge of the secondary structure of the protein.

The third level of complexity in the protein structure representation occurs with the all-atoms model.

Pedersen and Moult (1995, 1997) have used a GA successfully to perform ab initio folding of small fragments of proteins 12–22 residues long. The force field terms in the fitness function were parameterized by a potential of mean-force analysis of experimental structures. A gene was a string of φ, ψ and χ angles for each residue. The fitness function was evaluated after an extensive annealing of the side chains in close proximity to the crossover point. The mutation operator was not used while crossover points were weighted towards the positions in the peptide chain that had higher variability than in the current population.

Cui et al. (1998) described a GA-based method to compute native structures of proteins from their primary sequence. The conformational search space was restrained by eleven types of frequently occurring supersecondary structures, defined as regions in the φ/ψ plane. The residues were then assigned to one of these regions using an artificial neural network model (Sun et al., 1997). Side chain conformations were selected from a rotamers library. The potential function used had two contributions, namely a hydrophobic interaction and a van der Waals interaction term. In the implemented GA, each gene encoded the set of φ, ψ, and χ angles for every residue. The initial population was built by randomly selecting the backbone and side chains within the constrained regions imposed by the supersecondary structure assignment. The recombination of the trial conformations were carried out with a one-point crossover acting only at the bounds of the genes. Two mutation operators were simultaneously used, one producing higher perturbations but only within the predefined constraints, and the other generating small variations of $\pm 5°$ for more local searches. Despite the simplicity of the potential function used, the results were quite impressive if compared with the topologies of five small model proteins (46–120 residues long).

Ring and Cohen (1994) used a GA to sample the conformational space of loop regions using an alphabet of tetrapeptide conformations. The alphabet accounts of only four conformations (ULZJ) that partially overlapping tetrapeptides can assume, and loops can be approximated by a sequence of these conformations. Hence, in the GA implementation each gene encoded one of the four available conformations, which was represented by a 2-bits binary number. The results showed that the quality of the loops obtained from the proposed procedure degraded with the length of the loop. In addition, the procedure required an extensive minimization step to fit the chain ends to the protein scaffold after the GA convergence.

Tuffery et al. (1991) used a GA procedure to tackle the problem of predicting the side chain packing when the backbone conformation is known. With this approach, the genes simply encode the χ angles of the side chains. A rotamers library was used to describe side chain conformations, and each gene encoded one rotamer for every side chain. Molecular mechanics energy terms were used in the fitness function. Tuffery et al. (1993) also compared the effectiveness of GAs and other techniques in side chain packing. This study analysed a test set of 14 proteins of varying size and demonstrated that GA is better for large proteins, while the SMD (heuristic sparse matrix driven) method outperformed GA for small proteins.

7. Conclusions

As has been shown in this chapter, GAs have been widely used to tackle the problem of searching the conformational space of a molecular system both in constrained and unconstrained conditions. The success of GAs in this specific field has been quite impressive. The fact that the choice of conformers is a crucial step in predicting the properties of any molecular system makes the GA a valuable tool for all the current molecular modelling applications, especially for those devoted to the design of new drugs. However, in spite of their conceptual simplicity, the fine-tuning of GA-based procedures is not a straightforward process. The optimization of GA parameters may be a cumbersome issue, mostly when dealing with larger molecular systems. There is a serious suspicion that most GA procedures have been undervalued because their parameter settings were far from being optimal. Nonetheless, GAs have shown their potential usefulness also in cases of complex search spaces suggesting that future improvements in their effectiveness may be expected from further work on their optimization protocols. Another important development perspective continues to be the hybridization of GAs with other computational methods. The combination of a global optimization method such as GA with a local search algorithm or other heuristic method, has proven to be extremely useful. Again to follow useful research directions in the performance exploration of more complex hybrid systems, better and probably more time-consuming optimization protocols will be required to deal with a growing number of parameters and design solutions.

References

Allen, F.H., Davies, J.E., Galloy, J.J., Johnson, O., Kennard, O., Macrae, C.F., Mitchell, E.M., Mitchell, G.F., Smith, J.M., Watson, D.G., 1991. The development of Versions 3 and 4 of the Cambridge structural database system. J Chem. Inf. Comput. Sci. 31, 187–204.

Allinger, N.L., 1977. Conformational Analysis.130.MM2. A hydrocarbon force field utilizing V1 and V2 torsional terms. J. Am. Chem. Soc. 99, 8127–8134.

Anfinsen, C.B., 1973. Principles that govern the folding of protein chains. Science 181, 223–230.

Baylay, M.J., Jones, G., Willet, P., Williamson, M.P., 1998. GENFOLD: a genetic algorithm for folding protein structure using NMR restraints. Protein Sci. 7, 491–499.

Barnum, D., Greene, J., Smellie, A., Sprague, P., 1996. Identification of common functional configurations among molecules. J. Chem. Inf. Comput. Sci. 36, 563–571.

Beckers, M.L.M., Buydens, L.M.C., Pikkemaat, J.A., Altona, C., 1997. Application of a genetic algorithm in the conformational analysis of methylene-acetal-linked thymine dimers in DNA: comparison with distance geometry calculations. J. Biomol. NMR 9, 25–34.

Blundell, T.L., Sibanda, B.L., Sternberg, M.J.E., Thornton, J.M., 1987. Knowledge-based prediction of protein structures and the design of novel molecules. Nature (London) 326, 347–352.

Böhm, H.J., 1992. LUDI: rule-based automatic design of new substituents for enzyme inhibitor leads. J. Comput.-Aided Mol. Des. 6, 593–606.

Böhm, H.J., 1994. The development of a simple scoring function to estimate the binding constant for a protein–ligand complex of known three-dimensional structure. J Comput.-Aided Mol. Des. 8, 243–256.

Böhm, H.J., Stahl, M., 2002. The use of scoring functions in drug discovery applications. In: Lipkowitz, K. B., Boyd, D. B. (Eds.), Chapter 2 in Reviews in Computational Chemistry, Vol. 17. Wiley-VCH, New York.

Bowie, J.U., Eisenberg, D., 1994. An evolutionary approach to folding small α-helical proteins that use sequence information and an empirical guiding fitness function. Proc. Natl Acad. Sci. USA 91, 4436–4440.

Brodmeier, T., Pretsch, E., 1994. Application of genetic algorithms in molecular modelling. J. Comput. Chem. 15, 588–595.

Brooks, B.R., Bruccoleri, R.E., Olafson, B.D., States, D.J., Swaminathan, S., Karplus, M., 1983. CHARMM: a program for macromolecular energy, minimization, and dynamics calculations. J. Comput. Chem. 4, 187–217.

Budin, N., Majeux, N., Caflisch, A., 2001. Fragment-based flexible ligand docking by evolutionary optimization. Biol. Chem. 382, 1365–1372.

Clark, K.P., Ajay, J., 1995. Flexible ligand docking without parameter adjustment across four ligand–receptor complexes. J. Comput. Chem. 16, 1210–1226.

Clark, M., Cramer, R.D. III, Van Opdenbosch, N., 1989. Validation of the general purpose TRIPOS 5.2 force field. J. Comput. Chem. 10, 982–1012.

Clark, D.E., Jones, G., Willet, P., Kenny, P.W., 1992. Pharmacophoric pattern matching in files of three-dimensional chemical structures: use of bounded distance matrices for the representation and searching of conformationally-flexible molecules. J. Mol. Graphics 10, 194–204.

Clark, D.E., Jones, G., Willet, P., Kenny, P.W., Glen, R.C., 1994. Pharmacophoric pattern matching in files of three-dimensional chemical structures: comparison of conformational-searching algorithms for flexible searching. J. Chem. Inf. Comput. Sci. 34, 197–206.

Crippen, G.M., 1978. Rapid calculations of coordinates from distance matrices. J. Comput. Phys. 26, 449–452.

Cui, Y., Chen, R.S., Wong, W.H., 1998. Protein folding simulation with genetic algorithm and supersecondary structure constraints. Proteins: Struct. Funct. Genet. 31, 247–257.

Dandekar, T., Argos, P., 1994. Folding the main-chain of small proteins with the genetic algorithm. J. Mol. Biol. 236, 844–861.

Dandekar, T., Argos, P., 1996. Identifying the tertiary fold of small proteins with different topologies from sequence and secondary structure using the genetic algorithm and extended criteria specific for strand regions. J. Mol. Biol. 256, 645–660.

Davis, L., 1991a. Handbook of Genetic Algorithms, Van Nostrand Reinhold, New York.

Davis, L., 1991b. Handbook of Genetic Algorithms, Van Nostrand Reinhold, New York, p. 385.

Dewar, M.J.S., Zoebisch, E.G., Healy, E.F., Stewart, J.J.P., 1985. Austin model 1, newly parameterized MNDO version. J. Am. Chem. Soc. 107, 3902–3909.

Dennis, J.E., Torczon, V.J., 1991. Direct search methods on parallel machines. SIAM J. Optimization 1, 448–474.

Dixon, J.S., 1993. Flexible docking of ligand to receptor site using genetic algorithm. In: Wermuth, C.G., (Ed.), Trends in QSAR and Molecular Modelling 92: Proceedings of the 9th European Symposium on Structure—Activity Relationships: QSAR and Molecular Modelling, ESCOM Science Publishers, Leiden, The Nederlands, pp. 412–413.

Fesik, S.W., 1991. NMR studies of molecular complexes as a tool in drug design. J. Med. Chem. 34, 2937–2945.

Flury, B., 1988. Principal Component Analysis and Related Multivariate Models, Wiley, New York, pp. 5–50.

Forrest, S., 1993. Genetic algorithms: principles of natural selection applied to computation. Science 261, 872–878.

Gehlhaar, D.K., Verkhivker, G.M., Rejto, P.A., Sherman, C.J., Fogel, D.B., Fogel, L.J., Freer, S.T., 1995. Molecular recognition of the inhibitor AG-1343 by HIV-1 protease. Conformationally flexible docking by evolutionary programming. Chem. Biol. 2, 317–324.

Goldberg, D.E., 1989a. Genetic Algorithms in Search Optimization & Machine Learning, Addison-Wesley, Reading, MA.

Goldberg, D.E., 1989b. Genetic Algorithms in Search Optimization & Machine Learning, Addison-Wesley, Reading, MA, pp. 170–175.

Hasel, W., Hendrickson, T.F., Still, W.C., 1988. A rapid approximation to the solvent accessible surface areas of atoms. Tetrahedron Comput. Meth. 1, 103–116.

Havel, T.F., 1991. An evaluation of computational strategies for use in the determination of protein structure from distance restraints obtained by nuclear magnetic resonance. Prog. Biophys. Mol. Biol. 56, 43–78.

Hermann, F., Suhai, S., 1995. Energy minimization of peptide analogues using genetic algorithms. J. Comput. Chem. 16, 1434–1444.

Hurst, T., 1994. Flexible 3D searching: the directed tweak technique. J. Chem. Inf. Comp. Sci. 34, 190–196.

Jin, A.Y., Leung, F.Y., Weaver, D.F., 1999. Three variations of genetic algorithm for searching biomolecular conformation space: Comparison of GAP 1.0, 2.0 and 3.0. J. Comput. Chem. 20, 1329–1342.

Jones, G., Willett, P., Glen, R.C., 1995a. A genetic algorithm for flexible molecular overlay and pharmacophore elucidation. J. Comput.-Aided Mol. Des. 9, 532–549.

Jones, G., Willett, P., Glen, R.C., 1995b. Molecular recognition of receptor sites using a genetic algorithm with a description of solvation. J. Mol. Biol. 254, 43–53.

Jones, G., Willett, P., Glen, R.C., 1996. Genetic algorthms for chemical structure handling and molecular recognition. In: Devillers, J., (Ed.), Chapter 9 in Genetic Algorithms in Molecular Modeling, Academic Press, London, pp. 211–242.

Jones, G., Willett, P., Glen, R.C., Leach, A.R., Taylor, R., 1997. Development and validation of a genetic algorithm for flexible docking. J. Mol. Biol. 267, 727–748.

Jorgensen, W.L., Tirado-Rives, J., 1988. The OPLS (optimized potentials for liquid simulations) potential functions for proteins, energy minimizations for crystal of cyclic peptides and crambin. J. Am. Chem. Soc. 110, 1657–1666.

Jorgensen, W.L., Maxwell, D.S., Tirado-Rives, J., 1996. Development and testing of the OPLS all-atom force field on conformational energetics and properties of organic liquids. J. Am. Chem. Soc. 118, 11225–11236.

Judson, R.S., 1997. Genetic algorithms and their use in chemistry. In: Lipkowitz, K. B., Boyd, D. B. (Eds.), Chapter 1 in Reviews in Computational Chemistry, Vol. 10. VCH Publishers, New York.

Judson, R.S., Colvin, M.E., Meza, J.C., Huffer, A., Gutierrez, D., 1992. Do intelligent configuration search techniques outperform random search for large molecules. Int. J. Quantum Chem. 44, 277–290.

Judson, R.S., Jaeger, E.P., Treasurywala, A.M., Peterson, M.L., 1993. Conformational searching methods for small molecules II. Genetic algorithm approach. J. Comput. Chem. 14, 1407–1414.

Judson, R.S., Jaeger, E.P., Treasurywala, A.M., 1994. A genetic algorithm based method for docking flexible molecules. J. Mol. Struct. (THEOCHEM) 308, 191–206.

Judson, R.S., Tan, Y.T., Mori, E., Melius, C., Jaeger, E.P., Treasurywala, A.M., Mathiowetz, A., 1995. Docking flexible molecules: a case study of three proteins. J. Comput. Chem. 16, 1405–1419.

van Kampen, A.H.C., Buydens, L.M.C., Lucasius, C.B., Blommers, M.J.J., 1996. Optimization of metrics matrix embedding by genetic algorithms. J. Biomol. NMR 7, 214–224.

van Kampen, A.H.C., Buydens, L.M.C., 1997. The ineffectiveness of recombination in genetic algorithm for the structure elucidation of a heptapeptide in torsion angle space. A comparison to simulated annealing. Chemom. Intell. Lab. Syst. 36, 141–152.

Kim, P.S., Baldwin, R.L., 1990. Intermediates in the folding reactions of small proteins. Annu. Rev. Biochem. 59, 631–660.

Kirkpatrick, S., Gelatt, C.D., Vecchi, M.P., 1983. Optimization by simulated annealing. Science 220, 671–680.

König, R., Dandekar, T., 1999. Improving genetic algorithms for protein folding simulations by systematic crossover. BioSystems 50, 17–25.

Kuntz, I.D., 1992. Structure-based strategies for drug design and discovery. Science 257, 1078–1082.

Lau, K.F., Dill, K.A., 1990. Theory for protein mutability and biogenesis. Proc. Natl. Acad. Sci. USA 87, 638–642.

Lawler, E., Wood, D., 1966. Branch and bound methods: a survey. Oper. Res. 14, 699–719.

Leach, A.R., 1991. In: Lipkowitz, K. B., Boyd, D. B. (Eds.), A Survey of Methods for Searching the Conformational Space of Small and Medium-Sized Molecules, vol. 2. VCH, New York.

Levinthal, C., 1968. Are there pathways to protein foldings. J. Chem. Phys. 65, 44–45.

Lybrand, T.P., 1995. Ligand–protein docking and rational drug design. Curr. Opin. Struct. Biol. 5, 224–228.

Lucasius, C.B., Kateman, G., 1994. Understanding and using genetic alghoritms. Part 2. Representation, configuration and hybridization. Chemom. Intell. Lab. Syst. 25, 99–146.

Majeux, N., Scarsi, M., Apostolakis, J., Ehrhardt, C., Caflisch, A., 1999. Exhaustive docking of molecular fragments on protein binding sites with electrostatic solvation. Proteins. Struct. Funct. Genet. 37, 88–105.

Majeux, N., Scarsi, M., Caflisch, A., 2001. Efficient electrostatic solvation model for protein-fragment docking. Proteins: Struct. Funct. Genet. 42, 256–268.

Markley, J.L., 1989. Two-dimensional nuclear magnetic resonance spectroscopy of proteins: an overview. Meth. Enzymol. 176, 12–64.

Martin, Y., Bures, M., Dahaner, E., DeLazzer, J., Lico, I., Pavlik, P., 1993. A fast approach to pharmacophore mapping and its application to dopaminergic and benzodiazepine agonists. J. Comput.-Aided Mol. Des. 7, 83–102.

McGarrah, D.B., Judson, R.S., 1993. Analysis of the genetic algorithm method of molecular conformation determination. J. Comput. Chem. 14, 1385–1395.

Mekenyan, O., Dimitrov, D., Nikolova, N., Karabunarliev, S., 1999. Conformational coverage by a genetic algorithm. J. Chem. Inf. Comput. Sci. 39, 997–1016.

Metropolis, N., Rosenbluth, A., Rosenbluth, A., Teller, A., Teller, E., 1953. Equation of state calculations by fast computing machines. J. Chem. Phys. 21, 1087–1092.

Meza, J.C., Judson, R.S., Faulkner, T.R., Treasurywala, A.M., 1996. A comparison of a direct search method and a genetic algorithm for conformational searching. J. Comput. Chem. 17, 1142–1151.

Mohamadi, F., Richards, N.J.G., Guida, W.G., Liskamp, R., Lipton, M., Caufield, C., Chang, G., Hendrickson, T., Still, W.C., 1990. MacroModel—an integrated software system for modeling organic and bioinorganic molecules using molecular mechanics. J. Comput. Chem. 11, 440–467.

Morris, G.M., Goodsell, D.S., Halliday, R.S., Huey, R., Hart, W.E., Belew, R.K., Olson, A.J., 1998. Automated docking using a Lamarckian genetic algorithm and empirical binding free energy function. J. Comput. Chem. 19, 1639–1662.

Muegge, I., Rarey, M., 2001. Small molecule docking and scoring. In: Lipkowitz, K. B., Boyd, D. B. (Eds.), Chapter 1 in Reviews in Computational Chemistry, Vol. 17. Wiley-VCH, New York.

Nicholls, A., Hönig, B., 1991. A rapid finite difference algorithm, utilizing successive over-relaxation to solve Poisson–Boltzmann equations. J. Comput. Chem. 12, 435–445.

Nissink, J.W.M., Murray, C., Hartshorn, M., Verdonk, M.L., Cole, J.C., Taylor, R., 2002. A new test set for validating predictions of protein–ligand interaction. Proteins: Struct. Funct. Genet. 49, 457–471.

Oprea, T.I., Marshall, G.R., 1998. Receptor based prediction of binding affinities. Perspect. Drug Disc. Des. 9–11, 35–61.

Orr, W.J.C., 1947. Statistical treatment of polymer solutions at infinite dilution. Trans. Faraday Soc. 43, 12–27.

Oshiro, C.M., Kuntz, I.D., Dixon, J.S., 1995. Flexible ligand docking using a genetic algorithm. J. Comput.-Aided Mol. Des. 9, 113–130.

Pak, Y., Wang, S.J., 2000. Application of a molecular dynamics simulation method with a generalized effective potential to the flexible molecular docking problems. J. Phys. Chem. B. 104, 354–359.

Patel, Y., Gillet, V.J., Bravi, G., Leach, A.R., 2002. A comparison of the pharmacophore identification programs: Catalyst, DISCO and GASP. J. Comput-Aided Drug Des. 16, 653–681.

Payne, A.W.R., Glen, R.C., 1993. Molecular recognition using a binary genetic search algorithm. J. Mol. Graphics 11, 74–91.

Pedersen, J.T., Moult, J., 1995. Ab initio structure prediction for small polypeptides and protein fragments using genetic algorithms. Proteins: Struct. Funct. Genet. 23, 454–460.

Pedersen, J.T., Moult, J., 1997. Protein folding simulation with genetic algorithms and a detailed molecular description. J. Mol. Biol. 269, 249–269.

Pearlman, D.A., Case, D.A., Caldwell, J.W., Ross, W.S., Cheatham, T.E. III, DeBolt, S., Ferguson, D., Seibel, G., Kollman, P., 1995. AMBER, a package of computer programs for applying molecular mechanics, normal mode analysis, molecular dynamics and free energy calculations to simulate the structural and energetic properties of molecules. Comput. Phys. Commun. 91, 1–41.

Plackett, R.L., Burman, J.P., 1946. The design of optimum multifactorial experiments. Biometrika 33, 305–325.

Priestle, J.P., Paris, C.G., 1996. Experimental techniques and data banks. In: Cohen, N. C., (Ed.), Chapter 5 in Guidebook on Molecular Modeling in Drug Design, Academic Press, New York, pp. 139–149.

Rarey, M., Kramer, B., Lengauer, T., Klebe, G., 1996. A fast flexible docking method using an incremental construction algorithm. J. Mol. Biol. 261, 470–489.

Ring, C.S., Cohen, F.E., 1994. Conformational sampling of loop structures using genetic algorithms. Isr. J. Chem. 34, 245–252.

Rooman, M.J., Kocher, J-P A., Wodak, S.J., 1991. Prediction of protein backbone conformation based on seven structural assignements. J. Mol. Biol. 22, 1961–979.

Sanctuary, B.C., 2000. Structure determination by NMR spectroscopy. In: Mannhold, R., Kubinyi, H., Timmerman, H. (Eds.), Chapter 10 in Evolutionary Algorithms in Molecular Design. Methods and Principles in Medicinal Chemistry, Vol. 8. Wiley-VCH, Weinheim.

Sanderson, P.N., Glen, R.C., Payne, A.W.R., Hudson, B.D., Heide, C., Tranter, G.E., Doyle, P.M., Harris, C.J., 1994. Characterization of the solution conformation of a cyclic RGD peptide analogue by NMR spectroscopy allied with a genetic algorithm approach and constrained molecular dynamics. Int. J. Pept. Protein Res. 43, 588–596.

Solis, F.J., Wets, J.B., 1981. Minimisation by random search techniques. Math. Oper. Res. 6, 19–30.

Stewart, J.J., 1990. MOPAC: a semiempirical molecular orbital program. J. Comput.-Aided Mol. Des. 4, 1–105.

Sun, S., 1993. Reduced representation model of protein structure prediction: statistical potential and genetic algorithms. Protein Sci. 2, 762–785.

Sun, S., 1995. A genetic algorithm that seeks native states of peptides and proteins. Biophys J. 69, 340–355.

Sun, S., Thomas, P.D., Dill, K.A., 1995. A simple protein folding algorithm using a binary code and secondary structure constraints. Protein Engng 8, 769–78.

Sun, Z., Rao, X., Peng, L., Xu, D., 1997. Prediction of protein supersecondary structures based on the artificial neural network method. Protein Engng 10, 763–769.

Sun, Z., Xia, X., Guo, Q., Xu, D., 1998. Protein structure prediction in a 210-type lattice model: parameter optimization in the genetic algorithm using orthogonal arrays. J. Protein Chem. 18, 39–46.

Taketomi, H.H., Ueda, Y.G.N., 1975. Studies on protein folding, unfolding, and fluctuations by computer simulation. Int. J. Pept. Protein. Res. 7, 445–459.

Taylor, J.S., Burnett, R.M., 2000. DARWIN: a program for docking flexible molecules. Proteins: Struct. Funct. Genet. 41, 173–191.

Taylor, R.D., Jewsbury, P.J., Essex, J.W., 2002. A review of protein–small molecule docking methods. J. Comput-Aided Mol. Des. 16, 151–166.

Tuffery, P., Etchebest, C., Hazout, S., Laverly, R., 1991. A new approach to the rapid determination of protein side-chain conformations. J. Biomol. Struct. Dyn. 8, 1267–1289.

Tuffery, P., Etchebest, C., Hazout, S., Laverly, R., 1993. A critical comparison of search algorithms applied to the optimization of protein side-chain conformations. J. Comput. Chem. 14, 790–798.

Unger, R., Moult, J., 1993. Genetic algorithms for protein folding simulations. J. Mol. Biol. 231, 75–81.

Vedani, A., Zbinden, P., Snyder, J.P., Greenidge, P.A., 1995. Pseudoreceptor modeling: the construction of three-dimensional receptor surrogates. J. Am. Chem. Soc. 117, 4987–4994.

Verkhivker, G., Appelt, K., Freer, S.T., Villafranca, J.E., 1995. Empirically free energy calculations of ligand–protein crystallographic complexes: I. Knowledge-based ligand–protein interaction potentials applied to the prediction of HIV-1 protease binding affinity. Prot. Engng 8, 677–691.

Vieth, M., Hirst, J.D., Dominy, B.N., Daigler, H., Brooks III, C.L., 1998. Assessing search strategies for flexible docking. J. Comput. Chem. 19, 1623–1631.

Wallqvist, A., Jerningan, R.L., Covell, D.G., 1995. A preference-based free-energy parameterization of enzyme–inhibitor binding: application to HIV-1 protease inhibitor design. Protein Sci. 4, 1881–1903.

Wang, J., Hou, T., Chen, L., Xu, X., 1999. Automated docking of peptides and proteins by genetic algorithm. Chemom. Intell. Lab. Syst. 45, 281–286.

Weiner, S.J., Kollman, P.A., Case, D.A., Singh, U.C., Ghio, C., Alagona, G., Profeta, S. Jr, Weiner, P., 1984. A new force field for molecular mechanical simulation of nucleic acids and proteins. J. Am. Chem. Soc. 106, 765–784.

Weiner, S.J., Kollman, P.A., Nguyen, D.T., Case, D.A., 1986. New force field for simulations of proteins and nucleic acids. J. Comput. Chem. 7, 230–252.

Wehrens, R., Pretsch, E., Buydens, L.M.C., 1998. Quality Criteria of Genetic Algorithms for Structure Optimization. J. Chem. Inf. Comput. Sci. 38, 151–157.

Wehrens, R., Pretsch, E., Buydens, L.M.C., 1999. The quality of optimization by genetic algorithms. Anal. Chim. Acta 388, 265–271.

Westhead, D.R., Clark, D.E., Murray, C.W., 1997. A comparison of heuristic search algorithms for molecular docking. J. Comput.-Aided Mol. Des. 11, 209–228.

CHAPTER 5

MobyDigs: software for regression and classification models by genetic algorithms

Roberto Todeschini, Viviana Consonni, Andrea Mauri, Manuela Pavan

Milano Chemometrics and QSAR Research Group, Department of Environmental Sciences, P.za della Scienza, I-20126 Milan, Italy

1. Introduction

Genetic algorithms (GAs) are an evolutionary method widely used for complex optimisation problems in several fields such as robotics, chemistry and QSAR (Goldberg, 1989; Wehrens and Buydens, 1998). A specific application of GA is variable subset selection (GA-VSS) (Leardi et al., 1992). Since complex systems are described by several variables, a major goal in system analysis is the extraction of relevant information, together with the exclusion of redundant and noisy information. However, an exhaustive search of all possible solutions is not feasible.

In regression and classification modelling the most relevant variables with respect to the specific problem of interest are searched for by different selection strategies. GAs perform this selection by considering populations of models generated through a reproduction process and optimised according to a defined *objective function* related to model quality.

In the following, the GA strategy for VSS as implemented in the MobyDigs software (Todeschini, 2002) is presented.

The procedure is based on the evolution of a *population* of models, i.e. a set of ranked models according to some objective function. In GA terminology, each population individual is called *chromosome* and is a binary vector, where each position (a *gene*) corresponds to a variable (1 if included in the model, 0 otherwise). Each chromosome represents a model given by a subset of variables.

Once the objective function to optimise is defined, the model population size P (e.g. $P = 100$) and the maximum number L of allowed variables in a model (e.g. $L = 5$) have to be defined; the minimum number of allowed variables is usually assumed equal to one. Moreover, the user must define both the *crossover probability* and the *mutation probability*.

Data Handling in Science and Technology, Volume 23
ISSN: 0922-3487

DOI: 10.1016/S0922-3487(03)23005-7

At this point GA evolution starts, based on three main steps:

1. *Random initialisation of the population*. The model population is built initially by random models with a number of variables between 1 and L. The value of the selected objective function of each model is calculated in a process called *evaluation*. The models are then ordered with respect to the selected objective function—model quality (the best model is in first place in the population, the worst at position P).
2. *Crossover*. From the actual population, pairs of models are selected (randomly or with a probability function of their quality). Then, from each pair of selected models (*parents*), a new model is generated, preserving the common characteristics of the parents (i.e. variables excluded in both models remain excluded, variables included in both models remain included) and mixing the opposite characteristics according to the crossover probability. If the generated offspring coincides with one of the individuals already present in the actual population, it is rejected; otherwise, it is evaluated. If the objective function value is better than the worst value in the population, the model is included in the population, in the place corresponding to its rank; otherwise, it is no longer considered. This procedure is repeated for several pairs.
3. *Mutation*. After a number of crossover iterations, the population proceeds through the mutation process. This means that for each individual of the population every gene is randomly changed into its opposite or left unchanged. Mutated individuals are evaluated and included in the population if their quality is acceptable. This process is controlled by mutation probability which is commonly set at low values; thus allowing only a few mutations and new individuals not too far away from the generating individual.

Unlike the classical GA, crossover and mutation steps are kept disjoint in MobyDigs approach. Population crossover and mutation are alternatively repeated until a stop condition is encountered (e.g. a user-defined maximum number of iterations) or the process is ended arbitrarily.

An important characteristic of the GA–VSS method is that it provides not a single model but a population of acceptable models; this characteristic, sometimes considered a disadvantage, makes the evaluation of variable relationships with response from different points of view possible.

MobyDigs is a software which extends the genetic strategy based on the evolution of a single population of models to a more complex genetic strategy based on the evolution of more than one population. These populations evolve independently from each other and, after a number of iterations, can be combined according to different criteria, thus obtaining a new population with different evolutionary capabilities.

Models can be optimised by different statistical parameters to measure their quality. Moreover, the genetic parameters that control the population evolution can be changed during the model searching. Mutation and crossover probabilities are tailored by this strategy and different criteria to put the variables in quarantine are proposed.

Once the best models from one or more optimised populations are obtained, bootstrap techniques can be used for further validation, response predictions and leverage values can be obtained easily, and consensus models defined.

Finally, a new distance measure between models is adopted to check the similarity/diversity among the final models.

2. Population definition

The GA–VSS method produces not just one solution, but a set of acceptable solutions, a population achieved through an evolutionary process. Each solution corresponds to a model constituted by some of the original available variables.

As GA-VSS is a stochastic optimisation method, there is no guarantee that the whole model space will be explored adequately or, therefore, that the best solution can be found. This is particularly relevant when there is a huge number of original variables available, the space of all the possible models being too large to be explored in a reasonable time. Moreover, if the evolutionary process converges towards the region of a single model space, the final population models differ in only a few variables and appear very similar, and thus provide the same information on the studied response. Therefore, to avoid these drawbacks, the independent process and optimisation of more than one population is the strategy we propose in MobyDigs in order to explore the model space deeply and maintain maximum diversity among the final models selected. This multipopulation approach reveals to be particularly useful in QSAR when there are different logical sets of molecular descriptors, encoding diverse information and there is the need to preserve diverse sources of information. This approach enhances the GA exploration ability.

To set the initial population the original set of variables is split into several subsets, each variable subset constituting the specific genetic fingerprint of each population. The initial partition does not allow a variable to be in more than one population.

MobyDigs deals with disjoint initial populations (up to 10) settled by a random splitting of the original variables or a user-defined selection. When there is no logical variable partition, random splitting into different populations is suggested; this allows the finding of local optima that can then be joined into a single new population for a final search for the best absolute models.

3. Tabu list

When a huge number of original variables are available, there is a very high probability that some will be completely useless for modelling purposes due to their distributional properties, high correlation with other variables or their being completely uncorrelated with the studied response. Thus, it is suggested that a preliminary rough screening of the original variables be made to reduce the number of variables undergoing analysis. Variables not fulfilling some statistical requirements are excluded from all the populations and included in the so-called *Tabu list*.

Variables in the Tabu list take no part in the evolutionary process, but they can be recovered at any time.

Five different criteria define Tabu list variables, and the user can choose one or more of these criteria to recognise the variables, putting them on the Tabu list if at least one criterion is fulfilled.

The first four criteria analyse one variable at a time, while the fifth criterion evaluates the pair correlations of the independent variables.

The criteria for the Tabu list are:

1. The Q^2 value of the univariate regression model (x, y) is less than a user-defined threshold (default 0).
2. The R^2 value of the univariate regression model (x, y) is less than a user-defined threshold (default 0.1).
3. The kurtosis of the variable is greater than a user-defined threshold (default 8).
4. The standardised entropy of the variable is less than a user-defined threshold (default 0.3).
5. The correlation between two variables (x, x) is greater than a user-defined threshold (default 0.9).

The first two criteria catch those variables that, alone, do not have an acceptable linear relationship with the Y response; in particular, Q^2 is the leave-one-out cross-validated explained variance and R^2 is the determination coefficient.

The third criterion evaluates the variable distributional properties. Unlike standard deviation, which depends on the numeric scale of the variable, kurtosis represents the 'peaknedness' of the variable, taking values between -1 and infinity. For example, kurtosis equalling -1, 1.8 and 3 represents typical values, independent of the variable scale, for bimodal, uniform and Gaussian distributions, respectively. Values higher than 3 correspond to leptokurtic distributions; very high values represent a peak distribution. Therefore, a variable with a high kurtosis value is constant or near-constant and will be included in the Tabu list.

The fourth criterion is based on the Shannon definition of standardised entropy (values between 0 and 1), which represents the quantity of information encoded by the variable. The lower the value, the lower the information content of the variable. Like the kurtosis criterion, the entropy-based criterion allows the detection of constant or near-constant variables.

The fifth criterion calculates all the pair correlations between variables belonging to the same population and, for each pair of variables with a correlation greater than the threshold value, includes one of them in the Tabu list.

4. Random variables

The addition of artificial random variables to the original ones is a way of checking whether or not the evolutionary procedure is selecting random models (Jouan-Rimbaud et al., 1996), i.e. models with at least one variable correlated randomly with response.

In fact, when simulated random variables start to appear in the evolving model population it means that the allowed maximum model size can no longer be increased since optimal complexity has been reached.

Random variables, both normally and uniformly distributed, can be generated automatically (up to 200). Random variables generated with a Y response correlation higher than the threshold (up to 0.1) are rejected and substituted by new ones; in fact, our definition of random variables precludes their encoding of useful information about response.

5. Parent selection

In GA terminology, parent selection can be defined as a genetic operator choosing an individual from the current population to be used as a parent. The different operators can select any single individual.

The simplest operator is *random selection* of the individual, in which quality is not taken into account . However, the most common operator is the so-called *roulette wheel* (RW) which is biased towards the best individuals, where the chance of an individual selected is a function of its quality (or rank). In this case the concept of quality survival comes into play by applying *selection pressure*. Additional pressure can be introduced by using the RW operator several times to produce a *tournament selection* of a subset of individuals: the best individual is then chosen as the selected parent.

MobyDigs allows the user to modulate the selection pressure by a user-defined parameter B, taking a value of 0–1. A B value equal to zero settles the random selection, while a B value equal to 0.5 gives the classical RW. For B values decreasing from 0.5 to 0, the importance of the individual quality in parent selection is gradually smoothed. By increasing B values from 0.5 to 1, the number of RW repetitions increases from 1 to 5, thus allowing a tournament selection (Table 1).

6. Crossover/mutation trade-off

The GA search is characterised by two evolutive processes, crossover and mutation, which are responsible for the generation of new individuals. In MobyDigs such processes are controlled by a user-defined parameter T that performs a trade-off between crossover and mutation, taking values between 0 and 1.

Table 1
Different parent selection operators on the basis of the selection pressure defined by the B parameter

B parameter	RW repetitions	Selection
$B = 0$	1	Random
$B < 0.5$	1	RW smoothed
$B = 0.5$	1	RW
$0.5 < B < 1$	2, 3, 4,..., 5	Tournament

(a) $T = 0$ enhances the pseudo-deterministic part of the GA procedure. The new models are searched for only by the crossover procedure (genetic part of the GA procedure) among the models actually present in the population (no new available variables can enter a model);
(b) $T = 1$ enhances the stochastic part of the GA algorithm. The new models are only searched for by the mutation procedure (evolutionary part of the GA procedure).
(c) $0 < T < 1$: both crossover and mutation are taken into account; in particular, for $T = 0.5$ the role played by the two processes is equally balanced.

The crossover consists of selecting pairs of individuals (*parents*) and combining them in order to generate new individuals (*offsprings*). The proposed GA exploits the selected pair of parents twice for each generated offspring. This preserves the common characteristics of the parents and mixes the opposite characteristics according to crossover probability.

Let Parent 1 and Parent 2 be the selected parents:

Parent 1: 0 1 **0** 0 **1** 1 0 0
Parent 2: 0 1 **1** 0 **0** 1 0 0

Each offspring derived from these two parents will preserve their common genetic part, being a chromosome like 0 1 **?** 0 **?** 1 0 0.

Offspring generation is performed by using one parent at a time and analysing its changeable genes by comparing a random number with the crossover probability (*unbiased uniform crossover*). For each variable included in one parent but not in the other, a number is randomly extracted and compared with the crossover probability: if the random number is lower than the crossover probability, then the variable is included if not present in the parent ($0 \rightarrow 1$) or excluded if present in the parent ($1 \rightarrow 0$) otherwise it remains unchanged.

In the MobyDigs approach, duplicate individuals are not allowed in order to avoid any elitist aspect, i.e. the replication of the same individuals in the population. Therefore, each pair of parents should have the maximum probability of producing offsprings that differ from the parents they are derived from. This can be achieved by using a crossover probability equal to 0.5.

The crossover probability is calculated as the following:

$$P^X = \begin{cases} 0.5 & \text{for} \quad 0 \leq T \leq 0.5 \\ 1 - 0.25^{(1-T)} & \text{for} \quad 0.5 < T \leq 1 \end{cases} \tag{1}$$

where T is the crossover/mutation trade-off parameter.

Mutation is a mechanism that produces, by a completely random process, new genetic material during population evolution. For each individual present in the population, p random numbers are tried, p being the number of individual genes, and one at a time each is compared with the defined mutation probability: each gene remains unchanged if the corresponding random number exceeds the mutation probability, otherwise, it is changed from 0 to 1 or vice versa. Low values of mutation probability allow only a few mutations, thus obtaining new chromosomes not too different from the generating chromosomes.

When several candidate variables were allowed (for example, 1000) and models with a low number of selected variables are required (for example, 4–6 variables per model), we observed that mutation probability should be distinguished as two different kinds of mutation, with two different values. An *input mutation probability* ($P_{\mathrm{IN}}^{\mathrm{MUT}}$) is aimed at controlling the selection of variables which are actually outside the model and which can be put into the model; an *output mutation probability* ($P_{\mathrm{OUT}}^{\mathrm{MUT}}$) aimed at controlling the exclusion of variables actually present in the model.

In the MobyDigs approach, the input mutation probability has to be low in order to avoid the selection of very high number of variables, which would exceed the maximum number of allowed variables in a model. On the contrary, the output mutation probability has to be relatively high in order to allow the exclusion of variables actually present in the model. The contemporary use of two different kinds of mutation probability seems to allow an efficient evolutionary mutation step.

$$P_{\mathrm{IN}}^{\mathrm{MUT}} = \left[\frac{1}{p} + \frac{L-k}{p}\right]T \tag{2}$$

$$P_{\mathrm{OUT}}^{\mathrm{MUT}} = \left[\frac{1+k}{L+k}\right]T \tag{3}$$

where L is the maximum user-defined model size, p the total number of available variables, k the actual model size and T the trade-off parameter.

Note that for a saturated model ($k = L$), the input mutation probability takes the minimum value proportional to $1/p$, while for the most simple model ($k = 1$), it takes the maximum value proportional to L/p. Unlike the input mutation probability, the output mutation probability does not depend on the total number of available variables p and increases with the number k of variables present in the model.

Figs. 1–3 show the trends of the input and output mutation probability, together with the crossover probability for different values of the crossover/mutation trade-off parameter (T).

Fig. 1 shows the crossover, input and output mutation probability values for a population with a total number of available variables equal to 10, model size equal to 1 and maximum user-defined model size equal to 5. It can be observed that in this case, i.e. the simplest model, the input mutation probability is greater than the output mutation probability, as is reasonable.

Fig. 2 shows crossover, input and output mutation probability values for a population with a total number of available variables equal to 10, model size equal to 4 and maximum user-defined model size equal to 5. It can be observed that in this case, a quite saturated model, the output mutation probability is greater than the input mutation probability.

Fig. 3 shows crossover, input and output mutation probability values for a population with a total number of available variables equal to 100, model size equal to 4 and maximum user-defined model size equal to 5. Comparing Fig. 3 with Fig. 2, it can be highlighted that crossover probability as well as output mutation probability, does not

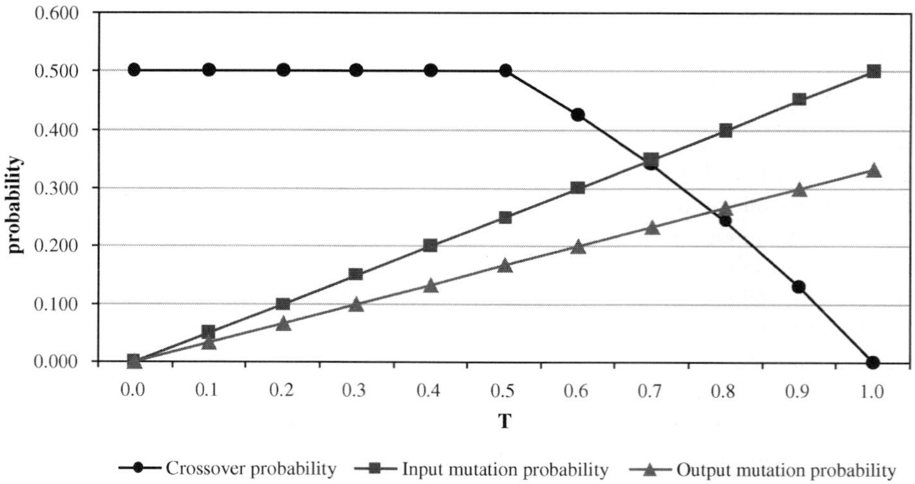

Fig. 1. Crossover and mutation probability trends ($p = 10$; $k = 1$; $L = 5$).

depend on the total number of available variables, while the input mutation probability changes with respect to the available variables.

7. Selection pressure and crossover/mutation trade-off influence

The influence on population evolution of selection pressure (B) and crossover/mutation trade-off (T) was verified on a sample data set constituted by 209 PolyChloroBiphenyls

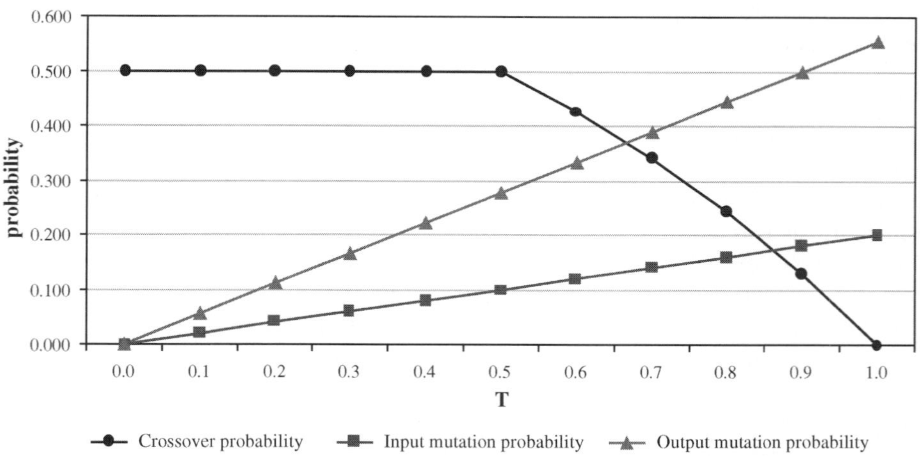

Fig. 2. Crossover and mutation probability trends ($p = 10$; $k = 4$; $L = 5$).

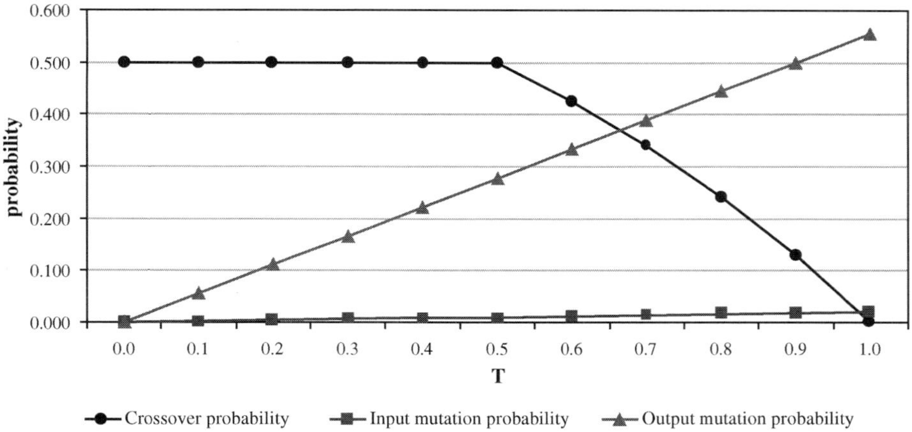

Fig. 3. Crossover and mutation probability trends ($p = 100$; $k = 4$; $L = 5$).

(PCB) described by 522 molecular descriptors calculated by the software DRAGON (Todeschini et al., 2003a) and using the melting point as response variable. Regression models based on descriptor subsets were calculated by Ordinary Least Squares (OLS) method and leave-one-out Q^2 was chosen as the objective function to search for the best models.

The study was performed in nine points corresponding to different pairs of T and B values (Fig. 4). The points were chosen in such a way as to explore the model population evolution by different strategies starting from a purely random search

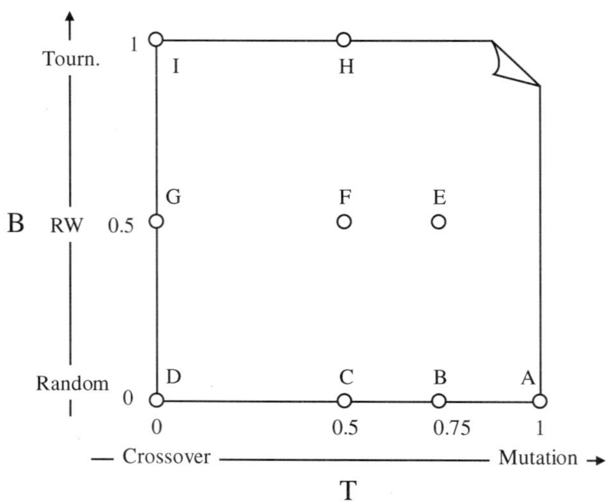

Fig. 4. Design points for the study of T and B parameters.

without crossover (case A), and ending with a full crossover procedure based on a statistically biased selection towards a few best models present in the population (case I).

Calculations were performed on populations of 50 individuals, allowing model size from 1 to 5 variables, without preserving the best models for each size and ending the evolution procedure at 10,000 iterations. The average quality (Q^2_{LOO}) of the population and the quality of the best model—evaluated with a step of 200 iterations—were considered to give an estimate of the population evolution.

Figs. 5 and 6 show the population average quality and the best model quality, respectively, of the nine populations evolved in the studied cases.

Cases D, G and I represent populations evolved only by crossover procedure, without mutations, with a parent selection performed randomly, by Roulette Wheel and tournament, respectively. These populations are quite static and are unable to find high quality models due to the absence of mutations; therefore, the absence of mutations should be selected only in the final step of the model searching in order to better explore a limited model space to find the absolute best model.

From case D to A parent selection is randomly performed and the mutation probability increases, while the crossover probability decreases. The population evolved according to case A (no crossover, only mutations) slowly finds high quality models, while the population evolved according to case C (crossover and mutation equally balanced) finds the high quality models quickly, but is more constrained to the model space initially found, which might be a local optimum.

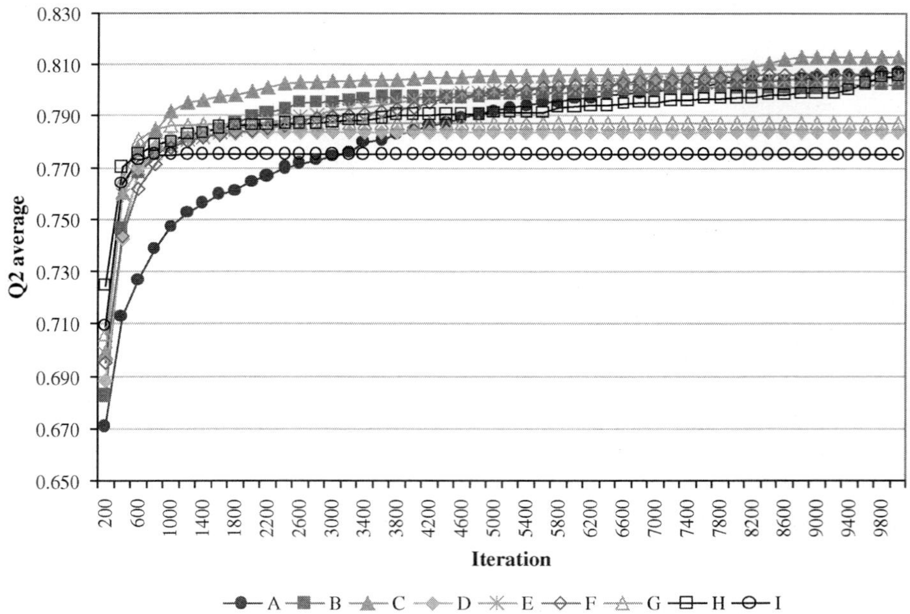

Fig. 5. Population average Q^2_{LOO} values for nine different $T-B$ values.

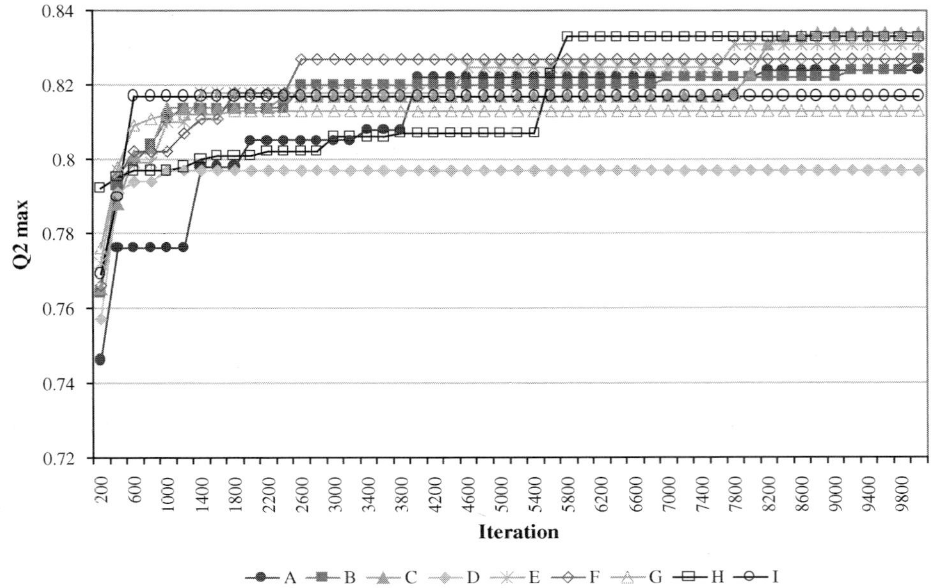

Fig. 6. Population best Q^2_{LOO} values for nine different $T-B$ values.

Case F corresponds to a population evolved with equal importance of the crossover and mutation probability and parent selection performed by RW. This corresponds to the classical GA model search.

8. RQK fitness functions

In searching for regression models by evolutionary methods, optimising only the leave-one-out explained variance, Q^2 has been demonstrated to be overoptimistic and unable to give optimal predictive models. In fact, the final selected models often turned out not to be as predictive as expected if more severe validation was applied. On analysing these models, it was found that chance correlation and random variables are frequently the cause of their lack of predictivity.

The RQK function is a new fitness function for model searching proposed to avoid unwanted model properties, such as chance correlation, presence in the models of noisy variables and other model pathologies that cause lack of model prediction power (Todeschini et al., 2003b). This is a constrained fitness function based on the Q^2_{LOO} statistics and four tests that must be fulfilled contemporarily. By using the RQK function in an evolutionary algorithm for optimal model population searching, one should maximise Q^2_{LOO} and accept models only if the following tests are satisfied:

1. $K_{XY} - K_X > \delta K$ (QUIK rule)
2. $Q^2_{\text{ASYM}} - Q^2_{\text{LOO}} < \delta Q$ (Asymptotic Q^2 rule)

3. $R^{\mathrm{P}} > t^{\mathrm{P}}$ (R^{P} rule)
4. $R^N > t^N$ (R^N rule)

Using the same tests, similar optimised model populations seem to arise, even maximising R^2 and R^2_{adj} instead of Q^2_{LOO} or minimising LOF.

Proposed in 1998, the *QUIK rule* (Todeschini et al., 1999) is a simple test that allows the rejection of models with high predictor collinearity, that can lead to chance correlation. The QUIK rule is based on the K multivariate correlation index (Todeschini, 1997) that measures the total correlation of a set of variables.

This rule is derived from the evident assumption that the total correlation in the set given by the model predictors X plus the response Y should always be greater than that measured only in the set of predictors X.

Therefore, the QUIK rule is: only models with the K_{XY} correlation among the $[X + Y]$-variables greater than the K_X correlation among the $[X]$-variables can be accepted, or

$$\text{if } K_{XY} - K_X < \delta K \rightarrow \text{reject the model}$$

where δK is a user-defined threshold (e.g. 0.01–0.05). The δK threshold can also be chosen equal to zero if a less severe constraint is needed. In any case, negative threshold values are not allowed, being theoretically unacceptable models with negative differences $K_{XY} - K_X$.

The QUIK rule has been demonstrated to be very effective in avoiding models with multi-collinearity without prediction power. On the other hand, this rule is not efficient enough to reject models with more than one random variable, as random variables are not usually correlated and, therefore decrease the total K_X correlation. In this case, even a low correlation of the Y response with the predictors may turn out significant with respect to the correlation among the predictors.

The second test is called the *Asymptotic Q^2 rule*. It arises from the widely accepted principle that a good model should have a small difference between fitting and predictive ability. In fact, marked differences between the R^2 and Q^2 values can be due to overfitting (high R^2 values) or to some not predictable samples (low Q^2 values) and do not guarantee future predictive ability of the model.

It has been demonstrated (Miller, 1990) that Q^2_{LOO} is asymptotically related to R^2 and, therefore, an asymptotic value of Q^2 can be calculated by the following expression:

$$Q^2_{\mathrm{ASYM}} = 1 - \left(1 - R^2\right)\left(\frac{n}{n - p'}\right)^2 \tag{4}$$

where n is the number of objects and p' the number of model parameters.

The asymptotic Q^2 rule is based on the comparison of the asymptotic and the actual Q^2_{LOO} value of the model:

$$\text{if } Q^2_{\mathrm{ASYM}} - Q^2_{\mathrm{LOO}} > \delta Q \rightarrow \text{reject the model}$$

The simplest threshold value is $\delta Q = 0$, but a more conservative value could, for example, be 0.005.

The R function based rules were proposed to avoid pathological model behaviour due to the presence of redundant or noisy variables in a regression model (Todeschini et al., 2003b).

Given a regression model with p variables, let R_{jY} be the absolute value of the correlation coefficient between the jth predictor and the Y response. In order to analyse the role of each variable in the model, we calculate the following quantity:

$$d = \frac{R_{jY}}{R} - \frac{1}{p} \tag{5}$$

We take as the reference a model with multiple correlation R equi-distributed on the p model variables. In this case, each variable contributes with $1/p$ to the multiple correlation R. Each contribution R_{jY}/R of the actual model can then be compared with the value $1/p$ referring to the reference model, in order to evaluate the role of the single variables in the model.

If in the model there are variables with high pairwise correlation R_{jY} with respect to the multiple correlation R, these should be redundant variables. Redundant variables are often, but not necessarily, highly correlated, and hence explain almost the same information in the response. In this case, by dropping all but one of the redundant predictors, the final model shows no loss in fitting with respect to the previous model and, furthermore, turns out much simpler.

In order to account for redundant variables the function R^{P} has been proposed. It is based on the positive differences $(R_{jY}/R - 1/p)$ calculated as the product of their transforms. More exactly, each positive difference is first scaled to its maximum value $(p - 1) / p$, then the complement to 1 is calculated as the following:

$$R_j^{\mathrm{P}} = 1 - \left(\frac{R_{jY}}{R} - \frac{1}{p}\right)\left(\frac{p}{p-1}\right) \quad 0 \le R_j^{\mathrm{P}} \le 1 \tag{6}$$

What is obtained in this way is a sort of penalty weight for modelling predictors. The value is low if the predictor has high correlation with the response, dropping to zero when R_{jY} equals R. The R^{P} function is then defined as:

$$R^{\mathrm{P}} = \prod_{j=1}^{p^+} R_j^{\mathrm{P}} \quad 0 \le R^{\mathrm{P}} \le 1 \tag{7}$$

where p^+ is the number of variables whose pairwise correlation over the multiple correlation is greater than the reference $1/p$.

The product of the penalty weights for modelling variables was chosen with the aim of having low values of the R^{P} function, when even just one variable in the model has a R_{jY} value very close to the multivariate coefficient. In fact, in this case, the product tends towards zero regardless of the value of the other terms, meaning that the other variables in the model are useless since they do not contribute to significantly increasing the multiple correlation; therefore, the model itself is considered too complex with respect to its quality. On the contrary, if each predictor explains the same fraction $1/p$ of the total multivariate correlation R, the R^{P} function is equal to one.

The R^P *rule* was finally settled as:

$$\text{if } R^P < t^P \rightarrow \text{reject the model}$$

t^P being a user-defined threshold ranging from 0.01 to 0.1 depending on the data. A suggested value for t^P is 0.05.

In order to catch the information related to all the predictors that, given their low correlation with response, could probably be random variables, the function R^N has been defined. It is the sum of the negative differences given by all the predictors whose ratio of R_{jY} over R is smaller than $1/p$:

$$R^N = \sum_{j}^{p^-} \left[\frac{R_{jY}}{R} - \frac{1}{p} \right] \qquad -1 < R^N \leq 0 \tag{8}$$

where the sum runs over the variables p^- giving a negative difference.

This quantity R^N accounts for different information with respect to R^P. It accounts for an excess of non-modelling or random variables and can be thought to be a measure of overfitting due to noisy variables. It takes the maximum value equal to zero when no supposed random variables are in the model.

Now, let us assume that all the non-modeling variables had almost the same low correlation with response equal to ε; then, each of such variables gives a contribution to R^N as:

$$\frac{\varepsilon}{R} - \frac{1}{p} = \frac{p\varepsilon - R}{pR} \qquad \varepsilon \ll R \tag{9}$$

The value of ε can be tuned by the user, depending on the knowledge of the Y response noise. Moreover, it can be assumed that no more that one noisy variable is allowed in the model, thus a threshold value for R^N can be estimated as:

$$t^N(\varepsilon) = \frac{p\varepsilon - R}{pR} \tag{10}$$

The choice of accepting at least one variable with low correlation with the response derives from the impossibility of knowing a priori if the variable is only noise or is useful in modelling residuals. Moreover, it seems reasonable that just one variable is allowed for residual modelling.

Finally, given the threshold value, the R^N *rule* has been settled as:

$$\text{if } R^N < t^N(\varepsilon) \rightarrow \text{reject the model}$$

Increasing the value of ε increases the value of the threshold, thus models with noisy variables will be rejected more easily.

9. Evolution of the populations

When VSS is applied to a huge number of variables, it is reasonable that several model populations based on different variable subsets are optimised independently. However, sooner or later the information encoded by the single populations should, in some way, be joined to find the absolute best models. To this end, the evolution procedure allows the merging of two or more populations into a new one, the duplication of an existing population and the transfer of one population to another.

While the first two options enable a new population to be added to the existing ones, the last option provides a modification of a selected population as it absorbs the information encoded into the transferred population.

The duplication of a selected population can be chosen to create a new population derived from the same variable subset, but optimised by a different setting of the genetic parameters controlling the evolution procedure.

Population junction and population transfer can be performed by adding together either all the original variables of the selected populations, or only those variables actually present in the population models. By adding all the original variables, the optimisation procedure allows better exploration of the model space of the new, or modified, population since all possible variable combinations are available. However, in this way all information related to the previous optimisation is lost and the number of all the possible models to be analysed increases significantly. This last drawback can be avoided in the population junction/transfer by adding only variables actually present in the population models; however, the loss of some of the original variables means that only a part of the model space will be explored.

In any case, to preserve the genetic information of selected populations the junction/transfer should be performed by allowing the migration of population individuals. The resultant population will then consist of joint population models ranked according to their quality; thus the population size is given by the sum of the two population sizes. If the new population size exceeds the maximum allowed size, the number of individuals is gradually and automatically reduced, excluding the worst ones.

Individual migration is allowed also when a new population is created by duplication. In this case the new population will be the same as the selected one, preserving its evolution level, and it will be allowed to evolve differently, according to the defined optimisation parameters. Otherwise the new population will be created by random initialisation of individuals from all the original variables of the selected population.

10. Model distance

When a more or less large set of possible models is obtained it becomes necessary to compare the models. A new measure of the distance between models has been proposed (Todeschini et al., 2003c), which allows an analysis of model populations. The distance proposed takes into account the correlation of variables within and between models and allows the finding of clusters of similar models, catching the most diverse models in such a way as to preserve maximum information and diversity.

Comparing two models means comparing two p-dimensional binary vectors where each position corresponds to a variable. The most common way to represent the relationships between two binary vectors, represented here by models A and B, is a

Table 2
Two-way table collecting variable frequencies between two binary vectors, represented by models A and B

		Model B		
		1	0	
Model A	1	a	b	p_A
	0	c	d	$(p - p_A)$
		p_B	$(p - p_B)$	p

two-way table as shown in Table 2. In Table 2, p is the total variable number, a the number of cases with 1 in the same position in both vectors, d the number of cases with 0 in the same position in both vectors, b the number of cases such that for a given position there is 1 in vector A and 0 in vector B, c the number of cases such that for a given position there is 1 in vector B and 0 in vector A. Therefore, b and c represent the number of variables not shared by the two models; b is the number of variables in model A but not in model B, and c the number of variables in model B but not in model A. The degree of similarity between the two models is in some way related to a and d, while their degree of diversity is related to b and c. The dimensionality of the models A and B is indicated by p_A and p_B, respectively.

The most common distance measure for two binary vectors $\mathbf{I}_A$ and $\mathbf{I}_B$ which are two models A and B each containing p_A and p_B variables, respectively, is the square Hamming distance d_H^2 defined as:

$$d_H^2 = b + c \tag{11}$$

where b and c are the numbers defined above.

The Hamming distance takes values in the range:

$$0 \leq \mathrm{DP} \leq d_H^2 \leq \mathrm{SP} \tag{12}$$

where

$$\mathrm{DP} = |p_A - p_B| \qquad \text{and} \qquad \mathrm{SP} = p_A + p_B \tag{13}$$

the minimum DP being equal to zero when the models coincide. Note that the Hamming distance can take only integer values.

It has been demonstrated that the Hamming distance usually overestimates the distance between two models, neglecting the variable correlations.

In order to measure model distances taking into account variable correlation *Model distance* has been proposed. Model distance calculation requires the identification of all the pairs of variables of a model having a correlation equal to one. Note that, if the models to be analysed have been searched for by variable selection, together with least squares regression, the case of pairs of variables in a model with a correlation equal to one is not possible. In any case all these redundant variables should definitely be excluded from the model, together with the common variables of the two models which are deleted for practical reasons. At this point, the number of diverse variables in the two models is calculated, this number being $b' + c'$ resembling that used for the Hamming distance even if the symbols b' and c' replace b and c since, in our case, a variable reduction could be made as explained above.

To make this step clearer, let us look at an example. Suppose a set of 10 ordered variables is given, let the model A ($\mathbf{I}_A$) be constituted by six variables ($p_A = 6$) and the model B ($\mathbf{I}_B$) by four variables ($p_B = 4$), with two common variables p_{AB} ($\mathbf{x}_3$ and $\mathbf{x}_9$), then their binary vector representation is:

$$\begin{aligned} \boldsymbol{I}_A &= [0 \quad 0 \quad 1 \quad 1 \quad 1 \quad 1 \quad 1 \quad 0 \quad 1 \quad 0] \\ \boldsymbol{I}_B &= [1 \quad 0 \quad 1 \quad 0 \quad 0 \quad 0 \quad 0 \quad 0 \quad 1 \quad 1] \end{aligned}$$

and the corresponding phenotypic representation:

$$\text{A}: \ x_3, x_4, x_5, x_6, x_7, x_9 \qquad \text{and} \qquad \text{B}: \ x_1, x_3, x_9, x_{10}$$

Let us now suppose that the variables x_4 and x_5 of the model A have a correlation equal to one, and the same for variables x_9 and x_{10} of the model B. Therefore, in both models one of the two variables, either x_4 or x_5 in model A and either x_9 or x_{10} in model B, has to be deleted together with the common variables. Then, the reduced models will be constituted by the following variables:

$$\text{A}: \ x_5, x_6, x_7 \qquad \text{and} \qquad \text{B}: \ x_1$$

It results that $b' = 3$ and $c' = 1$. For these reduced models the Hamming distance is equal to 4, while for the original models it would be 6.

The second step of the procedure deals with the evaluation of the correlation among the variables of the two reduced models. It involves the calculation of the cross-correlation matrix $\mathbf{C}_{AB}$, which contains the correlations between all the possible pairs of the variables of the two models. This matrix has b' rows, i.e. the number of variables in the reduced model A, and c' columns, i.e. the number of variables in the reduced model B. The counterpart of $\mathbf{C}_{AB}$ (size $b' \times c'$) is the cross-correlation matrix $\mathbf{C}_{BA}$ (size $c' \times b'$).

The cross-correlation matrix can be transformed into a symmetric matrix like the following:

$$\begin{aligned} \mathbf{Q}_A &= \mathbf{C}_{AB} \cdot \mathbf{C}_{BA} \quad (b', b') \\ \mathbf{Q}_B &= \mathbf{C}_{BA} \cdot \mathbf{C}_{AB} \quad (c', c') \end{aligned} \tag{14}$$

The no-zero eigenvalues of both matrices $\mathbf{Q}_A$ and $\mathbf{Q}_B$ coincide and the sum r_{AB} of the square root of these eigenvalues λ gives the desired information related to the inter-model variable correlation:

$$r_{AB} = \sum \sqrt{\lambda_j} \tag{15}$$

Finally, the *Model distance* d_M^2 is derived from the Hamming distance as follows:

$$d_M^2(A, B) = b' + c' - 2\sum \sqrt{\lambda_j} = b' + c' - 2r_{AB} \tag{16}$$

As is easily seen, if no preliminary variable reduction is carried out, i.e. $b' = b$ and $c' = c$, and no correlation exists between the two variable blocks, i.e. $r_{AB} = 0$, the model distance coincides with the Hamming distance.

The model distance satisfies the first two main postulates for a distance measure:

1. $d_{ij} = d_{ji}$
2. $d_{ii} = 0$

Moreover, it was observed that the Model distance does not always satisfy the triangle's inequality:

3. $d_{ij} + d_{jk} \geq d_{ik}$

thus belonging to the class of non-Euclidean distances.

11. The software MobyDigs

MobyDigs software has been developed for VSS in regression and classification analysis by GA. The implemented multivariate methods are OLS Regression and *K*-Nearest Neighbours (*K*-NN). MobyDigs software is a 32-bit application and can be run on Windows platforms. The programming language is Microsoft Visual Basic 6.0.

11.1. The data setup

Once the data has been loaded, the data setup menu allows the user to select the independent variables *X* and the response variable *Y*, which can be quantitative for regression analysis or categorical for classification analysis. In this last case only integer numeric variables are accepted.

The selected *X* variables are initially assigned to a single population; however, the user can split them into a chosen number of populations. This can be done by random or user-defined selection.

By default, all the objects are assigned to the training set except for those lacking values in at least one independent variable and the response. The former are automatically excluded from the analysis and the latter assigned to the prediction set. The user can modify, by hand, the object allocation by forcing an object's exclusion from the analysis or its belonging to the test set. Moreover, the object split into training/test sets can be performed by a user-defined binary variable or by cluster analysis.

The *X* and *Y* variables can be transformed before analysis. The available transformations are the logarithmic, inverse and square root. Moreover new *X* variables can be created automatically by selecting the second, third order and interaction terms.

Finally to fill missing values of independent variables the available options are random values or average values.

11.2. GA setup

Once the data have been loaded and the data setup performed the *GA setup* menu is activated. The options available in this menu are concerned with the GA parameters, the variable management and the choice of objective function.

The five main parameters for the GA are:

1. Population size: maximum number of models in a population (default: 50).
2. Maximum allowed model size: maximum number of variables in a model (default: 3).
3. Crossover and mutation trade-off: user-defined value of the T parameter which sets the values of the crossover and mutation probabilities (default: 0.5).
4. Number of retained models for each size: number of the best models for each size surviving in the population regardless their quality (default: 3). This option is important to save, in the final population, also the best models of lower complexity e.g. the first three models with one variable, the first three models with two variables, etc.
5. Selection pressure: user-defined value of the B parameter which sets the parent selection operator (default: 0.5).

At the beginning, the user-defined GA parameters hold for all the populations: however, they can be differentiated for each population in the evolution procedure.

An important characteristic of the software MobyDigs is the variable management, i.e. the set of strategies useful to rationalise the research process of the best models when a huge number of variables are available. Together with the splitting of the original variables in different populations, two tools are proposed in GA setup menu:

1. the Tabu list which catches variables that, according to the five criteria defined above, are possibly useless for modelling purposes;
2. the addition of randomly distributed variables, which test chance correlation during the evolution procedure. The user can add up to 100 random variables to each population with labels ZZNxx when normally distributed and ZZUxx when uniformly distributed (xx is an ID number associated to the random variable).

Several objective functions have been proposed in the literature (Frank and Todeschini, 1994) and are available in MobyDigs for evaluating the quality of population individuals. They differ in the degree of dependence on model complexity, some of them showing the maximum for optimal complexity and others the minimum. For regression analysis the user can select the following objective functions:

1. Q^2 leave-one-out

$$Q^2_{\mathrm{LOO}} = 1 - \frac{\mathrm{PRESS}}{\mathrm{TSS}} \tag{17}$$

2. R^2 adjusted

$$R^2_{\text{adj}} = 1 - \left(1 - R^2\right)\frac{n-1}{n-p'} \tag{18}$$

3. SDEP

$$\text{SDEP} = \sqrt{\frac{\text{PRESS}}{n}} \tag{19}$$

4. FF function

$$\text{FF} = \frac{R^2}{(1-R^2)}\frac{(n-p')}{p} \tag{20}$$

5. AIC Akaike information

$$\text{AIC} = \text{RSS}\frac{n+p'}{(n-p')^2} \tag{21}$$

6. FIT Kubinyi function (Kubinyi, 1994)

$$\text{FIT} = \frac{R^2}{(1-R^2)}\frac{(n-p')}{(n+p)^2} \tag{22}$$

7. LOF Friedman modified (Friedman, 1990)

$$\text{LOF} = \frac{\text{RSS}/n}{\left(1 - \frac{p(d+1)}{n}\right)^2} \tag{23}$$

8. RQK fitness function, defined above (Todeschini et al., 2003b)

In the formula above, R^2 is the determination coefficient, PRESS the predictive error sum of squares, TSS the total sum of squares, RSS the residual sum of squares, p the number of model variables, p' the number of model parameters, n the number of training objects and d is the LOF function smoothing parameter.

Note that for model searching the common coefficient of determination R^2 cannot be used monotonically increasing with the model complexity.

For classification analysis, the only objective function which has to be maximised is the Non-Error Rate (NER).

Together with the chosen objective function, the user can select another two parameters to be displayed during the evolution process. The displayed parameters are the above-mentioned objective functions plus the determination coefficient R^2 and the multivariate correlation index K (Todeschini, 1997) (see Fig. 7).

MobyDigs performs GA-based VSS either by a random initialisation of population individuals or a population initialisation searching all possible models. This last initialisation is feasible if, given the number of allowed model variables, the number of all the possible models is not too large; however, the user can, at any time during the initialisation, switch to the standard GA procedure. A suggested practice is starting the model searching by initialising the population by all possible models with one up to three variables.

11.3. Population evolution view

Once population optimisation is started, the user can follow the populations on two main screens (see Fig. 8). The first screen refers to all the evolving populations: for each population the best model is shown, described by the objective function value, two other selected fitness parameters, its size and variables, together with the maximum allowed number of model variables, the trade-off parameter T and the number of population individuals (models). The second screen shows the evolution of a selected population; all the population models are displayed, described by their objective function values, two other selected fitness parameters, model size and model variables.

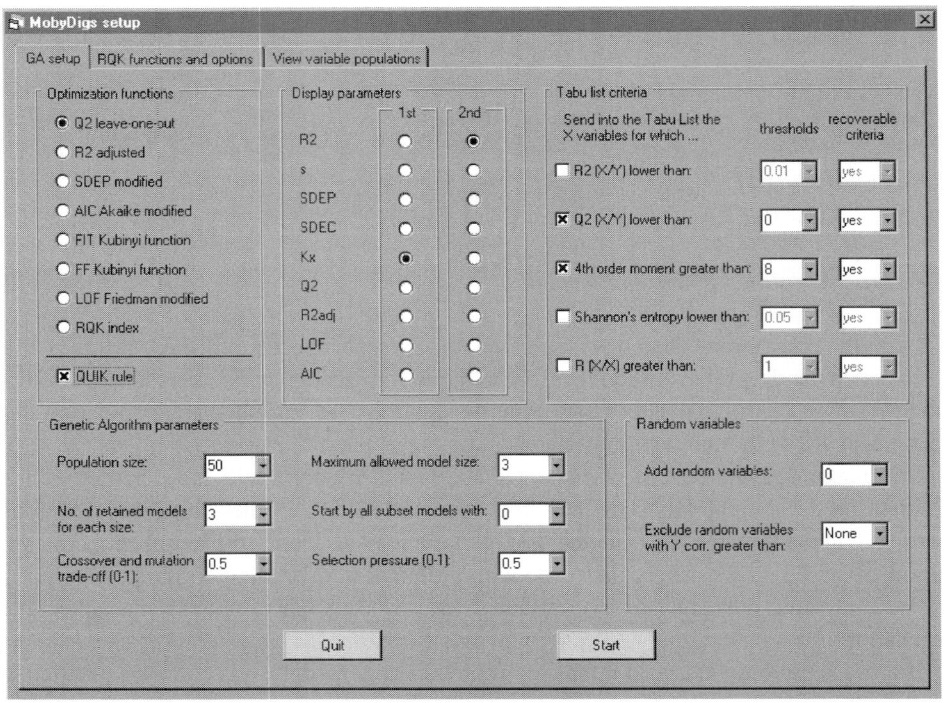

Fig. 7. MobyDigs GA setup.

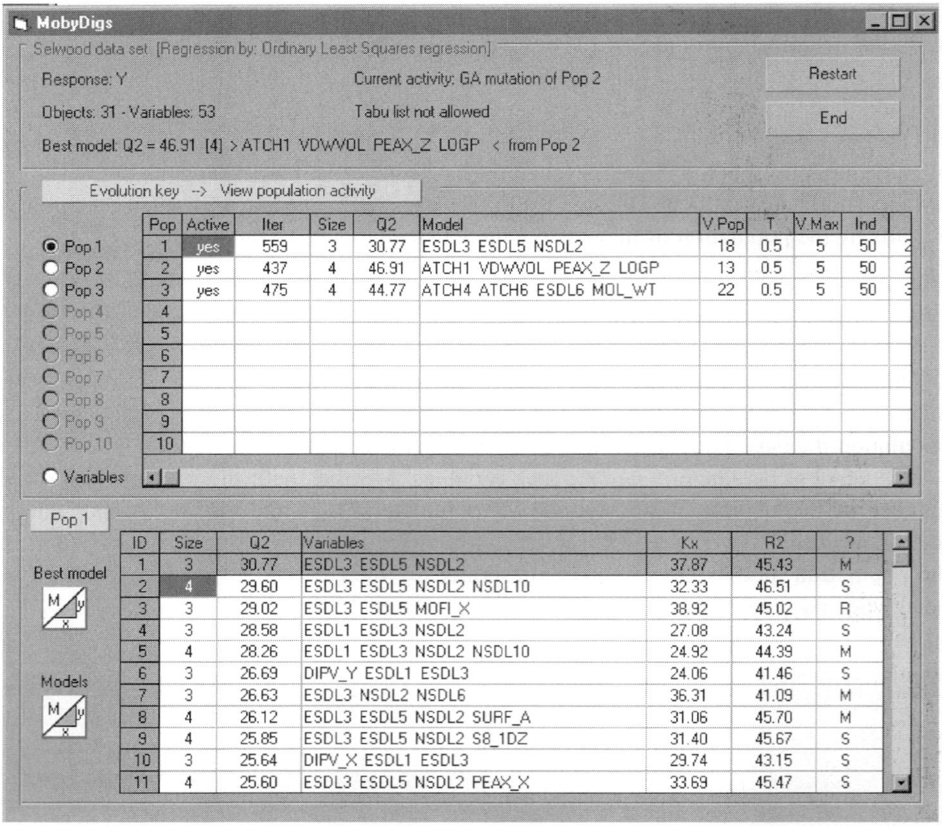

Fig. 8. MobyDigs population evolution view.

11.4. Modify a single population evolution

Once the population of interest is selected, the user can choose to make it inactive, if it has reached the desired evolution level, or to modify its genetic parameters to run through new evolution directions. The parameters are usually changed when the population has become so stable that no new individuals enter the population.

The maximum number of allowed variables in a model could be increased if the optimal model complexity has not yet been reached. As the objective functions available in MobyDigs take model complexity into account, the number of model variables can be progressively increased with little risk of exceeding optimal complexity. In fact, models with too many variables, i.e. too complex, should not enter the population.

If model searching has been initialised by small sized models, it is common practice to increase the model variable number for a satisfactory exploration of the model space, until the population becomes stable or the desired objective function has been obtained.

Since the coverage of the search space is influenced by crossover/mutation trade-off and selection pressure, these can be changed to augment the diversity of the population when it falls into a local optima and tends to be constituted by very similar individuals. In this case the crossover/mutation trade-off should be increased, while the selection pressure should be decreased to enhance the random process of GA and thus produce new genetic material in the population. On the other hand, when the population has been initialised favouring high diversity among individuals, the crossover/mutation trade-off should be decreased and the selection pressure increased to force the evolution towards a local, or hopefully absolute, optima.

Moreover, when the population is in the last optimisation stage, the Tabu list variables can be recovered to check if some significant variable has been prematurely excluded from the analysis.

11.5. Modify multiple population evolution

When more than one population has been optimised independently, MobyDigs allows the managing of these populations. The two options are:

A. Create a new population.
B. Transfer one population to another one.

All the active populations are shown in a view grid where the user can select those to be duplicated or merged.

A new population can be created by duplicating a selected population or merging two or more populations. The transfer of a population to another one allows the modification of a selected population by adding the information encoded in the transferring population, highlighted in the view grid.

Both these processes can be performed by checking one of the two following options:

1. Transfer of all the original variables.
2. Transfer only of actual population variables.

Moreover the condition *Allow migration of population individuals* can be settled as true (T) or false (F). When the condition is true, the same individuals as the parent population constitute the new population; when it is false the same variables (or a subset) only constitute it, but the new individuals are generated randomly.

Therefore, the following actions can be performed:

- A-1-T: a new population is created transferring all the individuals and original variables of the selected populations; if only one population has been selected, the new population is an exact duplication of it.
- A-1-F: a new population is created transferring all the original variables, but not the individuals of the selected populations, in such a way that the new population is randomly initialised.

- A-2-T: a new population is created transferring all the individuals of the selected populations and only the variables constituting them; if only one population has been selected, the new population is a duplication of it, but lacking all the original variables not present in the parent population.
- A-2-F: a new population is created transferring only the actual variables, but not the individuals of the selected populations, in such a way that the new population is randomly initialised.
- B-1-T: an existing population is modified by the inclusion of all the original variables and individuals of another population.
- B-1-F: the variable set of a population is increased with all the original variables of another population in such a way that the optimisation process can explore a larger model space.
- B-2-T: an existing population is modified by the inclusion of individuals, and therefore of variables, actually present in another population.
- B-2-F: the variable set of a population is increased with the variables actually present in another population in such a way that the optimisation process can explore a larger model space.

11.6. Analysis of the final models

Once the optimisation of the populations is concluded, the user has to choose the models to submit to the final analysis process. First of all, the user defines the total number k of best models to be processed and selects the populations from which the models are taken. Then, the final population will be constituted by:

1. *absolute best models*: the k absolute best models chosen among the models of all the selected populations;
2. *absolute best models of each population*: the k absolute best models of each selected population; therefore, the total number of final models is given by the number k multiplied by the number of selected populations;
3. *absolute best models + the best models for each size*: like case 1 with the addition of the best models for each model size chosen from among all the population models;
4. *absolute best models of each population + the best models for each size*: like case 2 with the addition of the best models for each model size taken from each selected population.

Note that the number of best models for each model size is that defined at the stage of GA setup, before any population optimisation has been started, and that duplicated models are always excluded from the final population.

The final population can be saved as is or analysed by switching to the menu *Population analysis*.

By this menu the final models, or only some of them, can be validated by techniques other than the leave-one-out that is automatically performed by the software. The available validation techniques are leave-more-out, bootstrap and external validation.

The *leave-more-out validation* is performed by a percentage of objects to be left out, in turn, from the training set defined by the user according to the strength of the required validation.

Bootstrap validation is performed by randomly generating training sets with sample repetitions and then evaluating the predicted responses of the samples not included in the training set. This procedure is repeated 5,000/10,000 times to evaluate Q^2_{BOOT}.

External validation requires the user to define a test set at the stage of data setup. The test objects require the final model variables and the observed response, and they do not take part in model development or internal validation. By using the selected model the values of the response for the test objects are calculated and the quality of these predictions is defined in terms of Q^2_{EXT}, which is defined as:

$$Q^2_{EXT} = 1 - \frac{\sum_i (y_i - \hat{y}_i)^2}{\sum_i (y_i - \bar{y}_{tr})^2} \tag{24}$$

where the sums run over the test set objects, y_i and $\hat{y}_i$ are the observed and predicted response of the ith object and $\bar{y}_{tr}$ is the average value of the training set observed response.

In the menu *Population analysis* the user can also calculate response values of all the data set objects by the option *Predictions*. This allows calculated response values to be obtained for training objects, and predicted values for test objects and objects lacking response in the data set, together with their leverage values useful for detecting influent objects and extrapolated responses. If more than one model has been selected in the final population then multiple predictions will be provided.

Consensus analysis is a strategy where the calculated or predicted responses are based on the average of the values obtained from more than one final selected model. Both simple and weighted average values are given, where the weighting scheme is inversely proportional to the leverage values of the models, thus giving less importance to the more extrapolated values.

The last tool provided by the *Population analysis* menu is concerned with the *similarity analysis of population models*, which is performed by the Model distance approach explained above.

Distances between all the selected final model pairs are calculated and the corresponding distance matrix is shown. Standardised distances can also be displayed. The button *MultiDimensional Scaling* (MDS) allows a graphical representation of the model distribution, distinguishing models derived from different populations by different colours.

11.7. Variable frequency analysis

This menu allows the analysis of variable frequency in the final population. In particular, the total number and percentage of variables present in the final models is given, together with a table showing the frequency of each variable in each original population calculated on the basis of only the final models.

Variable frequency analysis can be useful to detect those variable showing high importance in modelling a response.

11.8. Saving results

The final population models are saved by MobyDigs in a tabulated ASCII file where the models are listed according to the decreasing value of their quality. Each model is described by:

- model size;
- model variables;
- determination coefficient, R^2;
- explained variance in prediction by leave-one-out, Q^2;
- explained variance in prediction by leave-more-out, Q^2_{LMO}, explained variance in prediction by bootstrap, Q^2_{BOOT}, external explained variance, Q^2_{EXT} (optional);
- adjusted R^2_{adj};
- Akaike information AIC;
- multivariate correlation among the independent variables X, K_X;
- multivariate correlation among the independent variables X plus the response, K_{XY};
- standard deviation error in prediction, SDEP;
- standard deviation error in calculation, SDEC;
- Fisher test value, F;
- standard error, s;
- degrees of freedom, DF;
- average value of the training object leverages, AVH;
- coordinates of the models in the multidimensional scaling graph (optional);
- belonging population.

Other results that can be saved in MobyDigs are:

1. Responses with the corresponding leverage values calculated or predicted by all the selected models.
2. The average calculated/predicted responses (weighted and unweighted) and their minimum and maximum values for all the data set objects by consensus analysis.
3. The matrix constituted by the distances between all the model pairs.

MobyDigs constraints

Maximum number of objects: 3000
Maximum number of variables: 2000
Maximum number of variables in a model: 20
Maximum number of populations: 10
Maximum number of individuals (models) in each population: 100

References

Frank, I.E., Todeschini, R., 1994. The Data Analysis Handbook, Elsevier, Amsterdam, The Netherlands.

Friedman, J., 1990. Multivariate Adaptive Regression Splines, Technical Report No. 102, Laboratory for Computational Statistics, Department of Statistics, Stanford University, Stanford, CA.

Goldberg, D.E., 1989. Genetic Algorithms in Search, Optimization and Machine Learning, Addison-Wesley, Massachusetts, MA.

Jouan-Rimbaud, D., Massart, D.L., de Noord, O.E., 1996. Random correlation in variable selection for multivariate calibration with a genetic algorithm. Chemomet. Intell. Lab. Syst. 35 (2), 213–220.

Kubinyi, H., 1994. Variable Selection in QSAR Studies. I. An evolutionary algorithm. Quant. Struct.-Activity Relationships 13, 285–294.

Leardi, R., Boggia, R., Terrile, M., 1992. Genetic algorithms as a strategy for feature selection. J. Chemomet. 6, 267–281.

Miller, A.J., 1990. Subset Selection in Regression, Chapman & Hall, London, UK.

Todeschini, R., 1997. Data correlation, number of significant principal components and shape of molecules. The K correlation index. Anal. Chim. Acta 348, 419–430.

Todeschini, R., 2002. Moby Digs, rel. 1.0 for Windows; Talete srl: Milano, Italy.

Todeschini, R., Consonni, V., Maiocchi, A., 1999. The K correlation index: theory development and its applications in chemometrics. Chemom. Intell. Lab. Syst. 46, 13–29.

Todeschini, R., Consonni, V., Pavan, M., Mauri, A., 2003a. DRAGON, rel. 3.1 for Windows; Talete srl: Milano, Italy.

Todeschini, R., Consonni, V., Mauri, A., Pavan, M., 2003b. Detecting "bad" regression models: multicriteria fitness functions in regression analysis. Anal. Chim. Acta (submitted).

Todeschini, R., Consonni, V., Pavan, M., 2003c. A distance measure between models: a tool for similarity/diversity analysis of model populations. Chemomet. Intell. Lab. Syst. (in press).

Wehrens, R., Buydens, L.M.C., 1998. Evolutionary optimization: a tutorial. TrAC. Trends Anal. Chem. 17 (4), 193–203.

CHAPTER 6

Genetic algorithm-PLS as a tool for wavelength selection in spectral data sets

Riccardo Leardi

Department of Pharmaceutical and Food Chemistry and Technology, University of Genoa, via Brigata Salerno (ponte) – I-16147 Genova, Italy

1. Introduction

Spectral data consisting of hundreds and even thousands of absorbance values per spectrum can now be routinely collected in a matter of seconds. Methods such as Partial Least Squares (PLS) or Principal Component Regression (PCR), being based on latent variables, allow one to take into account the whole spectrum without having to perform a previous feature selection (Geladi and Kowalski, 1986; Thomas and Haaland, 1990). Owing to their capability of extracting the relevant information and producing reliable models, these full-spectrum methods were considered to be almost insensitive to noise. Therefore, until recently, it was commonly stated that no feature selection was required (Thomas and Haaland, 1990). In the last decade it has instead been gradually recognised that an efficient feature selection can be highly beneficial, both to improve the predictive ability of the model and to greatly reduce its complexity (Thomas, 1994). There is much empirical evidence now that variable selection is a very important step when using methods such as PLS or PCR (Jouan-Rimbaud et al., 1995; Bangalore et al., 1996; Broadhurst et al., 1997; Ding et al., 1998; Hasegawa et al., 1999).

In relatively simple cases, referring to materials with a limited number of components, spectroscopists can select some regions according to the knowledge of where these components are spectroscopically active and probably follow the Lambert–Beer law. When analysing much more complex materials, the wavelength selection based on spectroscopic considerations is much more difficult. The most important reason for this is that some components of the material may be unknown; furthermore, even in the case of known components, their spectral fingerprint can be changed by variable experimental conditions (e.g. temperature). Even if all the components (and their spectral fingerprints) are known, the relevant overlapping of the different fingerprints, the physical and chemical interactions among the components and other sources of deviations from the Lambert–Beer law can make this kind of selection extremely difficult.

Data Handling in Science and Technology, Volume 23
ISSN: 0922-3487

DOI: 10.1016/S0922-3487(03)23006-9

Computer-aided variable selection is important for several reasons. Variable selection can improve model performance, provide for more robust models and models that can transfer more readily and allow non-expert users to build reliable models with only limited expert intervention. Furthermore, computer-aided selection of variables may be the only approach for some models, for example, predicting physical properties from spectral data.

There are many approaches for variable selection.

'Univariate' approaches select those variables that have the greatest correlation with the response. In complex problems they give poor results, since they do not take into account the interrelationships among the variables. It has to be remembered that, together with the 'predictive' wavelengths (i.e. those wavelengths useful for directly modelling the relationship between spectral information and response), the 'synergic' wavelengths are also very important. The synergic wavelengths are those wavelengths that, though having a very limited relevance when used by themselves, become very important when added to the predictive wavelengths.

This drawback is partially overcome by the 'sequential' approaches. They select the best variable and then the best pair formed by the first and second and so on in a forward or backward progression. A more sophisticated approach applies a look back from the progression to reassess previous selections. The problem with these approaches is that only a very small part of the experimental domain is explored.

Recently, more methods specifically aimed at variable selection for multivariate calibration have been developed, namely Interactive Variable Selection (Lindgren et al., 1994), Uninformative Variable Elimination (Centner et al., 1996), Iterative Predictor Weighting PLS (Forina et al., 1999), Interval PLS (Nørgaard et al., 2000), significance tests of model parameters (Westad and Martens, 2000), and the use of genetic algorithms (Bangalore et al., 1996; Ding et al., 1998; Hasegawa et al., 1999; Leardi, 2000). A theoretical justification for variable selection in multivariate calibration has recently been offered along with a novel selection approach (Spiegelmann et al., 1998).

The selection of variables for multivariate calibration can be considered an optimisation problem. Genetic algorithms (GAs) applied to PLS have been shown to be very efficient optimisation procedures. They have been applied to many spectral data sets and are shown to provide better results than full-spectrum approaches (Leardi, 2000). The major concern with using GAs is the problem of overfitting. This problem has been recently addressed using a randomisation test (Jouan-Rimbaud et al., 1996; Leardi, 1996).

2. The problem of variable selection

As previously stated, one of the main problems during the examining of large data sets is the detection of the relevant variables (i.e. the variables holding information) and the elimination of the noise. Roughly speaking, each data set can be situated somewhere between the following two extremes:

(a) data sets in which each variable is obtained by a different measurement (e.g. clinical analyses);

(b) data sets in which all the variables are obtained by a single analytical technique (e.g. spectral data).

It can be easily understood that the goal of variable selection is quite different in the two cases.

For data sets of type (a) it is very worthwhile to try to reduce the number of variables involved in the model as much as possible, since this means shorter analysis time and lower costs. In such cases, even a small reduction of the predictive ability can be accepted, if it is counterbalanced by adequate time and cost saving.

For data sets of type (b) the main goal of variable selection is the elimination of noise, together with the reduction of the complexity of the model.

Apart from these quite obvious points, it has to be realised that spectra have another very relevant peculiarity; the spectral variables have a very high autocorrelation (i.e. variable n is very much correlated with variables $n - 1$ and $n + 1$). Hence, spectroscopists will never consider single wavelengths. Instead, their analysis will be in terms of spectral regions, each spectral region being formed by a certain number of adjacent variables. Therefore, when comparing different subsets of variables having the same predictive ability, one should take into account not only the number of retained variables, but also how many spectral regions are involved and how well they are defined.

One possible application of variable selection in spectral data sets is the setting of filter spectrometers, these instruments being much cheaper and faster than full-spectrum instruments (Lestander et al., submitted). It is easy to understand that in such cases the results of a variable selection are of practical use only if a small number of regions are detected.

Last but not least, the identification of spectral regions clearly involved in the modelling of the response can be a great help in the interpretation of the spectra (Leardi et al., 2002), because the spectroscopists will have very clear indications of which spectral features are the most relevant to the problem under study.

The procedure of variable selection, apparently so simple, is indeed very complicated and needs a careful validation to avoid the risk of overestimating the predictive ability of the selected model. In such cases, when using it on new data, one could be strongly deceived, discovering that it has no predictive ability.

This is mainly due to random correlations: if you try to describe 10 hypothetical objects with 100 random X variables and a random response, you will surely have some X variables perfectly modelling the response. This risk is, of course, higher when the variables/objects ratio is very high, as in the case of the QSAR data sets described in Chapter 5.

In the case of spectral data sets, this problem can be limited by reducing the number of variables. Owing to the relevant autocorrelation, the content of information of a new variable constructed from the average of a small number of adjacent original variables is very similar to that of the original variables themselves.

The risk of overfitting is also higher the longer the GA runs (i.e. the more models that are tested); a good solution consists of performing a large number of independent short runs, and obtaining the final model by taking into account the results of all the runs. By doing this, a much more consistent (and less overfitted) solution can be obtained.

3. GA applied to variable selection

As described in Chapter 1, in a data set in which each object is described by v variables, each chromosome is composed by v genes, each gene being formed by a single bit (Leardi et al., 1992).

As an example, with 10 variables, a set that only uses variables 1, 5, 8 and 9 will be coded as 1000100110.

The response will be the cross-validated variance explained by the selected set of variables.

While most GAs are intended to work in a continuous space, in our case the space under investigation is absolutely discontinuous: when working with v variables, it is like studying only the vertices of a v-dimensional hypercube. Several changes have thus been applied to the Simple Genetic Algorithm (SGA) to best adapt it to this specific purpose.

It has to be remembered that there is no 'ideal' GA structure; this heuristic is applicable to all sorts of problems. Therefore, the success in solving a particular problem is strictly dependent on how well the GA has been adapted to the problem itself.

As explained in Chapter 1, there needs to be a good balance between exploration and exploitation. If the search is too unbalanced towards exploration, the GA will become very similar to a random search. On the other hand, if exploitation would strongly prevail, then we would obtain something very similar to a 'classical' algorithm that finds a nearby optimum solution. In the case of variable selection, it is better to emphasise exploitation. This produces good results within a rather limited number of evaluations and therefore reduces as much as possible the risk of overfitting.

3.1. Initiation of population

According to the SGA the value of each bit of the initial population is determined by the 'toss of a coin'. Under this hypothesis an average of 50% of the variables would be selected in every initial chromosome.

Such a situation would lead to two main disadvantages:

- the computation of the response for a model containing a higher number of variables requires a much longer time;
- since PLS has a very low sensitivity to noise, the presence of a few 'good' variables in a chromosome (this will almost always happen if half of the variables are selected) may be enough to produce a good response, irrespective of the presence of some 'bad' variables; this will mean that the responses of different chromosomes will be very similar and therefore it would be much more difficult to get rid of the 'bad' variables.

By setting the probability of selection to n/v, n variables will be selected on average out of the v original variables. The selection of a much smaller number of variables will result in a much faster elaboration time and a greater ease in the detection of the relevant variables. This means that during the first stages 'bad' variables can be more easily discarded, since one of them can be enough to worsen the response of a chromosome in which only a few 'good' variables are selected.

Within each run, the combination of these small, highly informative 'building blocks' will lead to a gradual increase in the number of selected variables.

To avoid chromosomes that produce models containing too many variables, the maximum number of possible selected variables is also set. This constraint will also be enforced in subsequent phases.

3.2. Reproduction and mutation

To avoid any influence of the order of the variables, an unbiased uniform crossover is used. Since this mating operator promotes exploration, and the goal is to emphasise exploitation, it is used within a non-generational algorithm.

Mutation is then applied to the generated offspring. Then, the response of the two offspring is evaluated so that they can replace certain members of the population.

3.3. Insertion of new chromosomes

The quality, or fitness, of a subset of variables is determined both by the response it gives and by the number of features it uses. Thus it is rather important to know the best result obtained by using a certain number of variables. This chromosome is highly informative and deserves to be saved, regardless of its absolute position in the ranking of the responses. To do this, such a chromosome is 'protected' and cannot be eliminated from the population; its particular condition will end when another chromosome, using at most this number of features, gives a better result.

After evaluating the response of each offspring, it has to be decided whether to insert it into the population and, if so, what chromosome of the population to discard.

As stated before, a chromosome using v variables is protected when it gives the best response among all the chromosomes using at most v variables.

If the new chromosome is a protected one, then it will be a member of the population and the worst non-protected chromosome will be eliminated. If the new chromosome is non-protected, then it will be a member of the population only if its response is better than the response of the worst non-protected chromosome. In this case the worst non-protected chromosome will be eliminated.

By doing this, at each moment the population is composed of the best absolute chromosomes and of those chromosomes that are highly informative, since they give the best result with a limited number of variables.

Table 1 shows a simulated population ($N = 10$ chromosomes). In it, the following chromosomes are protected: #1 (highest response with $v \leq 7$), #4 (for $v \leq 5$), #7 (for $v \leq 4$), #8 (for $v \leq 3$), #9 (for $v \leq 2$) and #10 (for $v = 1$). Chromosome #5 is non-protected since, though it is the best for $v = 6$, it is dominated by chromosome #4 which gives a better result with a lower number of variables.

Table 1
Simulated population of 10 chromosomes

Chromosome #	Response	Number of selected variables
1	80.35	7*
2	78.32	7
3	75.47	8
4	70.32	5*
5	68.43	6
6	65.13	5
7	60.65	4*
8	50.43	3*
9	40.71	2*
10	30.09	1*

* protected chromosomes.

Let us assume four cases.

(a) The new chromosome has a response of 55.13 with $v = 4$; it is non-protected and its response is lower than the lowest response of a non-protected chromosome (65.13); the new chromosome is discarded.
(b) The new chromosome has a response of 68.19 with $v = 7$; it is non-protected and better than the worst non-protected chromosome (65.13); the chromosome enters the population and chromosome #6 is discarded.
(c) The new chromosome has a response of 72.14 with $v = 5$; it is protected and enters the population; chromosome #4 becomes non-protected and chromosome #6 is discarded.
(d) The new chromosome has a response of 42.11 with $v = 2$; it is protected and enters the population; chromosome #9 becomes non-protected and is discarded.

3.4. Control of replicates

The SGA does not check for the presence of 'twins', so that the same chromosome can be present more than once. In this algorithm this is not allowed and every new chromosome created is immediately checked; if it is found that a 'twin' has been previously created and evaluated, then this chromosome is discarded.

As a result of these modifications the population is always formed by highly informative combinations.

3.5. Influence of the different parameters

Four parameters (population size, probability of initial selection, maximum number of variables and probability of mutation) have to be defined.

The study of the influence of these parameters, performed on a very large number of data sets, showed that it is possible to obtain an architecture that is always very valid.

This is very important from the practical point of view, since it means that a parameter optimisation for each new data set is not required.

3.5.1. Population size

The lower the population size, the lower the exploration and the higher the exploitation, since the bias towards the best chromosomes being selected for crossover is greater; a population size of 30 chromosomes is a good compromise. Furthermore, a non-generational algorithm means that once a good chromosome has been found, it immediately enters the population, and can then be immediately picked up as a parent without having to wait for the evaluation of the other chromosomes of its same generation.

3.5.2. Probability of initial variable selection

A low value of this parameter allows one to explore many more combinations in the same amount of time, resulting in information that is much easier to interpret. The algorithm itself will build more complex combinations through the reproduction phase. A good value is the one that selects an average of five variables per chromosome (i.e. $5/v$, v the number of variables).

3.5.3. Maximum number of variables

The higher the number of selected variables, the higher the model complexity, and therefore the higher the time required for the evaluation of the response. A maximum value of 30 variables allows the algorithm to obtain very good models without making the computations too heavy.

3.5.4. Probability of mutation

The mutation step allows one to avoid deadlock situations and to 'jump" to new zones of the space. A very high value of this parameter disrupts the configuration of the chromosomes too much, with a low probability of obtaining a good result, while a very low value does not give this step the importance it deserves. A good compromise is the probability of 0.01 per gene.

3.6. Check of subsets

As stated in Section 3.4, the algorithm presented here requires that all chromosomes be unique. Since the number of variables contained within each chromosome can be from one to the maximum allowed, it is possible that one chromosome represents a subset of another. This is the case when the variables selected by chromosome *c2* are a subset of the variables selected by chromosome *c1*. If *c2* has a response higher than *c1*, the extra variables present in *c1* (but not present in *c2*) bring no information, and simply represent noise. By keeping chromosome *c1* in the population, we would reduce the diversity of the population (and therefore the exploration capability of the GA) without adding any supplementary information. As a consequence, chromosome *c1* is discarded.

3.7. Hybridisation with stepwise selection

Generally speaking, many common search techniques are characterised by a very high exploitation and a very poor exploration. This means that, given a starting point, they are able to find the nearest local maximum, but once they have found it, they get stuck on it. On the contrary, techniques such as GAs can have a very high exploration and a rather low exploitation. This means that they are able to detect several 'hills' leading to different local maxima, but it is not very easy for them to climb up to the maximum.

As described in Chapter 2, it is therefore rather intuitive to think that combining a GA with another optimising technique should produce a new strategy having both high exploration and high exploitation. The application of a different optimiser to one of the solutions found by the GA should lead to the identification of the local maximum near which the chromosome is lying (Hibbert, 1993).

As previously described, stepwise selection is one of the most commonly used techniques for variable selection. A GA's performance is improved by alternating it with cycles of backward stepwise selection. It is performed on the best chromosome that has not yet undergone a stepwise selection and if the backward elimination results in a better chromosome, this new chromosome will replace the 'original' one.

One cycle of backward elimination is performed every 100 evaluations and if the stop criterion is not a multiple of 100, a final cycle will also be performed.

4. Evolution of the GA

As previously stated, the presence of random correlations is surely the most important factor limiting a generalised and extensive use of GAs (Jouan-Rimbaud et al., 1996), and not taking it into account can lead to totally senseless models. For the same reason, the runs must be stopped very early. This means that only a minor part of a very complex search domain can be explored in a single run, and therefore the results of different runs can be rather different. A new approach is therefore needed by which the global information obtained in several runs can be exploited.

4.1. The application of randomisation tests

If the variables (mainly the y, or response, variable) are very noisy, or if a limited number of objects is present, or if the variables/objects ratio is very high, a GA cannot be used since it would model noise instead of information. To verify this, a randomisation test can be performed. In this verification, the order of the elements in the $\mathbf{y}$ vector is randomised, so that each row of the $\mathbf{X}$ matrix will correspond to a y that, though a real one, is not its own. Of course, in this case, there is no information in the data set and if some modelling can be performed, it means that noise is being modelled.

Several GAs are performed on randomised data sets (after each run a new randomisation of the $\mathbf{y}$ vector is applied); in each run, 100 chromosomes are evaluated and the best response is taken into account (if it is <0, then it is set equal to 0).

After having performed the required number of runs (100 is a good compromise between required time and consistency of the result), the average is computed. The better (more reliable) the data set, the lower this value. If it is very high, then it means that GA can find a good model even when no information is present. According to the results obtained on several real cases, with good data sets values <4 are obtained; as a rule of thumb, it can be said that GA can be safely applied till around 8.

4.2. The optimisation of a GA run

Another important decision is when to stop a GA run. When looking at the evolution of the response as a function of the number of evaluations, one can see very easily that, after the first evaluations in which it improves very fast, the improvement is much slower. In the case of a real data set, in presence of noise, it can be said that after having modelled the bulk of information (in the first evaluations), the GA starts refining the model until it will model noise. The danger in performing too many evaluations is to model noise.

To have an idea of when to stop, a series of R runs is performed ($R = 100$ is a good compromise); in the first $R/2$ runs the **y** vector is the original one, while in the second half of the runs the **y** vector is shuffled as in the randomisation test. After 100 runs, in each of which 200 chromosomes have been evaluated, a matrix $\mathbf{A}_{100,200}$ is obtained in which each element $a_{r,c}$ is the best result obtained during run r after having evaluated c chromosomes.

From it, a matrix $\mathbf{M}_{2,200}$ is obtained, in which the first row is the average of rows 1–50 of matrix **A**, while the second row is the average of rows 51–100.

A vector $\mathbf{d}_{1,200}$ is then computed, being the difference between the first and the second row of matrix **M**. When working with good data sets the values of **d** show a fast increase till a maximum value, after which a decrease or a plateau will follow. It can be said that the best moment to stop a GA run corresponds to the evaluation after which the maximum difference is obtained.

Such value will anyway be between 50 (the 30 chromosomes of the original population +20 offspring produced by GA, since with a lower value the effect of GA would be too limited and the results would be very similar to the results of a random search) and 200 (it has been noticed that a higher number of evaluations does not lead to any significant improvement of the quality of the model, in spite of a much higher computation time).

Generalising, it can be said that:

- the runs with the 'original' **y** show the ability of GA of modelling information + noise;
- the runs with the randomised **y** show the ability of GA of modelling noise;
- the difference vector corresponds to the ability of modelling information.

4.3. Why a single run is not enough

The result of a single GA run is usually a model in which only a very few variables are present. This means that the advantage of using PLS is not fully exploited. Furthermore, two opposite events have to be taken into account; on one hand non-relevant variables can be occasionally retained in the final model, while on the other side it can happen that some relevant variables are occasionally not included.

As previously stated, it has to be kept in mind that overfitting is the greatest risk in applying a GA. This risk increases as the number of tested models within a single run increases, since the probability of finding a model whose good performance is such only by chance (i.e. due to random correlations) becomes greater. Cross-validation is not a complete protection against overfitting, since the objects on which the performance of the models is tested are the same as those on which the variable selection is performed.

This basic consideration greatly influences the architecture of the GA. All the parameters are set in such a way that the highest exploitation is obtained, thereby meaning that the goal of the algorithm is to have a very fast increase in the response and therefore to have a very good solution in the very early stages of the process. This is the reason why a non-generational algorithm and a rather unusually limited population size have been applied. A drawback of this architecture is the fact that, since only a very small part of the domain is explored, the final result can be strongly influenced by the randomly generated original population, and therefore the variables selected in different runs can be substantially different. It is therefore worthwhile to perform a high number of different runs (e.g. 100) and to try to extract some 'global' information by taking into account the top chromosome of each run and computing the frequency with which each variable has been selected (Leardi and Lupiáñez González, 1998).

The final model is obtained following a stepwise approach in which the variables are entered according to the frequency of selections (i.e. in the model with n variables, the n most frequently selected variables are present), and evaluating the Root Mean Square Error in Cross-Validation (RMSECV) associated to each model. It can be noticed that usually the RMSECV decreases very fast, until it reaches a minimum or a plateau.

A crucial point is the detection of the number of variables to be taken into account. The model corresponding to the global minimum usually retains a rather high number of variables and very often it has the lowest RMSECV only due to some overfitting, without being significantly better than several other models. It can be said that the best model is the most parsimonious model among all the models which is not significantly different from the global optimum. This approach generally leads to models having a slightly lower Root Mean Square Error in Prediction (RMSEP) and a significantly higher 'definition' in terms of selected regions (less and/or smaller regions) than the models corresponding to the global minimum of RMSECV.

The following procedure is followed (Leardi et al., 2002):

- detect the global minimum of RMSECV;
- by using an F test ($p < 0.1$, d.o.f. = number of samples in the training set, both in the numerator and in the denominator) select a 'threshold value' corresponding to the highest RMSECV which is not significantly different from the global minimum;
- look for the solution with the smallest number of variables having a RMSECV lower than the 'threshold value'.

4.4. *How to take into account the autocorrelation among the spectral variables*

The techniques of feature selection usually assume that there is no autocorrelation among the variables. While this is true in the case of non-spectral data sets, as stated in

Section 2, it does not hold in the case of spectral data. This means that if wavelength n is selected as relevant, wavelengths $n - 1$ and $n + 1$ should also have a high probability of being selected.

The main feature of the previously described algorithm is the fact that, to further reduce the risk of overfitting, the final model is obtained from the results of 100 independent, very short GA runs, while usually the model is obtained from a single, very long run.

In this process, every single run actually starts from scratch without taking into account the results obtained in the previous runs. This approach, though ensuring complete independence of each run, is a waste of energy. Since the frequency with which the single wavelengths are selected in the best chromosome of each run can give valuable information about the relevance of the corresponding spectral region, it would be interesting if each run could somehow 'learn' from the information brought by the previous runs. By doing that, a new run could concentrate its efforts mainly on the most interesting regions, without completely discarding the possibility of a global exploration. It is also clear that the relevance of this information is higher if more runs have already been performed. A simple way to force the starting population of a new run towards the selection of certain variables consists in changing the vector of initial probabilities.

The first step of a GA is creation of the starting population. In it, each bit of each chromosome is given a random value. The probability of each variable being present in each chromosome of the starting population is

$$p = n/v$$

where n is the average number of 1s (i.e. selected variables) we want to be present in a chromosome and v is the total number of variables.

We can therefore imagine $\mathbf{p}$ to be a vector whose v elements have the same value.

The frequency of selection of the variables in the previously performed runs can be used to modify the vector $\mathbf{p}$ in such a way that the values of the elements corresponding to the most frequently selected variables are higher than the values of the least frequently selected variables:

$$p_i = n \frac{\text{sel}_i}{\sum_{j=1}^{v} \text{sel}_j}$$

where sel_j is the number of selections of variable j in the previous runs.

When creating the starting population at the beginning of a new run, each element of $\mathbf{p}$ is compared to a random number in [0,1]. If the random number is lower than the corresponding element in the vector $\mathbf{p}$, that bit will be set to 1 (i.e. variable present); otherwise it will be set to 0 (i.e. variable absent). Of course, the higher the value of p_i, the higher is the probability that variable i will be present in the chromosome.

Such a solution would give two main problems:

- it does not take at all into account the autocorrelation among adjacent wavelengths;
- the variables that have never been selected in a previous run would have $p = 0$.

The first problem is easily solved by applying to vector **p** a smoothing by using a moving average (window size 3), thereby obtaining a new vector **ps**. Owing to the high autocorrelation between spectroscopic variables, if variable v is thought to be relevant, the variables adjacent to it should also be relevant and therefore it is logical to increase their probability also.

The second problem is more complex, since one should also take into account the fact that the reliability of the pattern of the frequency of selections is a function of the number of runs already performed.

To do so, a weighted average between the 'original' probability vector in which the probability of each element is equal to n/v, and the 'weighted' probability vector is computed:

$$\mathrm{pf}_i = \frac{\frac{n}{v}(R - r) + \mathrm{ps}_i r}{R}$$

where pf_i is the final probability of variable i being present in a chromosome of the initial population, R the total number of runs to be performed, r the number of runs already performed and ps_i the probability of variable i after the smoothing.

This means that the weight of the previous runs, almost negligible at the beginning, becomes more and more relevant as the number of performed runs increases.

If 100 runs have to be performed, it will be 0 at the first run, 0.10 at the 11th run, 0.50 at the 51st run and 0.99 at the last run. As one can see, the probability associated with each variable, though sometimes very low, is never 0 and therefore each variable can always be present in a chromosome of the initial population.

In the case of 100 runs, 175 variables and five 1s per chromosome in the initial population on average, a variable that has never been selected has the following probability for the last run:

$$\mathrm{pf}_i = 5/175 \times 0.01 + 0 \times 0.99 = 0.00029$$

which, though very low, is not 0.

After the last run the plot of the frequency of selection may not be as smooth as one would expect from spectral data. Since it is not logical that the relevance of adjacent variables in a spectrum is very different, a smoothing by a moving average (window size 3) is also performed.

The final model is obtained by the stepwise approach previously described in which the variables are entered according to the smoothed value of the frequency of selection.

Owing to the previously reported modifications, the selected variables detect real spectral regions and the variability of the models obtained by several runs of the program on the same data set is quite limited. Unfortunately, it is always possible that some relevant spectral regions are not selected or, the other way round, some regions whose contribution is non-significant are included in the model.

To further reduce the variability of the model, the whole procedure is repeated several times (at least five). The regions selected by each repeat are then visually compared, together with the original spectrum. If a region has been selected in the majority of the

Table 2
Parameters of the GA

- Population size: 30 chromosomes
- On average, five variables per chromosome in the original population
- Regression method: PLS
- Response: cross-validated % explained variance (five deletion groups; the number of components is determined by cross-validation)
- Maximum number of variables selected in the same chromosome: 30
- Probability of mutation: 1%
- Maximum number of components: the optimal number of components determined by cross-validation on the model containing all the variables (no higher than 15)
- Number of runs: 100
- Backward elimination after every 100th evaluation and at the end (if the number of evaluations is not a multiple of 100)
- Window size for smoothing: 3

processes it means that it is significant, while the regions selected a limited number of times can be said to have been selected mainly by chance.

This sequence of operations cannot be fully automated, since the visual analysis comparing the regions selected by each trial with the spectrum is needed. In the case of very broad spectral features, it is possible that the regions selected by different trials seem to be different, since they cannot be overlapped. If all of them refer to the same spectral feature, they should be considered as equivalent and therefore taken into account in the final model.

The previously described procedure, though apparently quite complex, leads to a great improvement in the repeatability of the final model. Tests performed on several data sets showed that the models obtained by running the whole procedure several times are extremely similar in both the selected regions and the predictive ability; minor and non-significant differences can be found in the definition of the limits of the different regions.

Table 2 summarises the architecture of the GA.

5. Pretreatment and scaling

Three pretreatments (none, first derivative and Standard Normal Variate) together with three scalings (none, column centring and autoscaling) have been studied (Leardi, 2000).

As far as pretreatment is concerned, the best results are obtained when no pretreatment is used. For the first derivative this is probably due to the fact that its application increases the level of noise in the data. Though not a great problem with PLS itself, this can be very dangerous with a method such as the GA which is very sensitive to noise. Less clear is the reason why the GA produces worse results when SNV has been previously applied.

As far as scaling is concerned, the results obtained by the GA when no scaling has been applied are by far the worst. This could be due to the fact that since the major part of the variance is explained by the offset from the origin, the variations in the % CV variance (the response optimised by the GA) are very limited.

With any pretreatment the GA using autoscaled data is on average better than the GA on column-centred data. The reason for this behaviour is probably due to the fact that autoscaling, increasing the noise of the uninformative variables, makes them even worse and therefore less likely to be selected.

Globally, the best results have been obtained by applying the GA to autoscaled data without any pretreatment. Beyond producing, on average, the lowest RMSEP, the replications performed in such conditions were the ones having the lowest variability in terms of both the RMSEP and the selected variables.

6. Maximum number of variables

As a rule of thumb, it has been found that the performance of the algorithm decreases when more than 200 variables are used (Leardi, 2000). This is due to the fact that a higher variables/objects ratio increases the risk of overfitting and also due to the fact that the size of the search domain becomes too great. Of course, the number of wavelengths measured in a spectral data set is much greater than this. An acceptable number of variables is obtained by dividing the original spectrum in windows of equal size, each of them made by a number of adjacent wavelengths corresponding to the value obtained by rounding at the upper integer the ratio between the number of wavelengths and 200. Each new variable will be given the value corresponding to the average of the *x*s at these wavelengths. For instance, in the case of 1050 wavelengths, the width of the window will be 6 (1050/ $200 = 5.25$). 175 new variables will be obtained, with variable 1 being assigned the average of the *x*s of wavelengths 1–6, variable 2 being assigned the average of the *x*s of wavelengths 7–12, and so on. (Note: to avoid confusion, from now on the original variables will be referred to as ‘wavelengths’, while the new variables obtained by previously described procedure will be referred to as ‘variables’.)

This approach gives no problems with NIR spectra, being characterised by very broad peaks. FT-IR spectra instead have much thinner peaks, and therefore if the window size used for the averaging is not small enough it is possible to lose some fine features in the spectra that are potentially important to the calibration model. On these data sets, an iterative approach is applied which allows the window size to be reduced to a size compatible with the spectral features of interest (Leardi et al., 2002). A GA is first performed on the entire spectrum using the window size leading to less than 200 variables (if more than one response is associated to the same data, this procedure is run on all of them). The results of five independent GAs for each response are then visually examined and those portions of the spectra that are never chosen by the GA are removed. The analysis is then repeated using the remaining portions of the spectrum and thus allowing the window size to be reduced. This is repeated until either no more ‘unselected’ regions are found or there are less than 200 variables remaining. This procedure usually leads to a small (often non-significant) improvement in the predictive ability of the model, while greatly improving the definition of the selected regions (and therefore their interpretability) and significantly reducing the number of retained wavelengths.

7. Examples

In this section the results of the application of a GA for wavelength selection on two different data sets will be reported.

The first one (Soy) is a data set of NIR spectra of soy flour, on which three responses (moisture, oil and protein) have been measured (Forina et al., 1995). The spectra have been recorded from 1104 to 2496 nm with a step of 8 nm, for a total of 175 wavelengths. The 54 samples have been divided into a training set (40 samples) and a validation set (14 samples). The samples of the validation set have been selected in such a way that they are as representative as possible of the global data set (Pizarro Millán et al., 1998).

The second data set (Additives) is a data set of FT-IR spectra of polymer films, in which the amount of two additives (B and C) has been measured (Leardi et al., 2002). The spectra have been recorded from 4012 to 401 cm^{-1} with a step of 1.93 cm^{-1}, for a total of 1873 wavelengths. The samples were obtained from five production batches, with samples from batches 1–3 being used as calibration set and samples from batches 4 and 5 being used as validation set. For additive B, there are 42 calibration samples and 28 validation samples, while for additive C there are 109 calibration samples and 65 validation samples. Before the application of the GA, a pathlength correction has been applied. Since even under well-controlled situations the film thickness can vary slightly, a correction is made using a polymer peak in the spectrum. The pathlength normalisation factor is computed as the average peak height between 2662 and 2644 cm^{-1} (10 data points) minus a baseline value estimated as the average from 2459 to 2442 cm^{-1} (10 data points). The pathlength normalisation was computed in this manner because it is known that the peak height of the band at 2662–2644 cm^{-1} is solely related to the polymer and is thus directly proportional to the film thickness.

7.1. Data set Soy

7.1.1. Moisture

PLS on the whole spectrum produces a RMSEP of 1.12 (5 components).

According to the results of the optimisation of the GA, shown in Fig. 1, the stop criterion was set to 50 evaluations.

Fig. 2 shows the RMSECV as a function of the number of variables in the model (the results of the first GA program are displayed). It can be seen that the RMSECV decreases very fast after the first variables, then stabilises at a plateau and finally increases. The RMSECV in the 'critical' region are better displayed in Fig. 3 (a zoom of Fig. 2). The global minimum is obtained with 31 variables, but the RMSECV of the model with just four variables is not significantly different, and therefore this will be the model to be taken into account.

Fig. 4 shows the histogram of the smoothed frequency of selection. The two horizontal lines show the 'cutting levels' corresponding to the models with four (the upper line) and 31 (the lower line) variables. Fig. 5 shows the spectra and the variables selected in this first GA examination; the upper broken line corresponding to the regions selected by the four-variable model and the lower broken line corresponding to the regions selected by the 31-variable model.

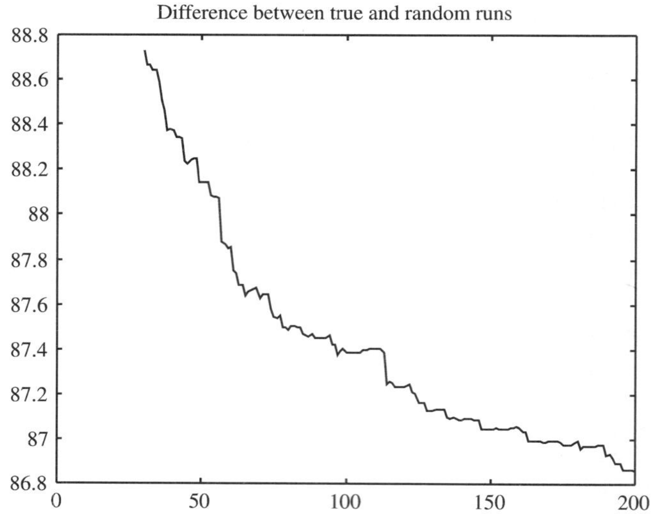

Fig. 1. Data set Soy, response Moisture: plot of the optimisation of the GA run.

To have more consistent results, nine more GA elaborations have been run. Fig. 6 shows the spectra and the regions selected in each of the 10 GA programs. This plot is used to construct the final model. It can be seen that two regions are consistently selected. The fact that they are slightly different for each run results from the stochastic nature of the algorithm, but it can be seen that the algorithm can bypass this problem converging onto very similar solutions. The final model is made by taking into account variables 108–113

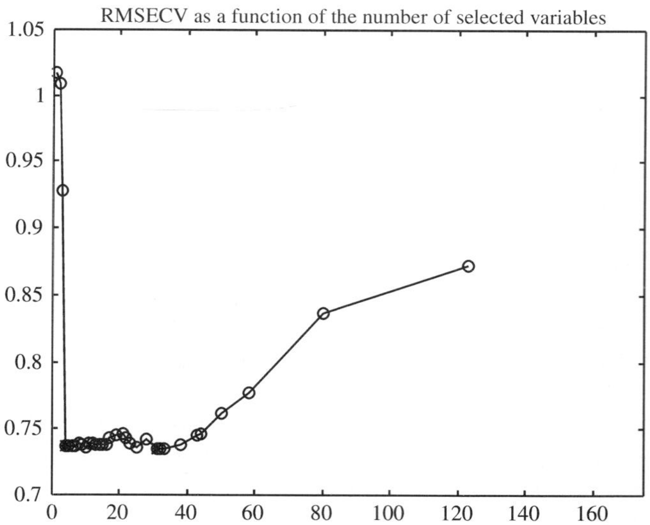

Fig. 2. Data set Soy, response Moisture (GA program 1): plot of the RMSECV as a function of the number of the selected variables.

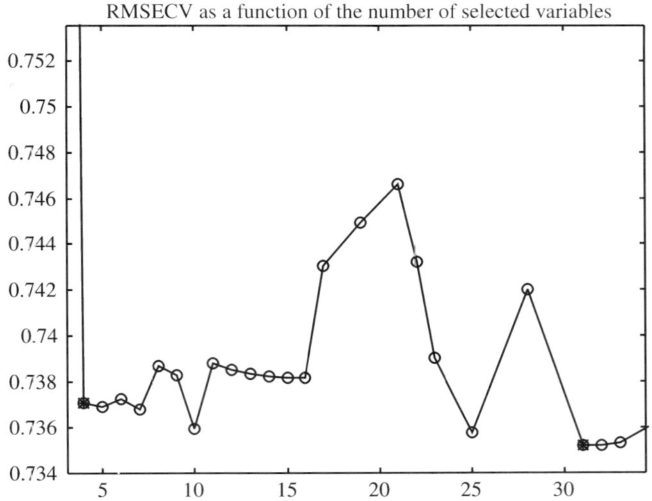

Fig. 3. Data set Soy, response Moisture (GA program 1): plot of the RMSECV as a function of the number of the selected variables.

and 124–129 (see Fig. 7). PLS on these two regions gives a RMSEP of 0.96, with just two components.

Summarising, the application of a GA to this response results in a model being both much more parsimonious (12 variables, 2 components vs. 175 variables, 5 components) and having a better predictive ability. Furthermore, the fact of having selected

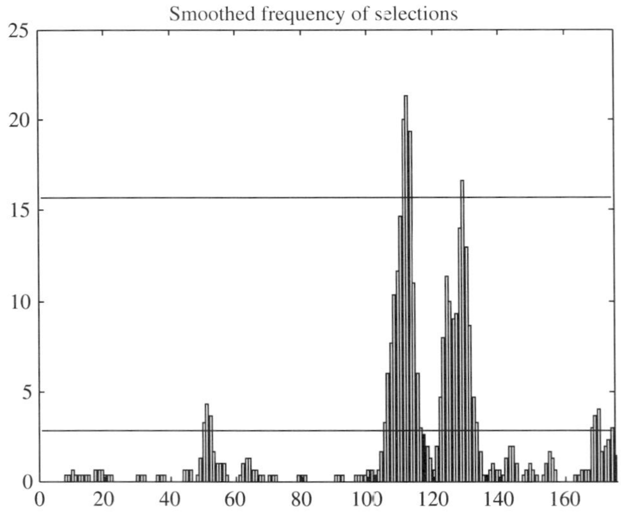

Fig. 4. Data set Soy, response Moisture (GA program 1): histogram of the smoothed frequency of selection. The two horizontal lines correspond to the 'cutting level' of the global RMSECV minimum (31 variables selected) and to the model selected according to the F-test (four variables selected).

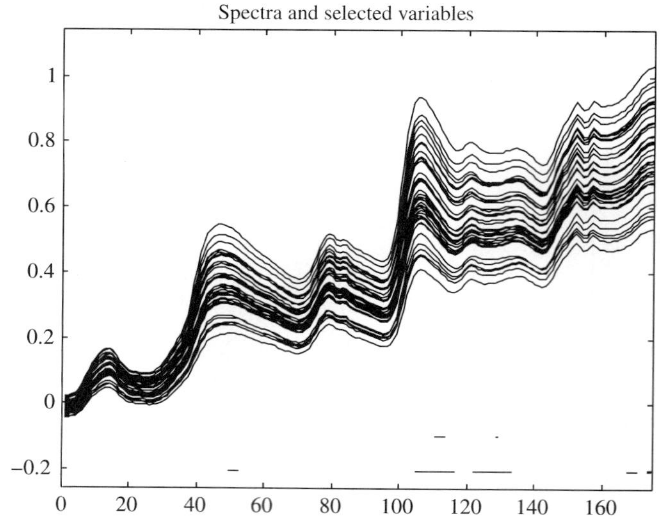

Fig. 5. Data set Soy, response Moisture (GA program 1): NIR spectra and selected wavelengths. The two broken lines correspond to the variables selected according to the global RMSECV minimum (31 variables selected) and to the F-test (four variables selected).

well-defined spectral regions can be a very relevant help in case a spectral interpretation is needed, or the good regions are sought for a filter instrument.

7.1.2. Oil

PLS on the whole spectrum produces a RMSEP of 1.29 (8 components).

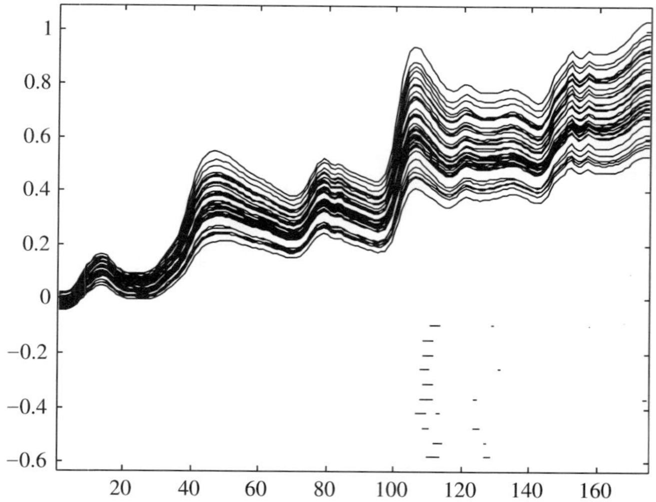

Fig. 6. Data set Soy, response Moisture: NIR spectra and selected wavelengths. Each broken line corresponds to the variables selected in each of the 10 GA elaborations.

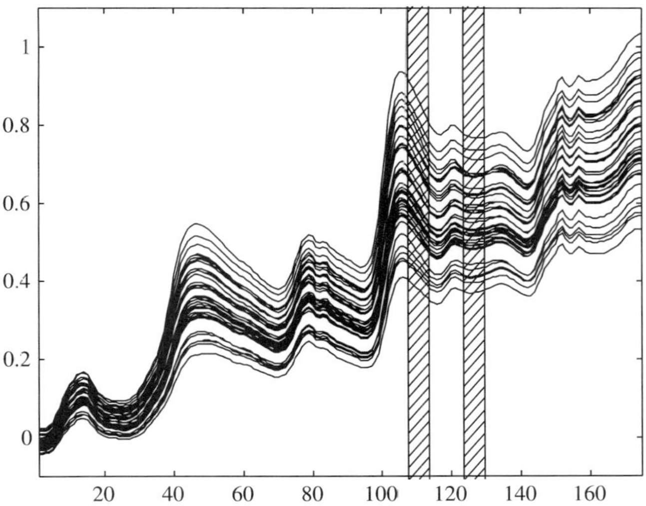

Fig. 7. Data set Soy, response Moisture: NIR spectra and selected regions in the final model.

According to the results of the optimisation of the GA, shown in Fig. 8, the stop criterion was set to 150 evaluations.

Fig. 9 shows the spectra and the regions selected in each of the 10 GA elaborations. It can be seen that four regions are consistently selected. The final model is made by taking into account variables 6–8, 14–17, 57–59 and 67–70 (see Fig. 10). PLS on these four regions gives a RMSEP of 1.07 (4 components).

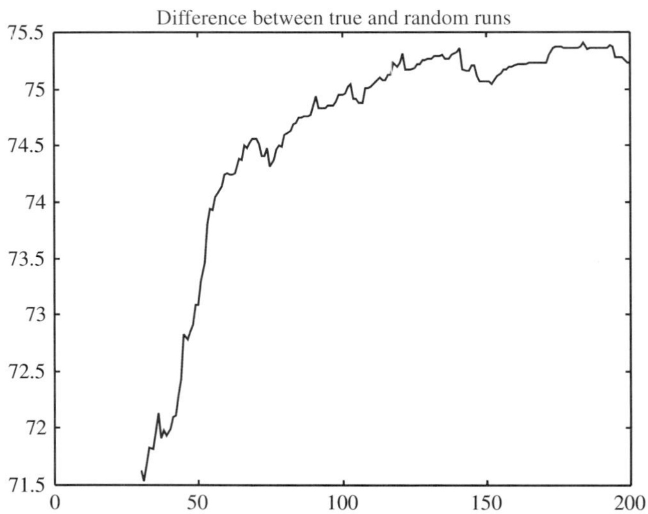

Fig. 8. Data set Soy, response Oil: plot of the optimisation of the GA run.

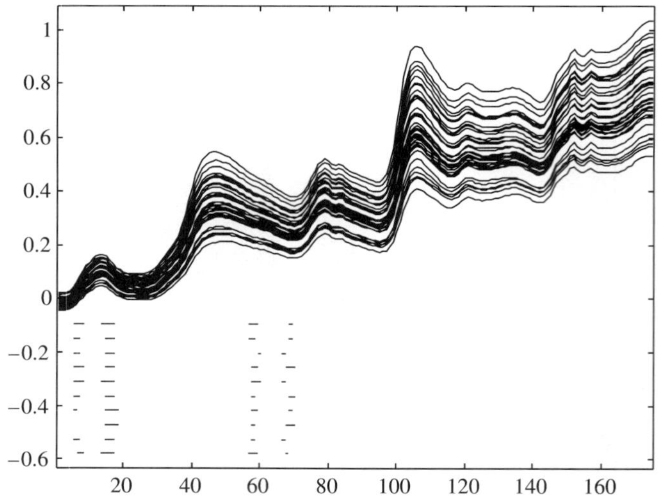

Fig. 9. Data set Soy, response Oil: NIR spectra and selected wavelengths. Each broken line corresponds to the variables selected in each of the 10 GA elaborations.

Also in this case, the application of GA results in a model being both much more parsimonious (14 variables, 4 components vs. 175 variables, 8 components) and with a better predictive ability.

7.1.3. Protein

PLS on the whole spectrum produces a RMSEP of 1.21 (8 components).

According to the results of the optimisation of the GA, shown in Fig. 11, the stop criterion was set to 100 evaluations.

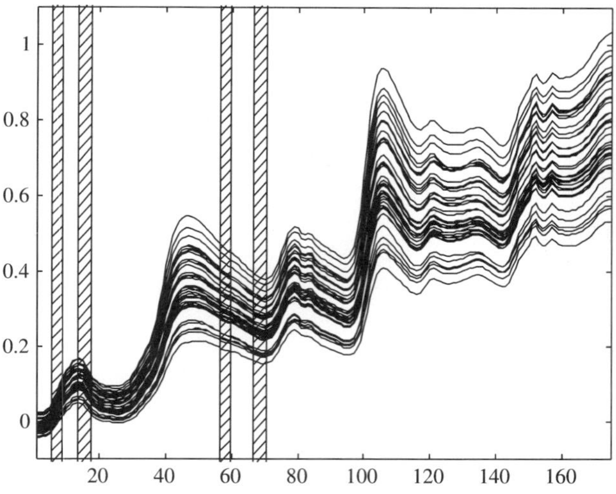

Fig. 10. Data set Soy, response Oil: NIR spectra and selected regions in the final model.

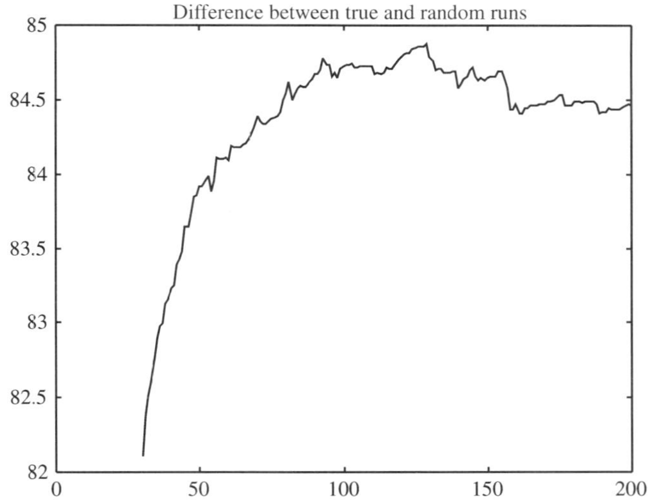

Fig. 11. Data set Soy, response Protein: plot of the optimisation of the GA run.

Fig. 12 shows the spectra and the regions selected in each of the 10 GA elaborations. It can be seen that two regions are consistently selected. The final model is made by taking into account variables 18–21 and 26–28 (see Fig. 13). PLS on these two regions gives a RMSEP of 1.21 (6 components).

In this case, the application of GA did not improve the predictive ability, but the resulting model is much more parsimonious (7 variables, 6 components vs. 175 variables, 8 components).

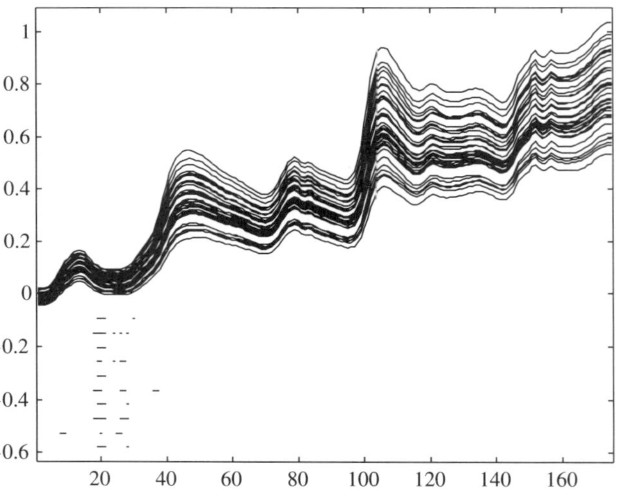

Fig. 12. Data set Soy, response Protein: NIR spectra and selected wavelengths. Each broken line corresponds to the variables selected in each of the 10 GA elaborations.

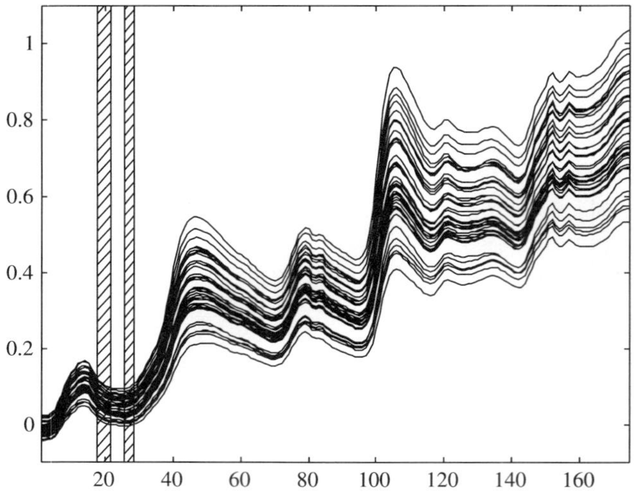

Fig. 13. Data set Soy, response Protein: NIR spectra and selected regions in the final model.

7.2. Data set Additives

For this data set, expert-selected models were already available and used. The regions were selected based on the knowledge about the spectroscopy of the additive, the polymers and the other additives present in the matrix. It is therefore interesting to compare the results of these models, obtained after hard work by a set of experts, with the results of an automated method. The comparison will take into account the predictive ability and the selected regions.

To this data set the iterative approach has been applied. A GA was first performed on the entire spectrum using a window size of 10. The following steps use a window size of 5, 4, 3 and finally 2. The results of this last step are used for the definition of the final model.

7.2.1. Additive B

Fig. 14 shows the average spectrum of the samples containing additive B. The model based on the region selected by the experts (178 wavelengths, highlighted in the figure) gives a RMSEP of 54 (11 components).

The first GA detected five different regions (see Fig. 15), for a total of 210 wavelengths, and its RMSEP was 48 (8 components).

It is interesting to see that the first region lies inside the experts' region, and that the four remaining regions were not taken into account by experts. The small improvement in the predictive ability of the model can therefore be ascribed by the synergistic effect of these regions. Further, it is important to notice that these results have been obtained with a window size of 19.3 cm^{-1} (average of 10 consecutive points), which many experts think is too large to give useful information.

Fig. 16 shows the regions that have been selected as the final model, obtained at the fifth step of the iterative GA (window size of two points).

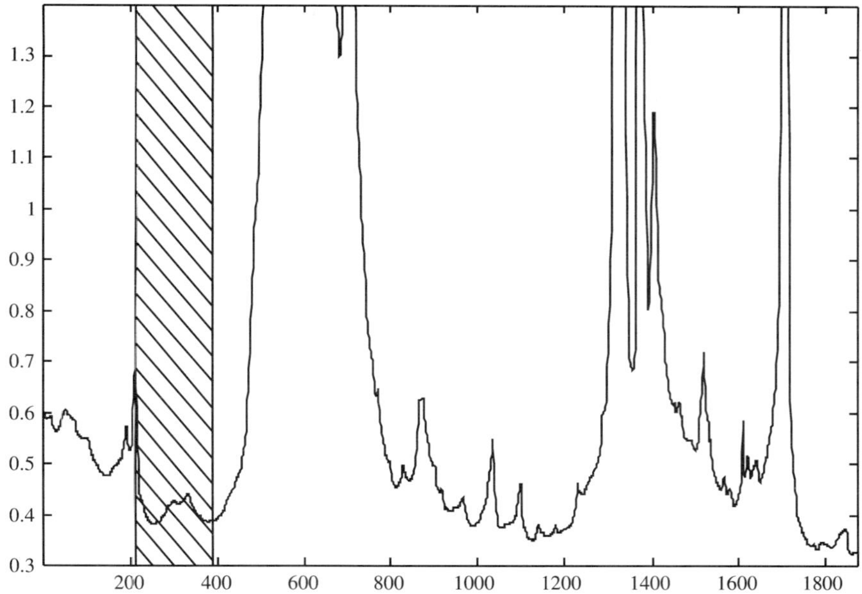

Fig. 14. Data set Additives, response Additive B: average FT-IR spectrum and region selected by experts.

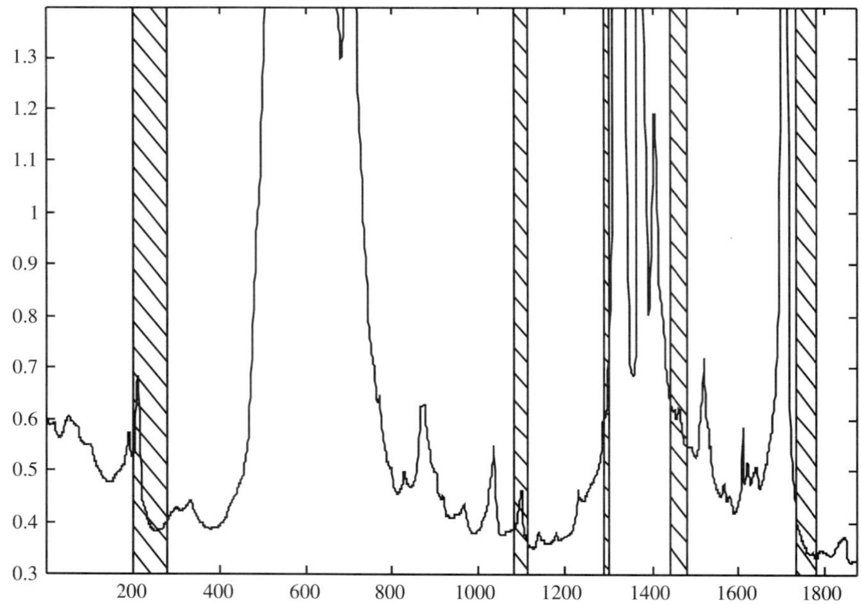

Fig. 15. Data set Additives, response Additive B: average FT-IR spectrum and regions selected by GA applied to the whole spectrum.

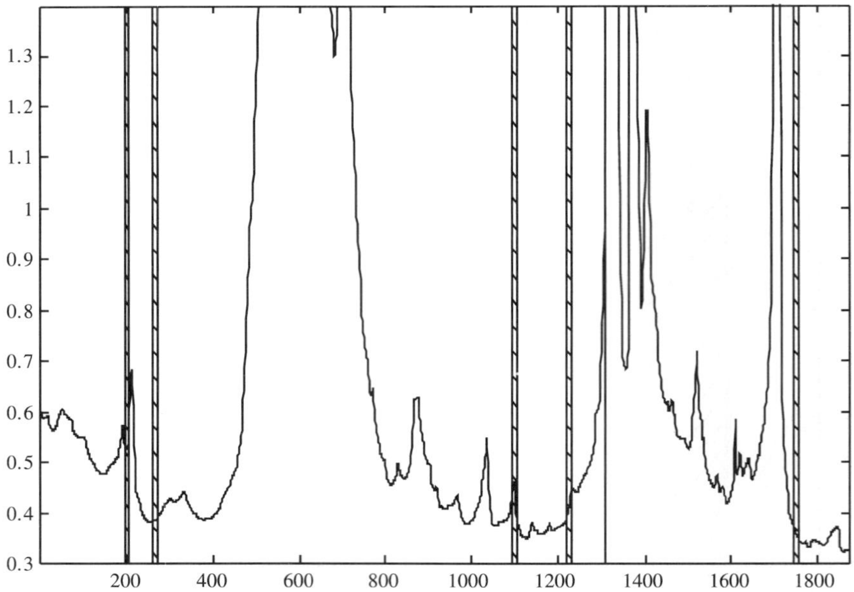

Fig. 16. Data set Additives, response Additive B: average FT-IR spectrum and regions selected by iterative GA.

This model takes into account wavelengths from six different regions, for a total of 60 wavelengths, and has a RMSEP of 48 (6 components). The predictive ability has not been further improved, but fewer wavelengths overall were selected. Inside the region indicated by the experts, two much smaller sub-regions have been selected, indicating a refinement of where the information is contained. The other regions appear to be related to the polymer. It is known that the catalyst 'health' influences the state of this additive, and of course also influences the polymer produced. So, it makes sense that polymer peaks would contribute to modelling this additive.

7.2.2. *Additive C*

Fig. 17 shows the average spectrum of the samples containing additive C (only the part of spectrum relevant to additive C is displayed). The model based on the region selected by the experts (37 wavelengths, highlighted in the figure) gives a RMSEP of 48 (12 components).

The first GA detected three different regions (see Fig. 18), for a total of 190 wavelengths, and its RMSEP was 48 (15 components).

In addition to the region identified by the experts, the GA selected two more regions. The second region was also known to be related to the additive, and at the time of the original modelling it was not clear to the experts whether this region should be included or not. The first region contains variables that might be important for modelling the offset (valley between peaks) and a polymer peak.

Fig. 19 shows the regions selected by the final model, obtained at the fifth step of the iterative GA (window size of two points).

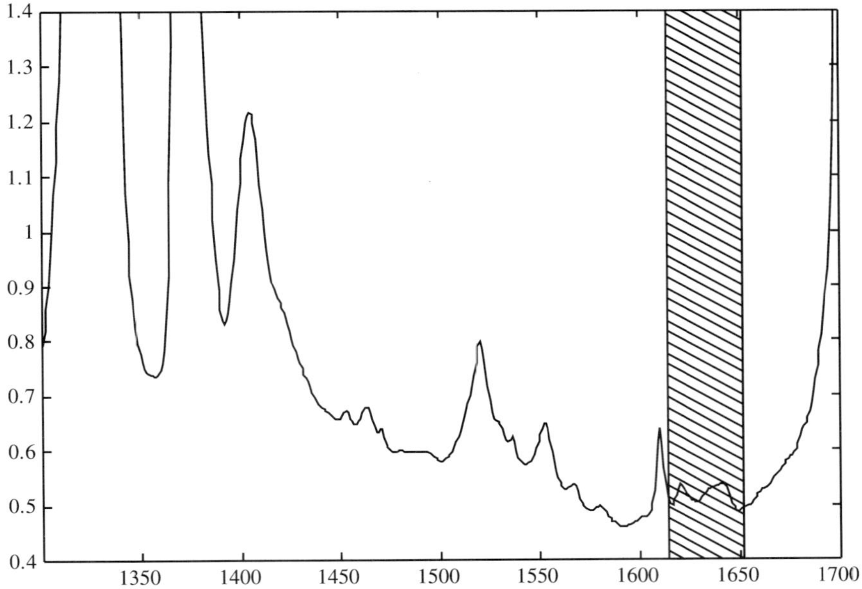

Fig. 17. Data set Additives, response Additive C: average FT-IR spectrum and region selected by experts (only the part of the spectrum related to Additive C is shown).

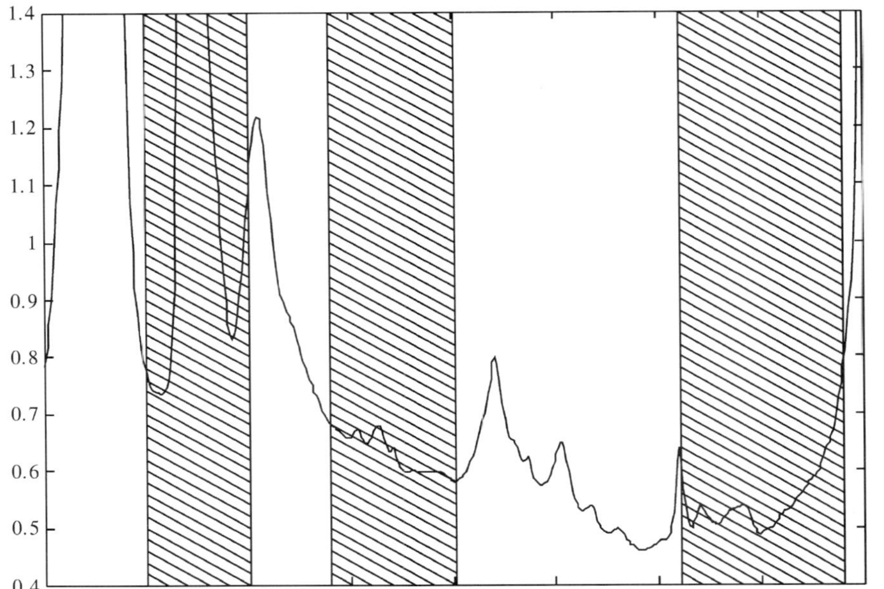

Fig. 18. Data set Additives, response Additive C: average FT-IR spectrum and regions selected by GA applied to the whole spectrum (only the part of the spectrum related to Additive C is shown).

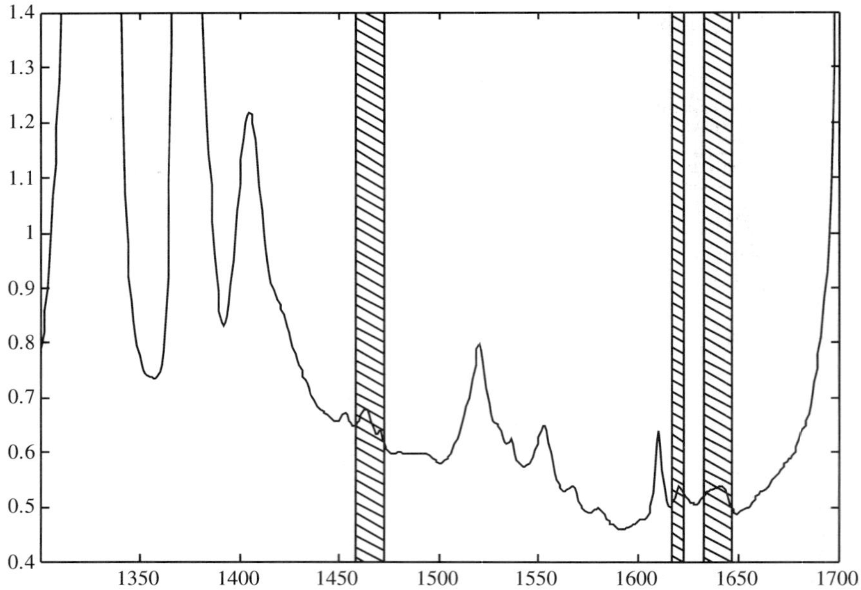

Fig. 19. Data set Additives, response Additive C: average FT-IR spectrum and regions selected by iterative GA (only the part of the spectrum related to Additive C is shown).

This model takes into account wavelengths from three different regions, for a total of 34 wavelengths, and has a RMSEP of 47 (12 components). As for additive B, the predictive ability has not been further improved, but fewer wavelengths overall were selected. The third region of the previous model has been refined by the iterative GA, and two sub-regions have been obtained. It is interesting to see that they are both inside the boundaries of the region selected by the experts and that they perfectly correspond to two small peaks, while the valley between them has not been included. Also the second region of interest was significantly reduced in size and corresponds very well to two small peaks, while the first region has not been selected. These results confirm that the application of the iterative approach provides better defined models, reducing the number of selected wavelengths.

8. Conclusions

It has been shown that a GA can select variables that provide good solutions in terms of both predictive ability and interpretability.

When comparing the selected variables with the models proposed by the experts, it can be seen that the GA models contain the suggested bands or part of the suggested bands, plus additional bands. In some cases these extra regions could be readily interpreted, in other cases they could not. It is logical to say that they contain relevant information since they decrease the RMSEP and they were consistently selected in independent GA elaborations.

It has to be emphasised that this method does not require any spectroscopic experience by the user. Using this approach, a non-expert will be able to efficiently construct reliable calibration models with little or no intervention by an expert. Further, this approach can aid the expert with difficult calibration problems where the variable selection is not obvious.

References

Bangalore, A.S., Shaffer, R.E., Small, G.W., 1996. A genetic algorithm based method for the selection of wavelengths and model size for partial least-squares regression and near-infrared spectroscopy. Anal. Chem. 68, 4200–4212.

Broadhurst, D., Goodacre, R., Jones, A., Rowland, J.J., Kell, D.B., 1997. Genetic algorithms as a method for variable selection in multiple linear regression and partial least squares regression, with applications to pyrolysis mass spectrometry. Anal. Chim. Acta 348, 71–86.

Centner, V., Massart, D.L., de Noord, O.E., de Jong, S., Vandeginste, B.M., Sterna, C., 1996. Elimination of uninformative variables for multivariate calibration. Anal. Chem. 68, 3851–3858.

Ding, Q., Small, G.W., Arnold, M.A., 1998. Genetic algorithm-based wavelength selection for the near-infrared determination of glucose in biological matrices: initialization strategies and effects of spectral resolution. Anal. Chem. 70, 4472–4479.

Forina, M., Drava, G., Armanino, C., Boggia, R., Lanteri, S., Leardi, R., Corti, P., Conti, P., Giangiacomo, R., Galliena, C., Bigoni, R., Quartari, I., Serra, C., Ferri, D., Leoni, O., Lazzeri, L., 1995. Transfer of calibration function in near-infrared spectroscopy. Chemom. Intell. Lab. Syst. 27, 189–203.

Forina, M., Casolino, C., Pizarro Millán, C., 1999. Iterative Predictor Weighting (IPW) PLS: a technique for the elimination of useless predictors in regression problems. J. Chemom. 13, 165–184.

Geladi, P., Kowalski, B.R., 1986. Partial least square regression: a tutorial. Anal. Chim. Acta. 185, 1–17.

Hasegawa, K., Kimura, T., Funatsu, K., 1999. GA strategy for variable selection in QSAR studies: enhancement of comparative molecular binding energy analysis by GA-based PLS method. Quant. Struct.–Activity Relat. 18, 262–272.

Hibbert, D.B., 1993. A hybrid genetic algorithm for the estimation of kinetic parameters. Chemom. Intell. Lab. Syst. 19, 319–329.

Jouan-Rimbaud, D., Walczak, B., Massart, D.L., Last, I.R., Prebble, K.A., 1995. Comparison of multivariate methods based on latent vectors and methods based on wavelength selection for the analysis of near-infrared spectroscopic data. Anal. Chim. Acta. 304, 285–295.

Jouan-Rimbaud, D., Massart, D.L., de Noord, O.E., 1996. Random correlation in variable selection for multivariate calibration with a genetic algorithm. Chemom. Intell. Lab. Syst. 35, 213–220.

Leardi, R., 1996. Genetic algorithms in feature selection. In: Devillers, J., (Ed.), Genetic Algorithms in Molecular Modelling, Academic Press, London, pp. 67–86.

Leardi, R., 2000. Application of genetic algorithm-PLS for feature selection in spectral data sets. J. Chemom. 14, 643–655.

Leardi, R., Lupiáñez González, A., 1998. Genetic algorithms applied to feature selection in PLS regression: how and when to use them. Chemom. Intell. Lab. Syst. 41, 195–207.

Leardi, R., Boggia, R., Terrile, M., 1992. Genetic algorithms as a strategy for feature selection. J. Chemom. 6, 267–281.

Leardi, R., Seasholtz, M.B., Pell, R., 2002. Variable selection for multivariate calibration using a genetic algorithm: prediction of additive concentrations in polymer films from Fourier transform-infrared spectral data. Anal. Chim. Acta. 461, 189–200.

Lestander, T.A., Leardi, R., Geladi, P., Selection of NIR wavelengths by genetic algorithms for determination of seed moisture content. J. NIRS (submitted).

Lindgren, F., Geladi, P., Rännar, S., Wold, S., 1994. Interactive Variable Selection (IVS) for PLS. Part 1: theory and algorithms. J. Chemom. 8 (1994), 349–363.

Nørgaard, L., Saudland, A., Wagner, J., Nielsen, J.P., Munck, L., Engelsen, S.B., 2000. Interval partial least squares regression (iPLS): a comparative chemometric study with an example from near-infrared spectroscopy. Appl. Spectrosc. 54, 413–419.

Pizarro Millán, C., Forina, M., Casolino, M.C., Leardi, R., 1998. Extraction of representative subsets by potential function method and genetic algorithms. Chemom. Intell. Lab. Syst. 40, 33–51.

Spiegelman, C.H., McShane, M.J., Goetz, M.J., Motamedi, M., Yue, Q.L., Coté, G.L., 1998. Theoretical justification of wavelength selection in PLS calibration: development of a new algorithm. Anal. Chem. 70, 35–44.

Thomas, E.V., 1994. A primer on multivariate calibration. Anal. Chem. 66, 795a–804a.

Thomas, E.V., Haaland, D.M., 1990. Comparison of multivariate calibration methods for quantitative spectral analysis. Anal. Chem. 62, 1091–1099.

Westad, F., Martens, H., 2000. Variable selection in near infrared spectroscopy based on significance testing in partial least squares regression. J. Near Infrared Spectrosc. 8, 117–124.

PART II

ARTIFICIAL NEURAL NETWORKS

CHAPTER 7

Basics of artificial neural networks

Jure Zupan

Laboratory of Chemometrics, National Institute of Chemistry, Ljubljana, Slovenia

1. Introduction

The research on artificial neural networks (ANNs) started almost 60 years ago with the pioneering work of McCulloch and Pitts (1943), Pitts and McCulloch (1947), and Hebb (1949). The reasons why it took until the work of Hopfield (1982) to gain full recognition can be at least partially explained by the work of Minsky and Papert (1989) in which they showed that perceptrons have serious limitations for solving non-linear problems. Their very good theoretical treatment of the problem diverted many scientists from working in the field. Hopfield (1982) shed new light on this topic by giving a new flexibility to the old ANN architecture through the introduction of non-linearity and feedback coupling of outputs with inputs.

Parallel to Hopfield's work, and even before then in the seventies and early eighties, research on ANNs proceeded, notably through Kohonen (1972). The interested reader can consult an excellent review of ANNs up to 1987 by Anderson and Rosenfeld (1989). This comprehensive collection of all the basic papers is accompanied by enlightening introductions and is strongly recommended for all beginners in the field.

As a response to the work on error backpropagation learning, which was published by Werbose (1982) and by Rumelhart and co-workers (1986), the interest in ANNs has grown steadily since 1986. Since then a number of introductory texts by Lippmann (1987), Zupan and Gasteiger (1991, 1993, 1999), Gasteiger and Zupan (1993), Despagne and Massart (1998), Basheer and Hajmeer (2000), Mariey et al. (2001), and Wong et al. (2002), to mention only a few, have been published.

Because ANNs are not one, but comprise a group of methods, there are many different situations in which they can be employed. Therefore, potential users have to ask what kind of task and/or sub-tasks are to be solved in order to obtain the final result. Indeed, in the solution of a complex problem many different tasks can be undertaken and one can complete many of them or another particular method using the possibilities offered by ANNs.

Data Handling in Science and Technology, Volume 23
ISSN: 0922-3487

DOI: 10.1016/S0922-3487(03)23007-0

2. Basic concepts

2.1. Neuron

Before actual ANNs are discussed and explained as problem-solving tools, it is necessary to introduce several basic facts about the artificial neuron, the way neurons are connected together, and how different data-flow strategies influence the setup of ANNs.

The way the input data are treated by the artificial (computer-simulated) neuron is similar in action to a biological neuron exposed to incoming signals from neighboring neurons (Fig. 1, left). Depending on the result of the internal transfer function, the computer neuron outputs a real signal y of value that is non-linearly proportional to the m-dimensional input. An appropriate form of the internal transfer function can transform any real-valued or binary input signal $\mathbf{X}_s = (x_{1s}, x_{2s}, \ldots, x_{ms})$ into the real-valued output between fixed minimum and maximum limits, usually between zero and one.

In the computer the neurons are represented as weight vectors $\mathbf{W}_j = (w_{j1}, w_{j2}, \ldots, w_{ji}, \ldots, w_{jm})$. When describing the functioning and/or transfer of signals within the ANNs different visualizations of neurons are possible: either as circles (Fig. 1, middle) or as column vectors (Fig. 1, right). Because any network is composed of many neurons, they are assigned a running index j and accordingly all the properties associated with the specific neuron j bear an index j. For example, the weights belonging to neuron j are written as $\mathbf{W}_j = (w_{j1}, w_{j2}, \ldots, w_{ji}, \ldots, w_{jm})$.

The calculation of the output value from the multi-dimensional vector $\mathbf{X}_s$ is carried out by the transfer function with the requirement that the outputted value is confined within

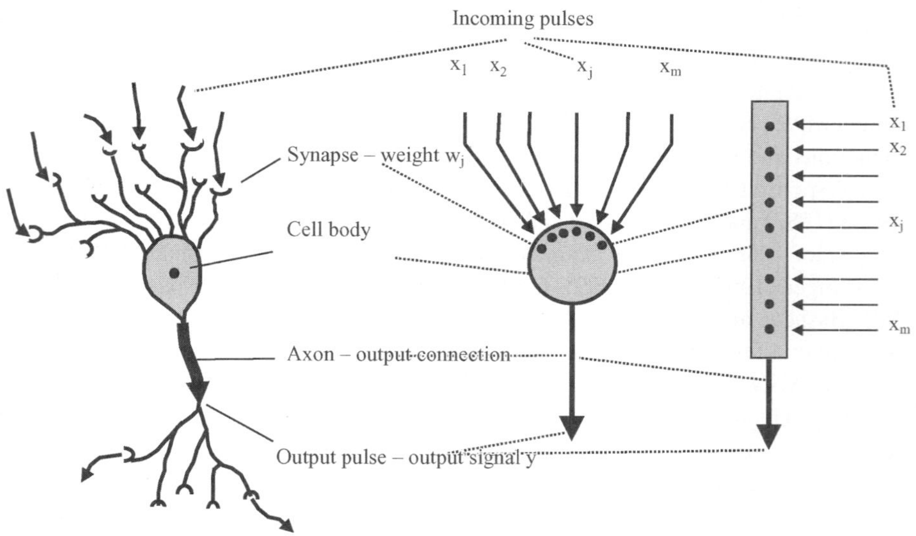

Fig. 1. Biological (left) and a computer-simulated neuron $\mathbf{W} = (w_1, w_2, \ldots, w_j, \ldots, w_m)$ in two visualizations: as a circle (middle) and as a column vector (right).

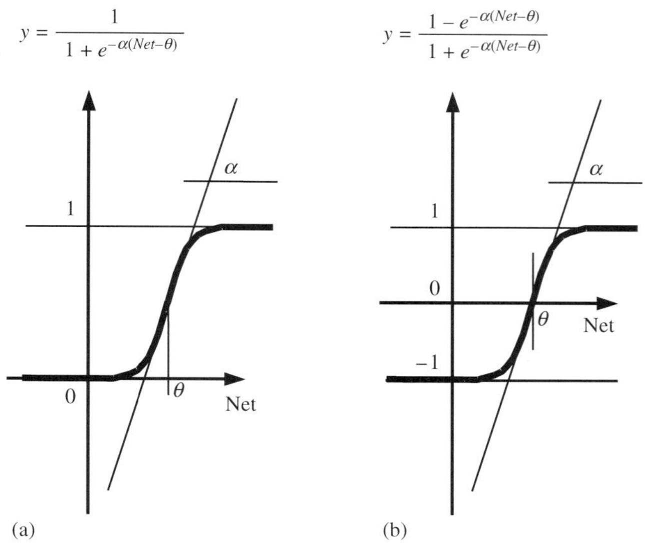

Fig. 2. Two different squashing functions: standard sigmoid (left) and tanh (right).

a finite interval. The transfer function usually has one of the two forms (Fig. 2)

$$\mathbf{y}_j = \frac{1}{1 + \mathrm{e}^{-\alpha_j(\mathrm{Net}_j - \theta_j)}} \tag{1}$$

$$\mathbf{y}_j = \frac{1 - \mathrm{e}^{-\alpha_j(\mathrm{Net}_j - \theta_j)}}{1 + \mathrm{e}^{-\alpha_j(\mathrm{Net}_j - \theta_j)}} \tag{2}$$

where the argument Net_j, called the net input, is a linear combination of the input variables:

$$\mathrm{Net}_j = \sum_{i=1}^{m} w_{ji} x_i \tag{3}$$

Once chosen, the form of the transfer function is used unchanged for all neurons in the network. What changes during the learning are the weights w_{ji}, the function parameters that control the position of the threshold value θ_j, and the slope α_j. The strength of each synapse between an axon and a dendrite (this means each weight w_{ji}) defines the proportion of the incoming signal x_i which is transmitted to the body of the neuron j.

By adding a dummy variable to the input the two mentioned parameters, α_j and θ_j, can be treated (corrected) during the learning procedure in exactly the same way as all other weights w_{ji}. Let us see how. First, the argument of the sigmoid function (Eq. 1) is combined with Eq. (3):

$$\alpha_j(\mathrm{Net}_j + \theta_j) = \sum_{i=1}^{m} a_j w_{ji} x_i + \alpha_j \theta_j$$

$$= \alpha_j w_{j1} x_1 + \alpha_j w_{j2} x_2 + \cdots + \alpha_j w_{ji} x_i + \cdots + \alpha_j w_{jm} x_m + \alpha_j \theta_j \tag{4}$$

and then for the new weights $a_j w_{ji}$ the same letter w_{ji} is used as before:

$$w_{j1}x_1 + w_{j2}x_2 + \cdots + w_{ji}x_i + \cdots + w_{jm}x_m + w_{jm+1} \tag{5}$$

At the beginning of the learning process none of the constants, α_j, w_{ji}, or θ_j, is known. Therefore, it does not matter if the products $\alpha_j w_{ji}$ and $\alpha_j \theta_j$ are written simply as one new unknown constant. This form requires the addition of one variable x_{m+1} to each input vector **X** obtaining $\mathbf{X}' = (x_1, x_2, \ldots, x_i \ldots, x_m, x_{m+1})$. The additional variable x_{m+1} is set to 1 in all cases; hence, one can write all augmented input vectors $\mathbf{X}'$ as $(x_1, x_2, \ldots, x_i \ldots, x_m, 1)$. This manipulation is made because one wants to incorporate the weight w_{jm+1} originating from the product $\alpha_j \theta_j$ into the new term Net_j containing *all* parameters (weights, threshold, and slope) in a unified form, which makes the learning process of the neurons much simpler. The additional input variable in $\mathbf{X}'$ is called the 'bias' and makes the neurons more flexible and adaptable. Eqs. (1) and (3) for the calculation of the output y_j of the neuron j are now simplified:

$$\mathbf{y}_j = \frac{1}{1 + \mathrm{e}^{-\mathrm{Net}_j}} \tag{1a}$$

and

$$\mathrm{Net}_j = \sum_{i=1}^{m+1} w_{ji} x_i \tag{3a}$$

It is important to realize that the inclusion of biases into neurons increases the number of weights by one into each of them. More weights (parameters) in the network (model) require more objects in the training set.

2.2. Network of neurons

ANNs are composed of different number of neurons. In chemical applications the size of the networks, i.e. the number of neurons, ranges from fewer than ten to tens of thousands. In the ANN the neurons can be placed into one, two, three, or even more layers of neurons. Fig. 3 shows a one-layer network in two variations. In both variations the network is designed to accept two variables x_1 and x_2 as input and to yield three outputs y_1, y_2, and y_3. In the upper left network (Fig. 3, upper, left), each neuron has two weights. Altogether there are six weights w_{ji} connecting the input variables i with the outputs j. The upper right network (Fig. 3, right) is designed to be more adaptive than the left one; hence, the *bias* input is added to the two input variables. Because one-layer networks are not able to adapt themselves to highly non-linear data, additional layers of neurons are inserted between the input and the output layers. The additional layers are called *hidden* layers. The hidden layer in Fig. 3 (lower) contains two neurons each having three weights: two for accepting the variables x_1 and x_2 and the additional one for accepting the bias input. Together with the weights that link the hidden layer with the output neurons, the network has 15 weights.

Most frequently used ANNs have one hidden layer. Seldom are two hidden layers and rarely (for very specific applications only) is a neural network with more than two hidden

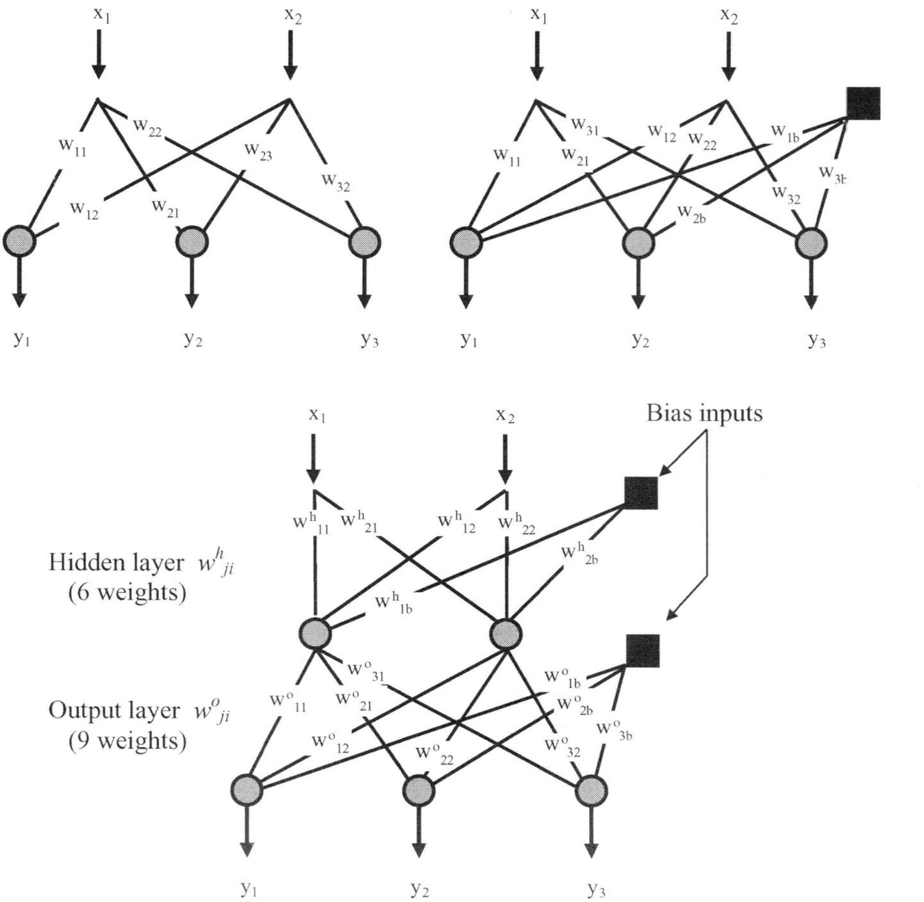

Fig. 3. Three neural networks. Above are two one-layer networks, the left one without, and the right one with the bias input and bias weight. Bias input is always equal to 1. Below is a two-layer neural network. The weights in different layers are distinguished by upper indices: '*h*' and '*o*' for *hidden* and *output*, respectively.

layers of neurons ever used. There is no recipe for the determination of either the number of layers or the number of neurons in each layer. At the outset only the number of input variables of the input vector $\mathbf{X}_s$ and the number of sought responses of the output vector $\mathbf{Y}_s$ are fixed. The factor that limits the maximal total number of neurons that can be employed in the ANN is the total amount of available data. The same rule as for choosing the number of coefficients for a polynomial model is applicable for determination of the number of weights in ANNs too. The number *should not* exceed the number of objects selected for the training set. In other words, if one wants to use the network shown in Fig. 3 with 15 weights at least 16 objects must be in the set used to train it. After the total number of weights is determined, the number of neurons in each layer has to be adjusted to it. Usually this is made by trial and error.

It is interesting to note that in spite of the fact that neural networks allow models with several outputs (like the ANNs shown in Fig. 3), for the modeling of real-valued output chemists rarely employ this advantage. Instead, they use as many different and separately trained networks as there are outputs. In the mentioned case probably three networks each with only one output would be made. The multi-output networks are regularly used for the *classification* problems where each output signals the probability for the input object to belong to a class associated with the particular output.

Some of the available ANN software programs already contain a build-in option, which automatically selects the optimal number of neurons in the hidden layer.

3. Error backpropagation ANNs

Error backpropagation learning is a supervised method, i.e. it requires a set of n_t input and target *pairs* $\{\mathbf{X}_s, \mathbf{R}_s\}$. With each m-dimensional object $\mathbf{X}_s$, an n-dimensional target (response) $\mathbf{R}_s$ must be associated. For example, m-component analysis of a product, m-intensity spectrum of a compound, or m experimental conditions of a chemical process, must be accompanied by n properties of the product, n structural features to be predicted, or n responses of the process to be controlled.

The generation of models by the error backpropagation strategy is based on the comparison between the actual output $\mathbf{Y}_s$ of the model and the answer $\mathbf{R}_s$ as provided by each known input/output pair ($\mathbf{X}_s$, $\mathbf{R}_s$). The method is named after the order in which the weights in the neurons are corrected during the learning (Fig. 4).

The weights are corrected from the output towards the input layer, i.e. in the *backward* direction to which the signals are propagated when objects are input into the network. The correction of the ith weight in the jth neuron of the layer l is always made according to the equation:

$$\Delta w_{ji}^{l} = \eta \delta_j^l \, \mathrm{out}_j^{l-1} + \mu \Delta w_{ji}^{l,\,\mathrm{previous}} \tag{6}$$

The first term defines the learning rate by the rate constant η (between 0.01 and 1.0) while the other term takes into account the momentum for learning μ. The superscript *previous* in the momentum term refers to the complete weight change from the previous correction. As the indices indicate, the term δ_j^l is related to the error committed by the jth neuron in the lth layer. The *input* signal out_j^{l-1} that causes the error is coming to the weight w_{ji} as the *output* from neuron i of the layer above it ($l-1$), hence the notation out_i^{l-1}. The term δ_j^l is calculated differently for the correction of weights in the hidden layers and in the last (output) layer. The detailed reasoning and mathematical background of this fact can be found in the appropriate textbooks (Zupan and Gasteiger, 1999):

$$\delta_j^{\mathrm{output}} = (t_j - y_j^{\mathrm{output}}) y_j^{\mathrm{output}} (1 - y_j^{\mathrm{output}}) \qquad \text{output layer} \tag{7}$$

$$\delta_j^{\mathrm{hidden}} = \left(\sum_{k=1}^{n_r} \delta_k^{\mathrm{output}} w_{kj}^{\mathrm{output}} \right) y_j^{\mathrm{hidden}} (1 - y_j^{\mathrm{hidden}}) \qquad \text{hidden layers} \tag{8}$$

The summation in Eq. (8) involves all n_r neurons in the level below the hidden layer for which the δ_j contribution is evaluated. The situation is shown in Fig. 5. The situation does

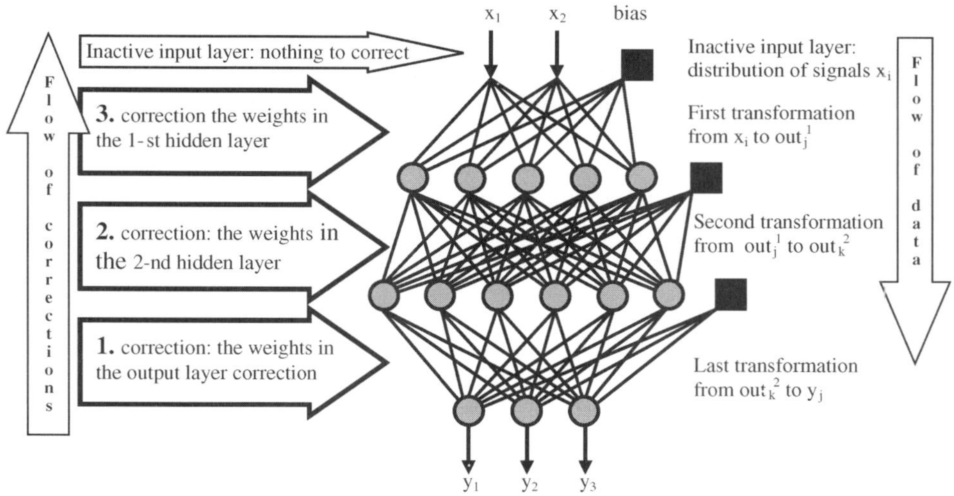

Fig. 4. Order of weight corrections (left-side arrow) in the error-backpropagation learning is opposite (backward) to the flow of the input signals (right-side arrow).

not change if the there is more then one hidden layer in the network. The only difference is that the index *output* changes to *hidden* and index *hidden* to *hidden-1*.

The momentum term in Eq. (6) symbolizes the *memory* of the network. It is included to maintain the change of the weights in the same direction as they were changed during the previous correction. A complex multivariate non-linear response surface has many

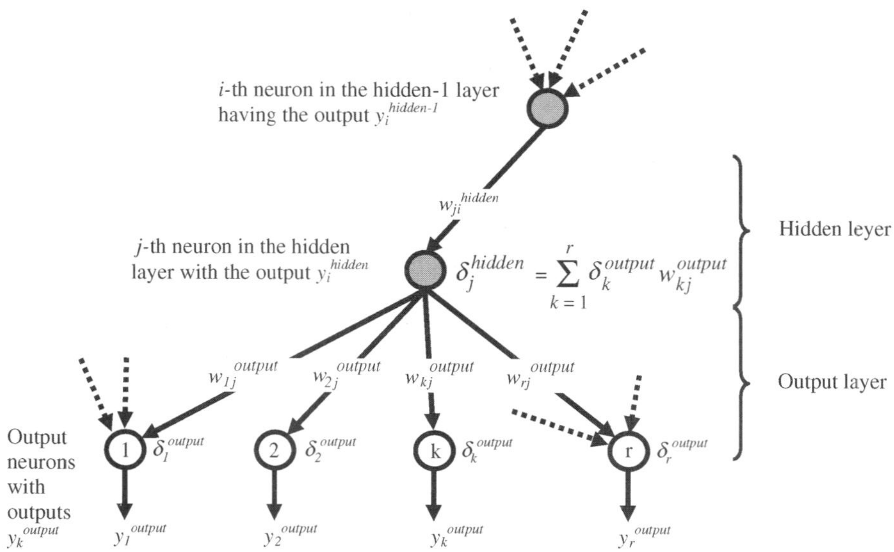

Fig. 5. The evaluation of the term δ_j^{hidden} in the hidden layer is made using the weighted sum of the δ_k contributions of one layer of neurons below.

traps or local minima and therefore, there is an imminent danger that the system will be captured by them if encountered. The momentum term is necessary because without it the learning term immediately reverses its sign if the system error starts increasing and the model trapped in the local minimum. By the inclusion of the momentum term the learning procedure continues the trend of weight changes for a little while (depending on the size of μ) trying to escape from the local minima. The momentum constant μ is usually set to a value between 0.1 and 0.9 and in some cases might vary during the learning.

4. Kohonen ANNs

4.1. Basic design

The Kohonen networks (see Kohonen 1972, 1988, 1995) are designed for handling the *unsupervised* problems, i.e. they do not need any targets or responses $\mathbf{R}_s$ for each object $\mathbf{X}_s$. In the absence of targets the weights in the Kohonen networks are learning to mimic the similarities with the objects. If the main concern in the error backpropagation networks is to train its weights to produce the answer closest to the response $\mathbf{R}_s$ of each individual object $\mathbf{X}_s$, then the main goal of the Kohonen layer is to train each neuron to mimic one or a group of similar objects and *to show the location* of the most similar neuron to an unknown object $\mathbf{X}$ which is input to the network. Therefore, the Kohonen network produces what is often called a *s*elf-*o*rganized *m*ap (SOM). A Kohonen type of network is based on a *single* layer of neurons ordered in a line or spread in a planar rectangular or hexagonal manner. Fig. 6 shows a rectangular Kohonen network.

Because the Kohonen ANNs are seeking an optimal topological distribution (positions) of the neurons with the respect to the objects, the *layout* of neurons, the *topology* of the neighborhood, and the actual *distances* of each neuron to its neighbors are very important. In a linear layout, each neuron has only *two* closest neighbors (one on each side), two second-order neighbors, etc. In the rectangular layout each neuron has *eight* immediate neighbors, sixteen second-order neighbors, twenty-four third-order neighbors, and so on, while in the hexagonal layout, there are only *six* first-order neighbors, twelve second-order neighbors, eighteen third-order neighbors, etc. (Fig. 7).

Although the topological distance between two neurons i and j in the, say, first neighborhood area is fixed (equal to 1), the distance $d(\mathbf{W}_i, \mathbf{W}_j)$ between the corresponding weight vectors $\mathbf{W}_i$ and $\mathbf{W}_j$ can differ considerably. Since each neuron influences its neighbors, the neurons on the borders and in the corners have less influence on the entire network than the ones in the middle of the plane or line. One can just ignore the problem (many computer programs do this), or alternatively, one can balance the inequality of the influence of particular neurons by 'bending' the line or plane of neurons in such a way that the ends or edges join their opposite parts. In the computational sense this means that neighbors of the last neuron in the line become the neighbors to the first one (Fig. 8, top). Similarly, in the planar rectangular layout the edge a of the neurons' layer is linked to the edge b, while the upper row c becomes the neighbor of the bottom row d (Fig. 8, middle). The situation of making a continuous plane in the hexagonal layer of neurons is solved by

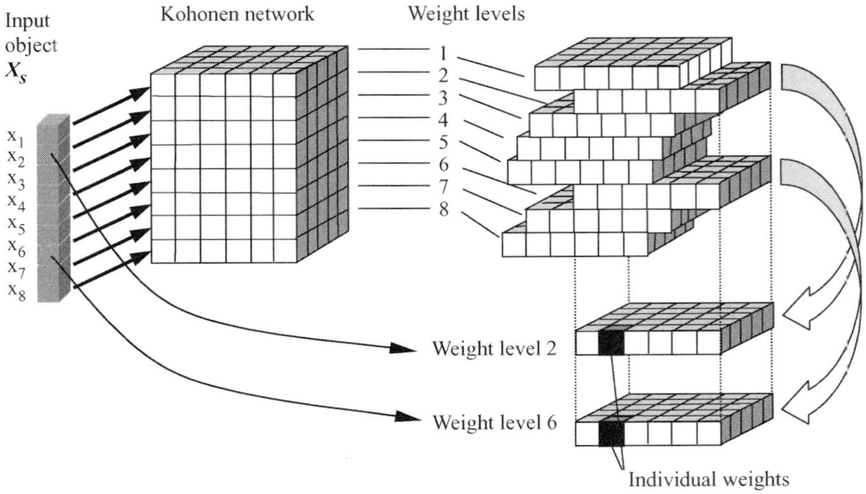

Fig. 6. The Kohonen network. The input object $\mathbf{X}_s$ and neurons are represented as columns; the complete network is a three-way matrix. The weights that accept the same variable, x_i, are arranged in planes or levels. Each weight is represented as a small box in a plane of weights (bottom, right).

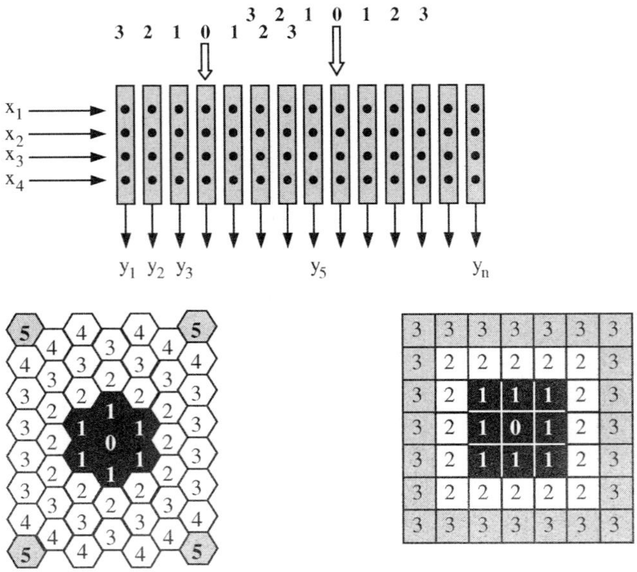

Fig. 7. In different Kohonen network layouts, neurons at the *same* topological distance (the same number) from the excited neuron $\mathbf{W}_e$, marked as '0', have different numbers of neighbors.

linking the *three* pairs of the opposite edges in the hexagon in such a way that they become the neighbors of their opposite edges (Fig. 8, below). This is equivalent to covering the plane with the hexagonal tiles. In the hexagonal Kohonen network restricted by the *toroid* boundary conditions, very interesting and informative patterns can emerge.

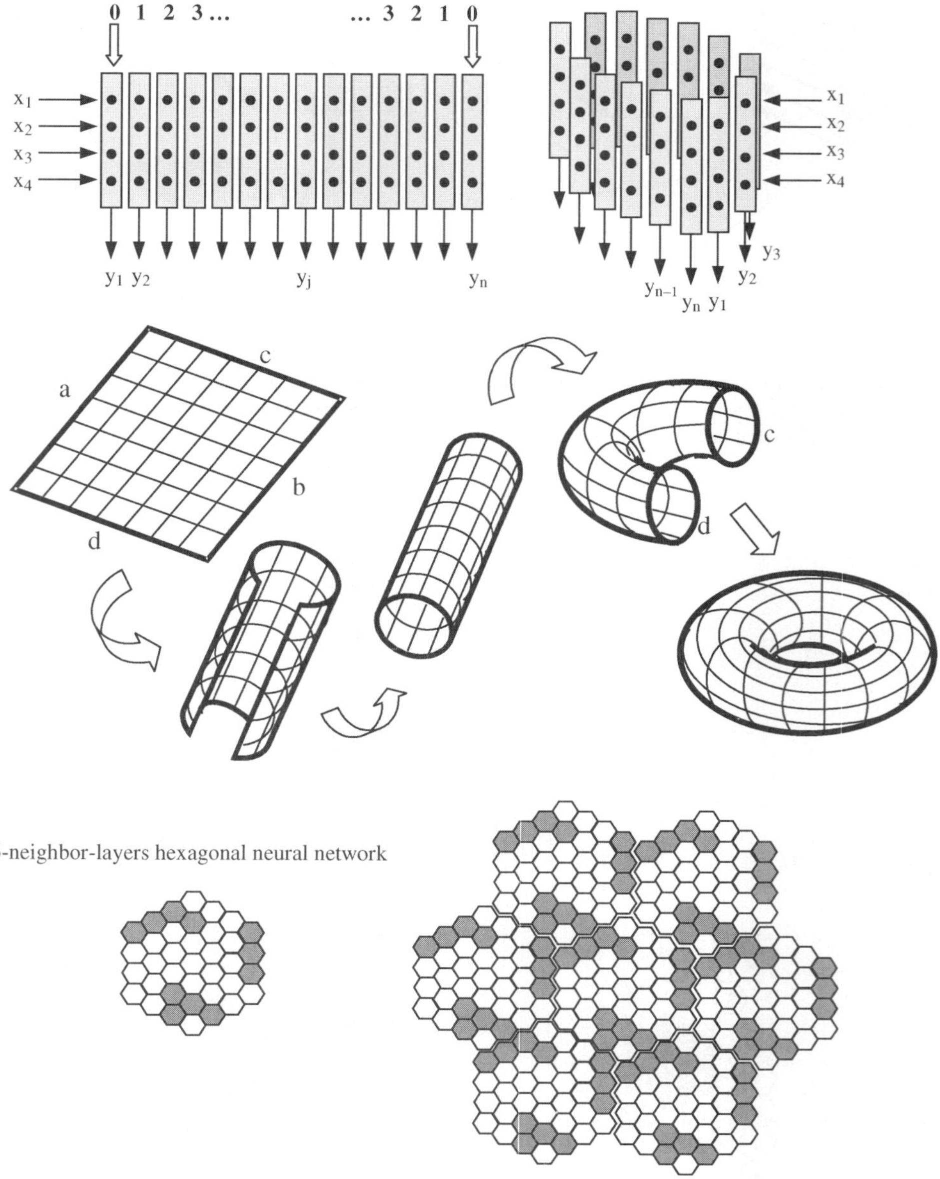

Fig. 8. Bending of the linear (above) Kohonen network into a circle and the rectangular one into a toroid (middle). A hexagonal neural network layer of neurons can be seen as a hexagonal tile. Tilling the plane with a single top map pattern can yield better information as obtained from the single one (below).

After an object enters the Kohonen network the learning starts with the selection of only one neuron $\mathbf{W}_e$ from the entire assembly of neurons. The selection can be based on the largest response among all neurons in the layer or on the closest match between the weights of each neuron and the variables of the object. The latter criterion of the similarity between the neuron's weights and the variables of the input object is employed in a vast majority of all Kohonen learning applications. The selected neuron is called the excited neuron $\mathbf{W}_e$. The similarity between the neuron j, represented as a weight vector $\mathbf{W}_j$ $(w_{j1}, w_{j2}, \ldots, w_{jm})$ and the object $\mathbf{X}_s = (x_{s1}, x_{s2}, \ldots, x_{sm})$ is expressed in terms of the Euclidean distance $d(\mathbf{X}_s, \mathbf{W}_j)$. The largest the distance, the smaller the similarity, and vice versa:

$$d^2(X_s, W_j) = \sum_{i=1}^{m} (x_{si} - w_{ji})^2 \tag{9}$$

$$W_e \leftarrow \min\left\{\sum_{i=1}^{m} (x_{si} - w_{ji})^2\right\} \qquad j = 1, 2, \ldots, e \ldots, N_{\text{network}} \tag{10}$$

Once the excited neuron $\mathbf{W}_e$ is found, the corrections which produces better similarity or smaller distance between $\mathbf{X}_s$ and $\mathbf{W}_e$ if the same object $\mathbf{X}_s$ is input to the network can be obtained. Again a very simple formula is used:

$$\Delta w_{ji} = \eta a(d_j)(x_{si} - w_{ji}^{\text{old}}) \tag{11}$$

Eq. (11) yields the correction of weights for all neurons at a certain distance around the excited neuron $\mathbf{W}_e = (w_{e1}, w_{e2}, \ldots, w_{em})$. The learning rate constant η is already familiar from the backpropagation learning, while the topological dependence is achieved via the factor $a(d_j)$. Additionally, through the function $a(\cdot)$ the shrinking condition is implemented. Namely, the neighborhood around the excited neuron in which the corrections are made must shrink as the learning continues

$$a(d_j) = 1 - \frac{d_j}{[d_{\max}(n_{\text{epoch}}) + 1]} \qquad d_j = 0, 1, 2, \ldots, d_{\max} \tag{12}$$

Parameter $d_{\max}(n_{\text{epoch}})$ defines the maximal topological distance (maximal neighborhood) to which neurons should be corrected. Neurons $\mathbf{W}_j$ that are more distant from the $\mathbf{W}_e$ than $d_{\max}$ are not corrected at all. Making $d_{\max}$ dependent on the number of learning epochs, n_{epoch}, causes the neighborhood of corrections to shrink during the learning. One epoch is the period of learning in which all objects from the training set pass once through the network. For the excited neuron $\mathbf{W}_e$ the distance d_j between $\mathbf{W}_e$ and $\mathbf{W}_e$ is zero, and then the term $a(d_j)$ becomes equal to 1. With increasing d_j within the interval $\{0, 1, 2, 3, \ldots, d_{\max}\}$ the local correction factor $a(d_j)$ yields linearly decreasing values from 1 to the minimum of $1 - d_{\max}/(d_{\max} + 1)$. The rate by which the parameter $d_{\max}$ decreases linearly with increasing current number of epochs of training n_{epoch} is given by the following equation:

$$d_{\max} = N_{\text{net}}\left(1 - \frac{n_{\text{epoch}}}{n_{\text{tot}}}\right) \tag{13}$$

At the beginning of learning ($n_{epoch} = 1, n_{epoch} << n_{tot}$), d_{max} covers the entire network ($d_{max} \sim N_{net}$), while at the end of learning ($n_{epoch} = n_{tot}$), d_{max} is limited only to the excited neuron $\mathbf{W}_e (d_{max} = 0)$. The parameter n_{tot} is a predefined maximum number of epochs that the training is supposed to run. N_{net} is the maximal possible topological distance between any two neurons in a given Kohonen network. Linking all three Eqs. (12–14) together, the correction of weights in any neuron $\mathbf{W}_j$ can be obtained:

$$\Delta w_{ji} = \eta \left(1 - \frac{d_j}{N_{net}\left(1 - \frac{n_{epoch}}{n_{tot}}\right) + 1} \right) (x_{si} - w_{ji}^{old}) \tag{14}$$

Additionally, during the training procedure, the learning constant η can be changed:

$$\eta = (\eta^{start} - \eta^{final})(1 - n_{epoch}/n_{tot}) + \eta^{final} \tag{15}$$

This can be easily implemented on the backpropagation networks as well.

4.2. Self-organized maps (SOMs)

After the training is completed the entire set of the training vectors $\mathbf{X}_s$ must be run through the network again. The last run is used for labeling all neurons that are excited by at least one object $\mathbf{X}_s$. The label can be any information associated with the object(s). The most usual labels are ID numbers, chemical names, structures, objects' class identification, values of a certain property, chemical structures, etc. Labeling of the neurons is stored in the so-called *top-map* or *self-organized* map. The top-map consists of memory cells (boxes) arranged in exactly the same manner as the neurons in the network.

The simplest top-maps show the number of objects that have excited each neuron (Fig. 9, top left). Such a map gives a two-dimensional frequency distribution of objects in the measurement space mapped on the plane of neurons. The map of the objects' distribution enables easier decisions for the planning of additional experiments, for the selection of representative objects, for the selection of subgroups, etc. Another possibility is to display the class memberships of the objects (Fig. 9, top right). If the representation of objects reflects the relevant information about each class, one expects that after the training the objects will be clustered into assigned classes. In general, it is expected that objects belonging to the same class will excite topologically close neurons. Due to the experimental errors, bad reparation of clusters, inadequate representation of objects, or any other reason, the top-map can show conflicting neurons, i.e. neurons excited by objects belonging to *different* classes. In Fig. 9 (top right), three such neurons are shown in black. Both the frequency distribution and the class membership can be shown on one map. A slightly more complex way of making a top-map is to display lists of ID numbers of objects that excite each neuron (Fig. 9, bottom). Because Kohonen mapping is non-linear there will almost always be some empty cells in the top-map (neurons not excited by any object) as well as cells representing neurons excited by many objects. The number of objects exciting various neurons can be highly unbalanced, ranging from zero to as much as large proportions of the entire population. Therefore, the quality of such a display

1	1	2	**3**			**4**
1	**6**		2	1		**5**
2		1	1	1		2
	1	1	1			1
1	1	1			1	1
2	1	2	1	1	1	2
1	1		1	2	2	**8**

A	A	A	A,C			B
A	A		A,C	C		B
A		C	C	C		B
	C	C	C			B
C	C	C			B	B
C	C	C,B	B	B	B	B
C	C		B	B	B	B

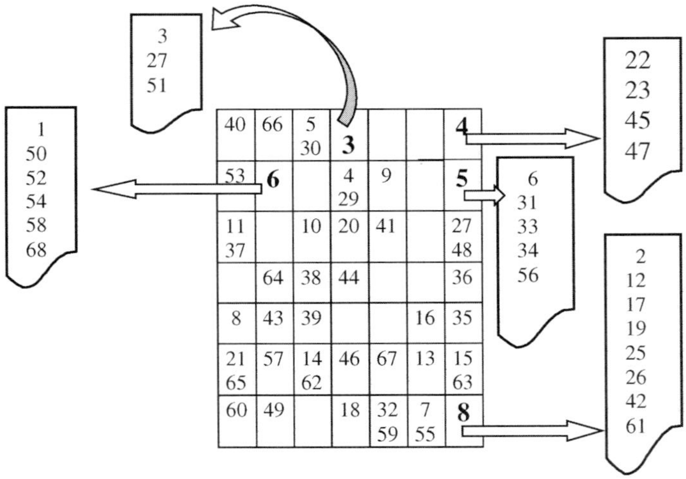

Fig. 9. Different top-maps of the same 7 × 7 Kohonen network. Frequency distribution of objects in the two-dimensional Kohonen map (top, left), distribution of objects according to the three class assignments (top, right), and lists of identification numbers of objects exciting the neurons.

depends on the program's ability to link each neuron in the Kohonen network to the corresponding object in the database.

All neurons $\mathbf{W}_j$ of the Kohonen network are m-dimensional vectors and an additional way to show the relations between the adjacent neurons is to calculate four distances between each particular neuron $\mathbf{W}_j$ and its four (non-diagonal) neighbors of the first neighborhood ring. The display of the results in topologically equivalent boxes can be made on the *double top-map* (Fig. 10, left). The combination of the double top-map and a class assignation can provide very rich information, such as the relation between the objects, between and within the clusters of objects, and the relative separation of different clusters at different positions in the measurement space. This last information is particularly important when the trained Kohonen network is used for the classification of unknown objects which excite the empty neurons, i.e. the neurons forming the gap between the clusters. Still another use of the SOM is formation of the U-matrix, which is obtained from the double top-map by substituting each cell with the average of the four

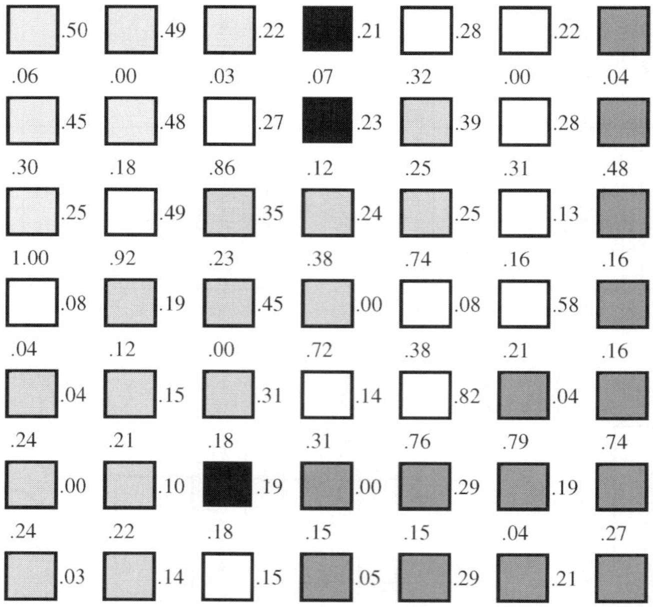

Fig. 10. Double top-map with the distances between the adjacent neurons (left) and the *U*-matrix (right). The numbers shown between the neurons (shaded boxes on left) are distances normalized to the largest one. The three 'empty' neurons are black. Each cell of the *U*-matrix contains the average distance to its four closest neighbors shown at left.

(three or two) distances to the neighboring neurons (Fig. 10, right). The *U*-matrix displays the reverse density of objects distributed in the space. The smaller the value, the denser are the neurons in the network. Such maps can serve for outlier detection.

Because in the Kohonen neural network the neurons are represented as columns it is easy to see that the weights w_{ji} accepting the same variable x_i are arranged as *planes* or *levels* (square, rectangular, or hexagonal). The term *level* or *plane* refers to the arrangement of *weights* within the layer (to be distinguished from the level) of neurons. The term level specifies a group of weights to which one specific component x_{si} of the input signal $\mathbf{X}_s$ is connected (see weight levels in Fig. 6). This means that in each *level* of weights only the data of one specific variable is handled. In the trained network the weight values in one weight level form a *map* showing the distribution of that particular variable. The combinations of various two-dimensional *weight maps* together with the top-map (specifying the class assignment, frequency distribution, or similarity) are the main source of the information in the Kohonen ANNs. Fig. 11 shows a simple example of how the overlap of two-dimensional weight maps together with the top-map information can give an insight into the relation between the properties of the object and the combination of input variables. The overlap of a specific class of samples (cluster of paint samples of quality class A on the top-map shown in Fig. 11) with the corresponding identical areas in the maps of all three variables defines the range of combinations of variables in which the paint with the quality of class A can be made.

.76	.85	.90	.93	.95	.97	.99
.63	.75	.84	.80	.85	.89	.95
.59	.66	.68	.65	.77	.75	.92
.47	.45	.52	.58	.60	.69	.85
.31	.33	.46	.42	.53	.61	.81
.16	.25	.30	.33	.46	.55	.75
.03	.09	.08	.22	.38	.50	.67

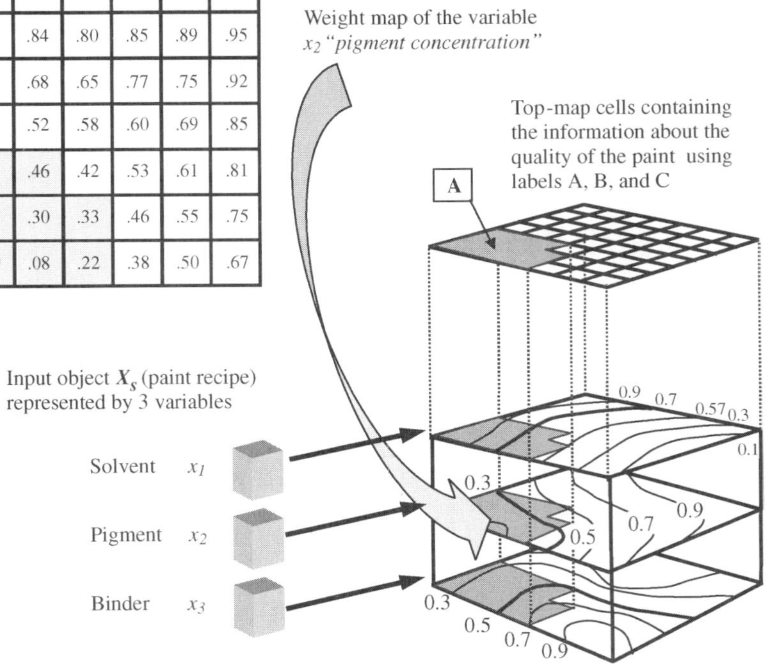

Fig. 11. Overlap of the weight maps with the part of the top–map (right) gives the information about the range of a specific variable (variable x_2 'the pigment concentration' in the shown case) for the corresponding class defined in the top-map (label 'A'). The weight map for the 'pigment concentration' (top, left) is from a real example of modeling coating paint recipes. Each cell of the weight matrix has its equivalent in the top-map. For a better visualization the weight planes are presented as maps with 'iso-variable' lines.

5. Counterpropagation ANNs

Counterpropagation neural networks were developed by Hecht-Nielsen (1987a,b, 1988) as the Kohonen networks augmented in such a way that they are able to handle the targets $\mathbf{R}_s$ associated with inputs $\mathbf{X}_s$. Counterpropagation ANNs are composed of two layers of neurons each having identical number and layout of neurons. The input or the Kohonen layer acts exactly in the same way as already discussed in the previous section. The second layer acts as a 'self-organizing' network of outputs. The number of weights in the Kohonen and in the output layer correspond to the number of input variables in the input vector $\mathbf{X}_s = (x_{s1}, x_{s2}, \ldots, x_{si}, \ldots, x_{sm})$ and the number of responses r_{sj} in the response vectors $\mathbf{R}_s = (r_{s1}, r_{s2}, \ldots, r_{sj}, \ldots, r_{sn})$, respectively (Fig. 12).

There are no fixed connections between the neurons of both layers in the sense that the signals from the Kohonen layer would be transmitted to the neurons in the output layer. The connection to the second layer of neurons is created each time *at the different location* only *after* the Kohonen layer processes the input signal. Still, *no* flow of data between the two layers is realized. The information that connects both layers of neurons is the *position*

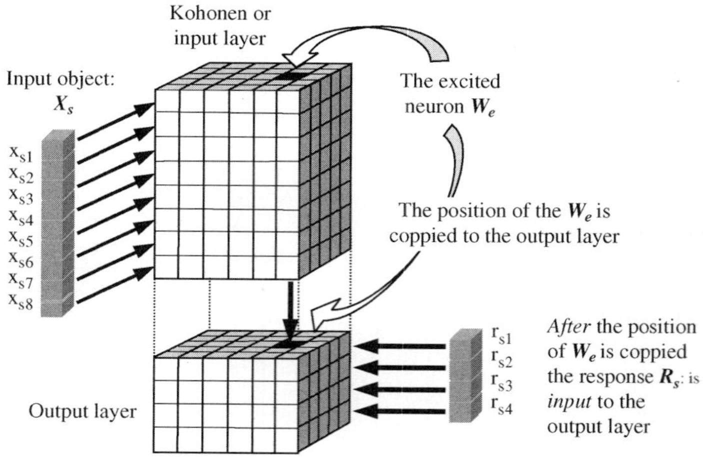

Fig. 12. A counterpropagation neural network is composed of two layers of neurons. Each neuron in the upper layer has its corresponding neuron in the lower output layer. Inputs $\mathbf{X}_s$ and responses $\mathbf{R}_s$ are *input* to the network during the training from opposite layers.

of the excited neuron $\mathbf{W}_e$ in the Kohonen layer that is copied to the lower one. This is the reason why the layout of neurons in both layers of the counterpropagation neural network must be identical.

After the excited neuron $\mathbf{W}_e$ in the Kohonen layer and its counterpart in the output layer, are identified, the correction of the weights is executed in exactly the same way as given by Eq. (11):

$$\Delta w_{ji}^{\text{Kohonen}} = \eta a(d_j)(x_{si} - w_{ji}^{\text{old, Kohonen}}) \tag{16}$$

$$\Delta u_{ji}^{\text{output}} = \eta a(d_j)(r_{si} - u_{ji}^{\text{old, output}}) \tag{17}$$

In the output layer the corrections of weights u_{ji}^{output} are made according to the target vectors $\mathbf{R}_s$, which are associated in pairs $\{\mathbf{X}_s, \mathbf{R}_s\}$ with the input vectors $\mathbf{X}_s$. The aim of the corrections in the output layer is similar to that in the Kohonen: to minimize the difference between the weights u_{ji} and the response r_{si}. In the counterpropagation ANN the response vectors $\mathbf{R}_s$ have exactly the same role as the $\mathbf{X}_s$. Because the response $\mathbf{R}_s$ enters the network in the same way as the object $\mathbf{X}_s$, but from the opposite, i.e. from the output side, this type of ANN is called counterpropagation. The complete training procedure of the counterpropagation network can be summarized in three steps.

- determination of the excited neuron in the Kohonen layer:

$$W_e \leftarrow \min\left\{\sum_{i=1}^{m}(x_{si} - w_{ji}^{\text{Kohonen}})^2\right\} \qquad j = 1, 2, \ldots, e \ldots, N^{\text{Kohonen}} \tag{18}$$

- correction of the weights in the Kohonen layer around the excited neuron W_e:

$$\Delta w_{ji} = \eta a(d_j)(x_{si} - w_{ji}^{\text{old}}) \tag{19}$$

- corrections of the weights in the output layer around the W_e position copied from the input layer:

$$\Delta u_{ji} = \eta a(d_j)(r_{si} - u_{ji}^{\text{old}}) \tag{20}$$

In Eqs. (19) and (20) the neighborhood function $a(d_j)$ and the learning rate η are the same as in Eqs. (12) and (15), respectively.

After the counterpropagation ANN is trained it produces the self-organized map of responses $\mathbf{R}_s$ accessible via the locations determined by the neurons excited through the training input objects $\mathbf{X}_s$. The input layer of the counterpropagation ANN acts as a pointer device determining for each query $\mathbf{X}^{\text{query}}$ the position of the neuron in the output layer in which the answer or prediction $\mathbf{Y}^{\text{query}}$ is stored.

This form of the information storage can be regarded as a sort of a *look-up table*. The most widely used forms of look-up tables are dictionaries, telephone directories, encyclopedias, etc. The disadvantage of the look-up table is that no information is available for a query that is not stored in the table. Another disadvantage is that in order to obtain an answer, the sought entry must be given exactly. If only one piece of the query (let us say one letter) is missing the information cannot be located even if given in the table. To a large extent the counterpropagation network, if used as a look-up table, overcomes these disadvantages. An answer is obtained for *every* question, provided that its representation (number of variables) is the same as that of the training objects. The quality of the answer, however, depends heavily on the distance between the query and the closest object used in the training. Nevertheless, any answer, even a very approximate one, can be useful in the absence of other information. Second, for the corrupted or incomplete queries which can be regarded as objects not given in the training set, the answer is always assured with the quality of the answer depending on the extent of corruption of the query.

It is important to realize that the answers are stored in *all* neurons in the output layer regardless of whether its counterpart neuron in the Kohonen layer was excited during the training or not. Hence, *any* input will produce an answer. The counterparts of the non-exited neurons in the output layer contain the 'weighted' averages of all responses. The individual proportional constant to this average is different for each object and is produced during the training. It depends not only on the responses, but also on the position of each neuron in the network, on the geometry of the network, and on all training parameters, i.e. the number of epochs, the learning rate, the initialization, etc.

In Kohonen and counterpropagation ANNs the training usually requires several hundred epochs to be completed, which is one to two orders of magnitude less than that required during the error backpropagation learning phase.

6. Radial basis function (RBF) networks

We have seen that both the Kohonen and the error backpropagation type of ANNs require intensive training. Corrections of weights are repeated many times after each entry in cycles (called epochs). In order to obtain a *self-organized map* with the Kohonen network, several hundred if not thousands of epochs are needed, while for producing a model by the error backpropagation learning, at least an order of magnitude more epochs are necessary. In contrast to these two learning methods, the radial basis function (RBF) network learning is *not* iterative. The essence of RBF networks is the conception that any function $y = f(\mathbf{X})$ can be approximated by a *set* of localized basic functions $\phi_j(\mathbf{X})$ in the form of a linear combination

$$y = \sum_{j=1}^{n} w_j \Phi_j(X) \tag{21}$$

where $\mathbf{X}$ represents an m-dimensional object $\mathbf{X} = (x_1, x_2, \ldots, x_m)$. For a more detailed description see Renals (1989), Bishop (1994), Derks (1995), or Walczak and Massart (1996). Once the set of basic functions $\{\phi_j(\mathbf{X})\}$ is determined, the calculation of appropriate weights w_j is made by standard multiple linear regression (MLR) techniques. As a localized function the Gaussian function for m-dimensional vectors $\mathbf{X}$ is mostly used:

$$\Phi_j(X) = A_j \exp\left[-\frac{(X - B_j)^2}{2C_j^2}\right] \qquad \text{for} \qquad \mathbf{X} = (x_1, x_2, \ldots, x_m) \tag{22}$$

Parameters A_j, B_j, and C_j are different and local for each function Φ_j. The parameters A_j are always omitted because the amplitudes can be incorporated in the weight vectors w_{ji}. The local point of the basis function Φ_j is $\mathbf{B}_j$, a vector in the same m-dimensional space as the input vector $\mathbf{X}$, while parameter C_j is the *width* or the *standard deviation* σ_j, which is in most cases set to a constant value σ equal for all functions Φ_j. Fig. 13 shows one- and two-dimensional cases of Φ_j for two different values of σ at two different positions $\mathbf{B}_j$. Because the parameter $\mathbf{B}_j$ is in the same space as all object $\mathbf{X}_s$, it will be written as $\mathbf{X}_j^c$.

The architecture of the RBF network is simple. It is similar to the error backpropagation ANN with one hidden layer. It has an *inactive* input layer, a hidden layer (consisting of p radial basis functions $\phi_j(\mathbf{X}_s - \mathbf{X}_j^c)$ with *non-linear* transfer), and one *linear* output layer. The connection between the nodes in different layers is shown in Fig. 14. The first layer serves only for the distribution of all variables x_i of the signal $\mathbf{X}_s$ to all nodes in the hidden layer, while the output layer collects all outputs from the hidden layer to one single output node. Of course, more output nodes $y_{s1}, y_{s2}, \ldots, y_{sn}$ could be easily incorporated, but here our attention will be focused to the RBF networks having only one output.

The first significant difference between the RBF and the sigmoid transfer function is that the input signal $\mathbf{X}_s$ is handled by the sigmoid transfer function as a *scalar* value Net_j (see Eqs. (1) and (3)), while in the RBF networks the $\mathbf{X}_s$ enters the RBF as a *vector* with m components:

$$\mathrm{out}_{sj} = \Phi_j(X) = \exp\left[-\frac{(X_s - X_j^c)^2}{\sigma^2}\right] \tag{23}$$

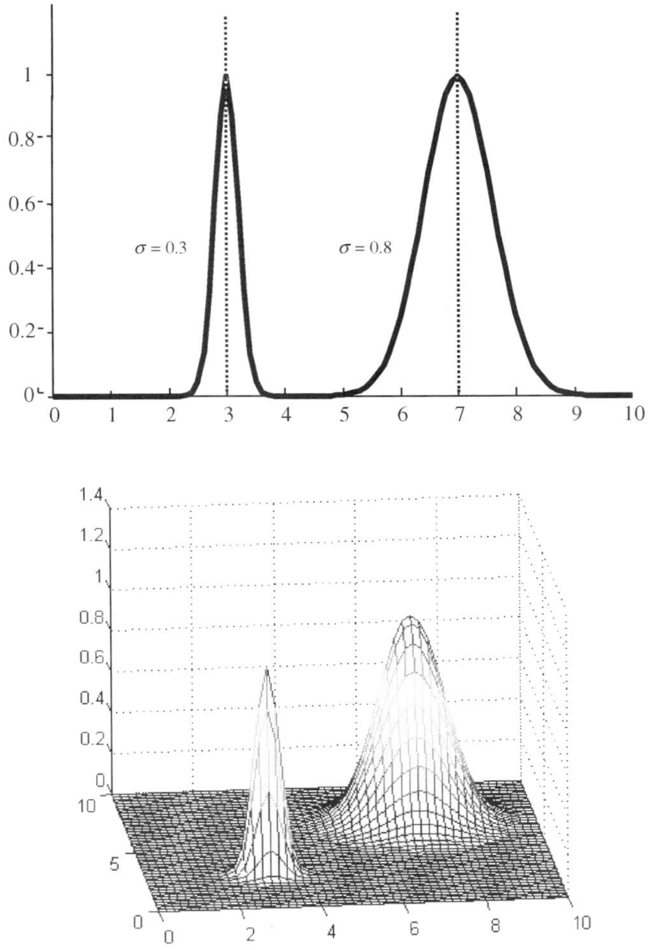

Fig. 13. Gaussian functions as localized radial basis functions (RBF) in one-dimensional (above) and two-dimensional space (below). In the upper example the centers B_1 and B_2 are equal to 3 and 7, while in the two-dimensional case below the centers $\mathbf{B}_1$ and $\mathbf{B}_2$ are (3,3) and (7,7), respectively. The widths of the left RBFs above and below are 0.3 while the right ones are 0.8.

The result out_j of the RBF Φ_j in the hidden layer is transferred to the output layer via the weight w_j, making the final output y_s a linear combination of all results from the RBFs and corresponding weights:

$$y_s = \sum_{j=1}^{r} w_j \mathrm{out}_{sj} + w_{\mathrm{bias}} = \sum_{j=1}^{r} w_j \Phi_j(X_s, X_j^c, \sigma) + w_{\mathrm{bias}} \tag{24}$$

The main concern in setting up the RBF network is the determination of the number of RBF functions r and finding all r local vectors $\mathbf{X}_j^c$ for each RBF. Although there is no single recipe, there are many ways to do this.

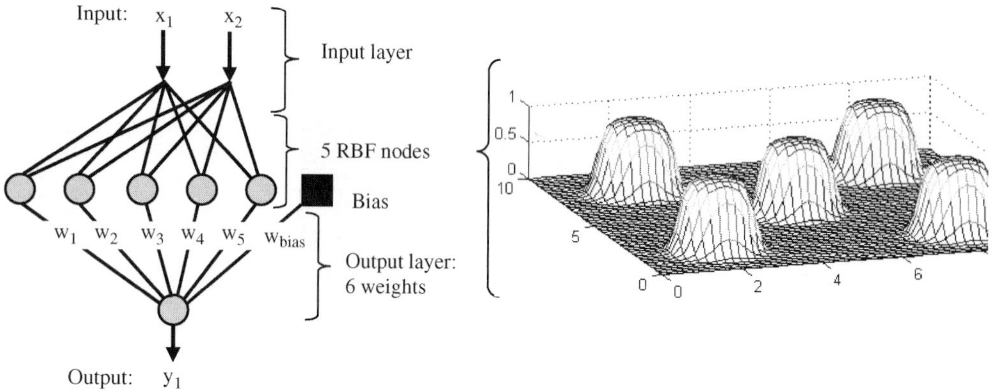

Fig. 14. An RBF network layout consists of three layers: input, hidden, and output. The hidden layer in the above example consists of five two-dimensional RBF functions (right). All RBFs have R and σ equal to 100 and 0.3, respectively, while the centers $\mathbf{X}_1^c$ to $\mathbf{X}_5^c$ have the coordinates (2,2), (2,8), (8,2), (8,8) and (5,5), respectively. An extra bias signal (black square) can be added to the RBF layer or not. If it is added it is transferred to the output via the weight w_{bias}.

A general strategy for the selection of centers $\mathbf{X}_j^c$ is to put them into such positions in the measurement space that they cover the areas within which the large majority of objects can be found. If one has enough indications that the data are distributed evenly in the measurement space a random selection of $X_j^c s$ values may be a good choice. The reasonable, but not necessarily the best way is to use a subset of the existing objects $\{\mathbf{X}_s\}$. On the other hand, for clustered or irregularly distributed data, more sophisticated methods have to be chosen. It is always advised to do some pre-processing of data by any clustering and/or separation technique. After the number of clusters in the measurement space are determined each cluster is supposed to provide one Φ_j. Hence the centers $\mathbf{X}_j^c$ are selected as averages or any other statistical vector representation of all objects in the clusters.

The final output y_s (we are talking about the RBF with only one output) is a linear combination of RBF outputs and weights. For each measurement $\mathbf{X}_s$ each RBF function Φ_j yields a different response out_{sj}. Hence for the sth measurement, r_s, the following relation between the output y_s and RBF outputs out_{sj} is obtained:

$$y_s = w_1\text{out}_{s1} + w_2\text{out}_{s2} + \cdots + w_n\text{out}_{sn} + w_{\text{bias}} = \sum_{j=1}^{n} w_j\text{out}_{sj} + w_{\text{bias}} \tag{25}$$

In order to obtain n weights w_j and the w_{bias} one must have at least $n+1$ equations:

$$\begin{aligned}
r_1 &= y_1 = w_1\Phi_1(\mathbf{X}_1) + w_2\Phi_2(\mathbf{X}_1) + \cdots + w_n\Phi_n(\mathbf{X}_1) + w_{\text{bias}} \\
r_2 &= y_2 = w_1\Phi_1(\mathbf{X}_2) + w_2\Phi_2(\mathbf{X}_2) + \cdots + w_n\Phi_n(\mathbf{X}_2) + w_{\text{bias}} \\
r_{p-1} &= y_{p-1} = rw_1\Phi_1(\mathbf{X}_n) + w_2\Phi_2(\mathbf{X}_n) + \cdots + w_n\Phi_n(\mathbf{X}_n) + w_{\text{bias}} \\
r_p &= wy_p = w_1\Phi_1(\mathbf{X}_{n+1}) + w_2\Phi_2(\mathbf{X}_{n+1}) + \cdots + w_n\Phi_n(\mathbf{X}_{n+1}) + w_{\text{bias}}
\end{aligned} \tag{26}$$

which in turn requires at least $n+1$ different input/output pairs $\{\mathbf{X}_s, r_s\}$. In the large majority of cases one has more measurements than weights ($p > n$). The above system of p equations with $n+1$ unknown weights w_j ($j = 1, 2, \ldots\, n, \text{bias}$), can be written in the matrix form:

$$[\mathbf{R}] = [\mathbf{\Phi}] \times [\mathbf{W}] + w_{\text{bias}} \tag{27}$$

where $[\mathbf{R}]$ and $[\mathbf{W}]$ are two $p \times 1$ and $n \times 1$ column matrices with elements r_i, and w_j, respectively, while $[\mathbf{\Phi}]$ is a $p \times n$ matrix with the elements out_j. The w_{bias} should be included in the matrix $[\mathbf{W}]$ as the $(n+1)$st element and, correspondingly, the matrix $[\mathbf{\Phi}]$ is augmented by one column of elements, all equal to 1. For the determination of the one-column matrix $[\mathbf{W}]$ containing all of the sought weights the following steps are followed:

$$\begin{aligned}
&[\mathbf{R}] = [\mathbf{\Phi}] \times [\mathbf{W}] \\
&[\mathbf{\Phi}]^{\mathrm{T}} \times [\mathbf{R}] = ([\mathbf{\Phi}]^{\mathrm{T}} \times [\mathbf{\Phi}] \times [\mathbf{W}] \\
&([\mathbf{\Phi}]^{\mathrm{T}} \times [\mathbf{\Phi}])^{-1} \times ([\mathbf{\Phi}]^{\mathrm{T}} \times [\mathbf{R}]) = [\mathbf{W}] \\
&[\mathbf{W}] = ([\mathbf{\Phi}]^{\mathrm{T}} \times [\mathbf{\Phi}])^{-1} \times ([\mathbf{\Phi}]^{\mathrm{T}} \times [\mathbf{R}])
\end{aligned} \tag{28}$$

Eq. (28) can be written in a more explicit form as follows:

$$\begin{bmatrix} w_1 \\ w_2 \\ \cdots \\ w_n \\ w_{\text{bias}} \end{bmatrix} = \left[\begin{bmatrix} \text{out}_{11} & \text{out}_{12} & \cdots & \text{out}_{1p} \\ \cdots & \cdots & \cdots & \cdots \\ \text{out}_{n1} & \text{out}_{n2} & \cdots & \text{out}_{np} \\ 1 & 1 & \cdots & 1 \end{bmatrix} \begin{bmatrix} \text{out}_{11} & \cdots & \text{out}_{n1} & 1 \\ \text{out}_{12} & \cdots & \text{out}_{n2} & 1 \\ \cdots & \cdots & \cdots & \cdots \\ \text{out}_{1p} & \cdots & \text{out}_{np} & 1 \end{bmatrix} \right]^{-1} \times \begin{bmatrix} \text{out}_{11} & \text{out}_{12} & \cdots & \text{out}_{1p} \\ \cdots & \cdots & \cdots & \cdots \\ \text{out}_{n1} & \text{out}_{n2} & \cdots & \text{out}_{np} \\ 1 & 1 & \cdots & 1 \end{bmatrix} \begin{bmatrix} r_1 \\ r_2 \\ \cdots \\ r_p \end{bmatrix} \tag{29}$$

One can easily see that in the evaluation of Eq. (28) or (29) the hardest numerical problem is the determination of the inverse matrix $([\boldsymbol{\Phi}]^{\mathrm{T}}[\boldsymbol{\Phi}])^{-1}$. This task can be achieved by various elaborate methods like Jacobean iteration. However, for a novice to the field, the easiest way to execute the above calculation including the inverse matrix calculation is with the help of the build-in matrix operation capabilities of the MATLAB® mathematical package, which runs on personal computers and on the Linux system.

7. Learning by ANNs

Learning, in the context of ANNs, means adapting the weights of the network to the specific set of data. If the learning is *supervised*, as in the case of the error backpropagation, the learning procedure is controlled by correcting the differences between the desired targets (responses) $\mathbf{R}_s = (r_{s1}, r_{s2}, \ldots r_{sj}, \ldots r_{sn})$ and the actual outputs $\mathbf{Y}_s = (y_{s1}, y_{s2}, \ldots y_{sj}, \ldots y_{sn})$ of the network.

A simple and statistically sound measure for the quality of the fit is the root-mean-square-error or RMSE, which measures the mentioned difference:

$$\text{RMSE}^{\text{supervised}} = \sqrt{\sum_{s=1}^{n_t} \sum_{j=1}^{n} \frac{(r_{sj} - y_{sj})^2}{n_t n}} \tag{30}$$

The aim of the training is to obtain the model that will give the smallest possible RMSE value. One has to be aware that the lowest RMSE is not the one that is recorded on the training set, but must be obtained on an independent set of objects *not used* in the training. To distinguish this final RMSE value from the RMSE value used in the training phase (in either the supervised or unsupervised learning procedure) this value is assigned the superscript test, $\text{RMSE}^{\text{test}}$. Besides the experimental error of the measured data there are other factors that influence the quality of the trained network, such as a the choice of the network design, the choice of the initial network parameters, statistical adequacy of the selected training set from the entire data collection, etc.

The testing with data not used in the training can be performed even before the training is completed. If the model is tested, say, after each of ten epochs of training, with the independent set of objects, one can follow the comparison between the $RMSE^{test}$ and $RMSE^{training}$ values obtained on the training data (Fig. 15). The behavior of both curves is predictable. At the beginning there is a longer period of training where both RMSE values are decreasing, with the $RMSE^{test}$ values always *above* the $RMSE^{training}$ one. Later on, the $RMSE^{training}$ is still decreasing, but the $RMSE^{test}$ value may reach its minimum. If this happens, i.e. the $RMSE^{test}$ curve shows a minimum at a certain number of epochs and from that point starts to increase, an indication of the *over-training* effect is obtained.

The over-training effect is a phenomenon caused by the fact that the model has too many weights which, after a certain period of learning, start to adapt themselves to the noise. This is an indication that the chosen layout of the neural network may not be the best one for the available data set. The model will not have a generalizing ability because it is too flexible, enabling the adaptation of weights and consequently the output(s) to all noise in the data. Such a network contains too many weights (degrees of freedom) and adapts to all small and non-essential deviations of responses $\mathbf{R}_s$ of the training set from the general trends that the data represent. It is advisable to find such a network where $RMSE^{test}$ would not have a minimum, but should steadily approach a certain limiting value (the difference between the minimal $RMSE^{test}$ and final $RMSE^{test}$). The gap should be as small as possible, but still larger than $RMSE^{training}$.

Even if the test and training set of objects are selected appropriately and the model with the lowest $RMSE^{test}$ is obtained, there is still a need for additional concern before giving the final model the credit it deserves. The problem is connected to the understanding of the

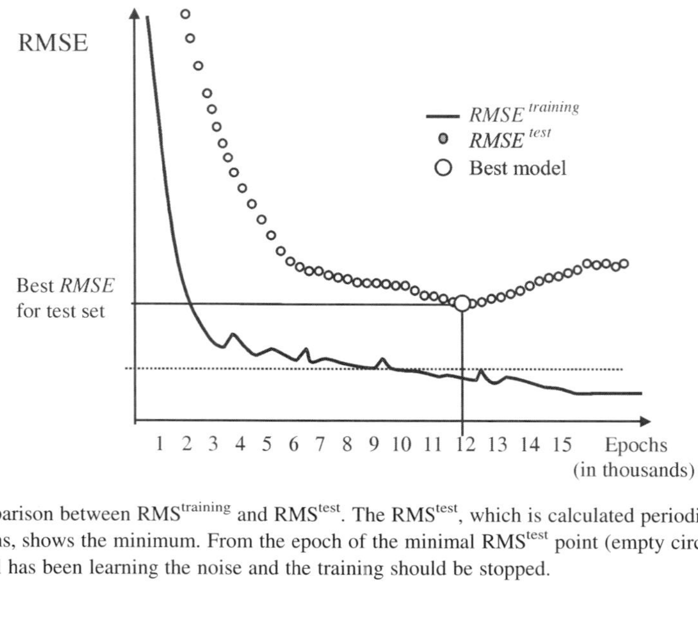

Fig. 15. Comparison between $RMS^{training}$ and RMS^{test}. The RMS^{test}, which is calculated periodically after each 10–100 epochs, shows the minimum. From the epoch of the minimal RMS^{test} point (empty circle) it is evident that the model has been learning the noise and the training should be stopped.

requirement that *'the model must be validated by an independent test set of data not used in the training procedure'*. The point is that the test set used to detect the over-training point is in a sense misused in the learning procedure. It has not been used directly for feedback corrections of weights, but it has nevertheless been used in the decision when and how to redesign the network and/or for the selection of new parameters (learning and momentum constants, initialization of weights, number of epochs, etc.). Such decisions can have stronger consequences and implications for the model compared to changes of weights dictated by the training data. Sometimes even a completely new ANN design is required. Considering such effects, it seems justified to claim that the test set was actually involved in the training procedure, which disqualifies it from making the final judgment about the quality of the model.

The solution of this situation is to employ a third set of data, the so-called *validation* set, and prove the quality of the model with it. The third test set should not be composed of objects used in either the training or the test phase. If the $\text{RMSE}^{\text{validation}}$ obtained with the third set is within the expected limits posed by the experimental error and/or the expectations derived from the knowledge of the entire problem, then it can be claimed that the neural network model was successfully completed.

In many cases one does not have enough data to form three independent different sets of data that will ensure reliable validation of the generated model. In such cases a 'leave-one-out' validation test can be employed. In this type of validation, one uses all available p data for generation of p models in the same layout of neurons in exactly the same way, with the same parameters, initialization, etc. The only difference between these p models is that each of them is trained with $p - 1$ objects, leaving one object out of the training set each time. The object left out from the training is used as a test. This procedure ensures that p models yield p answers y_j (predictions) which can be compared to the p responses r_j. The $\text{RMSE}^{\text{leave-one-out}}$ or the correlation coefficient (Massart et al. 1997) $R^{\text{leave-one-out}}$ between the predictions of the model y_j and responses r_j can be evaluated as an estimate of the reliability of the model:

$$R^{\text{leave-one-out}} = \frac{\sum_{j=1}^{p} y_j r_j - \frac{1}{p}\sum_{j=1}^{p} y_j \sum_{j=1}^{p} r_j}{\sqrt{\left[\sum_{j=1}^{p} y_j^2 - \frac{1}{p}\left(\sum_{j=1}^{p} y_j\right)^2\right]\left[\sum_{j=1}^{p} r_j^2 - \frac{1}{p}\left(\sum_{j=1}^{p} r_j\right)^2\right]}} \tag{31}$$

On the other hand, if the learning is *unsupervised*, as in the case of the Kohonen networks, the generation of the model is controlled by the monitoring of the difference between the input objects $\mathbf{X}_s = (x_{s1}, x_{s2}, \ldots x_{sj}, \ldots x_{sm})$ and the excited neurons, i.e. the corresponding weight vectors $\mathbf{W}_{es} = (w_{e,s1}, w_{e,s2}, \ldots w_{e,sj}, \ldots w_{e,sm})$. Because for each object $\mathbf{X}_s$ a different neuron is excited, two indices e and s are used to label the excited neuron:

$$\text{RMSE}^{\text{un-supervised}} = \sqrt{\sum_{s=1}^{n_t} \sum_{j=1}^{m} \frac{(x_{sj} - w_{e,sj})^2}{n_t m}} \tag{32}$$

Eq. (32) is formally similar to Eq. (30), but is quite different in its essence. While Eq. (30) calculates the difference between the *targets* and the *outputs*, Eq. (32) evaluates the difference between the *objects* and the *neurons* excited by the inputs. Because of a strong tendency of the unsupervised Kohonen method to learn noise, i.e. the tendency that the neurons adapt exactly to the input objects, one has to be careful when using Eq. (32) as a stop criterion. Its use is especially damaging if the Kohonen network has more neurons than there are objects in the training set (sparsely occupied Kohonen ANNs). Because in the mathematical sense the Kohonen ANN is not a model, the number of objects in the training set is by no means a restriction for the number of neurons used in the network. In many cases Kohonen ANNs contain many more neurons than there are input objects and in these cases the use of a stop criterion given by Eq. (31) is strongly discouraged. As was said before, it is much better that the Kohonen learning is carried out to a predefined number of training epochs, n_{tot}, although other *trial and error* stop criteria, like visual inspection of top-map clusters can be used.

8. Applications

In the last 20 years since the publication of Hopfield (1982) the number of ANN applications in various fields of chemistry has grown rapidly at a pace of more than 2500 publications a year. It is impossible to give a thorough review of this number of studies, so the interested reader must focus their attention to his or her specific interest. There are of course several reviews about the use of ANNs in chemistry by Zupan and Gasteiger (1991, 1999) and by Smits et al. (1994). There are also reviews in related fields such as different spectroscopic methods by Blank and Brown (1993), in biology by Basheer and Hajmeer (2000), etc., which provide initial information for these specific fields. In order to illustrate the large potential and great variety of possibilities where ANN methods can be applied, only several *types* of problems that can be tackled by ANNs will be mentioned and a few examples cited. The types of problems can roughly be divided into three groups: classification, mapping, and modeling.

8.1. Classification

In chemistry, classifications of many different kinds are sought quite often. The objects of classifications are multi-component analyses of merchandized goods, drugs, samples of criminal investigations, spectra, structures, process vectors, etc. On the output side the targets are products' origin, quality classes defined by the quality control standards, possible statuses of the analytical equipment at which the analyses have to be made, the presence or absence of fragments in the structures, etc. The classification problems are of either the *one-object-to-one-class* or the *one-object-to-several-classes* type. A common type of classification is the determination of the geographical origin of foods such as olive oils (Zupan et al., 1994, Angerosa et al., 1996), coffee (Anderson and Smith, 2002), or wine vinegars (GarciaParilla et al., 1997). Classification can be applied in the quality control of final products through their microstructure classification (Ai et al., 2003), monitoring the flow of processes via the control charts (Guh and Hsieh, 1999), etc. In general, for the multi-class classification, each object of the tested group is associated with several classes for which the network should produce the corresponding number of signals (each signal higher than the pre-specified threshold value) on the output. The final goal is to generate a network that will respond with the 'firing' of all output neurons that correspond to the specific class of the input object (see for example, Keyvan et al., 1997).

The prediction of the spectra–fragment relation is a typical multi-classification case. The chemical compounds are represented, for example, by the infrared spectra and the sought answers are the lists of structural fragments (atom types, length and number of the chains, sizes of the rings, types of bonds, etc.) that correspond to the structures. Structure fragments are coded binary, i.e. 1s and 0s, for the presence and absence of a particular fragment, respectively (Novic and Zupan, 1995; Munk et al., 1996; Debska and Guzowska, 1999; Hemmer and Gasteiger, 2000).

There are many more applications in the field of classification from planning of chemical reactions (Gasteiger et al., 2000) to classification of compounds using an electronic nose device (Boilot, 2003).

8.2. Mapping

Among all ANNs, Kohonen learning is best suited for the mapping of data. Mapping of data or projection of objects from m-dimensional measurement space into a two-dimensional plane is often used at the beginning of the study to screen the data or at the end of a study for better visualization and presentation of the results. Mapping is generally an applicable methodology in fields where permanent monitoring of multivariate data, for example chemical analyses accompanied by meteorological data, are required (Bajwa and Tian, 2001, Kocjancic and Zupan, 2000). Another broad field of mapping applications is the generation of two-dimensional maps of various spectra from the infrared made by Cleva et al. (1999) or Gasteiger et al. (1993), to NMR spectra analyzed by Axelson et al. (2002). In such studies the objective is to distinguish between different classes of the objects for which the spectra were recorded. Increasingly powerful personal computers with a number of easily applicable programming tools enable generation of colorized maps (Bienfait and Gasteiger, 1997). The maps of large quantities (millions and more) of multivariate objects can be obtained by special Kohonen ANN arrangements to check where and how the new (or unknown) objects will be distributed (Zupan, 2002). The properties of the excited neurons together with the vicinity of other objects of known properties provide the information about the nature of the unknowns (Polanco et al., 2001).

Besides the self-organizing maps (SOMs) produced by the Kohonen ANNs, mapping can be achieved by the error backpropagation ANNs as well. This so-called *bottle-neck* mapping introduced by Livingstone et al. (1991) uses the idea that the objects employed for the training as inputs can be considered *at the same time* as targets, hence, the training is made by the $\{\mathbf{X}_s, \mathbf{X}_s\}$ input/output pairs. Such a composition of the inputs and targets requires the network having m input and m output nodes. The mapping is achieved by inclusion of several hidden layers of which the middle one must have only 2 (two!) nodes. The outputs of these two nodes serve as (x, y)-coordinates for each object in the two-dimensional mapping area. The bottle-neck mapping has been recently used in ecological (Kocjancic and Zupan, 2000) and chemical applications as well (Thissen et al., 2001).

It has several advantages over the Kohonen maps, such as better resolution and continuous responses. Unfortunately, the training is very time consuming because adaptation of at least two-hidden layer error backpropagation ANNs on the input and output sides of m-dimensional input and output nodes, respectively, requires a large number of objects in the training set. This demand, together with the known fact that the error backpropagation network, compared to the Kohonen one, needs at least an order of magnitude (if not two) more epochs to be fully trained, shows that this valuable method has serious limitations and, unfortunately, in many cases cannot be applied.

8.3. Modeling

Modeling is the most frequently used approach in ANN applications. It is far beyond the scope of this chapter to give an account of all possible uses or even the majority of them. Models in the form $(y_1, y_2, \ldots, y_n) = M(x_1, x_2, \ldots, x_m, w_{11}, w_{12}, \ldots . w_{kn})$ can be built for virtually any chemical application where the relation between two multidimensional features $\mathbf{Y}_s$ and $\mathbf{X}_s$ represented as vectors is sought. For this modeling the error backpropagation, counterpropagation, or radial basis networks can be used. Probably the best-known field in chemistry in which modeling is the main tool is quantitative structure–activity relationship (QSAR) studies. Therefore, it should be of no surprise that scientists in this field have quickly included ANNs in their standard inventory (Aoyama, 1991). For more up-to-date information it is advisable to consider recent reviews by Maddalena (1998) and Li and Harte (2002). In connection with QSAR studies, it might be worthwhile pointing out the problem of uniformly coding chemical structures. The uniform structure representation is needed not only for the input to ANNs, but in any other standard modeling method as well. This mandatory form of coding scheme is again outside the scope of the present work. However, chemists should be at least aware that several possibilities for coding the chemical structures in a uniform representation exist. Some of them are explained in the ANN tutorial book by Zupan and Gasteiger (1999).

Modeling of a process or property is mainly carried out with the optimization of the process or properties in mind. The effectiveness of the optimization depends on the quality of the underlying model used as the *fitness* function (or part of it). Because the influence of the experimental variables on the properties or processes can be obtained by on-line measurements, there is usually enough experimental data to assure the generation of a reliable ANN model for quantitative predictions of responses for any combination of the possible input conditions. Once the model, be it an ANN or polynomial form, is available the optimization, i.e. the search for the state yielding the *best suited* response, can be implemented by any optimization technique from the gradient descent to simplex (Morgan et al., 1990) or genetic algorithm (Hilberth, 1993). Optimizations using ANNs and various genetic algorithms are intensively used in chemical and biochemical engineering (Borosy, 1999; Kovar et al., 1999; Evans et al., 2001; Ferentinos and Albright, 2003) and in many optimizations performed in high throughput analytical laboratories as described by Havlis et al. (2001).

9. Conclusions

As it was shown in the above discussion, ANNs have a very broad field of applications. With ANNs one can do classifications, clustering, experimental design, mapping, modeling, prediction of missing data, reduction of representations, etc. The ANNs are quite flexible for adaptation to different types of problems and can be custom-designed to almost any type of data representation, i.e. to real, binary, alphanumeric, or to mixed ones. In spite of all advantages of ANNs, one should be careful not to try to solve all problems using the ANN methodology, just because of its simplicity. It is always beneficial to try different approaches to solve a specific problem. Only one method, no matter how powerful it may seem, can fail easily.

This warning is important in solving problems where large quantities of multivariate data must be handled. In such problems the best solutions are not necessarily obtained immediately and they are far from self-evident, even when already obtained. For example, to obtain a good model based on a thousand objects represented in, say, 50-dimensional measurement space, hundreds of ANN models with different kinds of architecture and/or different initial and training parameters have to be trained and tested. Many times even polynomial models (non-linear in factors) must also be made and compared to the ANN models before the concluding decision can be made. The same is true for clustering of large multivariate data sets, as well as for the reduction of the measurement space. Reducing the number of variables (for example molecular descriptors) from several hundred to the best (or optimal) possible set of a few dozen ones is not an easy task. The best representation does not depend only on the property to be modeled and on the number and distribution of the available compounds, but on the choice of the modeling method as well.

In the experimental data normally used by the ANNs there is a lot of noise and entropy, the solutions are fuzzy, and they are hard to validate. The best advice to follow is to try several methods and to compare the results. It is important to validate the obtained results with several tests. Often the data do not represent the problem well and therefore, do not correlate well with the information sought. It can happen that at the beginning of the research even the user does not know precisely what he or she is looking for. It is important for the users of ANNs to gain the necessary insight into the data, to their representation, and to the problem before the appropriate method is selected. It has to be repeated that the proper selection of *the number* of data and *the distribution* of data in the measurement space are crucial for successful modeling. The proper distribution of data is not only essential for successful training, but for reliable validation as well. Potential users have to be aware of the fact that most of the errors in ANN modeling are made by the inadequate selection of the number and the distribution of the training and validation objects.

Many times, this procedure is not straightforward, bust must be accomplished in loops: after gaining the first 'final' results one gets better insight into the data and to the problem, which in turn opens new possibilities for different choices of parameters, design, and adjustment of the ANNs in order to achieve still better results and come a step closer towards a deeper understanding of the problem and the final goal.

Acknowledgments

This work has been supported by the Ministry of Education, Science, and Sport of Slovenia through Program grant P-104-508.

References

Ai, J.H., Jiang, X., Gao, H.J., Hu, Y.H., Xie, X.S., 2003. Artificial neural network prediction of the microstructure of 60Si2MnA rod based on its controlled rolling and cooling process parameters. Mat. Sci. Enign. (A), Struct. Mat. Prop. Microstruct. Process. 344 (1–2), 318–322.

Anderson, A.J., Rosenfeld, E., 1989. Neurocomputing. Foundation of Research, MIT Press, Cambridge, (Fourth Printing).

Anderson, K.A., Smith, B.W., 2002. Chemical profiling to differentiate geographic growing origins of coffee. J. Agricult. Food Chem. 50 (7), 2068–2075.

Angerosa, F., DiGiacinto, L., Vito, R., Cumitini, S., 1996. Sensory evaluation of virgin olive oils by artificial neural network processing of dynamic head-space gas chromatographic data. J. Sci. Food Agricult. 72 (3), 323–328.

Aoyama, T., Ichikawa, H., 1991. Neural networks applied to pharmaceutical problems. Chem. Pharm. Bull. 39 (2), 372–378.

Axelson, D., Bakken, I.J., Gribbestad, I.S., Ehrnholm, B., Nilsen, G., Aasly, J., 2002. Applications of neural network analyses to in vivo H-1 magnetic resonance spectroscopy of Parkinson disease patients. J. Magn. Res. Imaging 16 (1), 13–20.

Bajwa, S.G., Tian, L.F., 2001. Aerial CIR remote sensing for weed density mapping in a soybean field. Trans. ASAE 44 (6), 1965–1974.

Basheer, I.A., Hajmeer, M., 2000. Artificial neural networks: fundamentals, computing, design, and application. J. Microbiol. Meth. 43 (1), 3–31.

Bienfait, B., Gasteiger, J., 1997. Checking the projection display of multivariate data with colored graphs. J. Mol. Graph. Model. 15 (4), 203–218.

Bishop, C.M., 1994. Neural networks and their applications. Rev. Sci. Instrum. 65, 1803–1832.

Blank, T.B., Brown, S.D., 1993. Data-processing using neural networks. Anal. Chim. Acta 227, 272–287.

Boilot, P., Hines, E.L., Gongora, M.A., Folland, R.S., 2003. Electronic noses inter-comparison, data fusion and sensor selection in discrimination of standard fruit solutions. Sens. Act. (B), Chem. 88 (1), 80–88.

Borosy, A.P., 1999. Quantitative composition-property modelling of rubber mixtures by utilizing artificial neural networks. Chemom. Intell. Lab. 47 (2), 227–238.

Cleva, C., Cachet, C., Cabrol-Bass, D., 1999. Clustering of infrared spectra with Kohonen networks. Analysis 27 (1), 81–90.

Debska, B., Guzowska-Swider, B., 1999. SCANKEE - computer System for interpretation of infrared spectra. J. Mol. Struct. 512, 167–171.

Derks, E.P.P.A., Sanchez Pastor, M.S., Buydens, L.M.C., 1995. Robustuess analysis of radial base function and multilayered feedforward neural-network models. Chemom. Intell. Lab. Syst. 28, 49–60.

Despagne, F., Massart, D.L., 1998. Neural networks in multivariate calibration. Analyst 123, 157R–178R. (Tutorial Review).

Evans, J.R.G., Edirisingh, M.J., Coveney, P.V., Eames, J., 2001. Combinatorial searches of inorganic materials using the ink jet printer: science, philosophy and technology. J. Eur. Ceram. Soc. 21 (13), 2291–2299.

Ferentinos, K.P., Albright, L.D., 2003. Fault detection and diagnosis in deep-trough hydroponics using intelligent computational tools. Biosyst. Engng 84 (1), 13–30.

GarciaParrilla, M.C., Gonzalez, G.A., Heredia, F.J., Troncoso, A.M., 1997. Differentiation of wine vinegars based on phenolic composition. J. Agri. Food Chem. 45 (9), 3487–3492.

Gasteiger, J., Zupan, J., 1993. Angew. Chem., Neural Networks Chem. 105, 510–536. Gasteiger, J., Zupan, J., 1993. Angew. Chem. Intl. Ed. Engl. 32, 503–527.

Gasteiger, J., Li, X., Simon, V., Novic, M., Zupan, J., 1993. Neural Nets for Mass and Vibrational Spectra. J. Mol. Struct. 292, 141–159.
Gasteiger, J., Pfortner, M., Sitzmann, M., Hollering, R., Sacher, O., Kostka, T., Karg, N., 2000. Computer-assisted synthesis and reaction planning in combinatorial chemistry. Persp. Drug Disc. Des. 20 (1), 245–264.
Guh, R.S., Hsieh, Y.C., 1999. A neural network based model for abnormal pattern recognition of control charts. Comp. Ind. Engng 36 (1), 97–108.
Havlis, J., Madden, J.E., Revilla, A.L., Havel, J., 2001. High-performance liquid chromatographic determination of deoxycytidine monophosphate and methyldeoxycytidine monophosphate for DNA demethylation monitoring: experimental design and artificial neural networks optimisation. J. Chromat. B 755, 185–194.
Hebb, D.O., 1949. The Organization of Behavior, Wiley, New York, pp. xi–xix, 60–78.
Hecht-Nielsen, R., 1987a. Counter-propagation Networks. Appl. Optics 26, 4979–4984.
Hecht-Nielsen, R., 1987b. Counter-propagation Networks. Proceedings of the IEEE First International Conference on Neural Networks, (II), 19–32.
Hecht-Nielsen, R., 1988. Application of Counter-propagation Networks, Neural Networks 1, 131–140.
Hemmer, M.C., Gasteiger, J., 2000. Prediction of three-dimensional molecular structures using information from infrared spectra. Anal. Chim. Acta 420 (2), 145–154.
Hilberth, D.B., 1993. Genetic Algorithms in Chemistry, Tutorial. Chemom. Intell. Lab. Syst. 19, 277–293.
Hopfield, J.J., 1982. Neural networks and physical systems with emergent collective computational abilities. Proc. Natl Acad. Sci. 79, 2554–2558.
Keyvan, S., Kelly, M.L., Song, X.L., 1997. Feature extraction for artificial neural network application to fabricated nuclear fuel pellet inspection. Nucl. Technol. 119 (3), 269–275.
Kocjancic, R., Zupan, J., 2000. Modelling of the river flow rate: the influence of the training set selection. Chemom. Intell. Lab. 54 (1), 21–34.
Kohonen, T., 1972. Correlation matrix memories. IEEE Trans. Computers C-21, 353–359.
Kohonen, T., 1988. An Introduction to Neural Computing, Neural Networks 1, 3–16.
Kohonen, T., 1995. Self-Organizing Maps, Springer, Berlin.
Kovar, K., Kunze, A., Gehlen, S., 1999. Artificial neural networks for on-line optimisation of biotechnological processes. Chimia 53 (11), 533–535.
Li, Y., Harte, W.E., 2002. A review of molecular modeling approaches to pharmacophore models and structure–activity relationships of ion channel modulators in CNS. Curr. Pharm. Desi. 8 (2), 99–110.
Lippmann, R.P., 1987. An introduction to computing with neural nets. IEEE ASSP Mag. April, 4–22.
Livingstone, D.J., Hesketh, G., Clayworth, D., 1991. Novel method for the display of multivariate data using neural networks. J. Mol. Graph. 9 (2), 115–118.
Maddalena, D.J., 1998. Applications of soft computing in drug design. Expert Opin. Ther. Pat. 8 (3), 249–258.
Mariey, L., Signolle, J.P., Amiel, C., Travert, J., 2001. Discrimination, classification, identification of microorganisms using FTIR spectroscopy and chemometrics. Vibrat. Spectr. 26 (2), 151–159.
Massart, D.L., Vandeginste, B.G.M., Buydens, L.M.C., De Jong, S., Lewi, P.J., Smeyers Verbeke, J., 1997. Handbook of Chemometrics and Qualimetrics: Part A, Elsevier, Amsterdam, 221 ff.
McCulloch, W.S., Pitts, W., 1943. A logical calculus of the ideas immanent in nervous activity. Bull. Math. Biophys. 5, 115–133.
Minsky, M., Papert, S., 1989. Perceptrons, MIT Press, Cambridge.
Morgan, E., Burton, K.W.C., Nickless, G., 1990. Optimisation using the super-modified simplex method. Chemom. Intell. Lab. Syst. 8, 97–108.
Munk, M.E., Madison, M.S., Robb, E.W., 1996. The neural network as a tool for multi-spectral interpretation. J. Chem. Inform. Comp. Sci. 36 (2), 231–238.
Novic, M., Zupan, J., 1995. Investigation of infrared spectra-structure correlation using Kohonen and counterpropagation neural-network. J. Chem. Inform. Comp. Sci. 35 (3), 454–466.
Pitts, W., McCulloch, W.S., 1947. How we know universals: the perceptron of auditory and visual forms. Bull. Math. Biophys. 9, 127–147.
Polanco, X., Francois, C., Lamirel, J.C., 2001. Using artificial neural networks for mapping of science and technology: A multi-self-organizing-maps approach. Scientometrics 51 (1), 267–292.
Renals, S., 1989. Radial basis function network for speech pattern-classification. Electr. Lett. 25 (7), 437–439.

Rumelhart, D.E., Hinton, G.E., Williams, R.J., 1986. In: Rumelhart, D.E., MacClelland, J.L., (Eds.), Distributed Parallel Processing: Explorations in the Microstructures of Cognition, vol. 1. MIT Press, Cambridge, MA, USA, pp. 318–362.

Smits, J.R.M., Melssen, W.J., Buydens, L.M.C., Kateman, G., 1994. Using artificial neural networks for solving chemical problems (Tutorial). Chemom. Intel. Lab. Syst. 22, 165–189. Smits, J.R.M., Melssen, W.J., Buydens, L.M.C., Kateman, G., 1994. Chemom. Intel. Lab. Syst. 23, 267–291.

Thissen, U., Melssen, W.J., Buydens, L.M.C., 2001. Nonlinear process monitoring using bottle-neck neural networks. Anal. Chim. Acta. 446 (1–2), 371–383.

Walczak, B., Massart, D.L., 1996. Application of Radial Basis Functions - Partial Least Squares to non-linear pattern recognition problems: Diagnosis of process faults. Anal. Chim. Acta 331, 177–185.

Werbose, P., 1982. In: Drenick, R., Kozin, F., (Eds.), System Modelling and Optimization: Proceedings of the International Federation for Information Processes, Springer Verlag, New York, pp. 762–770.

Wong, M.G., Tehan, B.G., Lloyd, E.J., 2002. Molecular mapping in the CNS. Curr. Pharm. Design 8 (17), 1547–1570.

Zupan, J., 2002. 2D mapping of large quantities of multi-variate data, Croat. Chem. Acta 75 (2), 503–515.

Zupan, J., Gasteiger, J., 1991. Neural networks: A new method for solving chemical problems or just a passing phase? (a review). Anal. Chim. Acta 248, 1–30.

Zupan, J., Gasteiger, J., 1993. Neural Networks for Chemists: An Introduction, VCH, Weinheim.

Zupan, J., Gasteiger, J., 1999. Neural Networks in Chemistry and Drug Design, 2nd edn., Wiley-VCH, Weinheim.

Zupan, J., Novic, M., Li, X., Gasteiger, J., 1994. Classification of multi-component analytical data of olive oils using different neural networks. Anal. Chim. Acta 292 (3), 219–234.

Zupan, J., Novič, M., Ruisanchez, I., 1997. Kohonen and counterpropagation artificial neural networks in analytical chemistry. Chemom. Intell. Lab. Syst. 38, 1–23.

CHAPTER 8

Artificial neural networks in molecular structures—property studies

Marjana Novic, Marjan Vracko

Laboratory of Chemometrics, National Institute of Chemistry, Ljubljana, Slovenia

1. Introduction

In molecular structure–property/activity (QSPR/QSAR) modelling three central questions are addressed. To describe the framework of models one has to define first, the data set; second, the descriptors of molecular structures and property; and third, the methods of modelling and testing of models. In Section 2 we present a short overview of molecular descriptors where the reader may obtain an insight into the type of data used in QSPR/QSAR modelling. Section 3 gives a description of the architecture and learning strategy of a counter propagation neural network. This method is successfully applied in QSPR modelling, particularly if applied to diverse data sets and for treating complicated biological properties such as binding affinity, toxicity parameters, or carcinogenicity. A diverse data set means that the data set consists of structurally different compounds that are active due to different mechanisms. In such cases, the relationship between descriptors and properties is usually not linear and therefore deterministic models would hardly be feasible. The advantage of neural network models is in their ability to learn from existing data and to describe non-linear relationships. Furthermore, in Section 3 we describe how the counter propagation neural network can be used in QSPR/QSAR modelling. In Section 4 examples from toxicology and drug design are shown.

2. Molecular descriptors

In molecular structure–property studies, the question of molecular representation is central. For modelling purposes, a molecule is represented as a multidimensional vector. In other words, a molecule is a point represented in multidimensional space. An ideal representation should be unique, uniform, reversible, and invariant on rotation and traslation of molecues. Unique means that different structures give different representations; uniform means that the dimension and domain of representation are the same for

Data Handling in Science and Technology, Volume 23
ISSN: 0922-3487

DOI: 10.1016/S0922-3487(03)23008-2

all structures; and reversible means that one can construct the structure from the representation. No representation fulfils all requirements simultaneously, and it is not expected that we will find a general representation suitable for all models. Usually, the molecules are represented by descriptors, i.e. parameters obtained empirically or calculated from the molecular structure (Todeschini and Consonni, 2000). In recent years dozens of descriptors have been proposed. The information, or descriptors, about each of the chemical structures can be hierarchically ordered (Basak and Mills, 2001; Basak et al., 2001).

One-dimensional (constitutional) descriptors. To this class belong the discrete numerical descriptors that describe the basic molecular structure; for example, molecular weight, number of atoms, number of rings, etc.

Two-dimensional descriptors. These are related to the topological picture (two-dimensional picture) of molecules. This picture carries information on how the atoms are connected and the nature of the bonds. This picture can be represented by a graph ('structural formula') or with connectivity matrices. The pioneering work in this field was published in 1947 by Wiener, who was studying paraffin hydrocarbons. Recently, several dozens or hundreds of descriptors have been deduced from the topological picture of a molecule, known as topological indices. Several authors describe the historical development of topological indices as well as their personal views of the controversy surrounding the used of topological indices (Randić, 1998; Devillers and Balaban, 1999; Balaban, 2001). Since these indices are mostly deduced from graphs using advanced mathematical algorithms and neglecting the nature of atoms and bonds, they have a little or no physical or chemical meaning. On the other hand, they are successfully used in many QSAR studies (Clerc and Terkovics, 1990; Basak and Niemi, 1991; Basak et al., 1994; Vracko, 2000).

Three-dimensional descriptors. These descriptors are related to the three-dimensional picture of molecules, which is defined by the coordinates of all atoms in the molecule. The step from a two-dimensional to a three-dimensional description of molecules is crucial (Randić and Razinger, 1997). The three-dimensional structure is not clearly determined. It depends on the molecular environment, i.e. it is different in crystal structure, in solutions, or in vacuo. If it is theoretically determined, the structure depends on the computational method. However, the three-dimensional structures form the basis for a broad range of descriptors. Mass distribution descriptors and shape descriptors based on van der Waals atomic radii belong to this class (Ciubotariu et al., 2001).

Quantum chemical descriptors. These are derived from quantum chemical calculations. Results of quantum chemical calculations are molecular orbital energies and charge distribution (Karelson and Lobanov, 1996). HOMO and LUMO energies and orbital energies close to them are often taken as descriptors of the ionization or activation potentials of molecules. From electron densities, the charge distribution descriptors, such as electronic indices or charge surface area indices, can be determinated. HOMO and LUMO electron densities determine electron donor and acceptor sites. Further, two descriptors calculated from quantum chemical results that describe the electrons' behaviour are delocalizability and polarizability (Pires et al., 1997). Alternatively, the density of spectral states, which are constructed from all orbital energies in the valence region, can be used to represent molecules (Vracko, 1997; Vracko et al., 2000).

Receptor-dependent descriptors. All of the above-mentioned classes of descriptors are related solely to the molecular structures, but they do not carry any information about biological targets. With the receptor-dependent descriptors, we try to include information about the receptor in QSAR studies. One of these approaches is comparative molecular field analysis (CoMFA), where the molecular three-dimensional structures are optimized together with the receptor (Cramer et al., 1988). This approach is often applied in drug design or specific toxicology studies where the receptor is known. On the other hand, it is seldom applied in environmental toxicology because the receptor is not known or the same compound acts on different receptors.

Empirical descriptors. These are usually easily measurable physico-chemical quantities. Such descriptors are the octanol–water partitioning coefficient ($\log P$), Hammet and Taft substituents constants, dipole moment, and aqueous solubility (Hansch, 1969; Hansch et al., 1969). The disadvantage of these descriptors is that they are available only for existing chemicals, and even for these the data are scarce. Their advantage is that they are obtained experimentally and therefore related to the reaction mechanism. A classical example is $\log P$, which was developed on the assumption that transportation from the site of application of the drug to its site of action depends on the lipophilicity of the molecule.

Besides the descriptors given, there have been further attempts to encode empirical or theoretical three-dimensional molecular structures with functions. Examples are the three-dimensional-MoRSE code (Schuur et al., 1996), a spectrum-like representation (Vracko, 1997; Zupan and Novic, 1997; Zupan et al., 2000), and radial distribution functions (Hemmer and Gasteiger, 2000). Also, experimentally determined infrared, mass, or NMR spectra can be taken to represent a molecule (Bursi et al., 1999).

The field of molecular descriptors and molecular representations has become very active in recent decades. Several computer programs are available for the calculation of descriptors and representations (CODESSA; PETRA; POLLY; DRAGON). For model construction, it is only important to have a vector of parameters for each molecule that describes its structure. We must be aware that all the uncertainties and ambiguities in structures are incorporated into the models.

3. Counter propagation neural network

3.1. Architecture of a counter propagation neural network

A counter propagation neural network is a generalization of the Kohonen network, or self-organizing map. The self-organizing map is a topology-preserving map obtained by a projection from the multidimensional descriptor space to a two-dimensional grid of points (neurons). After this projection, the objects are arranged in a two-dimensional network in such a way that similar objects are located close to each other. In contrast, the mapping is not metric preserving and therefore the information on distances between objects in descriptor space is generally lost. One of the reasons why such maps are used is to visualize the data. The human mind is not able to analyse data in multidimensional space, but it can very efficiently analyse two-dimensional pictures.

The counter propagation neural network has two layers: the input or Kohonen layer and the output layer (Hecht-Nielsen, 1987; Dayhof, 1990; Zupan and Gasteiger, 1999). The Kohonen layer is a two-dimensional network of neurons, all of which are vectors W_j $(w_{j1},w_{j2},\ldots,w_{jm})$. Here, W_j refers to the neuron, $w_{j,i}$ terms are components of the vector (weights), and m is the dimension of the vector. The dimension of the vectors is equal to the dimension of the descriptor space. The output layer is located beneath the Kohonen layer. It has the same number of neurons as the Kohonen layer with the dimension equal to the number of output variables (Fig. 1). The spirit of a counter propagation neural network

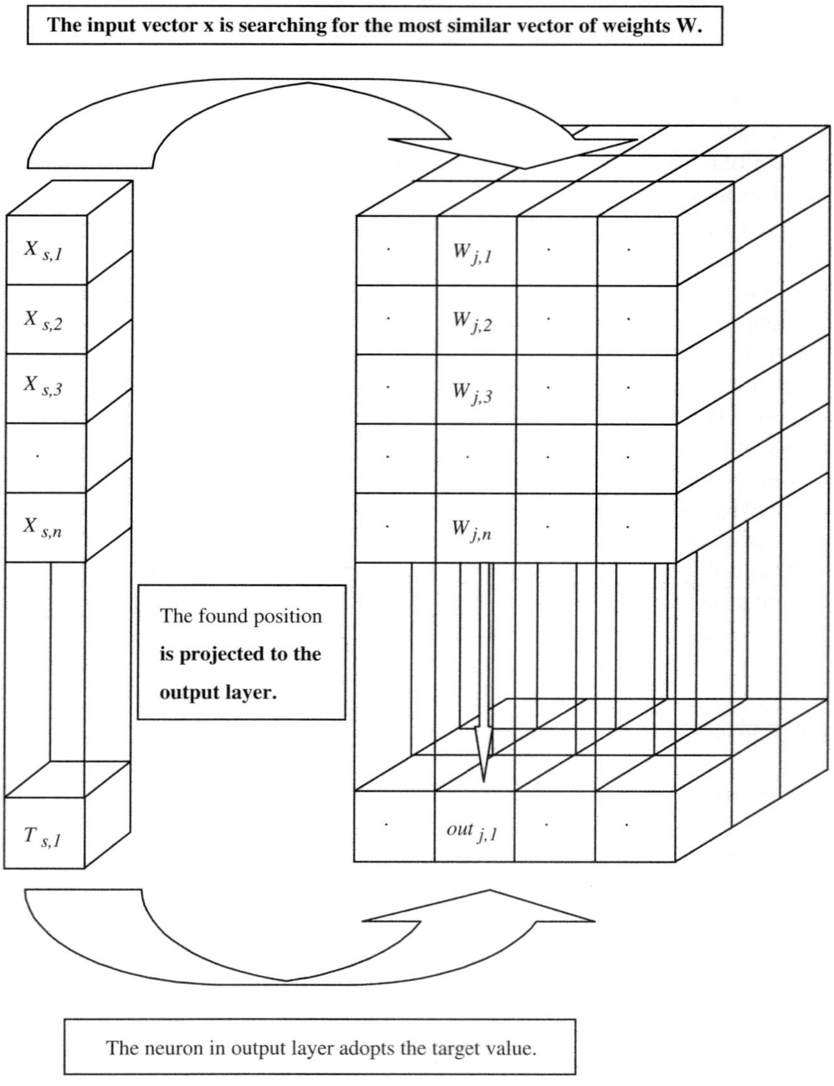

Fig. 1. The architecture and learning strategy of a counter propagation neural network.

lies in the learning procedure, which is in the Kohonen layer and different from the learning in the output layer.

3.2. Learning in the Kohonen and output layers

3.2.1. Learning in the Kohonen layer

The learning in the Kohonen (input) layer is done in the same way as in a Kohonen network. It is based on unsupervised competitive learning; 'the-winner-takes-all' strategy. This means that a vector of input variables (descriptors) is presented to all neurons. According to Eq. (1) the algorithm selects the neurons with weights closest to the input variables (the winning neuron):

$$\delta_j = \sum_{i=1}^{m} (x_{si} - w_{ji})^2 \quad \delta_c = \min(\delta_1, \delta_2, \ldots \delta_j, \ldots \delta_n) \quad \Rightarrow \quad W_c \tag{1}$$

In the next step the weights of the winning neuron, and all nearby neurons, are modified so that their weights become more similar to the input variables (Eq. 2).

$$w_{ji}^{\text{new}} = w_{ji}^{\text{old}} + \eta(t) \times b(d_c - d_j) \times (x_{si} - w_{ji}^{\text{old}}) \tag{2}$$

Parameter η determines the rate of learning; it is maximal at the beginning ($t = 1$, $\eta = a_{\max}$) and minimal at the end of the learning procedure ($t = t_{\max}$, $\eta = a_{\min}$). The function $b(\cdot)$ in Eq. (2) describes how the correction of the weights w_{ji} decreases with the increasing topological distance between the central neuron and the neuron being corrected. Index j specifies an individual neuron and runs from 1 to n. The topological distance of the jth neuron from the central one is defined according to the topology used for the distribution of neurons in the plane: in the rectangle net the central neuron has eight first neighbours ($d_c - d_j = 1$), 16 second neighbours ($d_c - d_j = 2$), etc. The minimal distance is zero ($j = c$, $d_c - d_j = 0$), which corresponds to the maximal correction function ($b = 1$). The maximal distance ($d_c - d_{\max}$) to which the correction is applied shrinks during the learning procedure. The correction function at maximal distance is minimal ($b = 0$). At the beginning the term $d_c - d_{\max}$ covers the entire network, while at the end, at $t = t_{\max}$, it is limited only to the central neuron. This process runs over all objects (one learning epoch) and it is repeated several times. The network continues this training until the weights are stabilized. The basic property of the trained network is that similar objects are located close to each other. It is important to emphasize that, in the arranging of objects, only the input variables were taken into account, regardless of the property (target) values.

3.2.2. Learning in output layer

The learning in the output layer occurs at the same time as the learning in the input layer. This step is a supervised learning since the target values are required for each input. The positions of objects are projected from input to output layer. The weights in the output layer are modified in a fashion comparable to the modification of the Kohonen layer. The weights at the central node and those in its neighbourhood are modified according to Eq. (3). In this way, the response surface is constructed:

$$\text{out}_{ji}^{\text{new}} = \text{out}_{ji}^{\text{old}} + \eta(t) b(d_c - d_j)(T_{si} - \text{out}_{ji}^{\text{old}}) \tag{3}$$

At the end of the learning process, the objects are organized in a two-dimensional grid with the response surface lying beneath.

3.3. Counter propagation neural network as a tool in QSAR

3.3.1. Prediction

A trained network can be used for prediction of output values for new 'unknown' objects. The prediction runs over two steps. In the first step the object is located in the Kohonen layer on the neuron with the most similar weight. In the second step, the position of that neuron is projected to the output layer, which gives the predicted output value. An advantage of the counter propagation is that we can follow the prediction. The objects close to the selected neuron determine the structure of the output layer in the neighbourhood of the selected neuron. We conclude that the objects close to the selected neuron determine the predicted value.

3.3.2. Clustering

Due to the basic property of Kohonen mapping that locates similar objects close to each other, the counter propagation network is a tool for clustering. Visual inspection of the map enables us to recognize the clusters and thus the similarity relationships within the data set (Zupan et al., 1994). Particularly interesting is the analysis of molecules that are located on the same neuron. The model recognizes such molecules as identical, or in other words their representation vectors are so similar that they cannot be discriminated by a model.

3.3.3. Outlier detection

Supose two compounds with very different activities are located on the same neuron. We conclude that one is an outlier. To realize which one is the outlier, we analyse the neighbourhood of the neuron. If the neighbouring neurons are occupied by active compounds the non-active compound is an outlier, and vice versa. The existence of such outliers is not a disaster; it simply means that our descriptors are not complete. We may consider extending the descriptor space or selecting the considered data set more carefully.

3.3.4. Training—test set division

To test the trained models, we sometimes divide the data set into two subsets: a training set used to build the model; and a testing set to test it. To obtain reasonable predictions for the test set, the training set must contain information on the entire descriptor space. The Kohonen map is divided into sub-parcels, selecting the objects for the training set from each sub-parcel equivalently. It is expected that such a training set possesses the information content of the entire set (Simon et al., 1993; Vracko and Gasteiger, 2002).

3.3.5. Classification

Sometimes, the output variable is not defined as a real number, but rather as an affiliation to a particular class. Such cases can be treated with a counter propagation neural network using a multidimensional output layer. If there are n possible classes, an n-dimensional output layer must be defined. The output variable of a compound belonging to the ith class is described with an n-dimensional vector with 'one' on the ith position and 'zero' on all other positions. The learning runs are as described above, only there are n different response surfaces. In the prediction of a class, one obtains an n-dimensional vector with components expressed as real numbers between zero and one. Two situations can occur. First, one component is essentially larger than the others. In such circumstances the predicted object unambiguously belongs to the identified class. Second, more than one component is approximately the same. In this case the object is set to several classes. In a very peculiar case all components are approximately equal. This means that the model cannot decide to which class the element belongs, and this is a very valuable result. We know a priori that the model is not able to describe this particular object (Vracko et al., 1999; Vracko, 2000).

3.3.6. Selection of descriptors

Visual inspection and comparison of individual levels in the input (descriptor) layer with the output layer can yield information on which descriptors correlate well with the output variable. This comparison means that two contour plots are inspected for overlapping extremes. The other possibility for descriptor selection is the 'modelling with the transposed data matrix'. In this modelling scheme a column in the network represents a descriptor and each individual input layer represents a molecule. After the training the obtained result is a map of descriptors. Descriptors, which are for all molecules similar, are located on the same neuron or collected together in clusters. By selecting one or some representatives from each cluster, one achieves a reduction in the number of descriptors (Roncaglioni, 2002).

4. Application in toxicology and drug design

4.1. A study of aquatic toxicity for the fathead minnow

The U.S. Environmental Protection Agency provided a data set of 568 organic compounds when referring to an acute aquatic toxicity dose after 96 h (LC_{50}) for the fathead minnow (*Pimephales promelas*) expressed as log(mol/l) (Russom et al., 1997; ECOTOX, 2000a; ECOTOX, 2000b; ECOTOX, 2000c). This is a large set of compounds belonging to different chemical classes making linear QSAR studies questionable. Here, an analysis of clusters and neighbours in the Kohonen network is more interesting than precise predictions of toxicity. Although the experimental errors are not reported, the data set provides homogeneous and reliable toxicological data.

A large number of descriptors was calculated using different software: (HYPERCHEM, CODESSA, PALLAS). Out of the hundreds of descriptors calculated only 150 gave a non-constant or non-missing value for all the objects. According to the CODESSA software, the descriptors can be classified into six categories: constitutional, geometrical, topological, electrostatic, quantum–chemical, and physico-chemical descriptors, such as log P (Katritzky et al., 1994; Karelson et al., 1999). All the descriptors and toxicity values were normalized between zero and one using a range-scaling procedure, thus maintaining the original distribution (Mazzatorta et al., 2002).

For the testing of models, the training/test set division of data was applied. In the final judgement of models, only statistical results for the test set were taken into account. General conditions for division of the data into training and test set is that both sets possess the same content of information. In this study the SphereExcluder program was used (Golbraikh, 2000). Basically, the division is based on distances between points in the descriptor space and on random sphere centre selection. The data set was divided into a training set (282 objects) and a test set (286 objects). Models were built with the counter propagation neural networks. Fig. 2 shows an overview of the results obtained by the developed model.

The model posesses the ability to predict the toxicity of the compounds in the training set. According to general experience, the recall ability of the counter propagation neural network is usually very good. The analysis of the network enables the selection of conflicting situations and outliers. In contrast, the model also has an acceptable ability to estimate the toxicity for compounds in the test set, which means that counter propagation neural networks are able to extract actual information and knowledge from the data set. In the next step some of the objects were selected as outliers and removed. The aim of identifying the outliers is not to achieve better statistical parameters, but to exclude the compounds for which their toxic character cannot be described with other compounds in the set. Two strategies for determining outliers were applied. First, we analysed the distances in the output layer. For the evaluation of the level of dissimilarity ($ds_{nx,ny}$) between neurons, we used an average of the difference between the weight of the neuron examined in the plane and its neighbourhood (Eq. 4):

$$ds_{nx,ny} = \left| \frac{w_{nx-1,ny-1} + w_{nx,ny-1} + w_{nx+1,ny-1} + w_{nx-1,ny} + w_{nx+1,ny} + w_{nx-1,ny+1} + w_{nx,ny+1} + w_{nx+1,ny+1} - 8*w_{nx,ny}}{8} \right| \quad (4)$$

Here $ds_{nx,ny}$ is the level of dissimilarity of the neuron in position (nx,ny); $w_{nx,ny}$ is the weight of the neuron in position (nx,ny). A level of dissimilarity >0.25 was considered as an

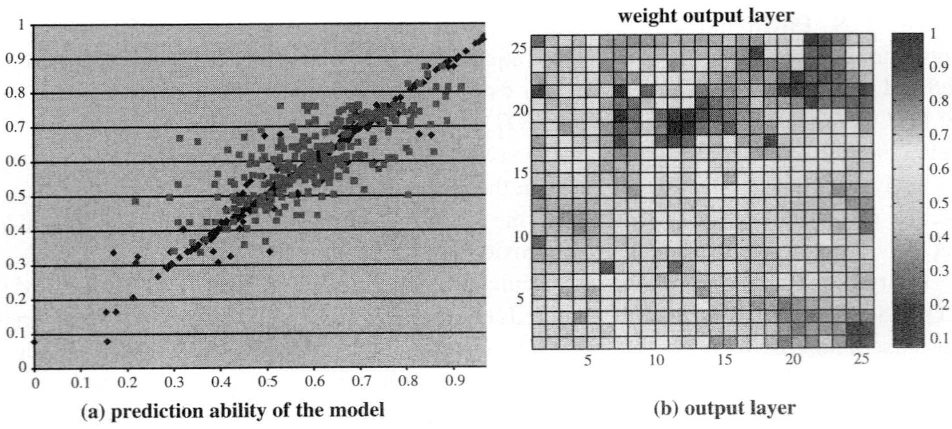

Fig. 2. Results for model 1. (a) The x and y axes show the experimental values and predicted values, respectively. Dark points represent the objects of the training set; bright points represent the object of the test set. (b) Output layer (bright represent high toxicity, dark represent low toxicity).

index of the presence of an outlier associated with that neuron. One compound (saccharin sodium salt hydrate) was selected as an outlier according to this criterion.

In the second strategy, we analysed the neurons occupied by two or more compounds. Such compounds are recognized as identical, which means that the corresponding descriptors are too similar to be discriminated by a neural network. If the compounds on the same neuron have similar properties, we obtain another confirmation of our model. If the properties differ essentially, there is a conflicting situation. One or more of the compounds on this neuron are outliers. Objects in the training set with an absolute error (AE) higher than 0.1 were analysed as possible outliers of the model. Nine objects were possible outliers. For further selection of outliers the neighbouring neurons were analysed. From this analysis four objects were clearly recognized as outliers. Fig. 3 shows two details from the output layer. For the objects associated with the neuron (10,21) it was not possible to determine an outlier because the closely neighbouring neurons did not present a clear trend. In other words, the neighbouring area did not give enough information to determine outliers. In this case, both objects were kept. On the other hand, the neighbourhood of the neuron (4,22) determines that the compound with a toxicity of 0.73 is an outlier. The same procedure was repeated for the test set. We found 11 compounds as possible outliers, 10 of which were recognized as outliers. After discarding the outliers, the training and testing of the model were repeated and analysed again (Model 2). In the analysis two other outliers were selected in the test set (Model 3). The statistical parameters of the models are shown in Table 1. More details of this study are reported in Mazzatorta et al., (2003).

4.2. *A study of aquatic toxicity toward Tetrahymena pyriformis on a set of 225 phenols*

The toxicity data for 225 phenols were taken from the literature (Schultz et al., 1997; Seward et al., 2001). The toxicity is expressed as growth inhibitory concentration against *Tetrahymena pyriformis* ($\log(IGC_{50}^{-1})$). The set of 225 phenols was used for training and testing of models. In addition, all models were validated with an external data set of 40 phenols. This emphasizes that these 40 compounds were never included in the modelling. The molecules were represented by 157 descriptors calculated by different software.

Table 1
Statistical parameters of models 1–3. R^2 (squared correlation coefficient), MAE (mean absolute error), RMSE (root mean squared error)

		No. of removed compounds	R^2	MAE	RMSE
Model 1	Training	/	0.953	0.016	0.032
	test	/	0.527	0.067	0.089
Model 2	Training	5	0.981	0.010	0.071
	test	10	0.567	0.065	0.084
Model 3	Training	/	0.981	0.010	0.071
	test	2	0.826	0.039	0.077

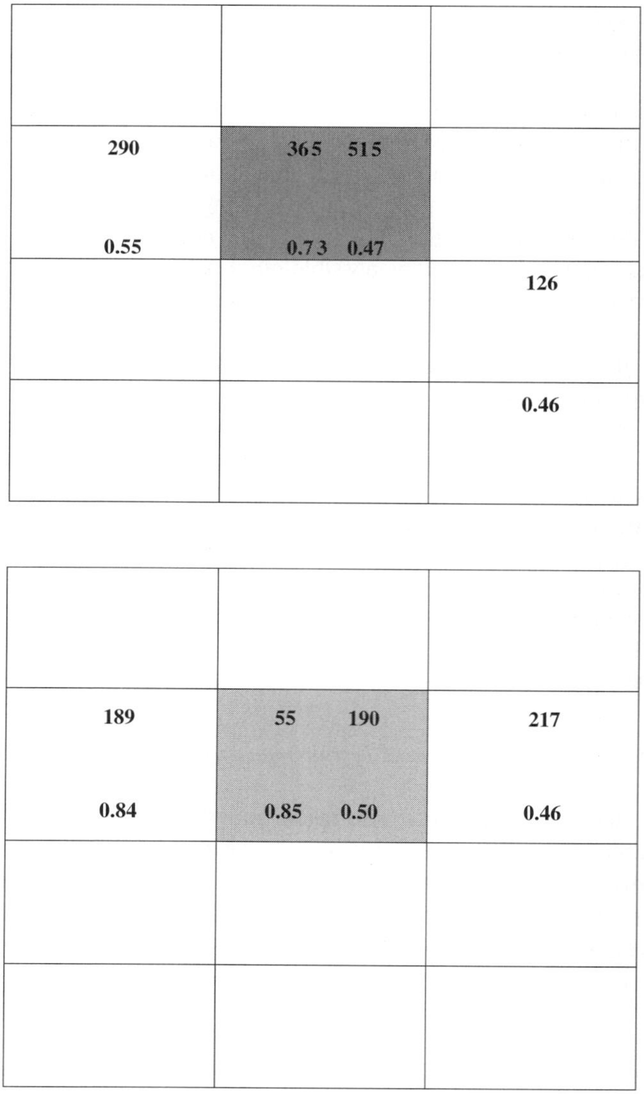

Fig. 3. Two details from the toxicity map. On the neuron (4,22) one outlier can be identified but none on the neuron (10,21).

First, the models were tested on recall ability to determine conflict situations and outliers. In the present study we define a conflict situation when the absolute difference in activity is larger than 0.4. In this situation, one molecule is an outlier, but the question is which one. To find an answer we analyse the neighbours in the network. If the closest neighbours are non-active, the active compound is an outlier and vice versa. In Table 3 we report the recall ability results and the predictions for the validation set of three models (A, B, C) (Szymoszek and Vracko, 2001).

Model A was built with 225 compounds. Table 2 shows groups of compounds that were recognized as identical. Three groups show a conflict situation. Group 3 has three compounds, which are structurally very similar (1,3,5-trihydroxybenzene, 1,2,4-trihydroxybenzene, and 1,2,3-trihydroxybenzene), but span the toxicity range from − 1.264 to 0.86. Two other conflict groups also contain structurally similar compounds (group 6 with ethyl-4-hydroxybenzoate and methyl-4-hydroxybenzoate, and group 9 with 4-nitrophenol and 3-nitrophenol). The groups 1, 2, 4, 5, 7, 8, 10, and 11 contain similar compounds that also show similar toxicity (the span in toxicity values within the group is lower than 0.4).

Model B was built without compounds 1,3,5-trihydroxybenzene, 1,2,4-trihydroxybenzene, ethyl-4-hydroxybenzoate, and 3-nitrophenol. There is still one conflict between the compounds catechol and resorscinol.

In model C, catechol was additionally removed, which shows no conflicts.

Statistical parameters (squared correlation coefficient R^2, and root mean square error, RMSE) for models A–C are given in Table 3. The statistical parameters for the training set validate the models. They improve from model A to model C.

In the next modelling experiments the set of 225 phenols was divided into training and test sets. The aim was to test the robustness of the models (Simon et al., 1993). Briefly, the entire Kohonen map is divided into sub-parcels selecting the objects for the training set from

Table 2
The groups of compounds recognized as identical by model A (bold values indicate groups of conflict; * denotes outliers)

Group	Molecule	Toxicity
1	4-chloro-2-methylphenol	0.701
	4-chloro-3-methylphenol (TE)	0.796
2	2,4-diaminophenol2HCl	0.127
	5-amino-2-methoxyphenol	0.45
3	1,3,5-trihydroxybenzene	− **1.264***
	1,2,4-trihydroxybenzene (TE)	**0.439***
	1,2,3-trihydroxybenzene (TE)	**0.85**
4	*m*-cresol(3-methylphenol)	−0.063
	p-cresol(4-methylphenol) (TE)	−0.184
5	2,4-dimethylphenol	0.0654
	2,3-dimethylphenol (TE)	0.122
	3,4-dimethylphenol (TE)	0.122
6	ethyl-4-hydroxybenzoate	**0.573***
	methyl-4-hydroxybenzoate	**0.084**
7	2,3,6-trimethylphenol	0.277
	2,3,5-trimethylphenol (TE)	0.36
8	3-hydroxy-4-methoxybenzylalcohol	−0.99
	4-hydroxy-3-methoxybenzylalcohol (TE)	−0.7
9	4-nitrophenol	**1.42**
	3-nitrophenol (TE)	**0.506**
10	4-*tert*-pentylphenol	1.229
	2-(*tert*)butyl-4-methylphenol (TE)	1.301
11	2-phenylphenol	1.0943
	4-phenylphenol (TE)	1.3928

Table 3
Statistical parameters for training, test, and validation sets for models A–C and I–V

Model	Set (number of objects)	R^2	RMSE
A	training (225)	0.983	0.110
	validation (40)	0.571	0.462
B	training (221)	0.989	0.100
	validation (40)	0.508	0.448
C	training (220)	0.998	0.094
	validation (40)	0.425	0.508
I	training (125)	0.999	0.005
	test (100)	0.665	0.450
	validation (40)	0.431	0.532
II	training (128)	0.998	0.026
	test (97)	0.633	0.466
	validation (40)	0.430	0.533
III	training (130)	0.999	0.032
	test (95)	0.681	0.436
	validation (40)	0.611	0.440
IV	training (132)	0.999	0.005
	test (93)	0.652	0.460
	validation (40)	0.458	0.520
V	training (134)	0.999	0.003
	test (91)	0.736	0.400
	validation (40)	0.465	0.516

each sub-parcel equivalently. It is expected that such a training set possesses the structural information content of the entire set. In our case the map of dimension 30 × 30 was divided into sub-parcels of dimension 5 × 5 selecting approximately 60% of molecules for training and 40% for the test set (model I). In the next steps two or three compounds from the test set were placed into the training set (models II–VI) in an attempt to improve the information content of the training set. The statistical parameters of the models are given in Table 3. The toxicity for the validation set of 40 compounds was predicted for models A–C (Table 4). Table 5 lists the compounds with predicted errors larger than 0.9. The last column in Table 5 shows the compounds of the training set which determined the predictions. Fig. 4a and 4b show two examples of predictions, one regarded as correct, the other as incorrect.

4.3. *Example of QSAR modeling with receptor dependent descriptors*

Receptor-dependent descriptors of molecular structure provide a good choice for QSAR studies if the structure of the receptor is known. Many successful applications are based on the CoMFA method in which the three-dimensional structure of a complex of ligand and receptor are optimized simultaneously (Cramer et al., 1988; Pitea et al., 1994; Bohm et al., 1999). Another possibility is to obtain the three-dimensional coordinates of the ligand–receptor complex from X-ray structure data. As a typical example of a QSAR study based on a neural network model using receptor-dependent descriptors based on X-ray structures, we will describe a study of an enzyme-binding neural network model for

Table 4
Validation set of 40 compounds (predictions for models A–C)

CAS	NAME	A	B	C	Exp.
611-99-4	4,4′-dihydroxybenzophenone	**1.512**	1.284	**1.849**	0.560
42019-78-3	4-chloro-4′-hydroxybenzophenone	1.056	1.088	0.982	1.596
581-43-1	2,6-dihydroxynaphthalene	0.857	**1.850**	1.236	0.834
117-99-7	2-hydroxybenzophenone	0.910	1.284	1.009	1.225
1143-72-2	2,3,4-trihydroxybenzophenone	1.024	1.384	1.009	0.879
835-11-0	2,2′-dihydroxybenzophenone	0.910	1.284	1.228	1.163
131-56-6	2,4-dihydroxybenzophenone	1.021	1.284	1.009	1.375
94-18-8	benzyl-4-hydroxybenzoate	1.283	1.284	**0.608**	1.547
2491-32-9	benzyl-4-hydroxyphenyl ketone	1.056	1.088	1.035	1.069
18979-50-5	4-propyloxyphenol	0.643	0.310	0.644	0.522
5471-51-2	4-(4-hydroxyphenyl)-2-butanone	−1.280	−0.059	−1.164	−0.497
131-55-5	2,2′,4,4′-tetrahydroxybenzophenone	1.024	1.384	1.009	0.959
136-36-7	resorcinol monobenzoate	1.512	1.516	1.136	1.109
28994-41-4	2-hydroxydiphenylmethane	1.349	1.349	1.349	1.311
1806-29-7	2,2′-biphenol	1.283	1.257	**1.849**	0.884
49650-88-6	2-(2-hydroxyethyl)resorcinol	−0.145	−1.121	−0.209	−0.868
101-18-8	3-hydroxydiphenylamine	1.465	1.257	1.500	1.014
70-70-2	4′-hydroxypropiophenone	−0.260	−0.223	−0.171	0.115
320-76-3	4-bromo-2-fluoro-6-nitrophenol	1.764	1.576	1.744	1.619
3947-58-8	2-bromo-2′-hydroxy-5′-nitroacetanilide	1.353	1.346	1.275	0.874
7693-52-9	4-bromo-2-nitrophenol	1.107	**0.603**	**0.643**	1.869
94-26-8	butyl-4-hydroxybenzoate	**0.327**	**0.290**	1.010	1.333
3264-71-9	4,6-dinitropyrogallol	−0.139	−0.133	−0.113	0.178
17696-62-7	phenyl-4-hydroxybenzoate	1.283	1.393	1.136	1.372
38713-56-3	nonyl-4-hydroxybenzoate	2.283	2.238	2.243	2.633
1131-60-8	4-cyclohexylphenol	1.289	1.227	1.247	1.558
5153-25-3	2-ethylhexyl-4′-hydroxybenzoate	**1.230**	**1.230**	**0.840**	2.507
6521-30-8	isoamyl-4-hydroxybenzoate	1.682	0.689	1.660	1.480
18979-53-8	*n*-pentyloxyphenol	0.890	0.711	0.701	1.360
86-77-1	2-hydroxydibenzofuran	1.465	1.465	1.279	1.523
29558-77-8	4-(4-bromophenyl)phenol	**1.349**	**1.349**	**1.349**	2.310
582-17-2	2,7-dihydroxynaphthalene	0.076	0.079	−0.209	0.554
575-44-0	1,6-dihydroxynaphthalene	0.521	0.301	−0.209	0.642
1689-64-1	9-hydroxyfluorene	1.465	1.377	1.213	0.844
1470-94-6	5-indanol	0.082	0.081	0.098	0.291
403-19-0	2-fluoro-4-nitrophenol	0.937	0.937	0.937	1.072
394-33-2	4-fluoro-2-nitrophenol	0.937	1.241	0.981	1.384
5460-31-1	2-methyl-3-nitrophenol	1.086	1.084	1.236	0.779
4920-77-8	3-methyl-2-nitrophenol	0.571	0.571	0.579	0.610
700-38-9	5-methyl-2-nitrophenol	0.571	0.571	0.643	0.586

prediction of the inhibitory effects of 18 chemicals towards human thrombin (Mlinšek et al., 2001).

Thrombin, a serine protease, controls thrombus formation. This blood coagulation system has to be controlled in many medical treatments. The inhibition of thrombus formation is achieved by the administration of anti-coagulant drugs, which bind to the thrombin-active site. In the search for new antithrombotic agents, a QSAR study of a series

Table 5
'Wrong predictions (difference is larger than 0.9)' calculated by models A, B and C. The last column shows the closest neighbours in the ANN, i.e. compounds that determinate the prediction

Compound	A	B	C	Experiment	Neighbours
4,4′-dihydroxybenzophenone	**1.512**	1.284	**1.849**	0.560	4-hydroxybenzophenone, phenylhydroquinone
2,6-dihydroxynaphthalene	0.857	**1.850**	1.236	0.834	methylhydroquinone
benzyl-4-hydroxybenzoate	1.283	1.284	**0.608**	1.547	ethyl-4-hydroxy-3-methoxyphenylacetate, 2-hydroxy-4-methoxybenzophenone
2,2′-biphenol	1.283	1.257	**1.849**	0.884	phenylhydroquinone
4-bromo-2-nitrophenol	1.107	**0.603**	**0.643**	1.869	4-methyl-2-nitrophenol
butyl-4-hydroxybenzoate	**0.327**	**0.290**	1.010	1.333	ethyl-4-hydroxybenzoate, methyl-4-hydroxybenzoate
2-ethylhexyl-4′-hydroxybenzoate	**1.230**	**1.230**	**0.840**	2.507	2,4,6-*tris*(dimethylaminomethyl)phenol, 4-(*tert*)octylphenol, 3,5-di(*tert*)butylcatechol
4-(4-bromophenyl)phenol	**1.349**	**1.349**	**1.349**	2.310	3-phenylphenol

of compounds showing an antithrombotic effect with known affinity constant K_i was started. The research was directed towards a model that would be used for predicting the K_i values for the binding of an inhibitor into the active site of thrombin. To construct such a model, the data set of known X-ray structures of inhibitor–enzyme complexes should be available together with their K_i values, while for a prediction of the binding activity of the new antithrombotic agent, one would need the X-ray structure of the complex with the new structure.

4.3.1. Data set

From the Brookhaven PDB database for thrombin structures, complexes of thrombin with 30 different ligands were selected. Only 18 of them were found to be reversibly bound inhibitors and thus supposedly active through the same biochemical mechanism (Banner and Hadvary, 1991; Maryanoff et al., 1993; Hilpert et al., 1994; Tabernero et al., 1995; Malley et al., 1996; Matthews et al., 1996; Krishnan et al., 1998; Steiner et al., 1998; Charles et al., 1999; Mochalin and Tulinsky, 1999). These 18 complexes were taken as the basis for the QSAR study. Their antithrombotic activity and the reference papers are given in Table 6. The molecular structures were diverse, from the smallest containing 18 to the largest containing 99 heavy atoms. Most of them were peptidomimetics, which are small peptide-like molecules that mimic the transition state of a substrate and work by competitively inhibiting the binding of the natural substrate—fibrinogen in our case. Binding the transition state will allow the peptidomimetic to competitively bind to the enzyme with a higher affinity than that of the natural substrate.

The majority of structures were peptidomimetics containing three or four structural units, one of which was a cyclic peptide derivative. Fig. 5, in which the thirteenth structure from Table 6 ($C_{25}H_{34}N_6O_5S$) is shown, illustrates what a peptidomimetic structure looks like. Neither the ligands alone nor the complexes with the receptor belong to a homologous series. However, after a careful analysis of the X-ray structural data of the series, it was

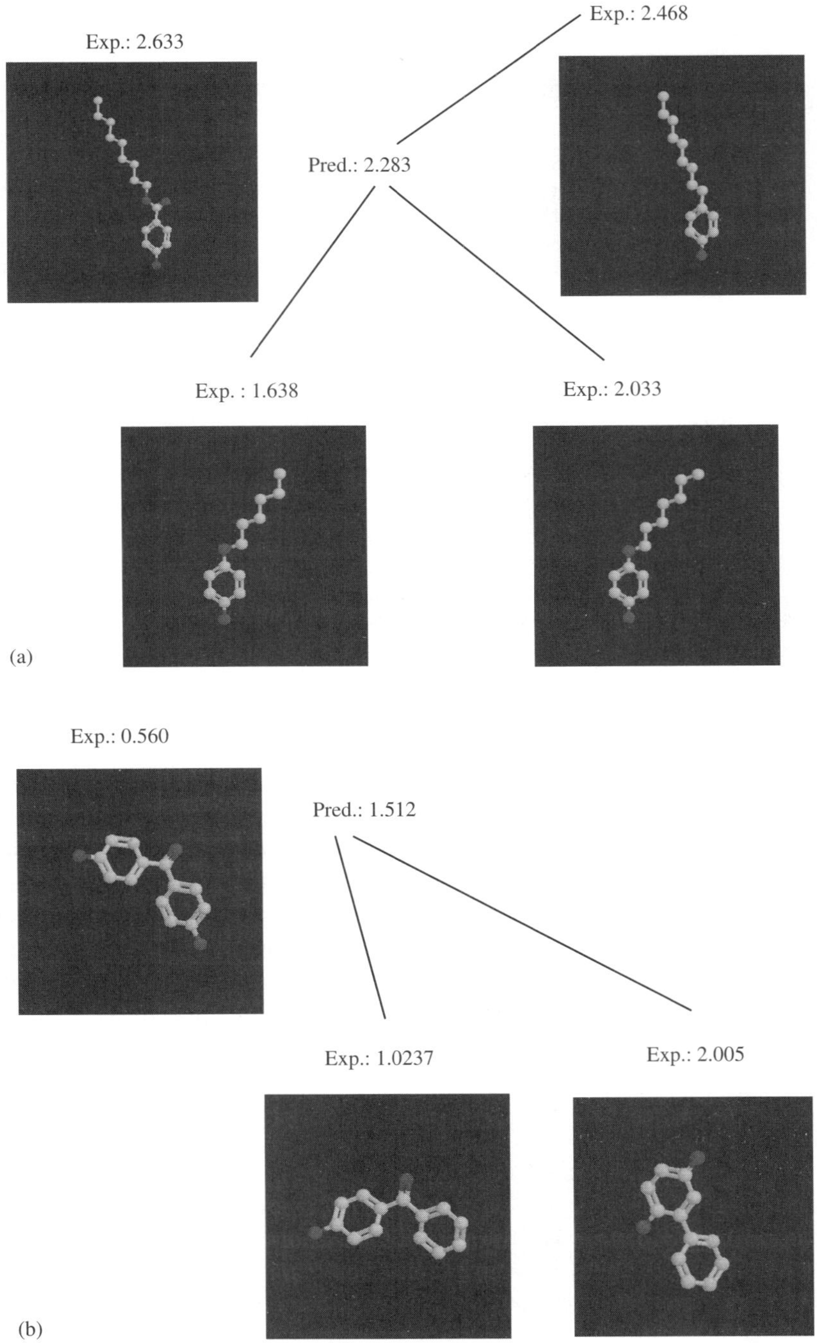

Fig. 4. (a) Prediction for nonyl-4-hydroxybenzoate on the basis of nonylphenol, 4-heptyloxyphenol, and 4-hexyloxyphenol. (b) 4,4′-dihydroxybenzophenone predicted from 4-hydroxybenzophenonen and phenylhydroquinone.

Table 6
List of the thrombin–inhibitor complex structures with references and binding constants K_i

No	PD code	K_i (nM)	pK_i	Reference
1[te]	7KME	4.0×10^4	−4.60	Mochalin and Tulinsky (1999)
2[pr]	8KME	8.0×10^3	**−3.90**	
3[tr]	1BB0	4.4×10^0	**−0.64**	Krishnan et al. (1998)
4[pr]	1BA8	1.0×10^0	**0.00**	
5[tr]	1CA8	3.2×10^{-1}	**0.49**	
6[tr]	1TMB	1.8×10^2	**−2.26**	Maryanoff et al. (1993)
7[* pr]	1A2C	0.5×10^0	**0.50**	Steiner et al. (1998)
8[te]	1A4W	1.2×10^3	**−3.08**	Matthews et al. (1996)
9[te]	1A5G	7.1×10^{-2}	**1.15**	Charles et al. (1999)
10[tr]	1A46	2.0×10^3	**−3.30**	
11[te]	1A61	2.4×10^0	**−0.38**	
12[te]	1B5G	1.0×10^1	**−1.00**	
13[tr]	1BMN	3.6×10^0	**−0.56**	Malley et al. (1996)
14[tr]	1BMM	7.9×10^1	**−1.90**	
15[tr]	1DWB	3.0×10^2	**−2.48**	Banner and Hadvary (1991)
16[tr]	1DWC	3.9×10^1	**−1.59**	
17[tr]	1DWD	6.6×10^0	**−0.82**	
18[te]	1HDT	1.72×10^1	**−1.24**	Tabernero et al. (1995)

[*] IC_{50} value available for this compound only. tr, te, pr indicate the compounds assigned for the training, test, or prediction set, respectively.

confirmed that all inhibitors were bound to the same site on the thrombin enzymatic surface.

There were two sources of uncertainty in comparisons of complex structures: first, the X-ray data were obtained with different crystallization procedures, and second, binding of the ligands might influence the structure of the thrombin active site differently. To avoid ambiguities, all enzyme structures were superimposed onto one reference structure. All inhibitors were at least in part anchored to the active site pockets of the three thrombins and were similarly oriented in the active site cavity. Comparing the complex structures, only a slight difference in the geometry of the active site (controlled by calculated RMSEP) was found, which implies that the conformation does not change significantly.

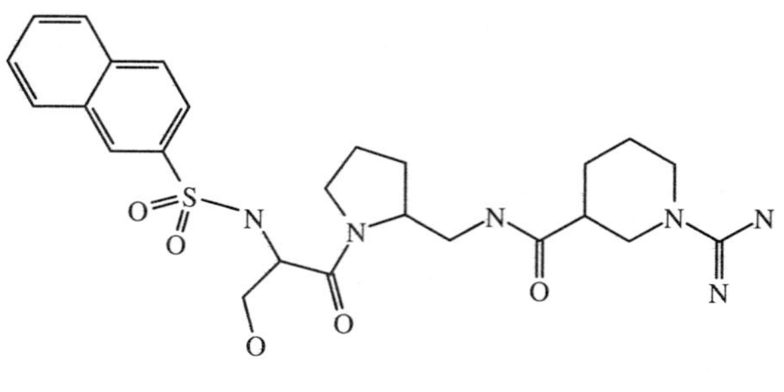

Fig. 5. Structure of one inhibitor, complex No. 13 from Table 6.

In the next step, the contact surfaces of the enzyme–inhibitor complex were determined. All the atoms in the active site less that 10 Å away from any atom of any of the inhibitors were selected and the van der Waals surface of these atoms was determined. Using 30–40 points per atom, the point density that is found to be sufficient for the smooth representation of the molecular electrostatic potential (MEP) values for each atom was obtained. There were $N = 322$ atoms on the contact surface and therefore up to 12,000 coordinate points were determined to express the contact surface for each thrombin–inhibitor complex. The complex structural descriptors were determined as a selected set of calculated MEP values.

4.3.2. Calculation of descriptors—MEP values

The molecular electrostatic potential at surface points obtained as described above was computed using a standard QM/MM method (Lee et al., 1998). Only the inhibitor atoms were treated quantum mechanically using the standard 6-31G basis set. The enzyme's electrostatic interactions were treated classically. The parameterized atomic charges of the enzyme atoms were included in the one electron Hamiltonian of the complex for the Hartree–Fock–Roothan iterative computation of the inhibitor wave function and electron density. In this way, the enzyme perturbs the density differently for each inhibitor. For each of the 322 atoms on the contact surface the 30–40 MEP values, obtained for all of the inhibitor–thrombin complexes, were ordered by size. For a unique equally dimensioned structural representation of each complex only two MEP values were chosen for every atom. If all the MEP values for a given atom were positive or all were negative, the two points with the highest values were chosen; if for a given atom the MEP values were of different signs (positive and negative) the highest positive and the highest negative value were chosen.

The described procedure for the computation of structural descriptors is an improved approach initiated by Schramm who constructed a spherical reference surface around the ligand placed in the enzyme active site (Kline and Schramm, 1995).

4.3.3. Model building

The structure representation vectors of the 18 inhibitors were composed of 644 descriptors—MEP values obtained in the procedure described above (X_i, $i = 1,\ldots,644$). Three of the inhibitors were separated from the rest for the external validation set (Table 6, nos. 2, 4 and 7), leaving 15 objects in the dataset. They served to train the model based on the counter propagation artificial neural network (Hecht-Nielsen, 1987; Dayhof, 1990; Zupan and Gasteiger, 1999). As described above, the learning strategy of a counter propagation neural network consists of two steps: (i) mapping of the inhibitors' representation vectors into the Kohonen or input layer, which is the unsupervised segment of the learning strategy, and (ii) forming the response surface on the basis of known targets or 'responses' connected to each input vector. In the first unsupervised part, each of the 644 dimensional input vectors was mapped onto a two-dimensional (5×5) Kohonen layer containing twenty-five 644-dimensional neurons ($W_{j,i}$, $j = 1,\ldots,25$, $i = 1,\ldots,644$). The mapping means that the position, given by the (X,Y) coordinate in the 5×5 plane corresponding to one of the 25 neurons ($j = j_c$, c denotes central), is determined. After the input vector is mapped, the weights of the neurons have to be modified, i.e. corrected in a way that they become more similar to the original input vector components as described in the previous section. During this stage, after each individual structure is input (called one cycle of the NN learning procedure), the supervised part of the learning was performed over the additional layer of neurons, the so-called output layer. In the output layer there were 25 one-dimensional neurons that were trained to become a response surface. This means that the weights of the output neuron ($O_{j,i}$, $j = 1,\ldots,25$, $i = 1$) were corrected in such a way that the weight at the position j_c became similar to the target value of the inhibitor mapped at this position. The procedure was repeated many times for all input structures (15 inhibitors). One epoch is completed after the mapping and correcting of the corresponding weights for all 15 structures. The predictions made by the resulting model on so-called raw data (prior to variable reduction) were tested with the leave-one-out cross-validation (LOO-CV) procedure, as shown in Fig. 6. During the leave-one-out test, the network was trained using the data for $n - 1$ inhibitors and the inhibition constant value for the nth inhibitor was predicted. This procedure was repeated n times yielding the list of n predictions.

As can be seen in Fig. 6, the error of predicted pK_i values for 15 inhibitors during model validation (LOO-CV) is large, with RMSEP = 6.54. This RMSEP is determined by Eq. (5)

$$\mathrm{RMSEP} = \sqrt{\frac{\sum_{i=1}^{N}(Y_i^{\mathrm{pred}} - Y_i)^2}{N}} \tag{5}$$

In order to improve the model, the 644-dimensional representation of the inhibitor–enzyme structure was reduced so that only the variables (descriptors) most relevant to the relationship between the complex structure and binding affinity were retained.

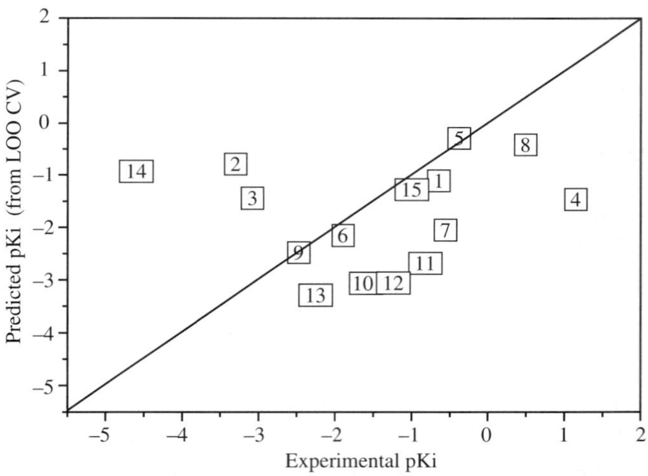

Fig. 6. Predicted versus known experimental binding affinities (p$K_\mathrm{i} = -\log(K_\mathrm{i})$) from leave-one-out cross-validation (LOO-CV) of the non-reduced inhibitor–enzyme interaction model based on CP-ANN.

4.3.4. Variable selection for model optimization

In the section 'Calculation of descriptors—MEP values', it was explained how the components of the 644-dimensional structure representation vectors describing the structure of the inhibitors were calculated. The reduction of the dimension of the structure representation vectors occurs by choosing the most relevant variables. This would lead to a model with the best predicting ability, and is beneficial for two reasons. First, removing a great number of variables found to be unimportant in the structure–activity relationship diminishes the so-called 'noise' in the model that is usually introduced by the irrelevant variables. Second, the analysis of the variables chosen to be the most relevant for the model informs us of the regions in the receptor's structure that are involved in the binding mechanism of the inhibitors from our data set.

The variable selection was done using the genetic algorithm (GA) (Smith and Gemperline, 2000) combined with the CP-ANN model. The predictive ability of the model that was obtained from the training set of compounds for each trial combination of variables obtained from GA was tested with a set of test compounds, namely the thrombin inhibitors. The resulting error in prediction (RMSEP) was the optimization criterion in the GA procedure. In order to carry out the subsequent genetic algorithm procedure, the inhibitors were divided into three groups: (i) a training set comprising nine inhibitors; (ii) a test set of six inhibitors; and (iii) the prediction (external validation) set. In the prediction set, there were three inhibitors chosen arbitrarily, except for no. 7 (see Table 6), which had an IC_{50} value instead of a K_i and was not the best choice with which to train the model. In Table 6 the assignation of compounds to the training, test and prediction set is given. Such a prediction set enabled us to test the quality of the final model. Furthermore, it should not influence either the determination of the CP-ANN model parameters, such as the number of training epochs, number of neurons, maximal and minimal learning rates, etc., or the next step of the procedure in which the selection of variables with GA is performed. It has to be stressed here that the statistical tests of the model are first done on the basis of the test set containing six inhibitors. However, this test is biased to the optimization procedure, because the prediction results of the test set are used to optimize model parameters and to select the variables. The three compounds excluded completely from the modelling procedure serve as an additional test and demonstrate how the property of a completely new compound could be predicted. Of course three compounds are not enough to obtain statistical parameters of the test, but excluding more compounds from the data set would jeopardize the quality of information needed for building the model. The lack of compounds with known binding affinities is a severe problem that modellers face all the time. A compromise between the number of compounds serving as a source of information for building the model and the number of compounds for validation of the model has to be accepted.

Once the compounds for the prediction set were removed, the remaining 15 compounds were divided into a training set and a test set, taking into account that the training set should contain as many compounds as necessary to encompass the entire information space, i.e. all structural and property coordinates. The selection of compounds for the training set was based on the results of leave-one-out cross-validation (LOO-CV), which measures the quality of the model itself but is too time consuming to be performed several thousand times in a GA procedure.

Genetic algorithm. A genetic algorithm is often used to treat various problems in which optimal values of a set of parameters, or an optimal reduced set (being a combination of original parameters), are sought. The origin of its name comes from the resemblance with the biological process. A genetic algorithm consists of three basic processes mimicking Darwinian evolution: crossover, mutation and survival of the fittest. In the crossover step, new chromosomes are generated by mixing fractions of old individuals, so-called parent chromosomes. Mutation is introduced to change randomly individual bits of the chromosome. In the last step, the survivors are chosen for the next generation, i.e. next cycle of the GA. The survivors are the chromosomes having the best criterion value determined by the fitness function. This last step should yield the lowest possible number of points on the enzymatic contact surface which gives a satisfactory prediction for K_i for each individual inhibitor in the data set. The length of the chromosome, i.e. the number of genes or bits, is determined by the number of input parameters. In our case the input parameters are 644 descriptors representing the electrostatic interaction of the inhibitor with 322 atoms of the enzyme.

The detailed description of the selection procedure implemented is given here. Initially 100 chromosomes of length 644 were randomly filled with discrete values 0 or 1. These discrete values are called bits, by analogy with computer terminology. A bit of value 1 at the position i in the chromosome says that the ith component (the MEP value in a point on the contact surface) is used in the reduced set of descriptors, while 0 means that the corresponding descriptor will be skipped. Each chromosome gave one pattern of descriptors accepted for the model. The counter propagation model designed according to the pattern of descriptors given by each chromosome was trained with nine training compounds and tested with six test compounds. The whole procedure was repeated for 100 chromosomes. The errors in the predicted values (RMSEP) of inhibition constants of compounds from the test set were taken as the criterion for the fittest chromosome. The best chromosomes were crossed over and mutated. In the new pool of 100 chromosomes, the new pattern representations were tested for quality. This procedure was repeated 900 times and thus, in each generation, a more representative pattern of contact points on the surface was obtained. Finally, the resulting optimized CP-ANN model incorporates the ability to predict the binding affinity of the inhibitor.

Optimal model with selected variables. A genetic algorithm was repeated from eight different random seeds for an initial random pattern of the bits in a pool of 100 chromosomes. In Table 7, the results of all eight GA runs are presented. The eight different reduced sets of variables were determined, consisting of 10–57 descriptors, yielding CP-ANN models that were able to predict the K_i values of test inhibitors with a correlation coefficient r between 0.951 and 0.969. Although the selected variables in all eight sets did not overlap as much as we would like, a careful examination of regions (residues) of the receptor connected to the sequence number of the selected descriptors (322 atoms determined 644 descriptors as described in the procedure of descriptors calculation) confirmed that the structural information from some regions were present in all reduced sets of variables. Not all residues were represented with equal frequency; the most frequent ones are: Tyr 60A, Pro 60B, Pro 60C, Trp 60D, Asn 60G, Phe 60H , Arg 73, Asn 95, Trp 96, Asn 98, Ser 171, Ile 174, Arg 175, Asp 189, Ala 190, Glu 217 and

Table 7
Eight different runs of GA yielding different bit patterns for variables accepted for the model, and regression coefficients for predictions of test compounds

Random seed no.	No. of bits (accepted variables)	r_{model}
1	57	0.957
2	52	0.965
3	56	0.969
4	55	0.954
5	56	0.953
6	35	0.951
7	45	0.961
8	10	0.965

Lys 224. The optimal model was built with 56 descriptors because the chromosome with the best-fit value of RMSEP = 0.35 had 56 genes turned to one. This suggests that 56 of the selected variables contain a large amount of information for predicting the biological activity of the compounds under investigation. The parameters for training the CP-ANN model were: 5 × 5 neurons of dimension 56, learning rates $\eta_{max} = 0.5$, $\eta_{min} = 0.01$, $N_{epochs} = 200$, triangular correction function, and a non-thoroidal condition for weights correction. The regression equation with regression coefficient $r = 0.97$ and standard error $S_d = 0.26$ of the predicted K_i versus experimental values for test compounds was:

$$K_i^{pred} = 0.85086\ (\pm 0.10923)\ K_i^{exp} - 0.16483\ (\pm 0.15149) \tag{6}$$

The predictions for the 9 training, 6 testing, and 3 non-biased compound prediction sets are demonstrated in Fig. 7a–c, respectively. The RMSEP of pK_i of the three compounds from the prediction set, which were never used in any optimization procedure during the presented modelling approach, is 1.24. This is a good prediction result, if we take into account that for one of the three compounds the IC_{50} value instead of K_i was available, which may differ from K_i by an order of magnitude. The model constructed on the basis of CP-ANN training and optimized by a GA procedure for variable selection enabled a correlation of the three-dimensional structure of an enzyme–inhibitor complex with its inhibition constant. The optimized model, if compared with the initial model obtained with the non-reduced set of descriptors, shows better performance for the six test compounds, while for the three compounds of the external validation set, the predictions were not significantly improved.

It was confirmed that the electrostatic potential calculated at the surface of the protein and inhibitor includes sufficient information on forces contributing to the enthalpy of the free energy of binding. The list of residues obtained by the variable selection method was found to correlate well with the current hypothesis on the importance of amino residues at the molecular surface of thrombin's active site (Jones-Hertzog and Jorgensen, 1997).

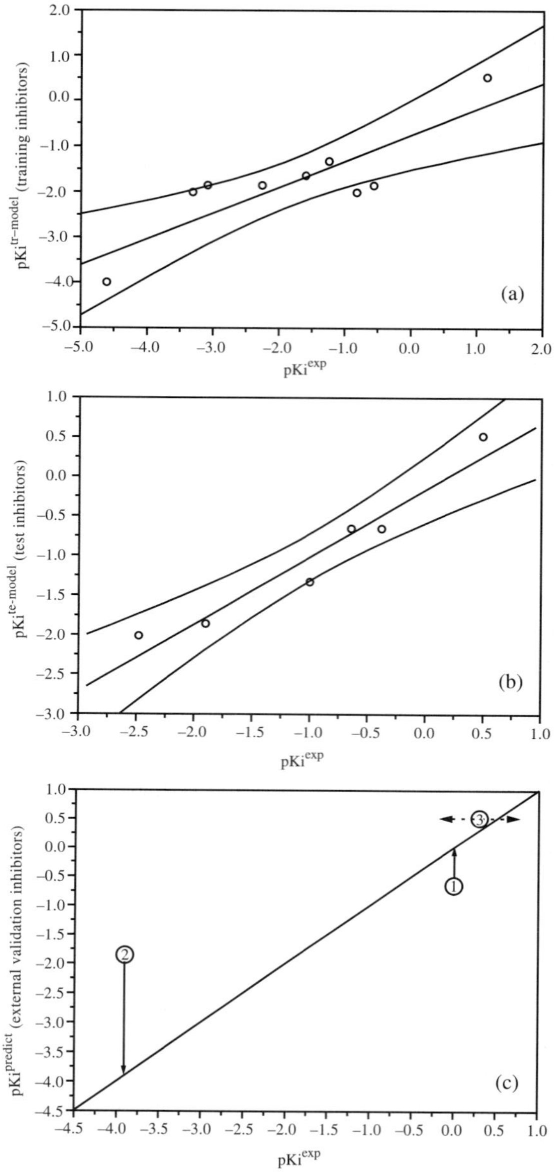

Fig. 7. Predicted versus experimental K_i values of (a) 9 training, (b) 6 testing, and (c) 3 non-biased compounds.

5. Conclusions

In this review we have presented the architecture and learning strategy of a counter propagation neural network and showed how it can be used in molecular structure–property modelling. Before the implementation of the modelling, several questions must be addressed. The first question is the selection of data used for modelling and for testing of models. A special challenge for researchers are large data sets with chemically diverse compounds, such as large databases containing different toxicology data. In this framework, visualization of data, screening of large databases, recognizing clusters, selection of outliers, and robust classification are more important tasks than a precise determination of activity. For these purposes, a counter propagation neural network was shown to be a suitable tool. The second question is the selection of descriptors. Nowadays, different softwares can calculate up to several thousand descriptors and it becomes difficult to understand the deep meaning of all of them, and consequently their relevance to the desired structure–property relationship. The selection of 'proper' descriptors remains a challenging task for an algorithm and computer program developer. A 'proper' descriptor should be correlated to the property and should also have a clear physical meaning, which enables insight into the mechanisms of the property. A short review on molecular descriptors and structure representations is given in the first part of this chapter. The third question is the testing and validation of models. The most reliable test is the prediction for an independent validation set, which was not included in the modelling. But such a set should not be completely independent of the training set; in fact, it must be a part of the domain of the training set. The selection of a proper test set analysing the descriptor domains of the training and test sets is also an important task. A complex computational approach to estimate biological properties of chemicals must treat these three essential questions. A further important question is also the quality of experimental biological data, because the quality of the model can never exceed the accuracy obtained for the data that served as the source of information for building the model. One has to be aware that by gathering the experimental data on the basis of a literature search, from different laboratories and databases, the experimental error, which has to be considered, enlarges considerably.

Acknowledgments

The research work described in this chapter was supported by the Ministry of Education, Science and Sport of the Republic of Slovenia (research programmes P1-0507 and P1-0508) and by the EU (projects COPERNICUS-EST CP94-1029, IMAGETOX HPRN-CT-1999-00015, 'Chemometrical Treatment of Toxic Compounds—Endocrine Disrupters' Marie Curie host fellowship, Contract No. HPMT-CT-2001-00240). The authors are grateful for this support.

References

Balaban, A.T., 2001. Personal view about topological indices for QSAR/QSPR Chapter 1. In: Diudea, M.V., (Ed.), QSPR/QSAR Studies by Molecular Descriptors. Nova Science Publishers, Inc., Huntington, CA.

Banner, D.W., Hadvary, P., 1991. Crystallographic analysis at 3.0Å resolution of the binding to human thrombin of four active site directed inhibirors. J. Biol. Chem. 266, 20085–20093.

Basak, S.C., Niemi, G.J., 1991. Predicting properties of molecules using graph invariants. J. Math. Chem. 7, 243–272.

Basak, S.C., Mills, D., 2001. Prediction of mutagenicity utilizing a hierarchical QSAR approach. SAR QSAR Environ. Res. 12, 481–496.

Basak, S.C., Bertelsen, S., Grunwald, G.D., 1994. Application of graph theoreical parameters in quantifying molecular similarity and structure–activity relationship. J. Chem. Inf. Comput. Sci. 34, 270–276.

Basak, S.C., Mills, D., Balaban, A.T., Gute, B.D., 2001. Prediction of mutagenicity of aromatic and heteroaromatic amines from structure: a hierarchical QSAR approach. J. Chem. Inf. Comput. Sci. 41, 671–678.

Bursi, R., Dao, T., Wijk, T.v., Gooyer, M.d., Kellenbach, E., Verwer, P., 1999. Comparative spectra analysis (CoSA): spectra as three-dimensional molecular descriptors for the prediction of biological activities. J. Chem. Inf. Comput. Sci. 39, 861–867.

Bohm, M., Sturzebecher, J., Klebe, G., 1999. Three-dimensional quantitative–structure–activity relationship analysis using comparative molecular field analysis and comparative molecular similarity indices analysis to elucidate selectivity differences of inhibitors binding to trypsin, thrombin and factor Xa. J. Med. Chem. 42, 458–477.

Charles, R.S., Matthews, J.H., Zhang, E., Tulinsky, A., 1999. Bound structures of novel P3-P1′ beta strand mimetic Inhibitors of thrombin. J. Med. Chem. 42, 1376–1383.

Ciubotariu, D., Gogonea, V., Medeleanu, M., 2001. Van der Waals molecular descriptors. Minimal Steric Difference. Chapter 10. In: Diudea, M.V., (Ed.), QSPR/QSAR Studies by Molecular Descriptors, Nova Science Publishers, Inc., Huntington, CA.

Clerc, J.-T., Terkovics, A.L., 1990. Versatile topological structure descriptor for quantitative structure/property studies. Anal. Chim. Acta 235, 93–102.

Cramer, R.D. III, Patterson, D.E., Bunce, J.D., 1988. Comparative molecular field analysis (Comfa). I. Effect of shape on binding of steroids to carrier proteins. J. Am. Chem. Soc. 110, 5959–5967.

CODESSA Semichem, 7128 Summit, Shawnee, KS, 66216.

Dayhof, J., 1990. Neural Network Architecture, An Introduction, Van Nostrand Reinhold, New York.

Devillers, J., Balaban, A.T., 1999. Topological Indices and Related Descriptors in QSAR and QSPR, Gordon and Breach, Reading, MA.

DRAGON 3.0, Milano Chemometrics and QSAR Research Group, © Talete srl.

ECOTOX, February 2000a. ECOTOXicology Database System Code List. By OAO Corporation Duluth Minnesota, Duluth, Minnesota, Prepared for US Environmental Protection Agency, Office of Research, Laboratory Mid-Continent Division (MED).

ECOTOX, February 2000b. ECOTOXicology Database System Data Field Definition. By OAO Corporation Duluth Minnesota, Duluth, Minnesota, Prepared for US Environmental Protection Agency, Office of Research, Laboratory Mid-Continent Division (MED).

ECOTOX, February 2000c. ECOTOXicology Database System User Guide. By OAO Corporation Duluth Minnesota, Duluth, Minnesota, Prepared for US Environmental Protection Agency, Office of Research, Laboratory Mid-Continent Division (MED).

Golbraikh, A., 2000. Molecular dataset diversity indices and their application to comparison of chemical databases and QSAR analysis. J. Chem. Inf. Comput. Sci. 40, 414–425.

Hansch, C., 1969. A quantitative approach to biochemical structure activity relationships. Acc. Chem. Res. 2, 232–239.

Hansch, C., Maloney, P.P., Fujita, T., Muir, R.M., 1969. Correlation of biological activity of phenoxyacetic acids with Hammett substituent constants and partition coefficients. Nature 194, 178–180.

Hecht-Nielsen, R., 1987. counter propagation networks. Appl. Optics 26, 4979–4984.

Hemmer, C.M., Gasteiger, J., 2000. Prediction of three-dimensional molecular structures using information from infrared spectra. Anal. Chim. Acta 420, 145–154.

Hilpert, K., Ackermann, J., Banner, D.W., Gast, A., Gubernator, K., Hadvary, P., Labler, L., Muller, K., Schmid, G., Tschopp, T.B., Waterbeemd, H.v.d., 1994. Design and synthesis of potent and highly selective thrombin inhibitors. J. Med. Chem. 37, 3889–3901.

HYPERCHEM 5.0, Hypercube Inc., Gainesville, Florida, USA.

Jones-Hertzog, D.K., Jorgensen, W.L., 1997. Binding affinities for sulfonamide inhibitors with human thrombin using Monte Carlo simulations with a linear response method. J. Med. Chem. 40, 1539–1549.

Karelson, M., Lobanov, V.S., 1996. Quantum-chemical descriptors in QSAR/QSPR studies. Chem. Rev. 96, 1027–1043.

Karelson, M., Maran, U., Wang, Y.L., Katritzky, A.R., 1999. QSPR and QSAR models derived using large molecular descriptors spaces. A review of CODESSA applications. Collect. Czech. Chem. C. 64, 1551–1571.

Katritzky, A.R., Lobanov, V.S., Karelson, M., 1994. CODESSA Comprehensive Descriptors for Structural and Statistical Analysis, Reference Manual, version 2.0, Gainesville.

Kline, P.C., Schramm, V.L., 1995. Pre-steady-state analysis of the hydrolytic reaction catalyzed by purine nucleoside phosphorylase. Biochemistry 34, 1153–1162.

Krishnan, R., Zhang, E., Hakansson, K., Arni, R.K., Tulinsky, A., 1998. Highly selective mechanism-based thrombin inhibitors: structures of thrombin and trypsin inhibited with rigid peptidyl aldehydes. Biochemistry 37, 12094–12103.

Lee, Y.S., Hodoscek, M., Brooks, B.R., Kador, P.F., 1998. Catalytic mechanism of aldose reductase studied by the combined potentials of quantum mechanics and molecular mechanics. Biophys. Chem. 70, 203–216.

Malley, M.F., Tabernero, L., Chang, C.Y., Ohringer, S.L., Roberts, D.G.M., Das, J., Sack, J.S., 1996. Crystallographic determination of the structures of human alpha thrombin complexed with BMS-186282 and BMS-189090. Protein Sci. 5, 221–228.

Maryanoff, B.E., Qui, X., Padmanabhan, K.P., Tulinsky, A., Almond, H.R., Andrade-Gordon, P., Greco, M.N., Kaufman, J.A., Nicolaou, K.C., Liu, A., Brungs, P.H., Fusetani, N., 1993. Molecular basis for the inhibition of human alpha thrombin by the macrocyclic peptide cyclotheonamide A. Biochemistry 90, 8048–8052.

Matthews, J.H., Krishnan, R., Costanzo, M.J., Maryanoff, B.E., Tulinsky, A., 1996. Crystal structures of thrombin with thiazole-containing inhbitors: Probes of the S1 binding site. Biophys. J. 71, 2830–2839.

Mazzatorta, P., Benfenati, E., Neagu, D., Gini, G., 2002. The importance of scaling in data mining for toxicity prediction. J. Chem. Inf. Comput. Sci. 42, 1250–1255.

Mazzatorta, P., Vračko, M., Jezierska, A., Benfenati, E., 2003. Modelling toxicity by using supervised Kohonen neural networks. J. Chem. Inf. Comput. Sci. 43, 485–492.

Mlinšek, G., Novič, M., Hodoščcek, M., Šolmajer, T., 2001. Prediction of enzyme binding: human thrombin inhibition study by quantum chemical and artificial intelligence methods based on X-ray structures. J. Chem. Inf. Comput. Sci. 41, 1286–1294.

Mochalin, I., Tulinsky, A., 1999. Structures of thrombin retro-inhibited with SEL2711 and SEL2770 as they relate to factor Xa binding. Acta Cryst allogr. D55, 785–793.

PALLAS 2.1, CompuDrug, Budapest, Hungary.

PETRA, Computer-Chemie-Centrum, University of Erlangen-Nuernberg.

Pires, J.M., Floriano, W.B., Gaudio, A.C., 1997. Extension of the frontier reactivity indices to groups of atoms and application to quantitative structure–activity relationship studies. J. Mol. Struct. (Theochem), 389, 159–167.

Pitea, D., Cosentino, U., Moro, G., Bonati, L., Fraschini, E., Lasagni, M., Todeschini, R., Davis, A.M., Cruciani, G., Clementi, S., 1994. 3D QSAR: the integration of QSAR with molecular modeling. In: Waterbeemd, H.v.d., (Ed.), Chapter 2 in Advanced Computer-Assisted Techniques in Drug Discovery, VCH, Weinheim.

POLLY 2.3, University of Minnesota, Duluth.

Randić, M., 1998. Topological indices. In: Schleyer, P.v.R., Allinger, N.L., Vlark, T., Gasteiger, J., Kollman, P.A., Schaefer III, H.F., Schreiner, P.R., (Eds.), The Encyclopedia of Computational Chemistry, Wiley, London.

Randić, M., Razinger, M., 1997. On characterisation of 3D molecular structures. In: Balaban, A.T., (Ed.), Chapter 6 in From Chemical Topology to Three-Dimensional Geometry, Plenum Press, New York.

Roncaglioni, A., 2002. Chemometrical Treatment of Toxic Compounds—Endocrine Disrupters, Marie Curie Host Fellowship, Contract No. HPMT-CT-2001-00240, report.

Russom, C.L., Bradbury, S.P., Broderius, S.J., Hammermeister, D.E., Drummond, S.J., 1997. Predicting modes of toxic action from chemical structure: acute toxicity in the fathead minnow (Pimephales promelas). Environ. Toxicol. Chem. 16, 948–967.

Schultz, T.W., Sinks, G.D., Cronin, M.T., 1997. Identification of mechanisms of toxic action of phenols to Tetrahymena pyriformis from molecular descriptors. In: Chen, F., Schüürmann, G., (Eds.), Quantitative Structure–Activity Relationships in Environmental Sciences —VII, SETAC Press, Pensacola, FL.

Schuur, J.H., Selzer, P., Gasteiger, J., 1996. The coding of three-dimensional structure of molecules by molecular transforms and its application to structure–spectra correlations and studies of biological activity. J. Chem. Inf. Comput. Sci. 36, 334–344.

Seward, J.R., Sinks, G.D., Schultz, T.W., 2001. Reproducibility of toxicity across mode of toxic action in the Tetrahymena population growth impairment assay. Aquat. Toxicol. 53, 33–47.

Simon, V., Gasteiger, J., Zupan, J., 1993. A combined application of two different neural network types for the prediction of chemical reactivity. J. Am. Chem. Soc. 115, 9148–9159.

Smith, B.M., Gemperline, P.J., 2000. Wavelength selection and optimization of pattern recognition methods using the genetic algorithm. Anal. Chim. Acta 423, 167–177.

Steiner, J.L.R., Murakami, M., Tulinsky, A., 1998. Structure of thrombin inhibited by Aeruginosin 298-A from blue-green alga. J. Am. Chem. Soc. 120, 597–598.

Szymoszek, A., Vracko, M., 2001. IMAGETOX, HPRN-CT-1999-00015, report.

Todeschini, R., Consonni, V., 2000. The Handbook of Molecular Descriptors. Series of Methods and Principles in Medicinal Chemistry, vol. 11. Wiley, New York.

Tabernero, L., Chang, C.Y., Ohringer, S.L., Lau, W.F., Iwanowicz, E.J., Han, W.C., Wang, T.C., Seiler, S.M., Roberts, D.G., Sack, J.S., 1995. Structure of a retro binding peptide inhbitor complexed with human alpha thrombin. J. Mol. Biol. 246, 14–20.

Vracko, M., 1997. A study of structure–carcinogenic potency relationship with artificial neural networks. The use of descriptors related to geometrical and electronic structures. J. Chem. Inf. Comput. Sci. 37, 1037–1043.

Vracko, M., 2000. A study of structure–carcinogenicity relationship for 86 compounds from NTP Data Base using topological indices as descriptors. SAR and QSAR in Environ. Res. 11, 103–115.

Vracko, M., Gasteiger, J., 2002. A QSAR study on a set of 105 flavonoid derivatives using descriptors derived from 3D structures. Internet Electron. J. Mol. Des. 1, 527–544.

Vracko, M., Novic, M., Zupan, J., 1999. Study of structure–toxicity relationship by a counterpropagation neural network. Anal. Chim. Acta 384, 319–332.

Vracko, M., Novic, M., Perdih, M., 2000. Chemometrical treatment of electronic structures of 28 flavonoid derivatives. Int. J. Quantum Chem. 76, 733–743.

Zupan, J., Gasteiger, J., 1999. Neural Networks in Chemistry and Drug Design, Wiley-VCH, Weinheim.

Zupan, J., Novic, M., 1997. General type of uniform and reversible representation of chemical structures. Anal. Chim. Acta 348, 409–418.

Zupan, J., Novic, M., Li, X., Gasteiger, J., 1994. Classification of multicomponent analytical data of olive oils using different neural networks. Anal. Chim. Acta 292, 219–234.

Zupan, J.M., Vracko, M., Novic, M., 2000. New uniform and reversible representation of 3D chemical structures. Acta Chim. Slov. 47, 19–37.

CHAPTER 9

Neural networks for the calibration of voltammetric data

Conrad Bessant, Edward Richards

Cranfield Centre for Analytical Science, Cranfield University, Silsoe, Bedford MK45 4DT, UK

1. Introduction

The instrumental simplicity of electroanalytical chemistry makes it an excellent basis for low cost, rapid and sensitive analytical devices. This has been demonstrated by the commercial success of electrochemical blood glucose biosensors currently being used by diabetics around the world. Such devices use glucose oxidase as a specific bio-receptor, in conjunction with electrochemical transduction. However, specific receptors do not exist for most other analytes, and the resulting lack of specificity makes it very difficult to quantify the analyte or analytes of interest because the relevant information is very deeply hidden within the acquired electrochemical data. Traditional chemometrics techniques such as partial least squares (PLS) can be used to produce multivariate calibration models for electrochemical systems, but the accuracy of such models can be severely limited due to complex processes occurring at the electrode surface. In this chapter a biologically inspired approach is introduced, employing neural networks, whose design is optimised using genetic algorithms (GAs). Using as examples a series of experiments conducted in our laboratories, we show that the technique produces accurate calibration models which surpass those produced by traditional chemometrics approaches.

2. Electroanalytical data

Although electrochemistry can be a very useful analytical tool, it is not used as widely as, for example, spectroscopy. The subject of electroanalytical chemistry is described in considerable detail elsewhere (Bard and Faulkner, 1980). In this section we will limit ourselves only to outlining some of the key electroanalytical techniques, and explaining why the data they generate can be difficult to calibrate.

Electroanalytical chemistry is a measurement science based upon oxidation and reduction reactions that occur at the surface of an electrode (Bard and Faulkner, 1980;

Data Handling in Science and Technology, Volume 23
ISSN: 0922-3487

DOI: 10.1016/S0922-3487(03)23009-4

Skoog et al., 1999). The principle behind all electrochemical methods is that electroreactive species undergo, or can be forced to undergo, a heterogeneous physical reaction at the surface of an electrode when integrated into an electrical circuit. For a reaction to occur, a current carrying electrical circuit must be formed using at least two electrodes in contact with a supporting electrolyte. This combination of electrodes and supporting electrolyte is known as an electrochemical cell and the physical reaction undergone by the species present in the cell at either electrode is the loss or gain of an electron, otherwise known as oxidation or reduction (Eq. (1)). The electrode at which the reaction occurs is referred to as the working electrode.

$$\mathrm{O} + ne \leftrightarrow \mathrm{R} \tag{1}$$

where O is the oxidised form of the standard system, R the reduced form of the standard system, and n the number of electrons per molecule.

There is a significant theoretical basis for electrochemistry, with many equations available to characterise the processes occurring at the electrode surface. For practical purposes, however, we can limit ourselves to the Cottrell equation (Eq. (2)) (Cottrell, 1902), which describes the relationship between the current, i, and time, t, when a constant disturbing potential is applied at a planar electrode of area A, submerged in an unstirred homogeneous solution containing excess supporting electrolyte and bulk concentration, C^*, of an electroactive species, O, with diffusion coefficient, D_O. Assumptions made about boundary conditions are that the solution is homogenised before the start of the experiment, the concentrations of regions sufficiently distant from the electrode remain unchanged throughout the experiment and that after the electrode surface potential transition has occurred the concentration of the electroactive species at the electrode surface is zero, i.e. the total concentration of species at the electrode surface has been completely reduced.

$$i(t) = \frac{nFA\sqrt{D_\mathrm{O}}C_\mathrm{O}^*}{\sqrt{\pi t}} \tag{2}$$

This relationship between the bulk concentration of a species and the measured current is the basis for the determination of the bulk concentration of a species.

2.1. Amperometry

The Cottrell equation forms the foundation for chronoamperometric techniques (Plambeck, 1982). These techniques use an applied disturbing potential that reduces or oxidises the species completely (in theory) at the electrode surface, and the current response is measured as a function of time. The technique is commonly named amperometry and there are many variants based upon it (Bard and Faulkner, 1980; Plambeck, 1982). Chronoamperometric techniques are useful for detection of species in flowing streams and are widely used after separation by liquid chromatography (Skoog et al., 1999).

2.2. Pulsed amperometric detection

Pulsed amperometric detection (PAD) (Hughes et al., 1981; Hughes and LaCourse, 1981, 1983) has been developed for the detection of aliphatic compounds. Although aliphatics are easily detected via amperometry, their breakdown products cause fouling of electrode surfaces and hence prevent further reactions at these surfaces (Johnson, 1986; LaCourse and Johnson, 1991). PAD removes working electrode fouling between amperometric measurements by the application of relatively large positive (oxidative) and negative (cathodic) potential pulses which oxidatively clean and then cathodically reactivate the working electrode surface ready for a subsequent measurement (Johnson and LaCourse, 1990; Johnson, 1986; LaCourse and Johnson, 1991). The use of a high pH (pH > 12) buffer, such as sodium hydroxide, has been found to significantly improve the sensitivity of detection for the PAD method (Neuburger and Johnson, 1987).

2.3. Voltammetry

Without a prior separation step, amperometry is non-specific in that all compounds that are electroactive at or below the applied potential will elicit a current response. Voltammetry (Bard and Faulkner, 1980; Plambeck, 1982) is one way of adding some specificity to electrochemical measurement.

In contrast to the constant potential applied in amperometry, in voltammetry a rapid linear potential sweep is applied. As the potential sweep progresses, species within the sample elicit a current response according to their specific oxidation or reduction potential. However, this response is limited by the rate of diffusion of the species to the electrode surface; so the response quickly falls away, resulting in the manifestation of peaks of current. When plotted as a function of applied potential, the current response takes the form of a series of peaks called a voltammogram, which is superficially similar to a spectrum obtained by spectroscopic techniques. As in spectroscopy, the height of each peak can indicate the concentration of a particular analyte.

2.4. Dual pulse staircase voltammetry

In dual pulse staircase voltammetry (DPSV), as in PAD, cleaning pulses are used to clean and prepare the working electrode surface between measurements. However, instead of a single potential measurement, a voltammetric scan is performed (Fung and Mo, 1995). Usually the potential steps are small (e.g. 10 mV) so that the waveform approximates a linear ramp. The potential profile applied in DPSV can be seen in Fig. 1.

The benefit of DPSV over normal voltammetry is that, like PAD, it is able to detect aliphatic compounds (Fung and Mo, 1995). The fact that DPSV includes a scanning element means that the resulting voltammograms comprise peaks of current, indicative of the analyte or analytes in the sample. Fig. 2 shows examples of DPSV voltammograms for glucose, fructose and ethanol. Superficially, these are analogous to spectra in spectrochemical analysis. However, it must be remembered that the voltammogram is the result of processes occurring at the electrode surface, and these processes can be very complex, making the resulting data difficult to analyse. In particular, it is important to be aware that,

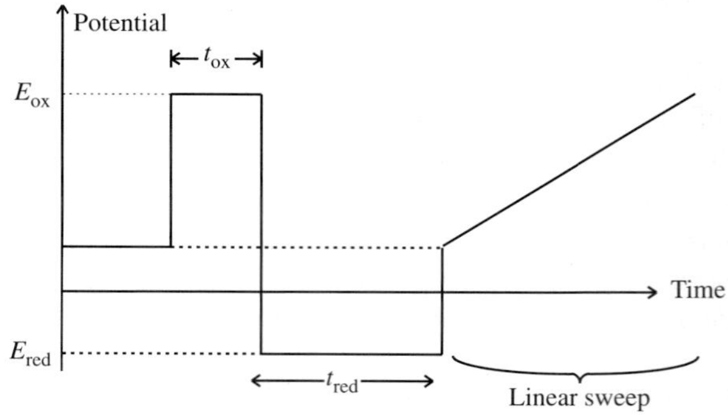

Fig. 1. DPSV waveform. E_{ox} is the oxidative cleaning potential, E_{red} the potential for cathodic reactivation of the electrode. The times t_{ox} and t_{red} indicate the duration for which each potential is applied.

as the voltammetric scan progresses, analytes are being broken down and their breakdown products are being adsorbed to the electrode surface. This reduces the effective surface area, and hence the current response of analytes which are oxidised at higher potentials. This limits the accuracy of traditional multivariate calibration techniques such as PLS.

Due to the information rich nature of DPSV results, and the fact that DPSV allows the detection of aliphatic compounds that are not readily detected spectroscopically, it is interpretation of DPSV data that forms the focus of this chapter.

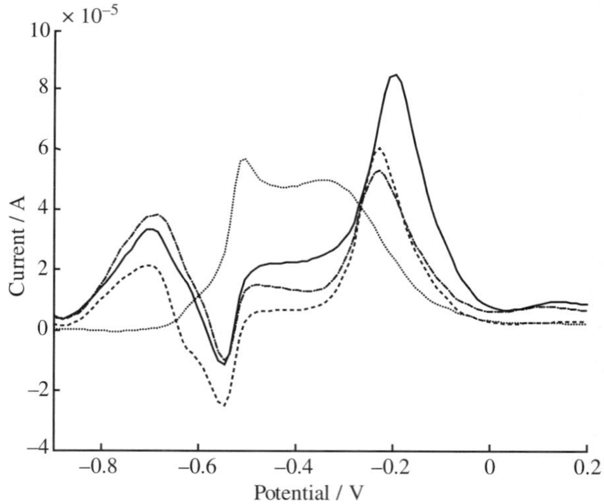

Fig. 2. Voltammograms obtained using DPSV for 0.72 mM glucose (–·–·–), 0.68 mM fructose (--), 12 mM ethanol (···) and a mixture of the three compounds (–). A platinum disc working electrode was used, with a Ag/AgCl reference and platinum wire counter. All samples were analysed in the presence of 0.1 M NaOH buffer solution.

2.5. *Representation of voltammetric data*

A voltammogram (whether acquired by DPSV or any other kind of voltammetry) naturally takes the form of a vector of current measurements, each one relating to a point on the potential sweep at which the current was measured. An example of such a matrix is shown on the right side of Fig. 3. When multiple experiments are performed, multiple vectors are acquired, and these can be concatenated into a matrix as shown in the figure. This matrix is referred to as a data set, of which several are required for a multivariate calibration, as will be described later in this chapter.

3. Application of artificial neural networks to voltammetric data

As artificial neural networks (ANNs) have been lucidly introduced elsewhere in this book (see Chapter 7), we will focus here only on how these networks are applied to voltammetric calibration.

Firstly, it is important to point out that from the many different ANN configurations available, it is the multilayer feed forward network with single hidden layer that stands out as the most suitable for multivariate calibration. Furthermore, it is widely accepted that back propagation of errors is the most effective training regime; so we will make no mention here of alternative ANN structures or training algorithms. In our work, we choose to use the resilient backpropagation algorithm (Riedmiller and Braun, 1993) for network training as it provides faster training than standard back propagation (Udelhoven and Schutt, 2000). Another invariant throughout this chapter is the choice of transfer functions. All the networks described in this chapter employ log sigmoidal neurons in the hidden layer, to give the network a non-linear modelling capability, and linear neurons in the output layer.

The feed forward ANN approach has been widely used for calibration in spectroscopy for some time (for example Long et al., 1990; Goodacre and Kell, 1993; Schulze et al., 1995; Zhang et al., 1997), and has been extended to electrochemistry in the last six or so years. One of the earliest applications of ANNs to electrochemical calibration was by Lastres et al. (1996) who used ANNs to remove interference caused by the formation of

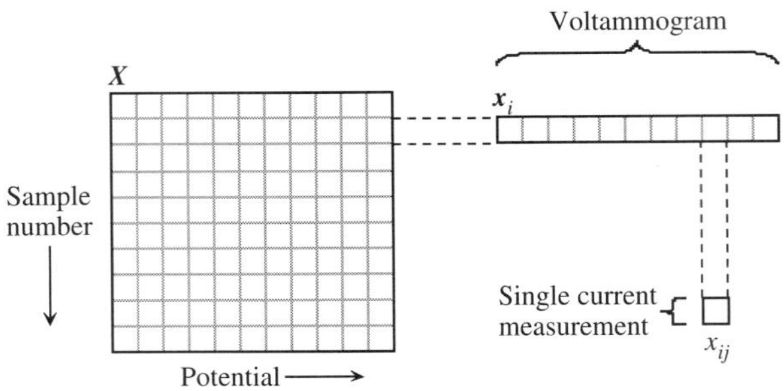

Fig. 3. Organisation of voltammetric data into a data matrix, $\boldsymbol{X}$.

intermetallic compounds in stripping voltammetry analysis of copper and zinc. Their final calibration model was capable of determining the concentrations of copper and zinc from a single voltammogram with reasonable accuracy. More recently, Cukrowska et al. (2001) have used ANNs to calibrate voltammograms to determine adenine and cytosine in mixtures that were experiencing interference from hydrogen evolution. In recent years, our laboratory has worked extensively on the development of ANNs for the calibration of voltammograms acquired using DPSV in mixtures of aliphatic compounds (Bessant and Saini, 1999; Richards et al., 2002). It is this work, and further developments of it that are described in this chapter.

Several papers have attempted to ascertain the suitability of ANNs over more traditional chemometrics techniques for electrochemical calibration (de Carvalho et al., 2000; Guiberteau Cabanillas et al., 2000; Bessant and Saini, 2000), and in all cases ANNs have been found to produce more accurate calibration models than techniques such as PLS. However, the literature usually highlights the difficulty in optimising neural networks compared to optimising PLS models.

3.1. Basic approach

Most aspects of the neural network calibration approach are common to other multivariate calibration methods. For example, the starting point is plenty of high quality data acquired from samples of known composition, collected according to a suitable experimental design. For the work described here, data was collected using an automated sample production and measurement system made specifically for this purpose in our laboratories. This system, described in detail elsewhere (Richards et al., 2003) permits accurate collection of voltammograms in a random order, and is sufficiently rapid to allow full factorial experimental designs to be carried out within sensible time scales.

Also in common with other calibration techniques, it is important to split the acquired data into three sets: a training set which is used to train the ANNs, a validation set which can be used to optimise the ANN training parameters, and a test set which contains completely unseen data and can therefore be used to carry out a blind test of any calibration model produced.

Once training data has been collected, thought needs to be given to how this is presented to a neural network. The simplistic approach is to use an ANN that has as many inputs as there are points in the voltammogram, and indeed this has been done in the past (Bessant and Saini, 1999). However, this is very inefficient as voltammograms typically comprise over 100 points, which necessitates a large input layer with a correspondingly large number of interconnections to the hidden layer, making it very difficult for network training algorithms to converge on the best network. An alternative is to use principal components analysis (PCA) (Martens and Næs, 1991; Otto, 1999) to compress the data prior to presentation to the ANN. In this way it is possible to reduce a voltammogram to just a handful of PC scores, thereby reducing the number of input neurons by the same amount. This is the preferred approach and is used throughout the following examples. However, it does mean that effort must be spent optimising the number of PCs, which, as we will see, is not as trivial a process as might be expected.

Another decision that needs to be made is whether one network should be used to determine the concentration of all analytes of interest from a voltammogram, or whether each analyte should have its own specifically trained network. Previous work in our laboratory has shown that having separate networks for each analyte invariably leads to more accurate calibration, and this approach is employed in the examples described below.

3.2. Example of ANN calibration of voltammograms

The example that follows involves a three component system of glucose, fructose and ethanol, as would be found during fermentation of grapes into wine. The aim is to calibrate the system such that the concentration of each of the three analytes can be determined from the voltammogram of an unknown mixture.

3.2.1. Hardware and software

All work described in this chapter was carried out in Matlab V6.1 (MathWorks Inc, USA). Neural network programs were written and trained in Matlab V6.1 using functions from the Neural Network Toolbox V4.01 (also from Mathworks).

3.2.2. Data

The training data set comprised four repetitions of a three-factor five-level experimental design with minimum concentrations of 0 M for all the analytes and maximum concentrations of 12, 0.68 and 0.72 mM for ethanol, fructose and glucose, respectively. The individual step concentrations used were 3, 0.17 and 0.18 mM for ethanol, fructose and glucose, respectively. The role of the training data set was to provide the ANN with data from which it could learn relationships between known concentrations of the three analytes and the resulting voltammograms obtained for those concentrations. Four repetitions of the experimental design were included in the training data set to enable natural variation in the sample preparation and measurement process to be expressed in the training data set. This was an important feature of the training data as the formation of a good general model required the natural variance to be represented in the model. Training on data with no natural variance expressed could have led to over-training of the model. The total number of voltammograms in the training set was 500.

A validation data set is essential when a calibration method requires the optimisation of model parameters, as is the case in ANNs as well as PLS (e.g. where latent variables need to be selected). The validation data set experimental design replicated the training data set experimental design and had, in addition, a three-factor four-level experimental design nested within it. The nested design had minimum concentrations of 1.5, 0.085 and 0.09 mM and maximum concentrations of 10.5, 0.595 and 0.63 mM for ethanol, fructose and glucose, respectively. The incremental concentration step sizes were as in the training set. The validation data set comprised three repetitions of the experimental design so that, like the training set, natural variance in sample production and measurement process was considered during the optimisation of the model. Using only one repetition of the validation data set could lead to a biased choice of modelling parameter based on the models' ability to predict well for one repetition rather than the general case. A total of 567 voltammograms were included in the validation set.

A test set is required to obtain a fully independent measure of a calibration model's accuracy. In the examples described below, the test set comprised a three-factor nine-level experimental design. The minimum concentrations for all the analytes was 0 M, and the maximum concentrations were 12, 0.68 and 0.72 mM for ethanol, fructose and glucose, respectively. The individual step concentrations were 1.5, 0.085 and 0.09 mM for ethanol, fructose and glucose, respectively. The test data set included all concentration permutations forming the training and validation sets as well as many more and hence provided a good thorough test of the model with previously unseen data which spanned the whole envelope of the concentration permutations for all the analytes. In total, there were 729 voltammograms in the test set, of which 536 were at concentration permutations that were not present in the training or validation sets.

3.2.3. Evaluation of network performance

To compare one network design against another, we need some kind of quantitative measure of network performance. Typically the overall error of a calibration model is presented as the root mean square error (RMSE) for the test set, as defined in Eq. (3) below.

$$\mathrm{RMSE} = \sqrt{\frac{\sum_{j=1}^{n} (y_j - \hat{y}_j)^2}{n}} \tag{3}$$

where y_j is the known concentration and $\hat{y}_j$ is the concentration predicted by the calibration model for sample j, and n is the total number of samples in the test set. To gain further insight into the behaviour of the calibration model, we chose to calculate separate errors for samples with the same concentrations used in training (we call this the replicate error, RMSR) and those at interpolation concentrations, i.e. concentrations between those used in training (the interpolation error, RMSI). This is useful because it allows us to easily identify overtrained networks (where RMSI is much greater than RMSR).

The two RMS errors were combined to provide a single measure of accuracy, the *distance RMS error*, RMSD, as described by the equation below.

$$\mathrm{RMSD} = \sqrt{\mathrm{RMSR}^2 + \mathrm{RMSI}^2} \tag{4}$$

The RMSD of the test set gives a good general indication of model capability, and minimisation of this error was the aim of all network optimisation. These measures were also used for evaluating models with the validation data set.

3.2.4. Network optimisation

There are three key parameters that need to be optimised in an ANN calibration model. These are:

- The number of epochs for which the network is trained.
- The number of principal components (PCs) used at the network input.
- The number of neurons in the hidden layer.

The most common approach to optimisation of ANN parameters is usually trial and error. However, this approach is inefficient and is very likely to miss the optimum network design. A more scientific approach is to use some kind of experimental design that allows networks to be tested across the design parameter space, and the optimum network pinpointed.

A serious problem when optimising neural networks is the issue of initial network weights. Random initial weights are widely used to provide scope for network training algorithms to find global error minima, but this means that two networks with identical design can produce very different calibration models once trained. Repeated evaluation of network designs is therefore essential in getting a clear idea of the capability of each design.

The most obvious way of rigorously optimising network parameters is a fully factored search through the parameter space. This steps through the possible parameter values in turn. The range of parameters investigated in this work is shown in Table 1.

For each permutation of network parameters, 10 separate networks were trained to reduce the effect of the random initial starting conditions. This resulted in a total of 32 480 neural network models being trained per analyte.

3.2.5. Number of training epochs

The effect of the number of training epochs on the trained neural network's ability to predict for the validation set are presented in Fig. 4. The figure shows how errors for ethanol calibration models vary when trained for a finite number of epochs. Similar results were obtained for the other two analytes, glucose and fructose. The number of epochs used to train models with minimum RMSR, RMSI and RMSD, for each analyte are summarised in Table 2.

All mean errors improved with increased training epochs; however, good low error networks were trained for all training epochs used. The appearance of good models in all epoch ranges is thought to be due to the random seeding of the initial weights—with so many network repetitions it is possible for a network to be seeded with weights that just happen to produce a reasonable model before training has even begun.

3.2.6. Principal components input to the network

Fig. 5 shows calibration errors for ethanol for neural networks trained using a variable number of PCs, when presented with the validation set. As the models were trained with increasing numbers of PCs, the average RMSR for the validation set can be seen to improve asymptotically. The greatest improvements were made between 2 and 7 PCs. Variance captured by the PCs showed that seven PCs was sufficient to capture over 99.9%

Table 1
Network parameter search space

Parameter	Minimum	Maximum	Step size
Training epochs	100	800	100
PCs	2	30	1
Hidden neurons	2	15	1

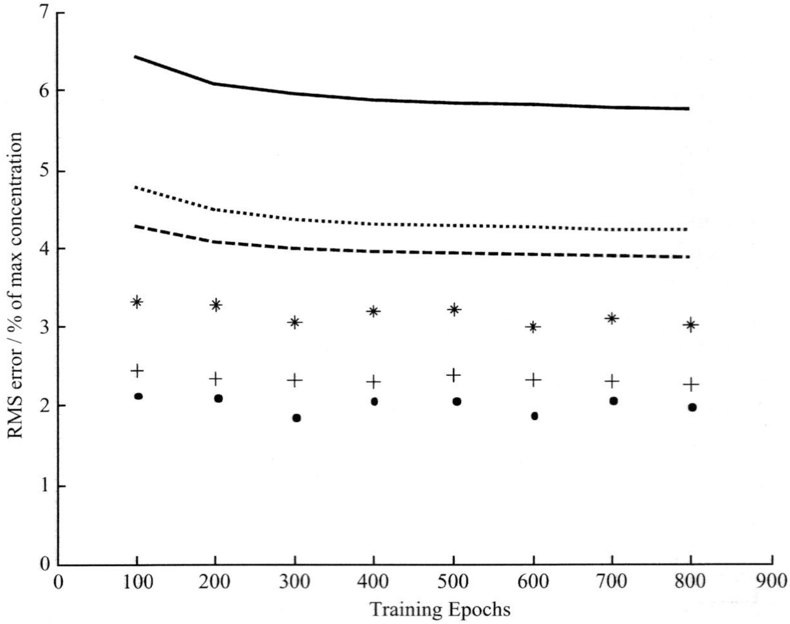

Fig. 4. Effect of number of epochs used to train neural network models for ethanol when presented with the validation set. The mean error is shown as a line and the minimum as a point, for RMSR (···, +), RMSI, (--,•) and RMSD (–, ∗).

of the variance in the data, so this was not unexpected. However, improvements in the mean RMSR are clearly seen when adding further PCs, suggesting that later PCs contain salient (presumably non-linear) information, a phenomena that has been observed elsewhere (Despagne and Massart, 1998). The average RMSR continued to fall to a minimum at 22 PCs then it began to rise slightly over the remaining PCs, as may be expected by the inclusion of PCs that only contain noise.

Generally the mean errors for all three analytes followed the same pattern, decreasing sharply as the number of PCs was initially increased, then the rate of decrease slowed before climbing again as the final PCs were added. The number of PCs that produced the lowest RMSR, RMSI and RMSD for each analyte are summarised in Table 3. Note that these figures are for the lowest individual network, which, in the case of ethanol, does not correspond to the lowest average errors on the traces in Fig. 6.

Table 2
Epochs used to train models with minimum prediction errors for each analyte

Analyte	Epochs for min RMSR	Epochs for min RMSI	Epochs for min RMSD
Ethanol	800	300	600
Fructose	800	700	400
Glucose	300	800	600

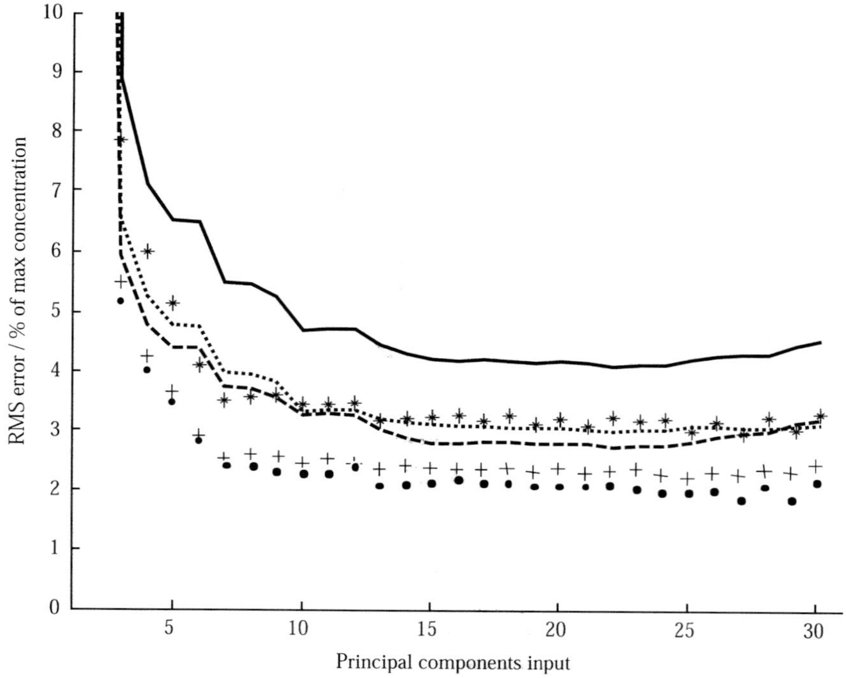

Fig. 5. Effect of number of PCs used to train neural network models for ethanol when presented with the validation set. The mean error is shown as a line and the minimum as a point, for RMSR (···, +), RMSI, (--,•) and RMSD (–, ∗).

Increasing the number of PCs used also increases the number of interconnections from the input layer to the hidden layer, thereby increasing the dimensionality of the error surface to be covered during training. As the search space increases, coverage of initial search space using randomly seeded initial weights becomes harder, especially where the topography of the error surface is complicated. This also provides a possible explanation why the average RMSEs deteriorated at higher PCs as finding the best random initial weights purely by chance became harder. It would, therefore, have been an advantage to perform more repetitions of networks containing many weights, to improve the chances of randomly finding suitable initial conditions for the network topology and parameters. This is something that can be achieved using a more targeted approach, as we will see in Section 4.

Table 3
Number of PCs used to train models with minimum prediction errors for each analyte

Analyte	PCs for min RMSR	PCs for min RMSI	PCs for min RMSD
Ethanol	25	29	27
Fructose	19	19	13
Glucose	17	15	17

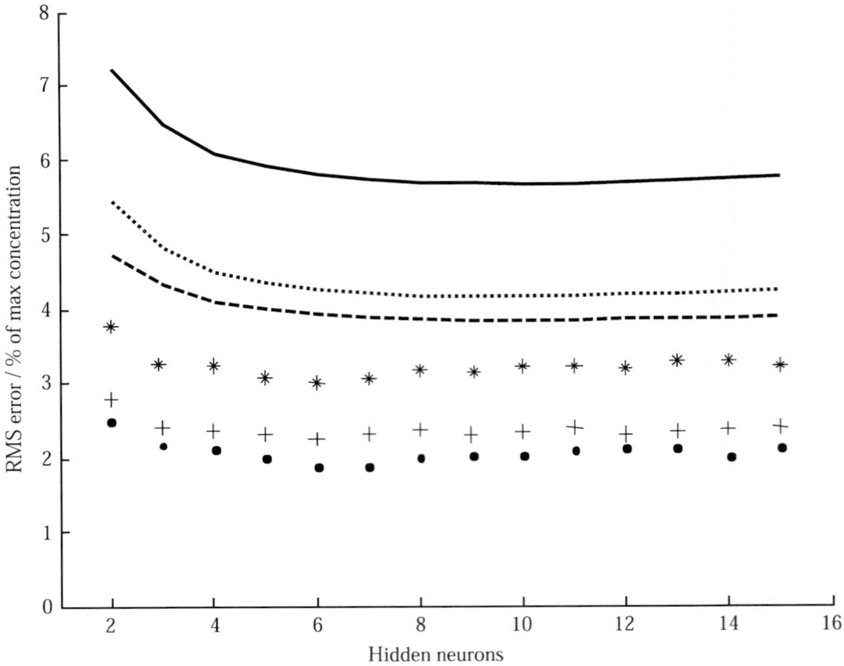

Fig. 6. Effect of number of hidden layer neurons used to train neural network models for ethanol when presented with the validation set. The mean error is shown as a line and the minimum as a point, for RMSR (· · ·, +), RMSI, (−−, ●) and RMSD (−, $*$).

3.2.7. Hidden neurons

Reducing the number of hidden neurons incorporated in an ANN restricts the relationships that can be modelled by limiting the number of degrees of freedom within the calibration model. Getting the correct balance between the number of neurons required to accurately model the system under study, and introducing over training by using too many neurons is one of the most difficult aspects of ANN design. Fig. 6 shows how the calibration errors for neural network models of ethanol varied depending on the number of hidden layer neurons. The average RMSR for the trained networks when presented with the validation set show that initially as the number of hidden neurons was increased from 2 to 10 the networks on average become more accurate. As the number of hidden neurons was increased further the average RMSR progressively increased, a sign that the networks with larger number of neurons were over trained. The best network prediction RMSR for the validation set occurred using a model employing six neurons in its hidden layer.

All the analytes followed the same pattern, with errors decreasing exponentially as the number of hidden neurons was initially increased, before climbing again as the final neurons were added. The number of hidden neurons used to train models with minimum RMSR, RMSI and RMSD for each analyte is summarised in Table 4.

Table 4
Number of hidden neurons used to train neural network models with the lowest prediction errors for each analyte

Analyte	Hidden neurons for min RMSR	Hidden neurons for min RMSI	Hidden neurons for min RMSD
Ethanol	6	7	6
Fructose	6	14	13
Glucose	8	12	6

3.3. Summary and conclusions

Having determined the best models for each analyte using the validation set, the test set was used to test those models to see how generally applicable they were, using previously unseen samples. The RMSEs can be seen for each analyte model in Table 5.

In summary, it can be seen that feed forward ANNs trained with resilient back propagation can produce low error calibration models for DPSV voltammograms obtained from a mixture of three aliphatic analytes. This is good news, but the process used to find the optimum ANNs proved slow and inefficient. The training of nearly 100,000 networks requires a lot of computer time, and unknowns such as the random initial weights and interactions between network parameters make it difficult to draw firm conclusions about what makes a good network.

4. GAs for optimisation of feed forward neural networks

One approach to optimising network designs more efficiently is to use GAs, which have been applied to a variety of problems in analytical chemistry (Leardi, 2001), and ANN optimisation in particular (Henderson et al., 2000). This section shows how GAs can be used to optimise ANNs for calibration of tertiary mixtures of the aliphatics analysed in Section 3.

4.1. Genes and chromosomes

The first step is to encode the network design into a binary format that is suitable for genetic evolution. The binary gene format has to be able to express any value in the

Table 5
Calibration errors for the best networks when tested using the test set

Analyte	RMSR% of max conc	RMSI% of max conc	RMSD	RMSE% of max conc
Ethanol	3.01	3.23	4.41	3.19
Fructose	2.80	2.65	3.85	2.68
Glucose	3.34	2.70	4.30	2.82

The replicate, interpolation and the standard RMSE are given as a percentage of the maximum concentration for each analyte.

Table 6
Parameter ranges and the corresponding ranges of binary genes chosen to represent the variables of interest

	Epochs	PCs	Hidden neurons
Scaled range	1–800	2–30	2–15
Binary range	$2^{10} = 0–1023$	$2^5 = 0–31$	$2^4 = 0–15$

variable range with the maximum number capable of being produced by the binary gene being greater than or equal to the maximum value in the variable range. Where the range of a variable was less than the maximum produced by the binary form of the gene, the range was scaled to fit the range of the binary gene. The format used is shown in Table 6.

The variables expressed in their binary form were concatenated into a single string, the chromosome, representing a network design to be evaluated. An example chromosome is shown below.

389 Epochs | 27 PCs | 12 Neurons

0111110001111100 1011

Ten different chromosomes were randomly generated to form the initial population.

4.2. Choosing parents for the next generation

The members of the next generation were chosen by employing the weighted roulette wheel method of selection (Goldberg, 1989). The weighted roulette wheel apportions the chromosomes from the previous generation to a segment of the roulette wheel, the area, A, of the segment being calculated as a function of the fitness of the individual chromosomes (Eq. (5)). The fitness function used was the RMSD distance error as defined earlier in this chapter.

$$A_i = \frac{\mathrm{RMSD}_{\mathrm{MAX}}^4 - \mathrm{RMSD}_i^4}{\sum_{k=1}^{10} \mathrm{RMSD}_{\mathrm{MAX}}^4 - \mathrm{RMSD}_k^4} \tag{5}$$

where RMSD_i is the best RMSD of networks trained using chromosome i and $\mathrm{RMSD}_{\mathrm{MAX}}$ is the best RMSD of networks trained for the worst performing chromosome. The RMSD values are all raised to the power of four to amplify the differences between the chromosomes, hence biasing the segment area apportioned towards the better chromosomes.

It should be noted that the weakest chromosome, the chromosome with the largest RMSD value, was always lost from the roulette wheel. The roulette wheel is only a method for describing how successive generations were biased towards the fitter chromosomes. In reality, segments were scaled so that the sum of all their ranges was 1. Nine parents were chosen from the roulette wheel by generating nine random numbers between 0 and 1.

Where a random number fell within a range of a segment, the chromosome represented by that segment was chosen to be a parent. The 10th chromosome, the *elite chromosome* (Brooks et al., 1996), was taken to be the chromosome that had the lowest RMSD over all the generations so far. The elite chromosome was used to prevent good chromosome attributes being lost in future generations due to poor random initial weights (Marshall and Harrison, 1991; Henderson et al., 2000). Ensuring that the best chromosome remained in the gene pool also allowed many different random initial weight combinations to be trialled, providing an opportunity for improved models to be trained by using different starting points on the error-surface (Hagan et al., 1996). Should a fitter chromosome be found in future generations, it would become the elitist chromosome.

Networks were allowed to evolve over 100 generations. This resulted in a total of 10,000 neural network calibration models being trained and evaluated per analyte, as 10 networks were generated for each chromosome (to allow a distribution of random initial weights).

4.2.1. Crossover

The parent chromosomes, chosen using the weighted roulette wheel, were randomly paired into five mating pairs. Crossover was performed between the mating pairs by swapping random length sections between the two chromosomes. Three separate types of crossover were used.

Type 1: A random value between 1 and 19 (the length of the chromosome) was chosen to provide the first cut in the chromosome. A second random value was generated between 1 and the remaining length of the chromosome to provide the end cut in the chromosome. The portions between the cuts for both the parents were exchanged to form two new chromosomes like this:

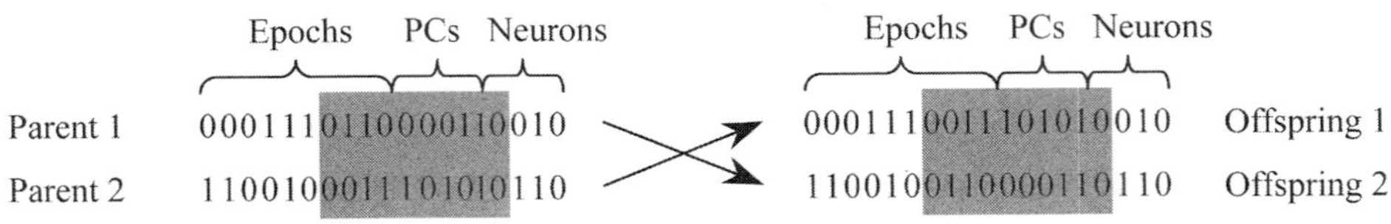

Type 2: This is essentially the same as type 1 except that the chromosome was split in two and crossover was performed in each half with the other parent:

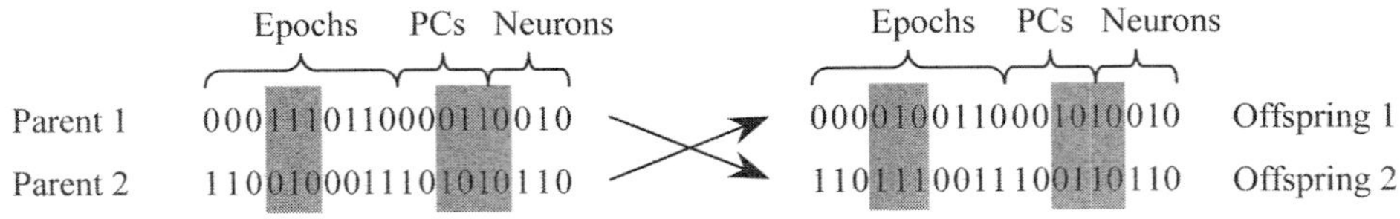

Type 3: This was the same as type 1 and 2, except that the chromosome is split into three sections and crossover is performed in each section between parents.

The type of crossover was chosen at random for each mating pair. Although the elite chromosome was paired with another parent chromosome it remained unchanged after crossover was performed.

4.2.2. Mutation

To prevent the GA converging too quickly in a small area of the parameter search space, mutation was applied. At random, a chromosome was chosen and, again at random, a bit within the chromosome was inverted. Mutations were performed 10 times in every generation. All except the elite chromosome could receive a mutation. The nine new offspring of the previous generation of chromosomes and the elite chromosome were then evaluated.

4.3. Results of ANN optimisation by GA

In this section we present detailed results obtained for ethanol only. The results for glucose and fructose followed a similar pattern, and are presented in Section 4.3.4.

4.3.1. Number of training epochs

Fig. 7 shows how the errors for neural network models trained for ethanol varied depending on the number of epochs that were chosen by the GA. Fig. 8 shows how many

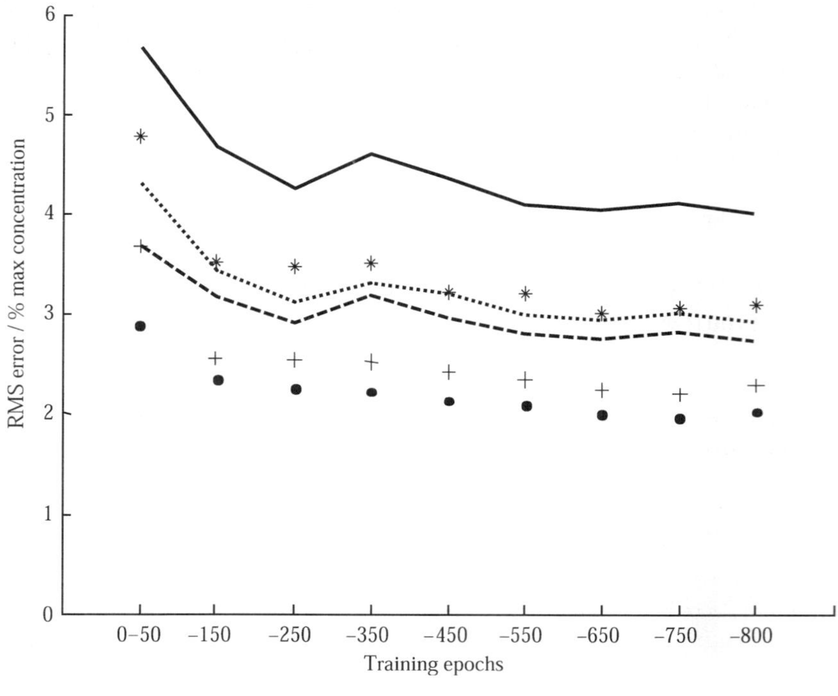

Fig. 7. Effect of number of epochs used to train neural network models for ethanol when presented with the validation set. The mean error is shown as a line and the minimum as a point, for RMSR (· · ·, +), RMSI, (––, ●) and RMSD (–, ∗).

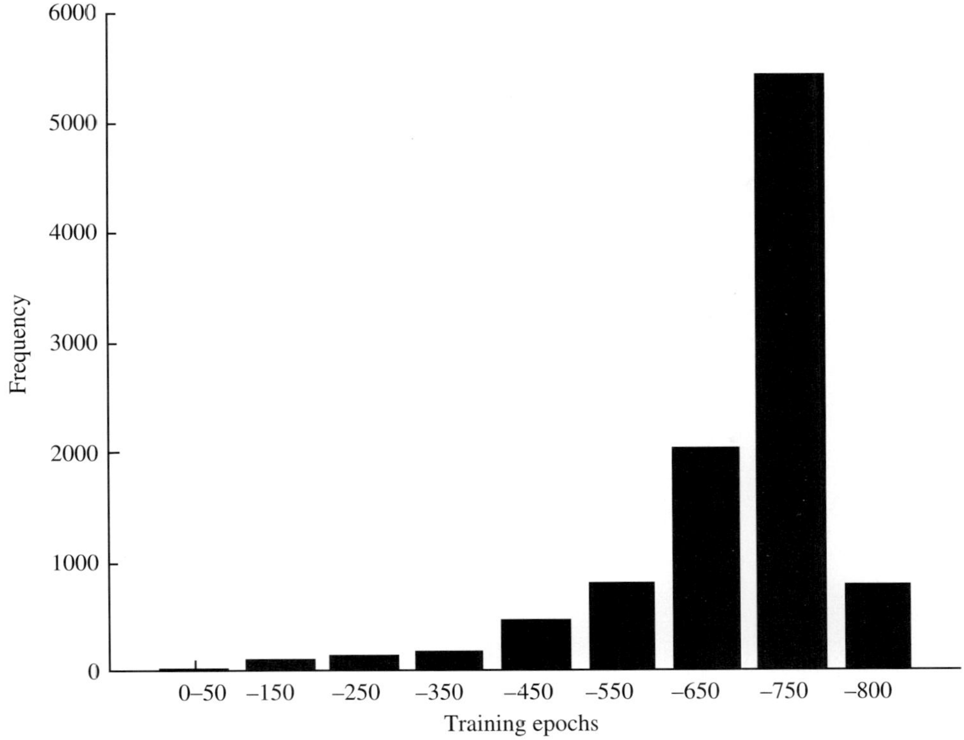

Fig. 8. Frequency for repetition of epochs chosen by the GA for modelling ethanol.

times networks were generated for each number of epochs from within the specified range. Note that, unlike the full factorial search of the network parameters, the GA could select the number of epochs over a continuous range between 1 and 800, but within these graphs the number of epochs are shown in discrete bands for simplicity. From the figure it can be seen that the RMSR and RMSI errors both followed similar trends, initially falling as the number of training epochs increased. Having found that networks trained for larger numbers of epochs produced better models, the lower epoch numbers were clearly neglected and the higher epoch ranges were concentrated upon by the GA. The lowest RMSD was produced by a model trained for 644 epochs.

As in the previous fully factored optimisation, it was very difficult to determine the optimal number of epochs to train a model for due to the randomly seeded initial weights. However, the GA approach reduces this problem to some extent in that it focuses on the most promising epoch range and is therefore able to perform many repetitions within those ranges, reducing the unpredictability caused by the random initial weights.

4.3.2. Principal components input to the network

Fig. 9 shows the average errors and the lowest errors produced by ANNs trained for ethanol using the number of PCs input to the networks specified by the GA. As seen for the

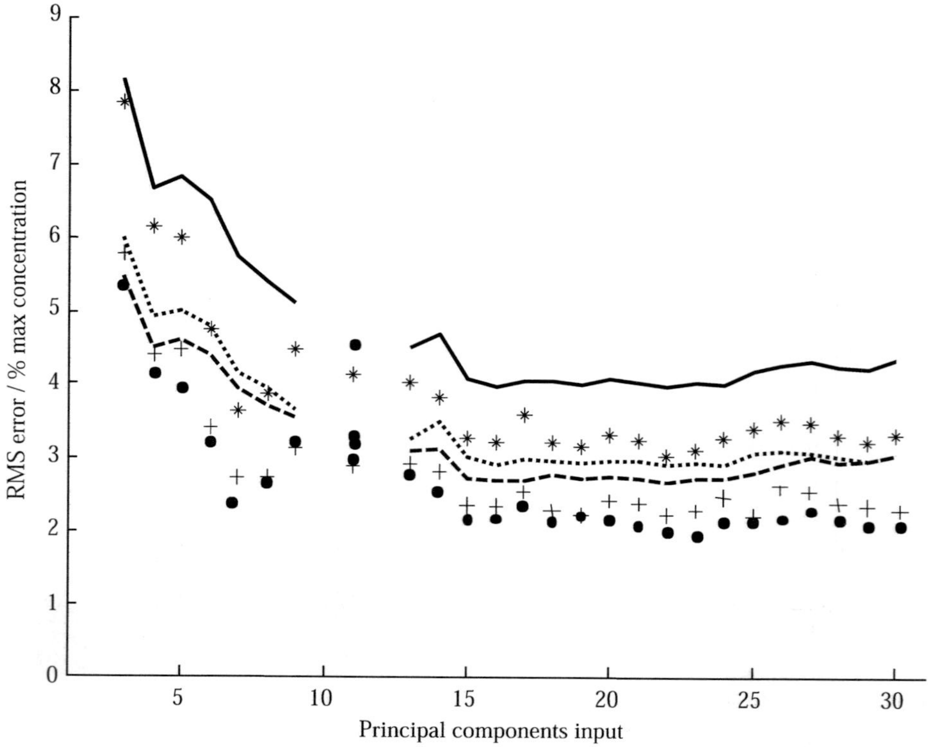

Fig. 9. Effect of the number of PCs input when training neural network models for ethanol when presented with the validation set. The mean error is shown as a line and the minimum as a point, for RMSR (· · ·, +), RMSI, (– –, ●) and RMSD (–, ∗).

fully factored parameter search approach, the errors initially fall as the number of PCs used was increased from 2 to 7. The average errors continue to improve up to the addition of the 15th PC. The average errors then remain relatively constant over the next eight additional PCs before beginning to rise. The network providing the best RMSD for the validation set was a model trained using 22 PCs.

The number of times that each PC was used to train a network is shown in Fig. 10. The GA can be seen to have favoured the PCs that produced good networks with almost half the networks trained using 22 PCs. Interestingly, some numbers of PCs were never evaluated; hence the gaps in the graphs. This could be a drawback of the GA approach under some circumstances, where the relationship between the fitness function and the parameter being optimised is more complex, but in this case it does not appear to be a problem as the missing points (10 and 12 PCs) are in a region of relatively high error.

The selection of additional PCs above those required to collect 99.9% of the variance in the data confirms the earlier results that suggested that there was useful non-linear information in the higher PCs. The GA chose many times to use the number of PCs which

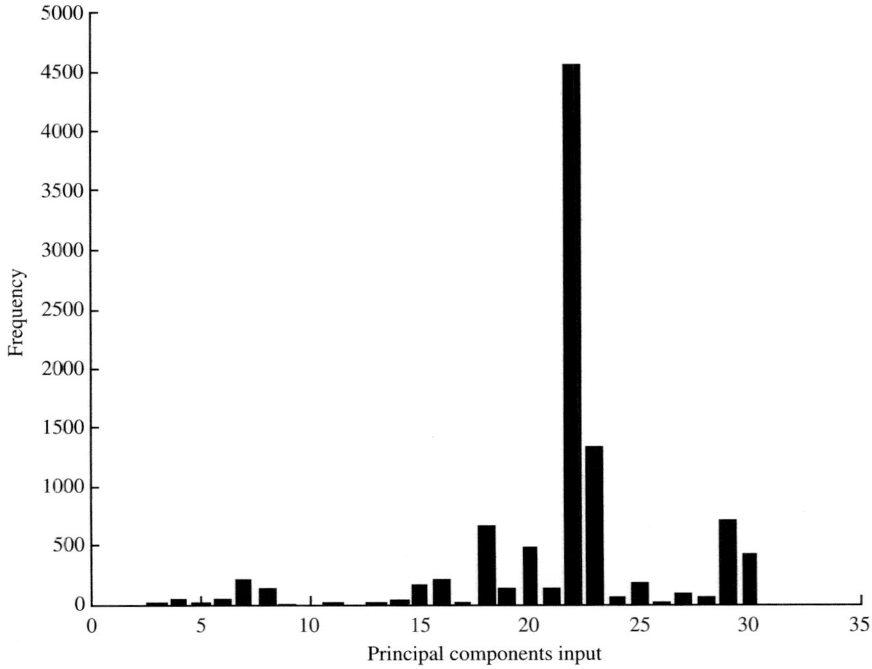

Fig. 10. Frequency for repetition of number of PCs used to train networks chosen by the GA for modelling ethanol.

had produced low error models and in doing so explored the error surfaces for models trained using that number of PCs thoroughly through the use of random initial weights.

4.3.3. Hidden neurons

Fig. 11 shows the ethanol calibration errors as a function of the number of hidden neurons, as determined using the GA approach. The average errors can initially be seen to fall as the numbers of neurons in the hidden layer were increased from 2 to 5. Subsequent additional neurons failed to improve the average errors beyond those obtained using five neurons. The individual network producing the lowest RMSD was trained using five hidden neurons.

Fig. 12 shows how often networks were trained using particular numbers of neurons in the hidden layer. Networks were predominantly trained using five neurons in the hidden layer. Networks were also trained many times using 3, 4, 8 and 12 neurons, for which networks with low errors were found.

For ethanol and glucose a relatively small number of hidden neurons, five and six, respectively, were required to create the neural network model with the lowest RMSD for the validation set. This suggests that the analyte interactions could be modelled relatively simply. Good prediction models were also trained using higher numbers of hidden

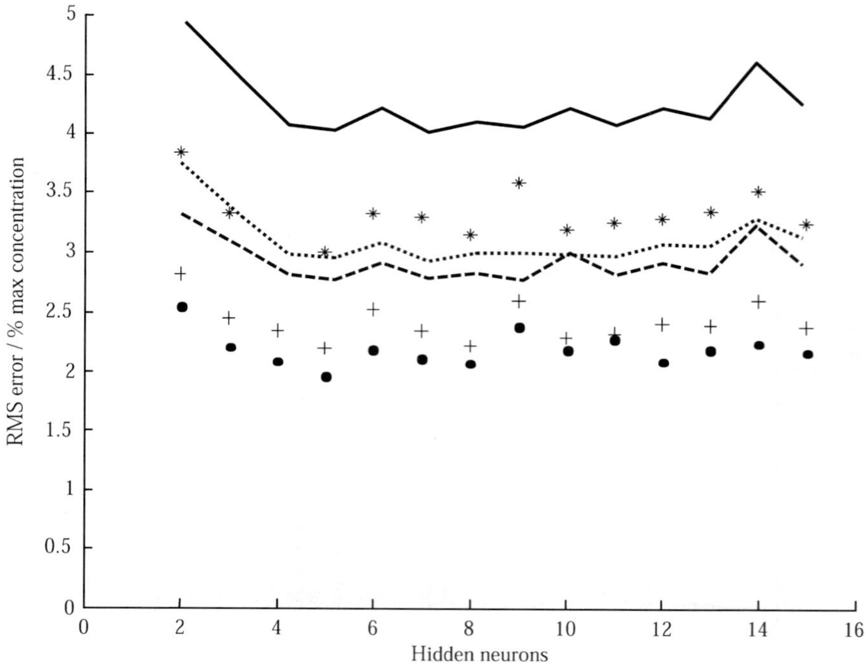

Fig. 11. Effect of the number of hidden neurons when training neural network models for ethanol when presented with the validation set. The mean error is shown as a line and the minimum as a point, for RMSR (· · ·, +), RMSI, (−-, ●) and RMSD (–, ∗).

neurons, for instance the best fructose model used 12 hidden neurons. One reason for this may be that by using a greater number of hidden neurons, areas of localised fitting were achieved rather than just over fitting the model to the training data (Lawrence and Giles, 2000). This suggests that the relationship between fructose and the other analytes was far more complicated to model. It also agrees with the findings of the fully factored parameter search.

4.3.4. Summary of optimisation using GAs

Table 7 shows a summary of the GA optimised ANNs trained for each analyte, which had the lowest RMSD when tested with the validation set. The generation column provides an insight as to when, within the total of 100 generations, the best models for each analyte were trained. The best model for glucose was trained relatively late in the GA parameter search; hence, extension of the numbers of generations beyond 100 used may increase the opportunity for an even better model to be found. This can be shown by the low number of times six neurons were used in the hidden layer. The best models for ethanol and fructose were found in the 88th and 64th generations.

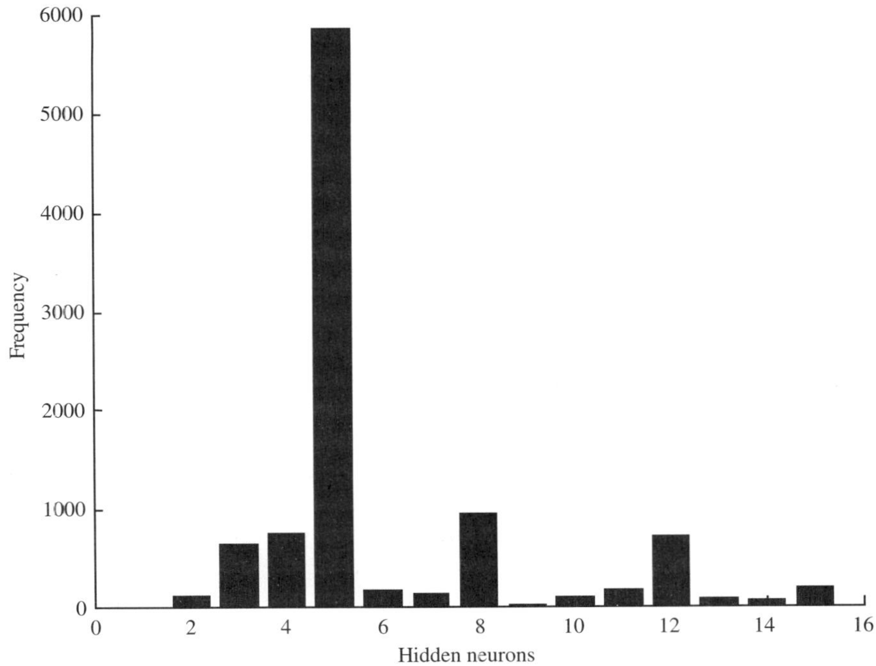

Fig. 12. Frequency of networks trained using a specified number of neurons in the hidden layer chosen by the GA for modelling ethanol.

Having determined the best models for each analyte using the validation set, the test set was used to see how general they were by using the previously unseen samples that made up the test set. The prediction errors can be seen for each analyte model in Table 8.

4.4. Comparison of optimisation methods

Both the fully factored and the GA parameter searches provided good general calibration models for the three analytes. However, when presented with the test set the best models produced by the GA outperformed those of the fully factored parameter search

Table 7
Calibration errors and parameter settings for the best individual networks found when making predictions for the validation set

Analyte	RMSR% of max concentration	RMSI% of max concentration	RMSD	Number of epochs	Number of PCs input	Number of hidden neurons	Generation
Ethanol	2.23	2.02	3.01	644	22	5	88
Fructose	2.34	2.18	3.20	784	19	12	64
Glucose	2.53	2.53	3.57	627	20	6	94

The RMSR and RMSI are given as a percentage of the maximum concentration for the analyte. The generation number indicates the generation when the best model was trained for each analyte.

Table 8
Calibration errors for the best models when tested using the test set

Analyte	RMSR % of max concentration	RMSI % of max concentration	RMSD	RMSE % of max concentration
Ethanol	2.84	3.18	4.26	3.12
Fructose	2.73	2.42	3.65	2.48
Glucose	2.71	2.71	3.83	2.71

The errors are given as a percentage of the maximum concentration for the analyte. The interpolation set comprised of any concentration mixture permutation that did not exist in the training set.

for all three analytes. Not only were better models produced using the GA, the total number of ANNs trained were also less than a third of those produced for the fully factored parameter search method, hence better models were found in a shorter time.

We suspect that the GA parameter search trained better models because far more repetitions of neural networks with general low error producing parameter designs were made, which in turn meant a more thorough exploration of the initial weight space within those parameter sets. The loss of good parameter sets was guarded against by the use of the elitist chromosome.

5. Conclusions

In this chapter we have seen that ANNs are a very powerful tool for calibrating voltammograms. Optimisation of network designs can be an issue, but the use of GAs to evolve accurate calibration networks clearly speeds this process up, allowing useful networks to be designed relatively quickly and with minimal human input. Although the examples have focussed specifically on the particularly challenging case of calibrating DPSV voltammograms of aliphatic mixtures, the techniques are equally applicable to data acquired using other types of voltammetry, and from other compounds. Indeed, we have successfully applied the approaches here to quantify a mixture of aliphatics and aromatics of interest in pharmaceutical production, and quaternions of aliphatic compounds, without major modification.

If attempting to employ the techniques described in this chapter, it should be noted that the examples used vastly greater amounts of data than would probably be necessary in a real calibration situation. The use of more efficient experimental designs would considerably reduce data acquisition time, probably with minimal impact on calibration error.

Acknowledgments

The authors would like thank Prof. Selly Saini, Head of Cranfield Centre for Analytical Science, for his assistance and guidance during the work described in this chapter.

References

Bard, A.J., Faulkner, L.R., 1980. Electrochemical Methods: Fundamentals and Applications, Wiley, New York.

Bessant, C., Saini, S., 1999. Simultaneous determination of ethanol, fructose and glucose at an unmodified platinum electrode using artificial neural networks. Anal. Chem. 71, 2806–2813.

Bessant, C., Saini, S., 2000. A chemometric analysis of dual pulse staircase voltammograms obtained in mixtures of ethanol, fructose and glucose. J. Electroanal. Chem. 489, 76–83.

Brooks, R.R., Iyengar, S.S., Chen, J., 1996. Automatic correlation and calibration of noisy sensor readings using elite genetic algorithms. Artif. Intell. 84, 339–354.

de Carvalho, R.M., Mello, C., Kubota, L.T., 2000. Simultaneous determination of phenol isomers in binary mixtures by differential pulse voltammetry using carbon fibre electrode and neural network with pruning as a multivariate calibration tool. Anal. Chim. Acta, 109–121.

Cottrell, F.G., 1902. Der reststrom bei galvanischer polarisation, betrachtet als ein diffusionsproblem. Z. Physik. Chem. 42, 385–430.

Cukrowska, E., Trnkova, L., Kize, R., Havel, J., 2001. Use of artificial neural networks for the evaluation of electrochemical signals of adenine and cytosine in mixtures interfered with hydrogen evolution. J. Electroanal. Chem. 503, 117–124.

Despagne, F., Massart, D.L., 1998. Neural networks in multivariate calibration. Analyst 123, 157R–178R.

Fung, Y.S., Mo, S.Y., 1995. Application of dual-pulse staircase voltammetry for simultaneous determination of glucose and fructose. Electroanalysis 7, 160–165.

Goldberg, D.E., 1989. Genetic Algorithms in Search, Optimization and Machine Learning, Addison-Wesley, Reading, MA.

Goodacre, R., Kell, D.B., 1993. Rapid and quantitative analysis of bioprocesses using pyrolysis mass spectrometry and neural networks: application to indole production. Anal. Chim. Acta 279, 17–26.

Guiberteau Cabanillas, A., Galeano Diaz, T., Mora Diez, N.M., Salinas, F., Ortiz Burguillo, J.M., Vire, J.C., 2000. Resolution by polarographic techniques of atrazine–simazine and terbutryn–prometryn binary mixtures by using PLS calibration and artificial neural networks. Analyst 125, 909–914.

Hagan, M., Demuth, H., Beale, M., 1996. Neural Network Design, PWS Publishing Company, Boston, MA.

Henderson, C., Potter, W., McClendon, R., Hoogenboom, G., 2000. Predicting aflatoxin contamination in peanuts: a genetic algorithm/neural network approach. Appl. Intell. 12, 183–192.

Hughes, S., LaCourse, D.C., 1981. Amperometric detection of simple carbohydrates at platinum electrodes in alkaline solutions by application of triple-pulse potential waveform. Anal. Chim. Acta 132, 11–22.

Hughes, S., LaCourse, D.C., 1983. Triple-pulse amperometric detection of carbohydrates after chromatographic separation. Anal. Chim. Acta 149, 1–10.

Hughes, S., Meschi, P.L., LaCourse, D.C., 1981. Amperometric detection of simple alcohols in aqueous solutions by application of a triple-pulse potential waveform at platinum electrodes. Anal. Chim. Acta 132, 1–10.

Johnson, D.C., 1986. Carbohydrate detection gains potential. Nature 321, 451–452.

Johnson, D., LaCourse, W., 1990. Liquid chromatography with pulsed electrochemical detection at gold and platinum electrodes. Anal. Chem. 62, 589–597.

LaCourse, W.R., Johnson, D.C., 1991. Optimization of waveforms for pulsed amperometric detection (p.a.d.) of carbohydrates following separation by liquid chromatography. Carbohydrate Res. 215, 159–178.

Lawrence, S., Giles, C., 2000. Overfitting and neural networks: conjugate gradient and backpropagation. Proceedings of the IEEE-INNS-ENNS International Joint Conference on Neural Networks, pp. 114–119.

Lastres, E., de Armas, G., Catasus, M., Alpizar, J., Garcia, L., Cerda, V., 1996. Use of neural networks in solving interferences caused by formation of intermetallic compounds in anodic stripping voltammetry. Electroanalysis 9, 251–254.

Leardi, R., 2001. Genetic algorithms in chemometrics and chemistry: a review. J. Chemom. 15, 559–569.

Long, J.R., Gregoriou, V.G., Gemperline, P.J., 1990. Spectroscopic calibration and quantitation using artificial neural networks. Anal. Chem. 62, 1791–1797.

Marshall, S., Harrison, R., 1991. Optimization and training of feed forward neural networks by genetic algorithms. Artificial Neural Networks, Second international conference on. 39–43. IEE, London.

Martens, H., Næs, T., 1991. Multivariate Calibration, Wiley, Chichester.

Neuburger, G.G., Johnson, D.C., 1987. Comparison of the pulsed amperometric detection of carbohydrates at gold and platinum electrode for flow injection and liquid chromatographic systems. Anal. Chem. 59, 203–204.

Otto, M., 1999. Chemometrics: Statistics and Computer Application in Analytical Chemistry, Wiley-VCH Verlag GmbH, Weinheim, Federal Republic of Germany.

Plambeck, J.A., 1982. Electroanalytical Chemistry: Basic Principles and Applications, Wiley, New York.

Richards, E., Bessant, C., Saini, S., 2002. Optimisation of a neural network model for calibration of voltammetric data. Chemom. Intell. Lab. Syst. 61, 35–49.

Richards, E., Bessant, C., Saini, S., 2003. A liquid handling system for the automated acquisition of data for training, validating and testing calibration models. Sensors and Actuators B: Chemical 88 (2), 149–154.

Riedmiller, M., Braun, H., 1993. A direct adaptive method for faster backpropagation learning: the RPROP algorithm. 1993 IEEE International Conference on Neural Networks 1, 586–591.

Schulze, H.G., Greek, L.S., Gorzalka, B.B., Bree, A.V., Blades, M.W., Turner, R.F.B., 1995. Artificial neural network and classical least-squares methods for neurotransmitter mixture analysis. J. Neurosci. Meth. 56, 155–167.

Skoog, D., West, D., Holler, F., Crouch, S., 1999. Analytical Chemistry. An Introduction, 7th edn, Thomson Learning, USA.

Udelhoven, T., Schutt, B., 2000. Capability of feed-forward neural networks for a chemical evaluation of sediments with diffuse reflectance pectroscopy. Chemom. Intell. Lab. Syst. 51, 9–22.

Zhang, L., Jiang, J.H., Ping, L., Liang, Y.Z., Yu, R.Q., 1997. Multivariate nonlinear modelling of fluorescence data by neural network with hidden node pruning algorithm. Anal. Chim. Acta 344, 29–39.

CHAPTER 10

Neural networks and genetic algorithms applications in nuclear magnetic resonance spectroscopy

Reinhard Meusinger[a], Uwe Himmelreich[b]

[a]*Technical University of Darmstadt, Institute of Organic Chemistry, Petersenstr. 22, D-64287 Darmstadt, Germany*
[b]*University of Sydney, Institute for Magnetic Resonance Research, Blackburn Bldg D06, Sydney NSW 2006, Australia*

1. Introduction

Over the decades, Nuclear Magnetic Resonance (NMR) spectroscopy has developed into one of the most powerful method in analytical chemistry and biochemistry, mainly for the determination of molecular structures. NMR spectroscopy is used in all disciplines of chemistry, pharmacy, biochemistry and biomedicine both in academic and industrial laboratories. The increase of NMR applications is mainly due to dramatic improvements in NMR instrumentation. This has, among others, resulted in a substantial increase in both the sensitivity and the resolution of NMR spectra. Superconducting magnet systems with ^{1}H NMR frequencies of up to 900 MHz (21.14 Tesla) now allow the elucidation of the structure of large proteins and polysaccharides and make NMR spectroscopy invaluable for the development of new drugs. Two-dimensional NMR methods and more recently three- and multi-dimensional experiments have allowed ever more complex molecules to be studied. Improved sensitivity due to cryogenic probes and micro- and nano-probes made it possible to study quantities of material as small as a few nanograms. For the observation of nuclei with low sensitivities such as ^{13}C or ^{15}N, so-called inverse detection techniques were developed. Together with gradient enhanced spectroscopy, increased sensitivity has also resulted in dramatically shortened time requirements for NMR experiments. Improved process control, the application of robotic sample preparation and advanced computation associated with modern NMR instruments have led to the development of fully automated systems. This enables unattended sample preparation, transfer, acquisition and processing; resulting in large sample throughput and continuous utilization of NMR spectrometer. Particularly successful developments for high sample throughput are flow-probes, which operate with capillary probes rather than systems that are based on NMR tubes. Flow NMR coupling techniques have been widely applied in combinatorial chemistry since the ease of automation. Other recent developments are

Data Handling in Science and Technology, Volume 23
ISSN: 0922-3487

DOI: 10.1016/S0922-3487(03)23010-0

NMR spectroscopy of tissue samples, body fluids and in vivo, resulting in spectra from complex multi-component samples.

The dilemma NMR spectroscopy was facing a decade ago, and is partly still facing, is the contradiction between the vast amount of automatically collected complex data and manual data analysis and interpretation. On the other hand, NMR spectroscopy was always closely linked with and dependent on computer technology so that computerized data analysis methods were quickly picked up by NMR spectroscopists. Applications of Neural Networks and Genetic Algorithms (GAs) in combination with NMR spectroscopy are numerous and driven by a variety of reasons.

The diversity of techniques being employed has resulted in the adaptation of a number of NMR methods for combinatorial chemistry purposes. High-throughput screening strategies were developed, which involve the acquisition of two-dimensional spectra from small organic molecules in only a few minutes. Libraries containing more than 10^5 compounds can be tested in less than one month. This includes the NMR analysis of reaction intermediates and products in solution or in the gel state and the analysis of ligands interacting with their receptors. An advantage of high-throughput NMR-based screening compared to conventional assays is the ability to identify high-affinity ligands for protein targets rapidly. So called structure–activity relationships ('SAR by NMR') are obtained from NMR spectra of small organic molecules that bind to proximal subsites of a protein. Traditional interpretation of NMR spectra, which involves manual interpretation of a large amount of various spectral parameters, became insufficient and too slow for such data, which include hundreds or thousands of compounds being synthesized per compound library. Rapid and automated methods for rapidly analysing NMR spectra were developed by using Neural Networks and self-organizing maps (Kalelkar et al., 2002).

Computerized data analysis and simulation methods also became increasingly important for the interpretation of NMR spectra from macromolecules in order to determine their structure. These compounds result in NMR spectra containing tens of thousands of signals that need to be assigned to functional groups in the molecule and interpreted in terms of a three-dimensional structure. Manual signal assignment often results in ambiguities and consequently in low-resolution structures. Such optimization problems are more successfully solved by computerized methods.

NMR spectroscopy is a valuable technique in biochemistry and clinical chemistry. In particular in vivo or in vitro ^{1}H NMR is used to study complex body fluids such as plasma, urine, and bile as well as different types of tissue. However, the derivation of diagnostic information from NMR spectra is often ad hoc when it is based on subjectively chosen signals. In addition, many clinicians are not familiar with NMR spectroscopy. Hence, a robust automated data interpretation is of advantage to make this method available for a broad range of diagnostic applications. Data interpretation is further complicated by the complex nature by which chemicals interact in biological systems and the fact that usually only a limited number of spectra is available as against the overabundance of spectral features. This has led to an interest in the classification of spectra using Artificial Neural Networks (ANNs) and the application of feature reduction methods for robust classification. Applications of ANNs and GAs predominantly include the screening and monitoring for metabolic diseases, the study of biochemical mechanisms associated with disease processes and the identification of potential drugs and their metabolites. ANNs and GAs are of interest for NMR data analysis because of their ability to systematically discriminate between complex data structures, without recourse to a priori assumptions about the underlying biochemistry.

2. NMR spectroscopy

In simple terms, NMR spectroscopy is the observation of an electric signal which is induced during the relaxation process of a previously excited nuclei with non-zero magnetic moments in a strong magnetic field. After a Fourier transformation the intensity of the observed signal is plotted as a function of the resonance frequency in a diagram called the NMR spectrum. These frequencies depend on the chemical environment of the atomic nuclei. The difference in the frequency of a particular nucleus from that of a reference nucleus is called the 'chemical shift'. The chemical shift was recognized early as a parameter of great utility in the solution of the central problem of chemical research—the determination of chemical structures. NMR spectroscopy is inherently a quantitative technique. The area under a resonance signal is directly proportional to the number of nuclei producing it. Provided that sufficient relaxation time was available between successive excitations, the area can be determined quantitatively by integration. Most ^{1}H NMR spectra are acquired under essentially quantitative conditions and their integrals provide the relative number of protons contributing to each resonance. A technique to obtain quantitative ^{13}C NMR data is the so-called inverse-gated decoupling experiment. The proton decoupling is applied during the data acquisition period only and the so-called nuclear Overhauser enhancement (NOE) is suppressed. The long and different relaxation times of carbon atoms can be shortened by addition of a paramagnetic relaxation reagent to the sample solution, like $Cr(acac)_3$. One important application of the integration of resonances is the determination of the quantitative composition of mixtures. Other important NMR parameters include the number of signals; the multiplicity of signals, which is an expression of magnetization transfer via chemical bonds; the NOE, which encodes for magnetization transfer through space and relaxation times, which influence the line widths. Some of these parameters are illustrated in Fig. 1. However, the chemical shift is by far the most important parameter for structure elucidation in chemistry. For ANNs, a segmentation or so-called 'binning' technique is frequently used in order to minimize the data size of the NMR spectrum. The binning technique creates a 'fingerprint' or overall profile of the entire ^{1}H or ^{13}C NMR spectrum, which can be used easily as input for an ANN, as shown in Fig. 1.

However, the binning may result in loss of information (resolution). Other feature space reduction methods may be more appropriate. Hereby, the entire data set (NMR spectrum) can be utilized as an input.

In summary, NMR spectroscopy is one of the most powerful methods for molecular structure analysis. It is sensitive for small structure modifications and produces quantitative and reproducible results with high accuracy. Additionally, large amounts of experimental data are available. Several databases exist, containing a large number of NMR chemical shift reference values and the appending information about the chemical

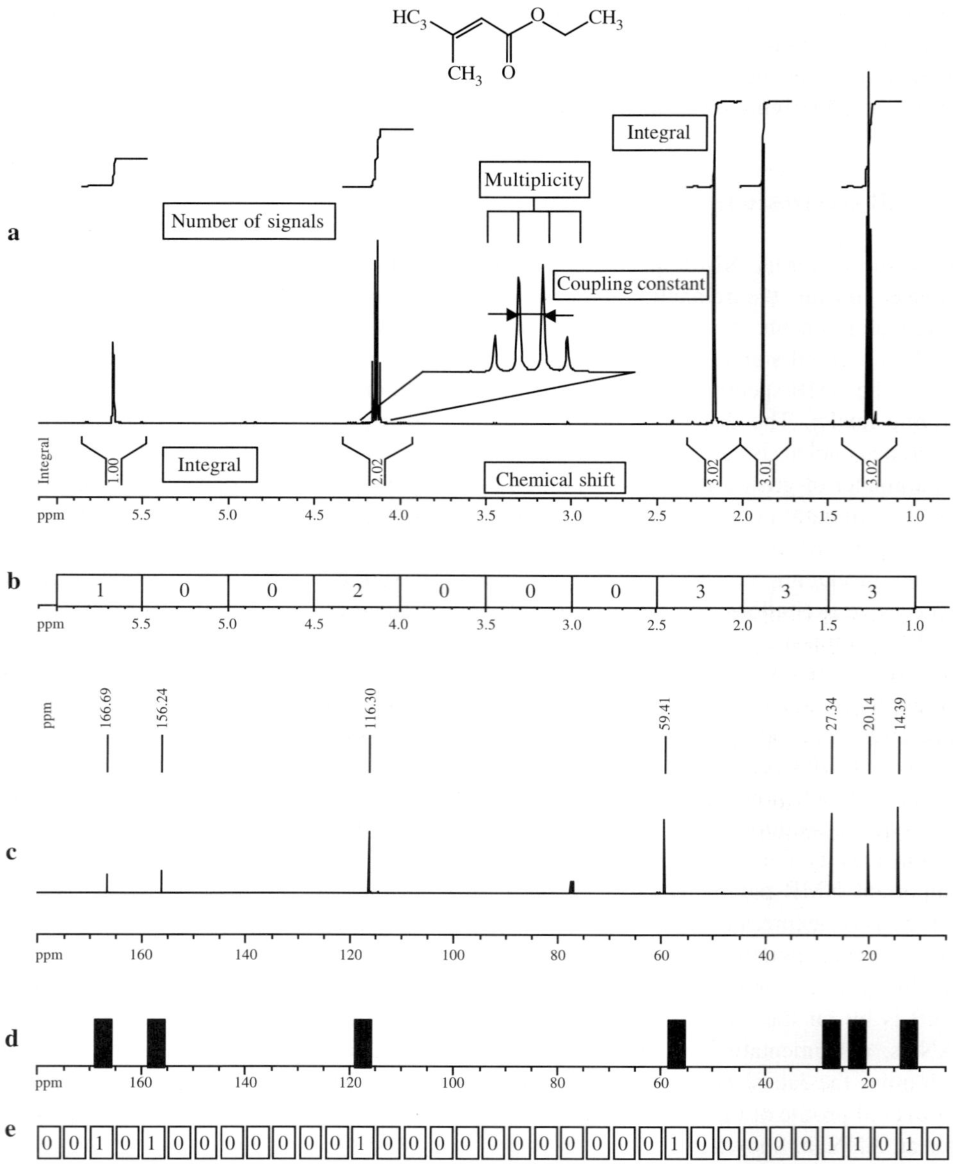

Fig. 1. NMR spectra of 3-methyl-2-butenoic acid ethylester (senecioic acid ethylester). (a) 500 MHz ^1H NMR spectrum. The basic NMR spectrum parameters are marked schematically. (b) The information from chemical shift and integration is summarized by the binning technique. Hereby, the integrated intensity (integral value) of the signals within defined spectral regions was registered, respectively. A uniform bin width of 0.5 ppm was used here. (c) The ^{13}C NMR routine spectrum. The three signals at around 77 ppm were caused by the solvent deuterochloroform. These signals were ignored during reduction of spectral information by (d) the uniform binning technique. Hereby, the number of signals in a bin of a defined width (e.g. 5 ppm) were counted as one. It should be noted that a theoretical signal intensity of 1 was supposed for each carbon instead of the experimental signal intensity. (e) The result is shown schematically in form of a 35-digit string, containing structural information of the compound in a most concentrated form.

environment for each individual nucleus. Such databases may form a basis for the identification of chemicals from multiple component mixtures based on NMR data. This is of particular advantage for biological and biomedical applications, where the entirety of NMR visible compounds (i.e. mobile relative to the NMR time scale) can be identified non-destructively and simultaneously. However, identification of individual chemicals is not always necessary for classification or correlation of NMR data with physical, chemical, biological or other properties, when ANNs or GAs are utilized. As NMR data exist in digitized form, input for computerized data analysis is uncomplicated. NMR spectroscopy represents not only an appropriate tool for the quantitative molecular structure description, but it is also convenient for the encoding of different chemical structures.

3. Neural networks applications

ANNs are particularly useful for problems where the relationship between input and output is complex and for large quantities of data, preferably distributed over a wide range of samples. These conditions are complied for NMR spectroscopy. Burns and Whitesides (1993) assessed a number of applications in NMR spectroscopy in one of the first reviews about ANNs in chemistry. Features of NMR spectra relate to specific structural groups often in a complex manner so that construction of rule-based interpretation systems is impractical. A typical example is the correlation between chemical structure and chemical shift. Simple linear approaches like increment systems are in most cases too inaccurate and ineligible for stereoisomers. In contrast to common statistical methods, ANNs are not restricted to linear correlation or linear subspaces. They are also an attractive computational tool to study the relationship between NMR chemical shift and a wide range of properties of a molecule. ANNs can be applied in NMR spectroscopy efficiently for:

- classification that supposes identification and clustering of objects, for example groups of chemical structures, physical properties or various diseases, and
- predictions that model functional correlation between experimental NMR parameters and quantifiable properties of the molecule.

Some other applications are known, which do not fit in this scheme. Kjaer and Poulsen (1991) as well as Corne and co-workers (Corne et al., 1992; Corne, 1996) classified cross-peaks of two-dimensional NMR spectra by ANNs. The ANNs functioned as a typical pattern recognition for which the actual NMR information is not correlated to structures. Hereby, ANNs were utilized to distinguish between authentic cross peaks and spurious peaks arising from noise and other artefacts. This is a complex problem since artefacts not only introduce spurious data with intensities comparable to real peaks, but may also be responsible for the corruption of genuine data. In another early application, Freeman (1992) explored several methods for designing pulse shapes of ‘soft’ radiofrequency pulses, used for the selective excitation in NMR spectroscopy. The frequency-domain excitation pattern of a nucleus from the pulse shape is calculated by the Bloch equations.

No analytical method exists for the reverse calculation of pulse shapes from excitation patterns. Instead, one can adjust a likely pulse shape with a simulated annealing (SA) technique in an attempt to find a pulse shape that provides a satisfactory excitation pattern. Apart from SA and an evolutionary algorithm, an ANN was also tested. Thereby, frequency-domain excitation patterns were calculated for a set of randomly generated pulse shapes. While most optimization programs are time-consuming because of their iterative nature and an asymptotic approach to the solution, the key advantage of ANNs is their very rapid operation. Freeman (1992) compared the protracted training process, in which a suitably representative range of pulse shapes has been studied, with the slow operating mode of a young child that learns to tie its shoelaces compared to automatic and rapid execution once the process was learned. Recently, Freeman (2003) recapitulated five ideas remote from magnetic resonance for the inspiration for new NMR experiments. Two of these are ANN and GA applications, used for discovering new radiofrequency pulse sequences.

Forshed et al. (2002) presented a method for the detection of impurities in pharmaceutical products. One particular example was 4-aminophenol as an impurity in ^{1}H NMR spectra of paracetamol. Because of interference between ^{13}C-satellites of the main product and proton peaks of the impurity and the non-linearity over a wide calibration range, ANNs were used for calibration. The mean error, which was obtained for the optimal calibration model was a hundred thousandth weight of 4-aminophenol per weight of paracetamol. This method was sufficiently sensitive for typical pharmaceutical impurity determinations.

Another type of application for ANNs is the reduction of the dimensionality of data sets. The accumulation of large experimental data sets can cause data complexity. Hoehn et al. (2002) demonstrated the efficiency of a Kohonen Self-Organizing Feature Map which discovered effectively the inherent structures of solid state NMR spectroscopic data for 72 polymers. An ANN was employed to unfold hidden information and to visualize similarities of the investigated compounds. This allowed a comparison of the multifaceted polymer structures by correlation of crosslinkage and NMR relaxation data that describe the structural features and dynamic behaviour of the polymers, which were not obvious beforehand.

3.1. Classification

Burns and Whitesides (1993) predicted that an ANN of sufficient size could model any relationship to any degree of precision. The fact that the network approximates a relationship from a limited number of examples with a finite number of units makes it more prudent to interpret the output as quality rather than quantity. They recommended that the simplest way to make this approach rigorous is to classify each example on the basis of the range in which its output falls, rather than to make use of the numerical value of the output itself.

ANNs have proved most useful for problems requiring classification of data or recognition of patterns. Pattern recognition is a term, which encompasses a wide range of techniques to classify data. Given a collection of objects characterized by a set of measurements made on each object, the goal is to find and predict a property of the objects that is not directly measurable. With NMR spectroscopy, the emphasis is more on discriminating between spectra from different classes of samples, and reducing the large numbers of spectral features in order to make the available information more accessible. Pattern recognition analysis provides an unbiased method of analysis, useful for both research and technical applications of NMR spectroscopy (Tate and Howells, 1998).

To our knowledge, the first analysis of NMR spectra by ANNs was published by Thomsen and Meyer (1989). They identified sugar alditols and complex oligosaccharides (Meyer et al., 1991) from their proton NMR spectra. The authors anticipated that ANNs can be used to identify the structures of any complex carbohydrate that was previously characterized and for which an NMR spectrum is available. Valafar and Valafar (2002) applied ANNs to the classification of a collection of one-dimensional ^{1}H NMR spectra of linked oligosaccharides, confirming the initial hypothesis. A so-called frequency to information transformation (FIT) was established as a novel method for information extraction from complex spectra. Amendolia et al. (1998) had extended the ANN application to mixtures of alditols. ^{1}H NMR spectra of six alditol binary mixtures and their relative concentrations were estimated.

In the last decade miscellaneous ANNs were applied to the assignment of NMR spectra from various types of compounds, from both pure substances and mixtures. One of the relatively simple application was presented by Isu et al. (1996). This group predicted the exo/endo branching of 38 bicyclic norbonane and norbonene derivatives using a three-layer ANN. Of these compounds, 25 were used for the learning process, and the remaining 13 for prediction. A single output neuron was sufficient, setting the output data for exo and endo isomers to be 0 and 1, respectively. The ^{13}C NMR chemical shift values of the seven ring carbons served as inputs in a fixed order. This procedure provided the correct assignment for all carbons in the NMR spectra, which is often difficult.

One of the most attractive fields of ANN applications is the determination of the sequence assignment for amino acids in biomacromolecules from their NMR spectra. Anand et al. (1993) as well as Hare and Prestegard (1994) first analysed the presence of various amino acids in protein molecules using NMR data and ANNs. High accuracies of 87 and 75%, respectively, were achieved for these classifications. Sharma et al. (1994) have employed an assignment tool based on the ^{1}H NMR chemical shift values of 20 amino acids. The ANN was trained with entries of typical NH and H_α chemical shifts and was capable to determine the amino acid types. As the ^{1}H- and ^{13}C NMR chemical shifts of proteins are known to correlate with their secondary structures, Choy et al. (1997) improved the accuracy of the amino acid recognition from predicted secondary structure information ANNs. Huang et al. (1997) took an approach to determine both, amino acid class and secondary structure from proteins, using different NMR data. In addition to the ^{1}H NMR chemical shifts, ^{15}N-chemical shifts and coupling constants were also used as input to a three-layer feed-forward network. Their model had the potential for further automation of NMR spectral analysis. Pons and Delsuc (1999) presented the publicly available RESCUE program as an assistance tool for the manual assignment of amino acids in polypeptides and proteins from ^{1}H NMR spectra. They trained two ANNs with a set of chemical shifts extracted from the BioMagResBank (BMRB) database, containing ^{1}H chemical shifts collected from over 1100 peptides and proteins. In a first step, chemically related amino acids were grouped together, while a second network separates particular amino acids of these groups. However, the ^{1}H chemical shifts were not corrected for reference, pH or temperature bias. These variations of chemical shifts for a given amino acid cannot be handled easily by an ANN. Therefore, the authors used an additional fuzzy logic layer encoding the chemical shift in the input layer using a grid on the chemical shift scale on which the position of each spectral line was coded. The error was reduced below 10% for certain cases.

Only ^{13}C NMR shifts were used by Fraser and Mulholland (1999) for a group classification of natural products. They prepared ^{13}C NMR data from more than 100 spectra of pure and impure limonoids and associated compounds isolated from numerous species within the Meliaceae family. The ^{13}C NMR input data were preprocessed by the binning technique to reduce the dimension of each data record without reducing information required for classification. Furthermore, the constructive 'Cascade Learning' method was used here to optimize the number of hidden neurons and speed up the learning rate. Previously, Munk and Madison (Munk et al., 1996) had investigated a profound effect on the performance of an ANN caused by the different number of hidden units. If too few units were used, the network would be ineffective in making discriminations. When more than the optimum number was used, the effectiveness of the network would not improve, but the required training time and the danger of overtraining would increase rapidly.

Kalelkar et al. (2002) presented recently a new, automated method for rapid analysis of NMR spectra from all compounds prepared in 96-well plates by rapid parallel syntheses. Their unsupervized ANN provided a 'bird's eye' view of the combinatorial plate by clustering NMR spectra and identifying outliers from such clusters. Hereby, the laborious process of analysing NMR spectra sequentially and manually can be reduced to the analysis of only outlier spectra. A particular advantage is that no additional input beyond the NMR spectra from an extensive databases of reagents was required. The 16 k data points of the origin of spectra were reduced by filtering and sub-sampling every 20th data point. Each run on a 96-well combinatorial plate took less than 60 s and provided a rapid quality control mechanism for combinatorial libraries.

Another application was developed by Ott (Ott et al., 2003) and Aranibar et al. (2001). They classified biochemical mechanisms for approximately 300 commercially available herbicides that are used in agriculture using an automated ANN analysis of ^{1}H NMR spectra from raw plant extracts. The ANN can simultaneously detect the modes-of-action of the various herbicides and other bioactive compounds, including under conditions where changes in sample characteristics are marginal. The different biochemical pathways were classified through one output unite for each of the three ANN layers. Only defined sections of the NMR spectra were used as the 1080 input unites. Although the herbicides and their metabolites were usually not visible in the NMR spectra due to their low concentrations, their spectral manifestation due to variations in the metabolite pools of the plant metabolites was well documented in the ^{1}H NMR spectra.

Munk and Madison (Munk et al., 1996) have shown that ANNs are also a useful tool for multispectral interpretation. They developed an ANN, which utilized data from infrared spectra, ^{13}C NMR spectra, and the molecular formulas of organic compounds for structure elucidation. It was shown that information from these disparate sources can principally be combined in a neural network. The numerical data were simply conjoined after scaling in a single input vector irrespective of its type or source. A database of 1560 compounds was used for training and testing a network to recognize substructural groupings with high accuracy. The trained network is part of the interactive computer-enhanced structure elucidation system SESAMI. The collective spectral properties of an unknown compound are directly reduced to a small number of compatible molecular structures with the assistance of such a comprehensive system. A sizeable ANN with a total of 512 input units was used: 128 for infrared data, 294 for ^{13}C NMR data and 90 for molecular formula information. The infrared as well as the ^{13}C NMR data were divided into intervals by the binning technique. Previously, the ^{13}C NMR data were grouped according to their signal multiplicity originating from the one-bond carbon–hydrogen couplings. Only the chemical shift regions in which signals of each multiplicity are normally found were divided into equal bins of 2 ppm in width, and each bin was assigned to an input unit. If at least one signal occurred in one of the 2 ppm intervals, a '1' was assigned as corresponding input unit; in the absence of a signal, a '0' was the input for that unit.

Applications addressing aspects of classification and property prediction will now be discussed. In 1992, Adler et al. (1992) applied an ANN for the prediction of carcinogenic properties of aminobiphenyls. This application was used for classification based on ^{13}C NMR chemical shifts and so-called electrotopological values. Two output neurons indicated the relation of the compound to either the carcinogenic (1) or non-carcinogenic category (0). This was the first publication in which ^{13}C parameters were used as an input for an ANN. A similar approach was taken by Isu et al. (1996) for the structure investigation of the endo/exo substitution of norbonane derivatives described above. Schaper et al. (1999) described non-linear relationships between the antituberculous activity of substituted xanthones and their ^{13}C NMR chemical shifts with a three-layer backpropagation ANN. The goal of these investigations was to find a correlation between the biological activity and physicochemical parameters. Since the biochemical mechanism is unknown, it is easier to test a large set of physicochemical parameters with regard to their influence on the activity and to reduce this initial parameter set stepwise. They started the QSAR study with a combination of lipophilic, steric and electronic descriptors in order to determine the importance of the different parameters. The input of their ANN consists of a total of 13 physicochemical parameter. Six ^{13}C NMR chemical shifts of aromatic carbons of the xanthone skeleton, the molar refractivities of all substituted carbons and the lipophilicity of the substituted derivatives were utilized. The target value was a so-called activity variable, consisting of the values 0 and 1. As a result, each compound was arranged in a class of biologically active (1) or inactive (0) compounds. For all strains of mycobacteria a correlation between the biological activity and certain ^{13}C NMR chemical shifts was observed, demonstrating the influence of the electronic properties of the substituents. In general, the authors obtained better results with this approach than with an Adaptive Least Squares Analysis (ALS), which was attributed to the non-linear relationships between physicochemical parameters and biological activity.

Axelson and Nyhus (1999) used probabilistic neural networks to classify crosslinked macroporous polymer particles either in porogen or crosslinker types on the basis of their NMR determined relaxation time data. Only two outputs were required, which dispensed output values of either 1 or 0.

All applications listed so far refer to single compounds. Meusinger and Moros (1996) demonstrated the possibility to implement this method also for complex mixtures. Gasoline constitutes a complex mixture. More than 140 unleaded gasoline samples, representing different grades and brands, from service stations in Germany and Austria were studied. The ^{1}H NMR spectra were acquired quantitatively and normalized against an internal standard. The chemical shift scale was not divided into regular intervals, but in definite chemical shift regions attributable to defined structure groups (see below). The possibility to determine the composition and the quality of a gasoline sample simultaneously from its ^{1}H NMR spectrum was investigated. Fifty-seven percent of the samples were used to train an ANN with the objective to estimate the gasoline grades 'Regular', 'Super' and 'Super-Plus'. This does not constitute a simple linear problem as each refinery blends their own mixture depending on available resources and current market prices. Additionally, some gasoline quality parameters differ from summer to winter season, which is also represented in the fuel composition. A multi-layer ANN with three output neurons was created. The ANN operated accurately, provided that representatives of all different grades and brands were part of the training data set.

3.2. Prediction of properties

Understanding interactions between the structure of molecules or the composition of molecular mixtures and their chemical, physical or biological properties is a main practical and theoretical aim. When these relationships are complex and of a non-linear nature ANNs were often chosen to estimate quantitative structure property (QSPR) or activity relationships (QSAR). A prerequisite is the unique property of ANNs—their ability to learn by observation. An elaborate review was published by Devillers (1996). Molecular structural information are encoded in NMR spectra through three basic spectroscopic parameter: number of signals, NMR chemical shifts and signal intensities. In contrast to the classification shown above, a defined property is quantified by a discrete computed value. ANN can assist here as a powerful tool for finding relationships between structural information and properties of single compounds or complex systems, even when a mathematical description of the structure–property relationships fails. This was demonstrated in a simple model—the computation of boiling points of low-molecular *n*-alkanes from the number of carbon atoms and the ^{13}C chemical shift of their methyl carbons (Meiler, 1998). While the correlation between carbon number of *n*-alkanes and their boiling points is well known, no relationship has been established for the ^{13}C shift of terminal carbons as yet. At first, the linear relations were computed for both properties (see Fig. 2).

Expectedly, an approximated linear relation was found for the relationship between number of carbons and boiling points. A relatively large regression coefficient of $R = 0.986$ verified the reliability of the results. In contrast, practically no correlation was found for the ^{13}C shifts of the methyl groups ($R = 0.733$). However, the combination of both entirely different values in an ANN led to a more robust result ($R = 0.9999$). Both numbers were then used as inputs for an ANN, which was trained with nine *n*-alkanes. The compound *n*-butane was used as cross-validation test compound. Its experimental boiling point ($BP_{exp} = -1°C$) was calculated correctly ($BP_{calc} = -0.97°C$). Similarly, good

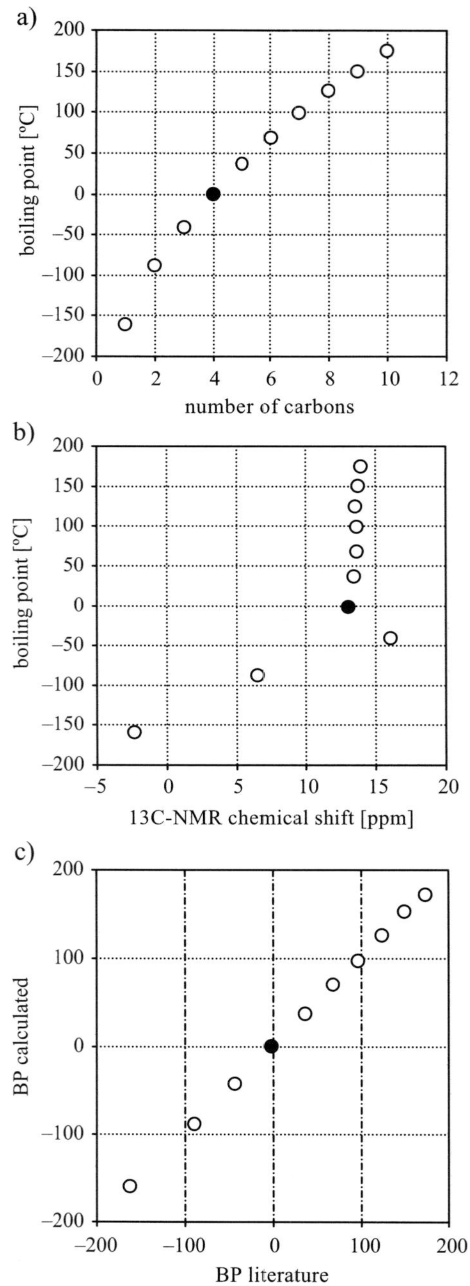

Fig. 2. Dependency of boiling points of low-molecular alkanes (C1–C10) (Meiler, 1998) (a) boiling points vs. number of carbon atoms plot from a linear regression analysis, (b) boiling points vs. ^{13}C chemical shift of methyl carbons plot from a linear regression analysis and, (c) calculated vs. experimental boiling points of ten alkanes. Number of carbons as well as ^{13}C shifts of methylgroups served as inputs of a neural network with one output neuron. *n*-Butane (●) was used for cross-validation.

results were achieved for this relative simple model using a multiple linear regression ($R = 0.997$). However, the boiling point of the test compound *n*-butane was calculated with a large deviation from the experimental value ($BP_{calc} = -5.2°C$), demonstrating the superiority of the ANN.

Meiler and Meusinger (1996) estimated boiling points for 150 *n*- and *iso*-alkanes solely from their ^{13}C NMR chemical shifts. Eighty-five alkanes and their corresponding boiling points were used for training of a three-layer ANN and the residual 65 were used as test set. All boiling points were taken from literature (Mihalic et al., 1992). The binning technique was used in order to minimize the number of input units. Hereby, the ^{13}C NMR chemical shift region was divided into the five subregions, the so-called bins: under 16 ppm, up to 24, 32, 40 ppm and, over 40 ppm, respectively. The carbon signals in each bin were added in consideration of isochronous nuclei that is nuclei with an equivalent chemical shift. Each molecular structure of the 150 hydrocarbons was represented by the simple set of five numbers. A single neuron acted as output, providing the learned or computed boiling point. This is schematically shown in Fig. 3.

The correlation coefficients *R* for experimentally determined and calculated values for the training and test datasets were $R_{train} = 0.995$ and $R_{test} = 0.997$ with standard errors *S* of 4.3 and 4.4, respectively. The standard errors for a multiple linear regression were $R_{train} = 0.991$, $S_{train} = 6.1$ and $R_{test} = 0.981$, $S_{test} = 10.6$. This study demonstrates the capacity of NMR spectroscopy for encoding macroscopic properties and their prediction. The strength of the small number of input units chosen here required a relative wide bin width of 8 ppm or more. Consequently, some different structures result in identical pattern. However, the average difference between boiling points of structures with similar NMR pattern was only 3.9 K. The same dataset of boiling points was used by Cherqaoui and Villemin (1994) to achieve comparable results by encoding the molecular structure with an artificial numeric *N*-tuple code ($R_{train} = 0.998$, $S_{train} = 2.6$ and $R_{test} = 0.983$, $S_{test} = 3.7$). However, they required the precise knowledge of the molecular structures to compute the boiling points with an ANN containing 10 input units.

In another application of an ANN for QSPR, Meusinger et al. (1999) determined the suitability of triazolopyrimidines as stabilizers in photographic silver halide materials from their ^{13}C NMR spectra. Triazolopyrimidines or similar compounds are often added to the photographic emulsion as so-called anti-fogging agents or anti-foggants for reduction of the unwanted blackening of unexposed grains in photographic emulsions, the so-called 'fogging' of silver halide photographic layers. ^{13}C NMR parameters were used for numeric coding of the chemical structures of 44 different substituted heterocycles. Assignment of observed ^{13}C NMR chemical shift values to the respective aromatic carbons was omitted, in order to avoid experimental uncertainty. Best results were achieved with a small feed-forward ANN including three hidden neurons, combining the experimentally determined relative fog value with the ^{13}C NMR chemical shifts of the five heterocyclic skeleton carbons and the sum of the carbons in side-groups. For some compounds with a good stabilizing effect, the calculated results differed unidirectionally from experimental values. This indicated a non-electronic effect, which is not covered by the ^{13}C NMR chemical shifts in the stabilizing mechanism. Another application of ANNs was developed by Ruan (1998) using the non-destructive and non-invasive NMR technique for evaluation of the internal quality of dried ginseng roots. The ginseng quality

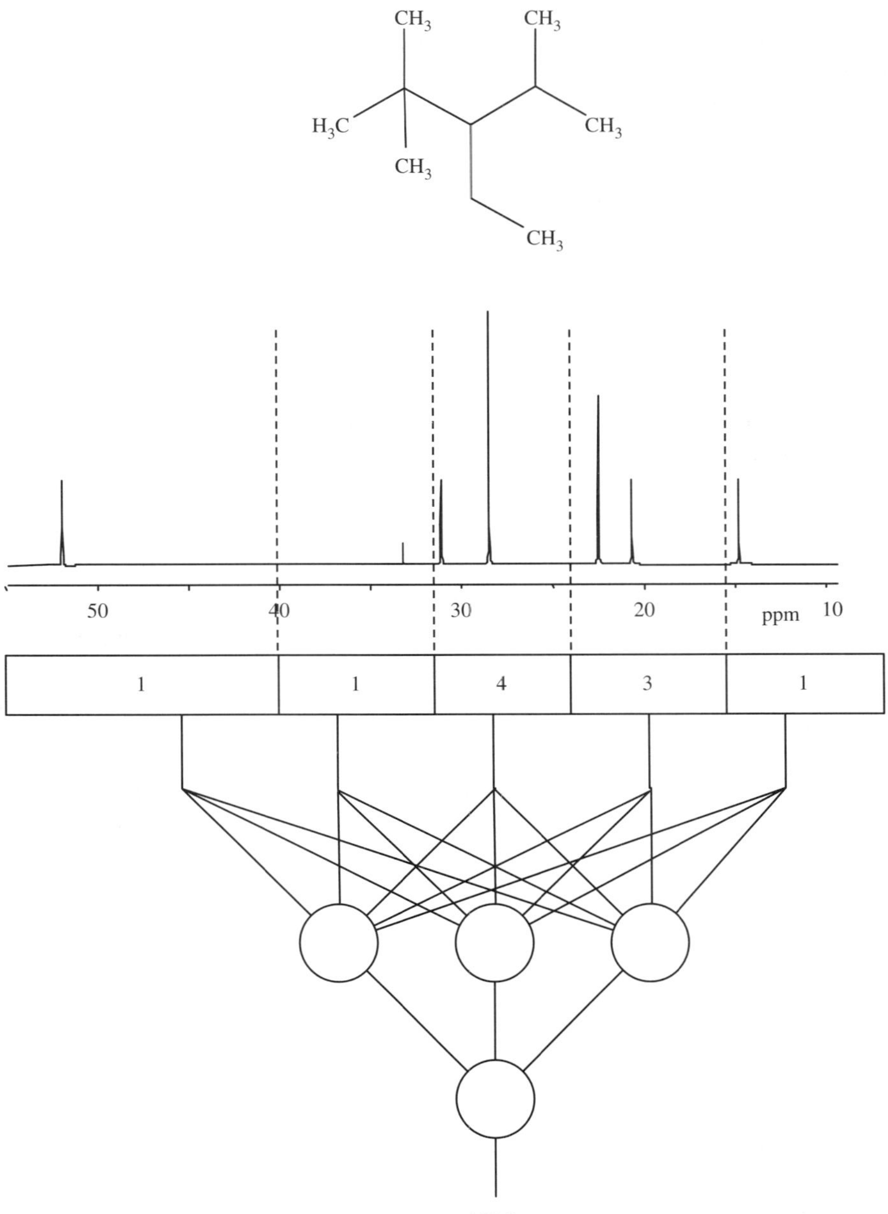

Fig. 3. A graphical representation of the boiling point prediction from the ^{13}C NMR spectrum of 3-ethyl-2,2,4-trimethylpentane. The molecular structure of the compound with a measured boiling point of 155.3°C was encoded in a five-digit pattern, which was obtained from the sum of the ^{13}C NMR signals in five defined subregions (bins). From this pattern, a corresponding boiling point of 158.7°C was computed by the trained neural network, whose neurons are pictured schematically as circles.

evaluation method needs to be non-destructive to keep the root's original shape and structure intact. Therefore, inspection by eyesight and backlighting are generally used to estimate the internal quality of the ginseng roots. Ruan determined the spin–lattice ($T1$) and spin–spin relaxation time ($T2$) of dried intact roots by low field ^{1}H NMR. These NMR parameters correlated with the quality parameters. The $T1$ and $T2$ values were used to train an ANN with the predetermined quality parameters, specific gravity and grade, based on a four-grade scale. The trained ANN was successfully tested for the evaluation of the quality of ginseng roots based on their NMR relaxation times.

Subsequently, a broad range of applications can be expected from the combination of NMR spectroscopy and ANNs in QSPR or QSAR studies. However, all applications published so far utilized poorly understood systems. In general, they can be distinguished into three main-groups:

- the prediction of chemical shift values for single nuclei in molecules;
- the characterization of composite hydrocarbon mixtures in oil chemistry, comprising the prediction of macroscopic properties of different oil products; and
- diverse applications in biomedical investigations.

3.2.1. Chemical shift determination

The chemical shift value combines two advantages for structural analysis: it is easily obtained from an experimental spectrum and, its dependence on chemical structure is well known. However, not all mechanisms influencing the ^{13}C chemical shifts are fully understood. In addition to its state of hybridization, the chemical shift of a carbon atom is mainly influenced by the kind and number of the bond carbon atoms and their distances to the observed carbon. ^{13}C chemical shifts can be influenced by another atom through electron interaction over covalent bonds or through space. In solution, the latter effect appears possibly as a 'solvent effect'. However, electron interaction through space is only important over short distances between observed and influencing atoms. The stronger effect is transmitted via covalent bonds.

The prediction of ^{13}C NMR spectra is one of the most intensely studied applications of empirical modelling. While theoretically offering the greatest predictive accuracy, ab initio or semi-empirical approaches are currently still too time consuming for large-scale use. The remaining empirical and most commonly used methods in predicting ^{13}C NMR chemical shifts are linear additivity relationships, database retrieval methods, molecular mechanics and empirical modelling techniques. Particularly the additivity rules, and substructure code approaches rely on the electronic and steric effects of substituents on the focal shift. This method is based on the assumption that the influence of different substituents on the chemical shift of an individual carbon atom can be defined simply by a set of constant values, the 'increments'. The chemical environment of a carbon atom is defined now by the kind and number of the neighbouring atoms or atomic groups and by their distances to the carbon atom considered. The chemical environment is comprehended by addition of all appropriate increments. The increments themselves were determined by multiple linear regression analysis using data sets of observed chemical shifts from structurally related compounds. In ^{13}C NMR spectroscopy, these substituent induced chemical shift (SICS) effects were first characterized for alkanes. However, the increments are structure class dependent and available only for some selected substance classes, like alkenes, substituted benzenes, naphthalenes and pyridines. The main advantage of these increments is their simple application and short computation time. Otherwise, not to consider possible interactions between several substituents is an obvious disadvantage of this approach. Consequently, the structure analyst must decide between the possibilities of either a more precise prediction requiring inexpedient computing time or rapidly available information afflicted with a larger uncertainty. Several attempts were undertaken to describe the relationship between the chemical environment of a carbon atom and its ^{13}C NMR chemical shift rapidly and precisely, using ANNs. The overall precision of the parametric additive model for alkanes already established in the 70s by Lindemann and Adams (1971) is approximately 0.8 ppm. All subsequent prediction methods must be at least as precise as this approach.

The substituent–shift relationship is biased by a dependent shift variable and independent structure variables. Molecular structures, which are graphical displays, are difficult to encode as direct inputs for ANNs. Therefore, different approaches were developed and tested to compare input strategies for structural parameters. The associated shift value is the network-output for all cases. In most modelling studies that use atom-based descriptors, different types of descriptors are distinguishable: electronic, topological and geometrical.

Approximately 30 studies have been published on the prediction of NMR chemical shifts using ANNs. Kvasnicka and co-workers (Kvasnicka, 1991; Kvasnicka et al., 1992) as well as Anker and Jurs (1992) and Doucet et al. (1993) predicted ^{13}C chemical shifts of monosubstituted benzenes, keto-steroid carbons and saturated hydrocarbons, respectively. Kvasnicka et al. used 11 mainly electronic descriptors to encode the basic physical and chemical nature of substituted functional groups using parameters like lone electron pairs, sum of main quantum numbers and number of hydrogen atoms attached to the next atom. These descriptors were divided in three categories analogous to α-, β- and γ-effects with respect to the carbons of the benzene ring. Weaker influences like δ- and ε-effects were neglected. Anker and Jurs used 24 electronic and geometric structural descriptors, which were reduced by stepwise linear regression to the 13 most influential such as Hückel charges, van der Waals energies, and inverse cubed through-space atom–atom distances. This procedure was applied to various classes of organic compounds (Ball and Jurs, 1993). Ivanciuc and co-workers predicted ^{13}C chemical shifts of saturated hydrocarbons (Ivanciuc, 1995) as well as alkenes (Ivanciuc et al., 1996) using electronic descriptors like the non-bonded van der Waals energy, the focal electrostatic energy and the degree of the carbon atoms. These descriptors were calculated by modeling that was comparable in complexity to ANNs. Doucet et al. (1993) preferred a simpler topological description for the ^{13}C chemical shift prediction of compounds in the alkane family. The environment of each carbon was concentrically described in terms of discrete and ordered atoms, which provided individual contributions to the chemical shift. Svozil et al. (1995) predicted ^{13}C chemical shifts of alkanes with topological descriptors, corresponding to so-called embedding frequencies of rooted subtrees. Hereby, the atoms in a structural formula were assigned to a vertex in a molecular graph. The environment of a chosen atom (a non-equivalent vertex in the tree) was given by single entries in a descriptor vector, the so-called root. This was determined by 13 descriptors that were used as inputs for a neural network. In a similar approach, Panaye et al. (1994) simulated the ^{13}C shifts for methylsubstituted cyclohexanes. They required only 12 structural descriptors, specifying the six possible ring positions and the respective axial or equatorial orientation of the methyl groups with respect to the resonating ring carbon. However, these procedures were limited to groups or classes of largely homogeneous molecular structures like hydrocarbons, which possess no other heteroatoms.

West (1993) used another mapping technique when he predicted ^{31}P NMR chemical shifts using ANNs. He represented the molecules as graphs, with the graph nodes corresponding to atoms and the arcs to bonds. The input parameters for the ANN were derived in a two-step process in consideration of the different phosphorus coordination classes in the initial step. To standardize the size of compounds, he defined a larger template graph, which contained a fixed number of nodes. By translating the molecular graph onto this template a new graphical representation for a compound was obtained. This template representation is known as molecular abstract graph space template (MAGS). In a further step of translation, substituents were arranged according to their extended connectivity values. The element symbols were represented by physical data like the electronegativity for this substructure code. The method can be modified to allow the network itself to derive a set of optimum substitution values, specific to different subclasses. This will allow an automatic derivation of parameters similar to those used in additivity rules for existing topological databases. This approach is evocative of a structure coding method in large structure–spectra databases. Bremser (1978) described a procedure, which is used to date in some NMR databases, the *h*ierarchically *o*rdered *s*pherical description of *e*nvironment (HOSE) code. Focussing on a randomly chosen atom in a molecule, all other atoms of this molecule were considered as members of spheres. Thereby the number of spheres surrounding the atom of interest is identical to the number of bonds between the focal point and the atoms combined in these spheres. The influence of substituents decreases with increasing numbers of bonds to the carbon atom of interest and also with increasing sphere numbers. This is shown in Fig. 4 for an olefinic carbon in the isopropenyl side chain of the monocyclic terpenoid carvone (Fig. 4a).

The ^{13}C NMR chemical shift of the carbon number 8 in carvone depends on neighbouring atoms as far as five spheres away (Fig. 4b). The substructure inside the first sphere is equivalent to isobutene (2-methylpropene). For this a chemical shift of 141.4 ppm was observed. In the second sphere two more carbons were added. In this case, the substructure is equivalent to 2,3-dimethylbutene, showing a ^{13}C chemical shift value of 151.2 ppm for the carbon atom of interest. Further on, the chemical shift of the observed carbon is modified from 147.2 ppm in the 3-ethyl-2-methylpentene (sphere III) to 146.6 ppm in the target molecule carvon (sphere V). An experimental NMR spectrum for the substructure in sphere IV was not available. Only the experimental ^{13}C NMR spectrum of carvone is shown in Fig. 4c. The signal of the quaternary carbon number 8 at 146.6 ppm is marked. For that carbon atom, the dedicated HOSE code is given in the following manner: =CCC(,CC/C,C/=OC,&=C/,C). Hereby ‘C’ and ‘O’ symbolize carbon and oxygen atoms, respectively, whereas ‘(’ and ‘/’ flag the separation between the spheres. The comma is used as a separator within one sphere and ‘&’ symbolizes the ring closure. The advantage of the HOSE code is the numerical and computer readable form of structure description. With this coding, the respective chemical shifts can be determined from a

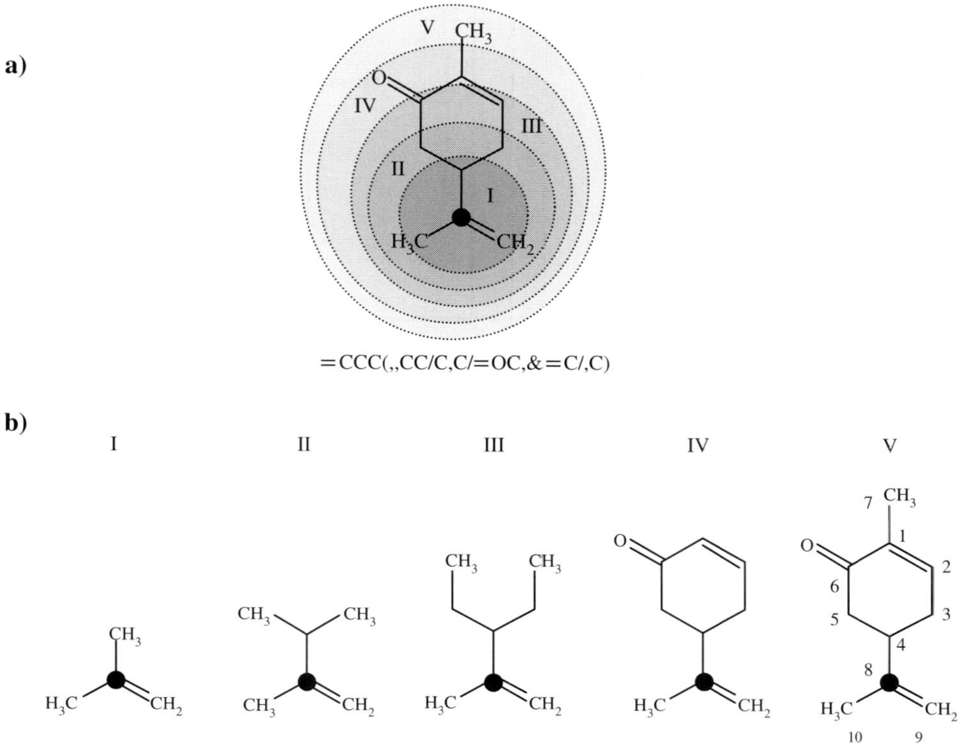

	I	II	III	IV	V
$^{13}C_{exp}$	141.4 ppm	151.2 ppm	147.2 ppm		146.6 ppm
$^{13}C_{calc}$	142.4 ppm	150.1 ppm	147.3 ppm	147.1 ppm	146.9 ppm

Fig. 4. Spherical coding of the chemical environment of a carbon and prediction of its ^{13}C NMR chemical shifts considering individual spheres of the cyclic natural compound carvone as example. (a) Five spheres (roman numbers) were required for complete environmental description of carbon 8 (●). The appropriate HOSE code for this carbon is also given. (b) The individual structural fragments representing these spheres. From the bottom-up, the respective experimental ^{13}C NMR chemical shift values of the observed carbon (●) reflect their dependency from the molecular framework. These values were reproduced by the ANN predicted chemical shifts (Meiler et al., 2000). (c) Experimental ^{13}C NMR spectrum of carvone in $CDCl_3$ (top) and the DEPT-135 spectrum (bottom) of the same sample for multiplicity determination of protonated carbons (CH_2 groups result in positive signals, CH and CH_3 groups in negative signals, quaternary carbons are disabled).

corresponding database. The ^{13}C NMR spectra databases are statistical tools to establish the relationships between NMR spectroscopic parameters (chemical shifts, intensities, multiplicities) and the chemical environment of individual carbon atoms to predict either chemical structures or spectra. ^{13}C NMR spectra databases can be utilized for:

- prediction of NMR parameters (chemical shifts) for given molecular structures;
- verification of existing assignments of NMR signals to carbon constitution;

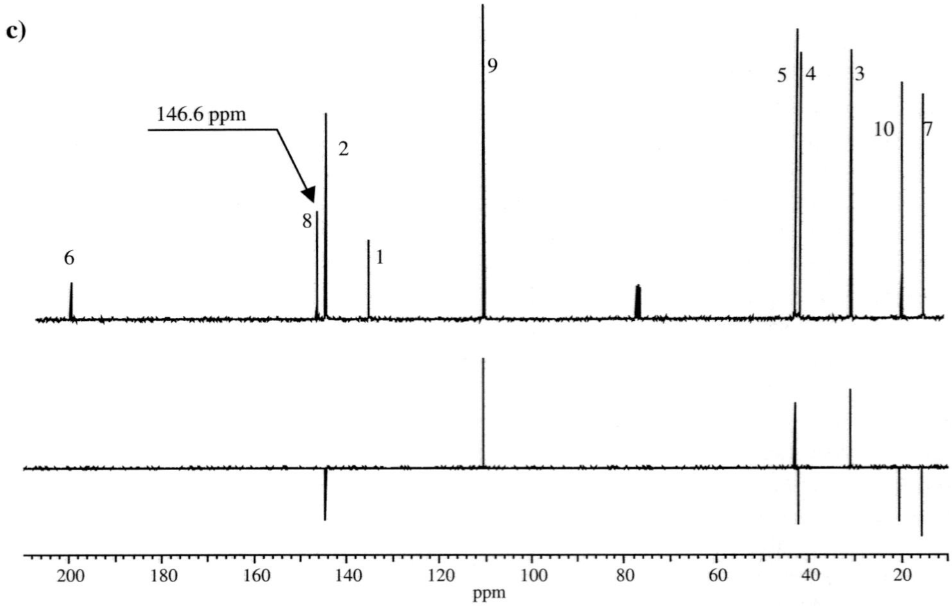

Fig. 4 (*continued*)

- determination of one or more possible molecular structures corresponding to a given ^{13}C NMR spectrum.

However, the results depend strongly on the quantity and quality of the available database entries. As shown above, ANNs are a rapid and accurate tool to calculate ^{13}C NMR chemical shifts of organic compounds. The computation time is 1000 times faster than using comparable accurate database predictions of chemical shifts. This is an advantage of ANNs for screening and checking the database entries. For this purpose, ANNs were applied for the first time in the 90s by Robien (2000). Meiler et al. (2000) exploited the advantage of ANNs to extract the principal information from a large NMR database (SpecInfo, 2000). The resulting parameter file, containing the condensed information from this database was integrated in the PC-program 'C_shift', which allows a rapid estimation of ^{13}C NMR chemical shift values with a precision comparable to the original database. Predictions are possible for any proposed molecular structure consisting of the covalently bond elements carbon, hydrogen, nitrogen, oxygen, phosphorus, sulphur and the halogens. The mean deviation was as low as 1.8 ppm with a computation time as short as known from increment calculations. For this purpose 526,565 carbon atoms from the SpecInfo database were distinguished in nine different atom types (saturated methyl, methylen, methin and quaternary carbons, protonated and non-protonated olefinic carbons, acetylenic carbons and protonated and non-protonated aromatic carbons). Nine different ANNs were trained with the spherically encoded chemical environments (representing 90% of these atoms) and their appropriated ^{13}C NMR chemical shifts. For each individual carbon atom, the atom type and the chemical environment were encoded numerically with descriptors analogous to the HOSE code. The number of the applied descriptors had to be as small as possible for computational reasons; but they must feature small differences in molecular structures clearly. Twenty-eight descriptors were determined for 28 different atom types (nine for the carbons (see above), seven for nitrogen, three each for oxygen and sulphur, two for phosphorus and four for the halogens). Two further descriptors were introduced in order to encode the number of all hydrogen atoms located in an individual sphere and the influence of ring formation on the ^{13}C NMR chemical shift (sum descriptor encoding the number of ring closures). The chemical environments of the carbons are described by arranging the atoms in the five spheres I–V as shown in Fig. 4 and counting the occurrence of every atom type in each sphere. All atoms in further distances than the five spheres were projected in an additional 'sum sphere'. Consequently, 30 numbers encode one sphere, and 180 numbers are necessary for the complete description of the environment of an individual carbon. In addition, this description method was extended by a so-called 'π-contact area' in order to consider conjugated π-electronic systems. This resulted in two sets of descriptors for the environment, one for all atoms and a second for the conjugated atoms in the π-contact areas only. A total number of 360 descriptors were used as input vector for the ANN of an individual carbon atom. For each of the nine atom types, representing a carbon atom, an autonomous ANN was constructed and trained. The individual ^{13}C NMR chemical shift value represent the output. The program was able to determine the chemical shift of all carbons in covalently bond molecular structures containing above-named chemical elements. The available molecules from the database were randomly subdivided into three sets. Ninety percent of data was used for the training. A second data set containing 7% of the data was used for simultaneous monitoring. The iterative training process was stopped if the deviation for this monitoring data set increased. Finally, the third data set of randomly selected molecules (3%) was used as an independent test set. The predicted ^{13}C chemical shifts are shown for five highlighted carbons in Fig. 4b. When considering all spheres, a chemical shift value of 146.9 ppm was computed for the carbon number 8 in carvon.

It should be mentioned, that results obtained directly from a database might be more accurate, as the ANN does not search for a congruently reference substance. However, the ANN allows a precise and rapid prediction of a large number of ^{13}C NMR spectra simultaneously, as needed for high-throughput NMR and screening of substance and spectroscopic libraries. This was shown recently by validation of structure proposals from the two-dimensional NMR-guided computer program COCON (*Co*nstitutions from *con*nectivities) using the ANN-assisted ^{13}C NMR chemical shift prediction (Meiler et al., 2002). COCON uses connectivity information from two-dimensional NMR spectra to generate all possible structures of a molecule, which agree with this information for a given molecular formula (Lindel et al., 1997). The primary COCON output was safely reduced to less than 1% of its original size, without losing the correct structure proposal by this shift prediction. A data set containing marine natural products was thoroughly studied (Meiler et al., 2002). For the heterocyclic system ascididemin ($C_{18}H_{11}N_3O$), which contains five condensed rings, a data set was generated that contained 14 ^{1}H,^{1}H-COSY and 35 ^{1}H,^{13}C-HMBC correlations. COCON suggested 28,672 possible structures. In order to find a small number of solutions that are identical or similar to the correct solution, additional data in form of C, C correlations or a rapid method to evaluate all structural proposals was required. The ANN is able to calculate 5000 ^{13}C chemical shifts per second. The calculation of all chemical shifts was performed in 103 s using a PC Pentium II. The ^{13}C chemical shifts deviations between the experimental and the theoretical values were calculated in less than 8 min for all 28,672 structures. The correct structure of the ascididemin was ranked as 25th, which was within the first 0.1% of all structural proposals. A disadvantage of this ANN application is the inability to consider stereochemical aspects of molecular structure. The ANN was not able to distinguish between stereoisomeric compounds such as diastereomers. This includes the inability to distinguish between diasterotopic atoms in an asymmetric compound like the diasterotopic methylcarbons in an isopropyl group. Robien and co-worker (Robien, 2000) developed an ANN, which also utilized stereochemical information based on a ^{13}C database containing 518,200 compounds. This improved the prediction quality for the assignment of some diastereomers. This evaluation of stereochemical interactions does not employ three-dimensional coordinates. Therefore, it is confined to olefinic and cyclic systems.

Le Bret (2000) combined ANNs and GAs to predict ^{13}C chemical shifts using 8300 carbons from the literature. The so-called genetic neural network with a genetic variable selection was successfully tested, but is very demanding for computation time. Each fitness computation involves teaching the model to converge, which takes a few minutes. For several thousand individuals and 100 generations, more than 3000 h were required for fitness computation. Backpropagation was used exclusively.

3.2.2. Prediction of macroscopic properties of complex mixtures

Numerous applications for the prediction of diverse properties of miscellaneous mixtures using the combination of NMR and ANN were published over the last 10 years. In many cases, a spectrum of a mixture consists of the sum of the spectra from the pure compounds with superimposed noise. For these cases, the degree to which the spectrum of each component contributes to the spectrum of the unknown sample can be calculated easily. An ANN is of advantage when interactions between the components cause deviations from linearity. The majority of products in the petrochemical industry are such complex mixtures. However, the knowledge of accurate composition is not the most important question here, but the molecular characteristics of composite hydrocarbon mixtures can be associated with the macroscopic properties of the oil products. Understanding relationships between the structure and composition of molecular mixtures and their chemical properties is of interest to industrial research. Both sophisticated spectroscopic detection methods and advanced chemometric evaluation techniques are required. The combination of NMR and ANN is a promising approach. Basu et al. (1998) predicted the biodegradability of mineral base oils from their chemical composition and viscosity. Hereby, the chemical composition was determined by NMR and mass spectrometry. Two mathematical models were constructed, which served to screen the base oils before they were subjected to a 21-day biodegradability test.

Michon and Hanquet (Michon et al., 1997) and Colaiocco and Espidel (2001) estimated rheological properties of asphalt samples using ^{13}C and ^{1}H NMR descriptors as input for an ANN, respectively. Rheological properties like creep slope at low temperature and stiffness at high temperature are of importance for the determination of the asphalt performance grade. Naturally, a correlation between descriptors and properties is necessary in each case. This was reliably achieved for the viscosity index of 38 hydrocarbon mixture test samples, determined by Vaananen et al. (2002) using an ANN analysis of their two-dimensional NMR spectra. Only the ^{1}H and ^{13}C methyl resonance regions were used in order to reduce the data dimension. A high correlation coefficient between the measured and the estimated viscosity index of 0.903 was achieved with only two neurons in the hidden layer.

Another important property in petrochemistry is the octane number (ON) or octane rating, indicating the quality of gasoline. The higher the ON, the lower the tendency of gasoline to produce knocking in a combustion engine. For testing, an engine is usually calibrated by measuring the knocking ability of mixtures of *n*-heptane (ON = 0) and 2,2,4-trimethylpentane (ON = 100). The research octane number (RON) and the motor octane number (MON) are used to describe the knocking characteristics under different severities in the combustion chamber. Both must be measured in a standard test engine. This traditional method is expensive and time consuming. Correlation between properties and the composition and structural differences of the complex mixture would be of advantage. Caused by mutual interaction between the hydrocarbons, the anti-knock behaviour of the complex gasoline mixtures do not represent the sum of the behaviour of the individual compounds. Models have to take into account the non-linear blending characteristics of single components for the calculation of the ON. However, no robust model has been developed, which would allow the prediction of the knock-rating of complex mixtures based on their quantitatively determined chemical structural groups. ^{1}H NMR spectra of more than 300 gasoline samples, representing all major brands and grades available at German and Austrian service stations were recorded (Meusinger, 1996). Manual preparation of one sample, acquisition and analysis of the NMR spectrum requires approximately 30 min. Fig. 5 shows typical ^{1}H and ^{13}C NMR spectra of a Super gasoline sample.

Thirteen different structural groups were distinguished and assigned in the ^{1}H NMR spectra (Fig. 5a) and 90 groups in the ^{13}C NMR chemical shift ranges. The assignments for the structural groups were determined by one- and two-dimensional NMR experiments and database searches (Meusinger, 1996). The correlation between individual structural groups and the ON was studied by a common factor analysis. Hereby, the structural groups were determined quantitatively by integration of defined chemical shift ranges. As expected, the oxygenates and the substituted aromatics showed the highest loading, whereas the aliphatic and olefinic groups are negatively correlated with the RON. As already indicated, the protons in a similar chemical environment resulted in NMR signals with similar chemical shifts. Therefore, comparable structural elements of different molecules, independent of their class or substances assignment, appear in the same integral region. Assuming that such equivalent structural groups behave similarly during combustion they should also result in comparable contributions to the ON. For further QSPR calculations, it was necessary to eliminate the auto-correlation between similar individual groups. These auto-correlations were identified by a hierarchical cluster analysis. The structural groups concerned were simply combined. The experimental integral values were used as input for an ANN, whereas the corresponding

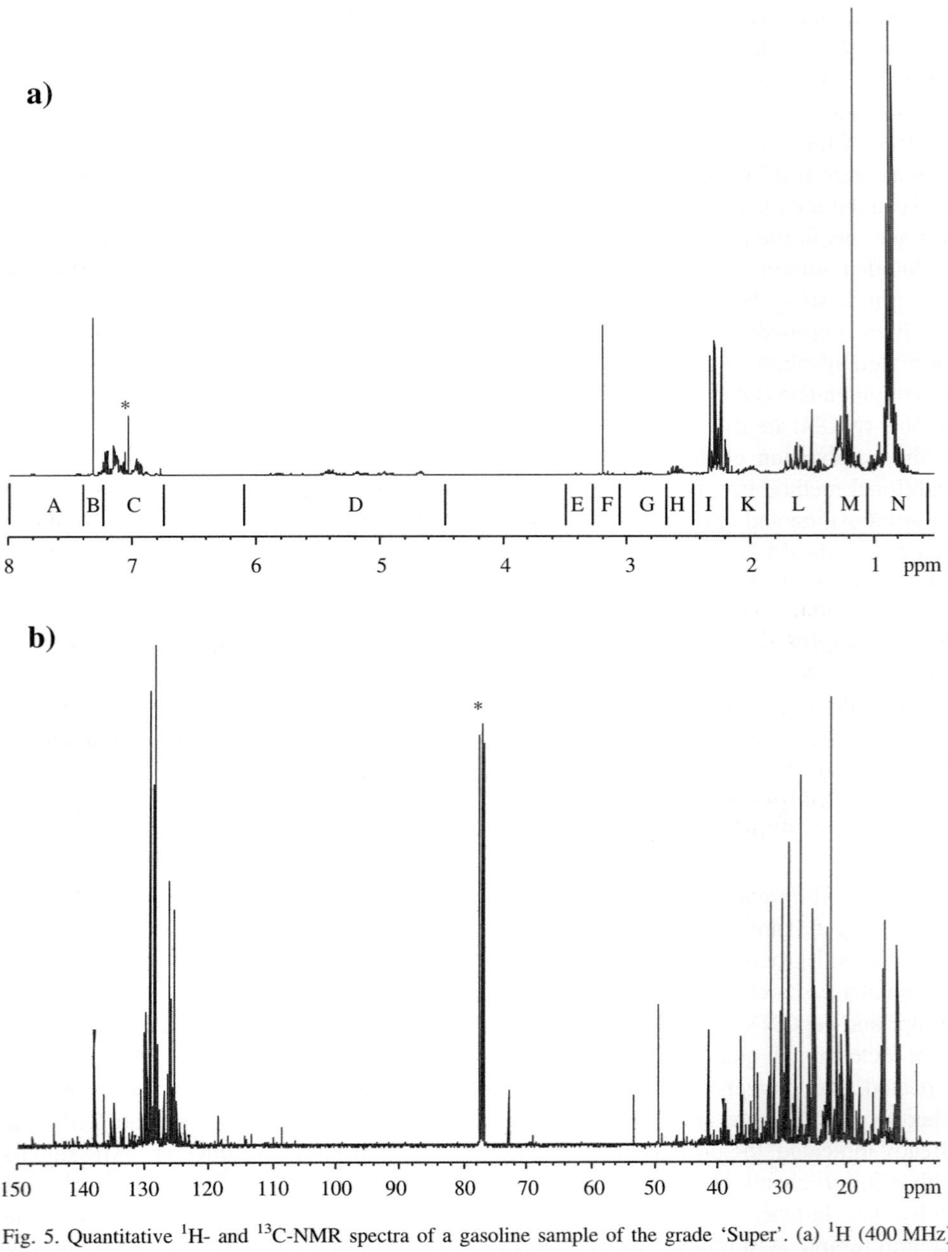

Fig. 5. Quantitative ^{1}H- and ^{13}C-NMR spectra of a gasoline sample of the grade 'Super'. (a) ^{1}H (400 MHz): 100 μl gasoline dissolved in 400 μl $CDCl_3$ (∗), 32 accumulations, repetition time 4 s, measuring time 2.1 min. 13 chemical shift sections, marked with letters A to N, were distinguished here, referring to different structural groups (Meusinger and Moros, 2001). (b) ^{13}C (100 MHz): 200 μl gasoline dissolved in 400 μl $CDCl_3$ (∗), 8000 accumulations, repetition time 6.2 s, measuring time 13.5 h. The 90 chemical shift sections represent different structural groups (not shown, see Meusinger and Moros (2001)).

RON represents the output. Different multi-layered backpropagation neural networks were constructed with variable numbers of neurons and hidden layers. The trained networks were tested with the data set already used for a multiple linear regression. The achieved result ($R_{test} = 0.984$) indicate that the relationships between structure and reactivity were best described by a three-layer ANN. While a ^{1}H NMR spectrum of small sample amounts is acquired in a few minutes, a routine application of ^{13}C NMR is not reasonable because of the very low sensitivity and time requirements for quantitative interpretation of the results. However, the spectral resolution is significantly larger for ^{13}C NMR data and more detailed structure information can be retrieved from ^{13}C NMR spectra. Therefore an ANN was also used for the determination of ONs of individual gasoline compounds from their ^{13}C NMR spectra (Meusinger and Moros, 2001). More recently the quantitative relationships between structural parameters of diesel fuels have been established with the diesel ignition delay characteristics, using ANN technique (Basu et al., 2003). The ^{1}H NMR spectra of 60 commercial diesel samples were studied using their structural characteristics and relative intensities of various regions in the NMR spectra were used as ANN inputs. The appropriate cetane number was determined by an ignition quality tester. The NMR spectra were divided into 18 regions representing paraffins, cyclo-alkanes, olefins and different types of aromatic compounds. To reduce the amount of data points, the data set was compressed to eight input parameters by training a primary neural network in which inputs and outputs were the same. The hidden layer of the developed primary network was used as the input and the cetane number as the output for the development of the final network. The primary network for data compression and the final network for cetane number prediction were appended together. The developed model, when tested on a validation data set, resulted in high correlation between the actual and predicted values of cetane number.

It was shown that NMR spectroscopy is a suitable method to describe chemical structures. Unfortunately, results obtained from the ANN calculations cannot be interpreted in a conventional manner. They encode complex structure–property relationships, but do not allow chemical or physical interpretation. However, they are well suited for the prediction of structure dependent properties (like ONs).

4. Genetic algorithms

Optimization problems associated with NMR spectra often involve the analysis of a very large search space. For some of these problems, a very sharp global minimum exists in the fitness landscape so that the risk of being trapped in locally optimal solutions is ever present if straight downhill techniques are used. Heuristic optimizations like GAs are probably the most suitable methods to overcome these two drawbacks.

GAs have been introduced by Holland (1975) and Goldberg (1989). Similar to biological evolution, only the fittest individuals (chromosomes) will survive a selection process (fitness function) resulting in new, modified generations (due to reproduction, crossover and mutation) that constitute better solutions to the optimization problem. A modification of GA is genetic programming (GP) (Koza, 1993) for which NMR spectra analysis will be discussed.

Applications of GAs and GP are found throughout all processes involved in the analysis of NMR spectra. They include:

(1) Data processing
(2) Assignment of NMR data (chemical shift, coupling constants and Nuclear Overhauser effects) to functional groups
(3) Conversion of NMR data to structural restraints and search for compounds that best represent a given data set
(4) Feature reduction of NMR data for classification
(5) Classification

In addition, Freeman (2003) has suggested utilizing GAs for the development of new pulse sequences in NMR spectroscopy. Hereby, the outcome could be defined by the required intensities, frequency ranges, phases, etc. These variables may form the genes that are initially arbitrary chosen. The Bloch equations would then be used to calculate the corresponding excitation pattern. The excitation pattern are mutated and crossed over. The resulting spectroscopic variables are judged against the target properties to select the most efficient pulse sequence. Freeman has also suggested that a strongly operator-supervised selection may occasionally result in abandoning the preconceived target and 'branching' into more complex developments. The success of this approach has been demonstrated by the development of a new pulse sequence for uniform excitation over a wide frequency band width (Wu et al., 1991).

4.1. Data processing

Compared to other spectroscopic methods, NMR spectroscopy is not very sensitive. Therefore, noisy data sets are always a problem, which causes errors in spectral quantification and signal recognition. Choy and Sanctuary (1998) combined GA with a priori knowledge for estimations of NMR spectra with very low signal-to-noise ratios. The detected NMR signal is usually transformed from the time domain into a frequency domain by fast Fourier transformation (FFT). The method applied by Choy and Sanctuary demonstrates that spectral simulation based on GA is superior to FFT and other estimation techniques in terms of signal recognition and quantification. The GA that was used by Choy and Sanctuary encoded the time dependence of the dumping factor and frequency as genes in chromosomes. Signal amplitudes and phases were derived separately based on linear fitting. The fitness of the chromosomes was tested on the basis that the least square error of the vectors of noiseless data should be minimized. The roulette wheel method (Goldberg, 1989), linear rank-based selection (Grefenstette and Baker, 1989), tournament selection (Goldberg and Deb, 1991) and truncation selection (Muehlenbein and Schlierkamp-Voosen, 1993) were used as selection schemes for reproduction; 10–50% of the chromosomes were selected for reproduction. The authors tested single-point crossover, two-point crossover, discrete recombination and intermediate recombination as crossover and recombination operators. A size of 100 populations and 100 generations was chosen. The intermediate recombination methods outperformed the others. Convergence was achieved after 50 generations. After incorporation of a priori knowledge like known dumping factors or frequencies, convergence was achieved quicker. GA outperformed other simulation methods in particular for NMR spectra with very low signal-to-noise ratios. Interestingly, the a priori knowledge of the frequency do not need to be precise in order to improve the performance of the GA.

Problems are more severe for in vivo NMR spectroscopy, where spectra are obtained from a living object. Usually, in vivo data suffer from poor signal-to-noise ratios, low resolution and consequently, overlapping signals (Gadian, 1995). Accurate quantification and determination of frequencies are essential for correct interpretation. Some models exist for data extraction that utilize iterative non-linear adaptation of a model function to the experimental data (Stephenson and Binsch, 1980; Webb et al., 1992). By minimizing the quadratic deviation between the measured data and the model a best solution was found. Weber et al. (1998) used a GA based optimization to overcome the dilemma of local minima. Comparison with simulated annealing (SA) revealed similarly good results for the quantification of in vivo NMR data that were superior to other methods that often found only local minima. SA showed hereby better reproducibility if calculations were repeated. GA, in contrast, found often deeper minima than SA. Weber et al. (1998) utilized model functions in the time domain that simulate NMR signals under consideration of amplitude, relaxation and frequency. The parameters that were optimized include scaling factors, frequency shift and dumping. The sum of the quadratic differences between the spectral data and the model function was used to assess the fitness of populations. Recombination was performed by either averaging of the parent genes or by random selection of genes. Random mutations were produced by either change of one decimal digit of a gene or by small decrease/increase of values. Up to 265 generations were produced. Convergence was achieved after less than 100 generations. Quantification was satisfying and reliable. Only discrimination of metabolites with very similar signal pattern required further investigation.

4.2. Structure determination

NMR spectroscopy is one of the most successful methods for the determination of chemical structures. Intermolecular NOEs provide information about distances between protons that are close in space, ${}^3J$-coupling constants provide information about dihedral angles and changes in chemical shifts provide information about functional group interactions. With the evolution of the field of protein NMR structure determination, the traditional method of qualitative analysis of these data, resulting in low resolution pictures of chemical structures, has developed into quantitative analysis of such data resulting in high-resolution structures.

Typically, NMR data are analysed by a two-step process that involves the assignment of the data to functional groups and conversion of this information into specific structural restraints, which will result in three-dimensional structures. If the accurate assignment of NMR data to structural restraints fails for even a small number this may result in failure to achieve high resolution structures or in distorted structure (Adler, 1996). It is therefore important to identify problem restraints in order to correct or remove them. Until 1996, this was often performed in an unsatisfying trial-and-error process. Pearlman (1996) has developed an automated objective method for structure refinement based on NMR data. The GA based program FINGAR utilizes a set of potential molecular conformations to refine relative weights of these confirmations based on comparisons with experimentally derived NOE distances and ^3J-coupling constants that encode for the dihedral angles. Hereby, the relative weights of the basic structures are calculated by minimization of the values for the fitness function that encodes for the weighted average of the energy over all base structures. The variables in FINGAR refinement are the basic set weights. Although, it has been demonstrated that FINGAR is very successful in the refinement of structures, the program still relies on correct assignment of NMR data.

An extended version of FINGAR addresses this problem and flags bad restraints during structure refinement (Pearlman, 1999). Relative weights of individual restraints and their weighted average value for NOE distances and ^3J-coupling constants were introduced in this version of the program. Fitness is compared between calculated and experimental values. The program is able to flag individual problem restraints. Several runs of the program may be required for a 100% success rate.

Other GA based approaches have been taken to address particular ambiguous NOE restraints and NOE violations. A large number of NOE signals cannot be assigned because of overlapping signals so that more than one pair of protons matches the chemical shifts. Many ambiguous NOEs can be resolved as some may contradict the low-resolution structure obtained by manual assignment. GAs have been developed to resolve such ambiguous NOEs and result in high-resolution structures due to unequivocal assignment of NOEs. Approaches taken vary between sequential assignment (Li and Sanctuary, 1997) and assignment of only ambiguous NOEs (Adler, 2000).

Adler (2000) treats each NOE as a single gene and possible assignments are treated as different traits as his program converges rapidly. NOEs will automatically be assigned. Each NOE will have more than one assignment. These NOE assignments will have an initially calculated probability of being correct based on the accuracy of the chemical shift match. Based on this first generation of different NOE assignments, initial structures are calculated using SA. The program GENNOE tests the self-consistency of a constraint list with the calculated structures. Structures are scored according to the number of NOE violations relative to the total NOE values. Revised probabilities of NOE assignments are directly calculated from the structure scores. As each gene is treated as a separate gene, GENNOE substitutes the direct calculation of the probabilities for the mathematical mating normally used in GA to generate the new generations. Adler also accounts for the possibility that a particular NOE cannot be correctly assigned (null_NOE) in order to prevent propagation of errors.

The program GENNOE analyses a large number of ambiguous NOEs in parallel, produces a self-consistent restraint set with a reduced amount of NOE violations and can assess the conformational flexibility of a preliminary structure. The program converges rapidly, resulting in a fast filter for ambiguous NOE assignments that may otherwise distort the calculated structure. Limitations found by Adler (2000) are that the results are path-dependent and a small proportion of misassigned NOEs can still be included in a self-constraint restraint set.

Hunter and Packer (1999) have used a GA to fit structures of supramolecular complexes with complexation induced chemical shift changes from NMR spectra. Complexation induced chemical shift changes were obtained from the difference in the chemical shift of all NMR signals of free molecules compared to the molecules in the supramolecular complex. During interaction of different parts of a molecule, complexation induced chemical shift changes provide information about interactions of functional groups that by themselves provide information about the structure of the complex. Computer programs are available that calculate chemical shifts for any given structure reliably (Iwadate et al., 1998; Williamson and Asakura, 1993). The difference between the calculated complexation induced chemical shift changes and the experimentally obtained chemical shift changes from the NMR spectra provide a quantitative measure of how well a given structure matches the experimental data set. A GA was used to minimize the root mean square difference between the experimental and the calculated values for all NMR signals, as a function of the relative position and orientation of the molecules of the complex and the internal torsion angles.

The fitness function was defined as $R_{expt}/R_{\Delta\delta}$ (with R_{expt} root mean square of the experimental values and $R_{\Delta\delta}$ the difference between calculated and experimental complexation induced chemical shifts). The chromosomes coded for a set of intermolecular variables (the three global rotations and translations for each molecule) and a set of intramolecular torsion angles. Each chromosome hereby represents a different structure with different internal torsion angles and different relative positions and orientations of the individual molecules in the complex. A rank-based generational GA with a population of typically 100 chromosomes was used by the authors. Single-point crossover and uniform mutation were used as reproduction operators. The mutation operator was allowed to mutate integers over the full range with an average of one mutation per chromosome. Rank-based selection was used to reduce premature convergence. In order to increase the speed of convergence, two strategies were applied: (1) restriction of the search space by introduction of interatomic distance constrains (user defined interatomic distances; experimental NOE constraints for distances and van der Waals clashes of distances) and (2) restart the GA once a reasonable structure has been found. Once a certain fitness value was reached the search space was halved in all dimensions, and the GA was restarted from the optimum structure of the previous run. This approach was repeated to reduce the search space at every restart and improve the resolution of the structures. The authors developed a software that has been successfully validated for the structure determination of medium size complexes. It remains to be seen how well the method will work with larger and more flexible molecules. It is of interest that with the parameters used by the authors, a space of 10^{38} conformations has been searched, a number too large to be successfully tackled by other means. In addition, complexation induced chemical shifts show large changes within relatively small conformational changes, resulting in a sharp spike in the fitness landscape, which may cause problems for other optimization techniques.

Similar approaches have been taken by others to calculate docking or binding sites. Evolutionary algorithms or GAs substantially speed up flexible docking by considerably narrowing the sampling of conformational space. However, this often results in a bias towards good hits found initially.

4.3. Structure prediction

In contrast to above-mentioned optimization of structures, structure prediction is aiming to make a best suggestion to identify an unknown compound. Meiler and Will (Meiler and Will, 2001, 2002) used experimental ^{13}C NMR spectra and a GA for automated structure elucidation of organic molecules. Using the molecular formula of the unknown compound, a set of random structures is created. The previously described ANN approach (Meiler et al., 2000) calculates the ^{13}C NMR chemical shifts of the random structures, which is then compared with the experimental ^{13}C NMR spectrum. Minimizing the root mean square deviation of the chemical shift deviations of all carbon atoms is the fitness criterion. The genetic code consisted of the numerical vector of all bonds between all possible atom pairs that are chemically reasonable and agree with the molecular formula. Recombination was performed by joining the two vectors of two parent molecules. Mutation was performed by randomly removing one bond and adding another bond to the molecule. In this way, the constitution of organic molecules could be optimized relative to an unknown organic sample consequently by combining the GA with the neural network based spectral prediction. An automated structure elucidation was possible for molecules with up to 14 non-hydrogen atoms. Also, by introducing a small list of forbidden fragments molecular structures with up to 20 non-hydrogen atoms could be determined using only their ^{13}C NMR chemical shifts. The method was implemented and tested successfully for different organic molecules with up to 20 non-hydrogen atoms. The time limiting step of the structure elucidation was the calculation of ^{13}C NMR spectra by the ANN. Currently considerable calculation time of up to 100 h is required for larger molecules using a PC.

4.4. Classification

Classification of samples by NMR spectroscopy is most often performed by pattern recognition methods, principle component analyses, linear statistical discrimination methods (e.g. linear discriminant analysis, LDA) or non-linear approaches (e.g. ANN). Somorjai et al. (1995) and Gray et al. (1998) have used GP for classification of NMR spectra from biological samples (see below also). GP produces rules that explicitly trace possible decision paths for classification problems. In contrast to GA, populations in GP consist of programs that are made up by simple arithmetic and logical operations (Koza, 1993). Programs that best classify a training set are retained. GP usually uses less mutation and more crossover as genetic operator (Koza, 1993). The fitness function for the classification problem was to minimize the number of samples minus number of correct assignment (Gray et al., 1998). In both applications, the first 10 (Somorjai et al., 1995) or 20 (Gray et al., 1998) principal components identified in the ^{1}H NMR spectra were retained for the analysis. Both authors compared the GP classification with classification by other methods (ANN). While GP usually performs as good as ANN, different individual misclassifications (errors) were identified for both methods, highlighting that both obtain solutions in different ways. This implies that combination of different types of classifiers will result in 'better' classifiers as suggested by Somorjai et al. (1995, 1996).

4.5. Feature reduction

A major complication of using NMR spectra for classification is the scarcity of samples compared to the vast amount of spectral features. As it is often not feasible to acquire NMR spectra from more samples, feature reduction methods are unavoidable. While conventionally principle component analysis (PCA) is applied for feature reduction (Howells et al., 1992, 1993), this has two disadvantages (Nikulin et al., 1998):

(1) Spectral features are usually not retained.
(2) The first principle components do not necessarily account for the most discriminatory features.

Nikulin et al. (1998) have introduced a GA based feature reduction method for classification based on NMR spectra that avoids both problems (see below also). The spectral features used in their approach are either the spectral data points or the first derivatives (Somorjai et al., 2002). The fitness function was the mean square error between the training set classification and the a priori class indicator. Populations are created by selecting a number of binary strings (chromosomes) that consist of a predefined number of desired subregions, which in turn consist of adjacent data points (e.g. subregions of variable length). These regions are then used to classify the training set and their fitness is assessed. Crossover was performed by conventional exchange of parts of parent chromosomes. Mutation was performed by a k-block mutation, whereby the size of the k-block decreases with increasing generations. For the first generation, a rapid but coarse exploration of the feature space is achieved, which is followed by fine-tuning due to the decreasing mutation block. For most problems, convergence was achieved after 50–100 generations. In addition to above-mentioned advantages, this optimal region selection (ORS) process is not restricted to a limited number of digitized integral regions, but uses the actual data points. The mapping procedure of the original attribute space then finds regions of variable width varying from single data points to adjacent data points spanning over up to half the chosen spectrum. This procedure accounts for variable line widths that are often found in inhomogeneous samples like NMR spectra from tissue biopsies or in vivo NMR spectra.

5. Biomedical NMR spectroscopy

Apart from some exceptions that have already been mentioned, applications of ANN and GA in biomedical NMR spectroscopy are mainly restricted to classification problems and feature reduction for classification.

In contrast to NMR spectroscopy in chemistry, biochemistry and pharmacy, biomedical samples are more complex. These samples are either extracts from biological material, biofluids (urine, blood plasma), biopsy samples or non-invasively collected in vivo NMR spectra. These data tend to be noisy and contain redundant information. Less favourable acquisition conditions often result in poor resolution so that it may become more difficult to resolve discriminatory peaks. Biological samples also represent a multi-component mixture resulting in overlapping signals. In addition, the problem of mixtures may be exhilarated by a dynamic situation where one has to deal with a large intensity range of the signals. These problems are further exacerbated by 'sampling errors' and 'expert errors'. During sample collection or for in vivo NMR spectroscopy, NMR spectra may represent a mixture of diseased tissue and surrounding 'normal' tissue. NMR spectroscopic applications in biomedicine aim to be superior to existing methods. However, NMR based classification relies on these frequently uncertain or unreliable traditional diagnoses during the training process. It is obvious that a powerful feature extraction method is required for successful classification of such samples.

Earlier toxicological studies on urine samples (Gartland et al., 1990), studies of urine samples from patients with metabolic disorders (Iles et al., 1984; Burns et al., 1992), tumour diagnosis on biopsy samples (Delikatny et al., 1993; Russell et al., 1994; Kurhanewicz et al., 1995; Rutter et al., 1995; Mackinnon et al., 1997) and in vivo tumour diagnosis (Gill et al., 1990; Negendank, 1992; Usenius et al., 1994; Preul et al., 1996) were partly very successful based on the identification of 'marker' metabolites. However, these classifications often focused on the distinctions between normal and diseased states. Classification between different diseased states or disease-grading is more complex.

A group at the University of London has successfully applied PCA and non-linear mapping techniques for the classification of NMR spectra from urine obtained from animal models and patients with different metabolic diseases and different diseases states (Gartland et al., 1990; Holmes et al., 1994; Holmes et al., 1998). Nicholson et al. (1999) have established the field of 'metabonomics', which was defined as "the study of the multivariate time-resolved metabolic changes in biofluids, cells and tissues to pathophysiological insult or genetic modification". Their approach and other unsupervised approaches were successful for many diseases, but significant contamination from less dominant classes occurred for some medical conditions. However, it was shown that the NMR data sets had intrinsic group structure. More successful and clinically more detailed results were obtained with a supervised approach using ANN or modelling of class analogy for the analysis of rat urine (Anthony et al., 1995; Holmes et al., 1998).

ANN have been successfully applied to a number of clinical problems that include: in vivo ^{31}P NMR spectra for the classification of tumours (Howells et al., 1993), ^{31}P NMR spectra for the evaluation of muscle diseases (Kari et al., 1995), ex vivo ^{1}H NMR spectra for the classification of thyroid neoplasms (Somorjai et al., 1995), ^{1}H NMR spectra for the characterization of blood plasma (AlaKorpela et al., 1996), ^{1}H NMR spectra from biopsies of brain tumours (Maxwell et al., 1998) and in vivo ^{1}H NMR spectra for the classification of brain tumours (Preul et al., 1996; Usenius et al., 1996; Poptani, 1999). The clear advantage of classification of NMR spectra by ANN is that neither biochemical prior knowledge nor any assignment of the signals is required. ANN results is more accurate classification of the samples than an analysis based on subjectively chosen signals.

Improved classification results were achieved by Holmes et al. (2001) by utilization of probabilistic ANNs compared to their previous approaches when applied to the classification of toxic responses in NMR spectra of rat urine. Unlike other ANNs, probabilistic ANNs incorporate all of the spectroscopic information while avoiding problems associated with overfitting. Holmes et al. (2001) have compared the classification using a probabilistic ANN (containing 1 hidden layer, 207 input and 10 output nodes) with multilayer perceptrons and a classical principal component approach.

The probabilistic ANN outperformed the other methods using a relatively large data set of 583 spectra in the training and 727 spectra in the validation set.

Hagberg (1998) has reviewed methods for classification of tumours, including in vivo MR spectra from patients. The majority of studies have used LDA or ANNs with or without feature reduction. Most studies that have used ANNs for tumour classification have used between 1 and 3 hidden layers with 3–60 nodes per layer. The robustness of classification can be improved by introduction of Gaussian noise in the training set (Branston et al., 1993). Only few examples are known for self-organizing ANNs for classification and feature reduction (Hagberg, 1998). It was found that LDA and ANN resulted in comparable classification. Considering that ANNs are more vulnerable to overfitting, LDA seem to be more robust. However, most of these data set contain only very few spectra compared to the number of features. One would expect more robust results if the entire sample space was sampled, hereby, avoiding overfitting. As expected, more accurate results were achieved if feature reduction methods were applied before training of the classifiers. Gerstle et al. (2000) used ANNs successfully for improved diagnosis of head and neck squamous cell carcinoma based on in vivo NMR spectra. The power but also the risks of ANNs were recently demonstrated by Axelson et al. (2002). This group compared Kohonen, backpropagation, probabilistic and radial basis function neural networks to distinguish between in vivo NMR spectra from patients with Parkinson disease and controls and to discriminate between patient subgroups. A GA was used for feature selection. Although no significant difference in metabolite ratios were detected between the different classes, the ANN was able to discriminate between these groups. However, the authors also noted that the number of cases in their study was rather low (four classes of 15, 11, 5 and 14 patients). It is highly likely that the low number has resulted in overfitting in this particular case. Cherniak et al. (1998) have used a two-stage feed-forward ANN to classify NMR spectra of polysaccharides isolated from 106 strains of *Cryptococcus neoformans*. They were able to identify eight chemotypes in this pathogenic yeast species reliably.

AlaKorpela et al. (1996) have compared an ANN analysis for the classification of plasma lipoprotein profiles with an unsupervised self-organizing map analysis using a Kohonen network (Kohonen, 1995). The accuracy of their analysis was comparable to the supervised approach. The clear advantage of the unsupervised approach is that a priori knowledge that may introduce expert bias is avoided. This may potentially reveal unexpected relationships that may not have been discovered otherwise. Kaartinen et al. (1998) used only a small subregion of the NMR spectra (methyl resonances; 22 data points between 0.72 and 0.89 ppm). This reduces the number of data points to the ones that are, based on a priori knowledge, the most informative. Hereby, the danger of overfitting might have been reduced.

However, for all the above studies, only a limited number of data was available compared to the vast amount of data points. Most of these studies utilized either none or a very small independent validation data set. These investigations can be regarded as very useful proof of concept studies. However, it needs to be seen if they will result in robust diagnostic applications.

Lisboa et al. (1998) have compared various classification strategies for the analysis of a data set containing five classes of tumour types. Non-parametric, statistical discrimination methods (LDA, nearest neighbour classification) were compared with dimensionality reduction using combinations of the attributes (PCA, partial least squares) and non-linear classifiers based on ANN (based on a Bayesian framework multi-layer perceptron) (Bishop, 1995). The PCA with leave-one-out cross-validation performed worse. The performance of the ANN based classification was similarly successful as statistical classifiers. For small data sets LDA may be a better choice for robust classification (Friedman, 1989).

A comparison of GP and ANN was performed by Gray et al. (1998) for the classification of two classes of brain tumours (see Chapter 4.4). The two methods resulted in similar accuracies when tested on a small independent validation set. The feature reduction (by PCA followed by varimax rotation) was more important for successful classification by GP than for ANN. Implementation of GP was more demanding than for ANN.

Somorjai et al. (1995) have suggested combining several independently developed classifiers to a consensus diagnosis that will improve reliability and robustness. They have combined an LDA classifier, an ANN that uses backpropagation with Gaussian noise added (Hertz et al., 1991) and a classifier based on GP (Koza, 1993) for the classification of thyroid neoplasms based on ^{1}H NMR spectra of biopsy samples. A consensus diagnosis was achieved using the medians of the individual classifications. The data set was split in training and test set, which was used for independent validation of the classification. The conceptional/methodological independence of the three classifiers was further confirmed by the fact that each misclassified different samples. There was no significant difference in the overall accuracy between the three classifiers on the test set. However, the consensus diagnosis gave better and more reliable results than any of the individual classifiers.

Somorjai et al. (1995) used PCA to reduce the dimensionality of their original data set as many of the other studies did. However, there are several problems associated with this type of preprocessing. As the principal components are linear combinations of the original features, original spectral features that could be correlated to biochemical background information of a particular disease are normally lost. More importantly, the principal components that account for most of the variability are not necessarily the most discriminating features of the NMR spectra. Nikulin et al. (1998) have taken a different approach, as explained in Chapter 4.5, by using a GA based optimal region selector (GA_ORS) for feature reduction. GA_ORS identifies the most discriminating features of the spectrum. Other advantages of this method are:

(1) All data points of a spectrum (>1000) are used as an input rather than broader chemical shift increments, resulting in larger resolution.
(2) GA_ORS produces most discriminatory subregions of variable size (width), which helps to restrict the optimal features to a low number.
(3) The feature reduction process retains the spectral identity.

Nikulin et al. (1998) originally combined the GA_ORS with LDA classification to discriminate between meningioma and astrocytoma based on in vivo ^{1}H NMR spectra and to distinguish between malignant and benign colorectal biopsy samples based on ex vivo ^{1}H NMR spectra.

Table 1
Comparison of different feature selection strategies for the classification of five yeast species based on their ^{1}H NMR spectra (see also Himmelreich et al., 2003)

Yeast species for that pair-wise classifiers were developed	Classification accuracy (%)		
	(a) Utilization of 16 spectral regions without feature selection	(b) Utilization of three principal components	(c) Utilization of the two most discriminatory regions (ORS)
C. albicans vs *C. glabrata*	92	96	100
C. albicans vs *C. krusei*	96	98	100
C. albicans vs *C. parapsilosis*	59	81	98
C. albicans vs *C. tropicalis*	71	78	100
C. glabrata vs *C. krusei*	85	92	100
C. glabrata vs *C. parapsilosis*	95	95	100
C. glabrata vs *C. tropicalis*	88	89	96
C. krusei vs *C. parapsilosis*	85	92	100
C. krusei vs *C. tropicalis*	88	94	98
C. parapsilosis vs *C. tropicalis*	64	82	100

The spectral region between 0.35 and 4.00 ppm was selected and divided into (a) 16 integral regions according to typical metabolite regions (see Bourne et al., 2001); (b) 73 regions of 0.05 ppm width and (c) 1500 individual data points. No additional feature selection was performed for (a); a principal component analysis was performed for (b); and a GA based ORS according to Nikulin et al. (1998) was performed for (c). Ten pair-wise LDA based classifier were developed to distinguish between the five yeast species using (a) all 16 spectral regions; (b) the three principal components accounting for most variability and (c) the two most discriminatory regions identified by the ORS. Classifiers were developed from NMR spectra of 876 cultures. The listed accuracies (correct identification with a class assignment probability larger than 0.75 divided by total number of isolates) were determined from a validation set, containing 133 isolates that were not part of the training process. The accuracies indicate that the three methods performed acceptably well for the classification of more distantly related isolates (Himmelreich et al., 2003) of the three groups (I) *Candida glabrata*, (II) *Candida krusei* and (III) *Candida albicans*, *Candida parapsilosis* and *Candida tropicalis*. LDA without previous feature selection resulted in unsatisfying results for classification between the more closely related species *C. albicans*, *C. parapsilosis* and *C. tropicalis* (approach (a)). LDA using the three principal components that account for most variability between these three species resulted in more accurate classification, but was outperformed by LDA using the two most discriminatory regions identified by the GA based ORS (Himmelreich et al., 2003).

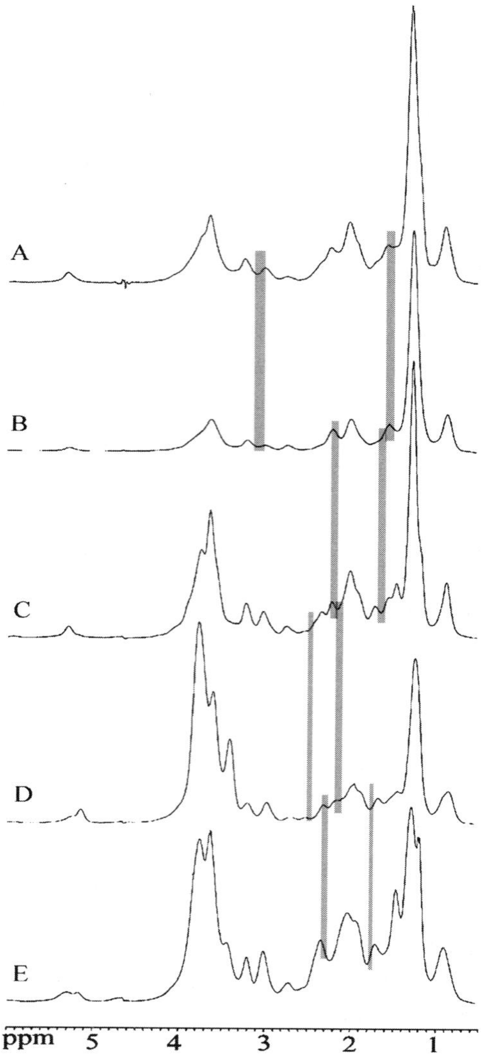

Fig. 6. ^{1}H NMR spectra of five *Candida* species. The bares between spectra indicate features selected by GA_ORS for four of the ten respective pair-wise classifiers. (A) *Candida tropicalis*, (B) *Candida albicans*, (C) *Candida parapsilosis*, (D) *Candida krusei* and (E) *Candida glabrata*. (A)–(C) are closely related species.

This method has been successfully applied (accuracies approaching 100%) to a wide range of clinical classification problems including diagnosis of breast cancer based on biopsy samples (Mountford et al., 2001), pathology of barrett's oesophagus (Doran et al., 2003), diagnosis of melanoma (Bourne et al., 2002), pathology of liver biopsies (Soper et al., 2002), diagnosis of squamous cell carcinoma (El-Sayed et al., 2002) and others.

In an application of this classification strategy to a classification problem in clinical microbiology, we were able to distinguish between five classes of closely related pathogenic yeast species (Himmelreich et al., 2003). The advantage of microbiological data sets is that, compared to human tissue or in vivo NMR spectra, additional samples can be obtained more easily from culture collections for validation of the methodology. In addition, microorganisms can be identified repeatedly using a wide range of independent methods, including molecular techniques. We have compared LDA based classification using GA_ORS for feature reduction with LDA based classification after PCA (Table 1). It is of interest that the more distantly related species (see Fig. 6) were classified with similar accuracy by both methods based on an independent validation data set containing 130 test spectra of new isolates. However, the more closely related species *Candida albicans* and *Candida parapsilosis* could not be distinguished from each other in 20% of the cases when LDA after PCA was used as a classification strategy.

In summary, applications of ANN and GA/GP for the classification of NMR spectra from biomedical samples is successful on relatively small data sets. This is a promising starting point for future applications of NMR spectroscopy in clinical and pathological diagnosis. For successful implementation, however, it is necessary to test various approaches on larger data sets.

It appears that the method for data analysis (statistical discrimination methods compared to non-linear approaches) is less critical than a sufficient and robust method of feature space reduction. The scarcity of samples is contrasted with an overabundance of spectral features. The theoretically and empirically suggested sample-to-attribute ratio of 10 to 1 (Fukunaga, 1990) is only met in a few of the above listed studies.

6. Conclusion

NMR spectra encode a large amount of information that can be utilized for classification and prediction. The full potential of NMR spectroscopy in chemical engineering, pharmaceutical research and biomedical applications has only been recognized after computerized data analysis methods have been applied to such data. Hereby, ANNs and GAs have shown promising results.

While ANNs based on the supervised learning paradigm perform reliably when trained on pattern with little variability in the total sample space, they do not have the ability to recognize new categories of pattern. Besides their potential based on the supervised learning paradigm, ANNs have also limitations in their applications due to their time-consuming iterations for training and their tendency to find local minima because of gradient descent algorithm. Unsupervised (self-organized) ANNs overcome some of the disadvantages of supervised ANNs, but often result in less accurate results.

GAs are very powerful in their ability to find global minima for optimization problems. It is expected that their importance will increase with more powerful computation. Currently, the application of GAs is often restricted due to their demanding computation.

A general problem with GAs and ANNs for the analysis of NMR data is the limited number of samples available compared to the number of features provided in NMR spectra. This results in many cases in overfitting and potential discreditation of these methods. With increasing data sets, it is expected that these data analysis methods will become more robust and therefore more important.

Acknowledgments

The authors would like to thank Jens Meiler and Ralf Moros for numerous PC calculations and some helpful suggestions. R.M. thanks the 'Fonds der Chemischen Industrie' for financial support. U.H. acknowledges the support by the National Health and Medical Research Council of Australia (NHMRC 153805).

References

Adler, M., 1996. Deviation versus violation plots: a new technique for assessing the self-consistency of NMR data. J. Biomol. NMR 8, 404–416.

Adler, M., 2000. Modified genetic algorithm resolves ambiguous NOE restraints and reduces unsightly NOE violations. Proteins 39, 385–392.

Adler, B., Ammon, K., Dobers, S., Winterstein, M., Ziesmer, H., 1992. Prediction of carcinogen properties of aminobiphenyls by computer-applications. Chem. Tech-Leipzig 44, 363–367.

AlaKorpela, M., Hiltunen, Y., Bell, J.D., 1996. Artificial neural network analysis of H-1 nuclear magnetic resonance spectroscopic data from human plasma. Anticancer Res. 16, 1473–1477.

Amendolia, S.R., Doppiu, A., Ganadu, M.L., Lubinu, G., 1998. Classification and quantitation of H-1 NMR spectra of alditols binary mixtures using artificial neural networks. Anal. Chem. 70, 1249–1254.

Anand, R., Mehrotra, K., Mohan, C.K., Ranka, S., 1993. Analyzing images containing multiple sparse patterns with neural networks. Pattern Recogn. 26, 1717–1724.

Anker, L.S., Jurs, P.C., 1992. Prediction of C-13 nuclear-magnetic-resonance chemical-shifts by artificial neural networks. Anal. Chem. 64, 1157–1164.

Anthony, M.L., Rose, V.S., Nicholson, J.K., Lindon, J.C., 1995. Classification of toxin-induced changes in H-1-NMR spectra of urine using an artificial neural-network. J. Pharmaceut. Biomed. 13, 205–211.

Aranibar, N., Singh, B.K., Stockton, G.W., Ott, K.H., 2001. Automated mode-of-action detection by metabolic profiling. Biochem. Bioph. Res. Co. 286, 150–155.

Axelson, D.E., Nyhus, A.K., 1999. Solid-state nuclear magnetic resonance relaxation times in crosslinked macroporous polymer particles of divinylbenzene homopolymers. J. Polym. Sci. Pol. Phys. 37, 1307–1328.

Axelson, D., Bakken, I.J., Gribbestad, I.S., Ehrnholm, B., Nilsen, G., Aasly, J., 2002. Applications of neural network analyses to in vivo H-1 magnetic resonance spectroscopy of Parkinson disease patients. J. Magn. Reson. Imaging 16, 13–20.

Ball, J.W., Jurs, P.C., 1993. Automated selection of regression-models using neural networks for C-13 NMR spectral predictions. Anal. Chem. 65, 505–512.

Basu, B., Singh, M.P., Kapur, G.S., Ali, N., Sastry, M.I.S., Jain, S.K., Srivastava, S.P., Bhatnagar, A.K., 1998. Prediction of biodegradability of mineral base oils from chemical composition using artificial neural networks. Tribol. Int. 31, 159–168.

Basu, B., Kapur, G.S., Sarpal, A.S., Meusinger, R., 2003. A neural network approach to the prediction of cetane number of diesel fuels using NMR spectroscopy. Energy Fuels (in press).

Bishop, C.M., 1995. Neural Networks for Pattern Recognition, Clarendon Press, Oxford.

Bourne, R., Himmelreich, U., Sharma, A., Mountford, C., Sorrell, T.C., 2001. Identification of *Enterococcus*, *Streptococcus* and *Staphylococcus* by multivariate analysis of proton magnetic resonance spectroscopic data from plate cultures. J. Clin. Microbiol. 39, 2916–2923.

Bourne, R., Thompson, J., De Silva, C., Li, L.X., Russell, P., Mountford, C., Lean, C., 2002. Proton MRS detects metastatic melanoma in lymph nodes. Proc. Intl. Soc. Mag. Reson. Med. 10, 2055.

Branston, N.M., Maxwell, R.J., Howells, S.L., 1993. Generalization performance using backpropagation algorithms applied to patterns derived from tumor H-1-NMR spectra. J. Microcomput. Appl. 16, 113–123.

Bremser, W., 1978. Hose—novel substructure code. Anal. Chim. Acta-Computer Tech. Optimization 2, 355–365.

Burns, J.A., Whitesides, G.M., 1993. Feedforward neural networks in chemistry–mathematical systems for classification and pattern-recognition. Chem. Rev. 93, 2583–2601.

Burns, S.P., Woolf, D.A., Leonhard, J.V., Iles, R.A., 1992. Clin. Chem. Acta 209, 47–60.

Cherniak, R., Valafar, H., Morris, L.C., Valafar, F., 1998. *Cryptococcus neoformans* chemotyping by quantitative analysis of H-1 nuclear magnetic resonance spectra of glucuronoxylomannans with a computer-simulated artificial neural network. Clin. Diagn. Lab. Immun. 5, 146–159.

Cherqaoui, D., Villemin, D., 1994. Use of a neural-network to determine the boiling-point of alkanes. J. Chem. Soc-Farad. Trans. 90, 97–102.

Choy, W.Y., Sanctuary, B.C., 1998. Using genetic algorithms with a priori knowledge for quantitative NMR signal analysis. J. Chem. Inf. Comp. Sci. 38, 685–690.

Choy, W.Y., Sanctuary, B.C., Zhu, G., 1997. Using neural network predicted secondary structure information in automatic protein NMR assignment. J. Chem. Inf. Comp. Sci. 37, 1086–1094.

Colaiocco, S.R., Espidel, J., 2001. Asphalt's performance grades estimation using artificial neural networks and proton nuclear magnetic resonance spectroscopy. Vis. Tecnol. 9, 17–24.

Corne, S.A., 1996. Artificial neural networks for pattern recognition. Concept Magn. Res. 8, 303–324.

Corne, S.A., Johnson, A.P., Fisher, J., 1992. An artificial neural network for classifying cross peaks in 2-dimensional NMR-spectra. J. Magn. Reson. 100, 256–266.

Delikatny, E.J., Russell, P., Hunter, J.C., Hancock, R., Atkinson, K.H., van Haaften-Day, C., Mountford, C.E., 1993. Proton MR and human cervical neoplasia: ex vivo spectroscopy allows distinction of invasive carcinoma of the cervix from carcinoma in situ and other preinvasive lesions. Radiology 188, 791–796.

Devillers, J., 1996. Neural networks in QSAR and drug design. In: Devillers, J., (Ed.), Strength and Weakness of the backpropagation neural network in QSAR and QSPR studies, Academic Press, London, pp. 1–46.

Doran, S.T., Falk, R.L., Somorjai, C.L., Lean, C.L., Himmelreich, U., Philips, J., Russell, P., Dolenka, B., Nikulin, A.E., Mountford, C.E., 2003. Pathology of Barrett's esophagus by proton magnetic resonance spectroscopy and a statistical classification strategy. Am. J. Surg. 185, 232–238.

Doucet, J.P., Panaye, A., Feuilleaubois, E., Ladd, P., 1993. Neural networks and C-13 NMR shift prediction. J. Chem. Inf. Comp. Sci. 33, 320–324.

El-Sayed, S., Bezabeh, T., Odlum, O., Patel, R., Ahing, S., MacDonald, K., Somorjai, R.L., Smith, I.C.P., 2002. An ex vivo study exploring the diagnostic potential of H-1 magnetic resonance spectroscopy in squamous cell carcinoma of the head and neck region. Head Neck-J. Sci. Spec. Head Neck 24, 766–772.

Forshed, J., Andersson, F.O., Jacobsson, S.P., 2002. NMR and Bayesian regularized neural network regression for impurity determination of 4-aminophenol. J. Pharmaceut. Biomed. 29, 495–505.

Fraser, L., Mulholland, D.A., 1999. A robust technique for the group classification of the C-13 NMR spectra of natural products from Meliaceae. Fresenius J. Anal. Chem. 365, 631–634.

Freeman, R., 1992. High-resolution NMR using selective excitation. J. Mol. Struct. 266, 39–51.

Freeman, R., 2003. Beg, borrow, or steal. Finding ideas for new NMR experiments. Concept Magn. Res. 17A, 71–85.

Friedman, J.H., 1989. Regularized discriminant analysis. J. Am. Stat. Assoc. 84, 165–175.

Fukunaga, K., 1990. Introduction to Statistical Pattern Recognition, Academic Press, Boston.

Gadian, D.G., 1995. NMR and Its Applications to Living Systems, Oxford University Press, Oxford.

Gartland, K.P.R., Sanins, S.M., Nicholson, J.K., Sweatman, B.C., Beddell, C.R., Lindon, J.C., 1990. Pattern recognition analysis of high resolution 1 H NMR spectra of urine. A nonlinear mapping approach to the classification of toxicological data. NMR Biomed. 3, 166–172.

Gerstle, R.J., Aylward, S.R., Kromhout-Schiro, S., Mukherji, S.K., 2000. The role of neural networks in improving the accuracy of MR spectroscopy for the diagnosis of head and neck squamous cell carcinoma. Am. J. Neuroradiol. 21, 1133–1138.

Gill, S.S., Thomas, D.G.T., van Bruggen, N., Gadian, D.G., Peden, C.J., Bell, J.D., Cox, I.J., Menon, K.D., Iles, R.A., Bryant, D.J., Coutts, G.A., 1990. Proton MR spectroscopy of intracranial tumors: in vivo and in vitro studies. J. Comput. Assist. Tomogr. 14, 497–504.

Goldberg, D.E., 1989. Genetic Algorithms in Search, Optimization, and Machine Learning, Addison-Wesley, Reading, MA.

Goldberg, D.E., Deb, K.A., 1991. A comparative analysis of selection schemes used in genetic algorithms, Foundations of Genetic Algorithms, Morgan Kaufmann Publ., San Mateo, CA, pp. 69–93.

Gray, H.F., Maxwell, R.J., Martinez-Perez, I., Arus, C., Cerdan, S., 1998. Genetic programming for classification and feature selection: analysis of H-1 nuclear magnetic resonance spectra from human brain tumour biopsies. NMR Biomed. 11, 217–224.

Grefenstette, J.J., Baker, J.E., 1989. How genetic algorithms works: a critical look at implicit parallelism. Paper presented at Third International Conference on Genetic Algorithms.

Hagberg, G., 1998. From magnetic resonance spectroscopy to classification of tumors. A review of pattern recognition methods. NMR Biomed. 11, 148–156.

Hare, B.J., Prestegard, J.H., 1994. Application of neural networks to automated assignment of NMR-spectra of proteins. J. Biomol. NMR 4, 35–46.

Hertz, J., Krogh, A., Palmer, R., 1991. Introduction to the Theory of Neural Computation, Addison-Wesley, Redwood City, CA.

Himmelreich, U., Somorjai, C.L., Dolenko, B., Cha Lee, O., Daniel, H.-M., Murray, R., Mountford, C.E., Sorrell, T.C., 2003. Rapid identification of chemical characterization of *Candida* species using nuclear magnetic resonance spectroscopy and a statistical classification strategy. Appl. Environ. Microbiol. 69, 4566–4574.

Hoehn, F., Lindner, E., Mayer, H.A., Hermle, T., Rosenstiel, W., 2002. Neural networks evaluating NMR data: An approach to visualize similarities and relationships of sol-gel derived inorganic-organic and organometallic hybrid polymers. J. Chem. Inf. Comp Sci. 42, 36–45.

Holland, J.H., 1975. Adaptation in Natural and Artificial Systems, University of Michigan, Ann Arbor, MI.

Holmes, E., Foxall, P.J.D., Nicholson, J.K., Neild, G.H., Brown, S.M., Beddell, C.R., Sweatman, B.C., Rahr, E., Lindon, J.C., Spraul, M., Neidig, P., 1994. Automated data reduction and pattern recognition methods for analysis of 1H Nuclear Magnetic Resonance spectra of human urine from normal and pathological states. Anal. Biochem. 220, 284–296.

Holmes, E., Nicholls, A.W., Lindon, J.C., Ramos, S., Spraul, M., Neidig, P., Connor, S.C., Connelly, J., Damment, S.J.P., Haselden, J., Nicholson, J.K., 1998. Development of a model for classification of toxin-induced lesions using H-1 NMR spectroscopy of urine combined with pattern recognition. NMR Biomed. 11, 235–244.

Holmes, E., Nicholson, J.K., Tranter, G., 2001. Metabonomic characterization of genetic variations in toxicological and metabolic responses using probabilistic neural networks. Chem. Res. Toxicol. 14, 182–191.

Howells, S.I., Maxwell, R.J., Griffiths, J.R., 1992. Classification of tumor H1-NMR spectra by pattern-recognition. NMR Biomed. 5, 59–64.

Howells, S.L., Maxwell, R.J., Howe, F.A., Peet, A.C., Stubbs, M., Rodrigues, L.M., Robinson, S.P., Baluch, S., Griffiths, J.R., 1993. Pattern recognition of 31P magnetic resonance spectroscopy tumour spectra obtained in vivo. NMR Biomed. 6, 237–241.

Huang, K., Andrec, M., Heald, S., Blake, P., Prestegard, J.H., 1997. Performance of a neural-network-based determination of amino acid class and secondary structure from H-1–N-15 NMR data. J. Biomol. NMR 10, 45–52.

Hunter, C.A., Packer, M.J., 1999. Complexation-induced changes in H-1 NMR chemical shift for supramolecular structure determination. Chem-Eur. J. 5, 1891–1897.

Iles, R.F., Hint, A.J., Chalmers, R.A., 1984. Clin. Chem. 30, 426–432.

Isu, Y., Nagashima, U., Aoyama, T., Hosoya, H., 1996. Development of neural network simulator for structure-activity correlation of molecules (NECO). Prediction of endo/exo substitution of norbornane derivatives and of carcinogenic activity of PAHs from C-13-NMR shifts. J. Chem. Inf. Comp. Sci. 36, 286–293.

Ivanciuc, O., 1995. Artificial neural networks applications. 6. Use of non-bonded van der Waals and electrostatic intramolecular energies in the estimation of C-13-NMR chemical shifts in saturated hydrocarbons. Rev. Roum. Chim. 40, 1093–1101.

Ivanciuc, O., Rabine, J.P., CabrolBass, D., Panaye, A., Doucet, J.P., 1996. C-13 NMR chemical shift prediction of sp(2) carbon atoms in acyclic alkenes using neural networks. J. Chem. Inf. Comp. Sci. 36, 644–653.

Iwadate, M., Asakura, T., Williamson, M.P., 1998. The structure of the melittin tetramer at different temperatures—an NOE-based calculation with chemical shift refinement. Eur. J. Biochem. 257, 479–487.

Kaartinen, J., Hiltunen, Y., Kovanen, P.T., Ala-Korpela, M., 1998. Application of self-organizing maps for the detection and classification of human blood plasma lipoprotein lipid profiles on the basis of H-1 NMR spectroscopy data. NMR Biomed. 11, 168–176.

Kalelkar, S., Dow, E.R., Grimes, J., Clapham, M., Hu, H., 2002. Automated analysis of proton NMR spectra from combinatorial rapid parallel synthesis using self-organizing maps. J. Comb. Chem. 4, 622–629.

Kari, S., Olsen, N.J., Park, J.H., 1995. Evaluation of muscle diseases using artificial neural network analysis of 31P MR spectroscopy data. Magn. Reson. Med. 34, 664–672.

Kjaer, M., Poulsen, F.M., 1991. Identification of 2D H-1-NMR antiphase cross peaks using a neural network. J. Magn. Reson. 94, 659–663.

Kohonen, T., 1995. Self-Organizing Maps, Springer, Heidelberg.

Koza, J.R., 1993. Genetic Programming, MIT Press, Cambridge, MA.

Kurhanewicz, J., Vigneron, D.B., Nelson, S.J., Hricak, H., MacDonald, J.M., Konety, B., Narayan, P., 1995. Citrate as an in vivo marker to discriminate prostate cancer from benign prostatic hyperplasia and normal prostate peripheral zone: detection via localized proton spectroscopy. Urology 45, 459–466.

Kvasnicka, V., 1991. An application of neural networks in chemistry—prediction of c-13 NMR chemical-shifts. J. Math. Chem. 6, 63–76.

Kvasnicka, V., Sklenak, S., Pospichal, J., 1992. Application of neural networks with feedback connections in chemistry—prediction of C-13 NMR chemical-shifts in a series of monosubstituted benzenes. Theochem.-J. Mol. Struct. 96, 87–107.

Le Bret, C., 2000. A general C-13 NMR spectrum predictor using data mining techniques, Sar and Qsar in Environmental Research, Overseas Publishers Association N.V., Amsterdam, pp. 211–234.

Li, K.B., Sanctuary, B.C., 1997. Automated resonance assignment of proteins using heteronuclear 3d NMR. Backbone spin systems extraction and creation of polypeptides. J. Chem. Inf. Comp. Sci. 37, 359–366.

Lindel, T., Junker, J., Kock, M., 1997. COCON: from NMR correlation data to molecular constitutions. J. Mol. Model 3, 364–368.

Lindemann, L.P., Adams, J.Q., 1971. Carbon-13 nuclear magnetic resonance spectrometry. Chemical shifts for the paraffins through C9. Anal. Chem. 43, 1245–1252.

Lisboa, P.J.G., Kirby, S.P.J., Vellido, A., Lee, Y.Y.B., El-Deredy, W., 1998. Assessment of statistical and neural networks methods in NMR spectral classification and metabolite selection. NMR Biomed. 11, 225–234.

Mackinnon, W.B., Barry, P.A., Malycha, P.L., 1997. Fine-needle biopsy specimens of benign breast lesions distinguished from invasive cancer ex vivo with proton MR spectroscopy. Radiology 204, 661–666.

Maxwell, R.J., Martinez-Perez, I., Cerdan, S., Cabanas, M.E., Arus, C., Moreno, A., Capdevila, A., Ferrer, E., Bartomeus, F., Aparicio, A., et al., 1998. Pattern recognition analysis of H-1 NMR spectra from perchloric acid extracts of human brain tumor biopsies. Magn. Reson. Med. 39, 869–877.

Meiler, J., 1998. Untersuchung von Struktur-Eigenschafts-Beziehungen für die Spezifität von Serin-Proteasen gegenüber Polypeptiden mittels NMR-Spektroskopie und Neuronaler Netze, Diploma, University of Leipzig, Leipzig.

Meiler, J., Meusinger, R., 1996. Use of neural networks to determine properties of alkanes from their 13C-NMR spectra. In: Gasteiger, J., (Ed.), Software Development in Chemistry, Gesellschaft Deutscher Chemiker, Frankfurt, pp. 259–263.

Meiler, J., Will, M., 2001. Automated structure elucidation of organic molecules from C-13 NMR spectra using genetic algorithms and neural networks. J. Chem. Inf. Comp. Sci. 41, 1535–1546.

Meiler, J., Will, M., 2002. Genius: a genetic algorithm for automated structure elucidation from C-13 NMR spectra. J. Am. Chem. Soc. 124, 1868–1870.

Meiler, J., Meusinger, R., Will, M., 2000. Fast determination of C-13 NMR chemical shifts using artificial neural networks. J. Chem. Inf. Comp. Sci. 40, 1169–1176.

Meiler, J., Sanli, E., Junker, J., Meusinger, R., Lindel, T., Will, M., Maier, W., Kock, M., 2002. Validation of structural proposals by substructure analysis and C-13 NMR chemical shift prediction. J. Chem. Inf. Comp. Sci. 42, 241–248.

Meusinger, R., 1996. Gasoline analysis by H-1 nuclear magnetic resonance spectroscopy. Fuel 75, 1235–1243.

Meusinger, R., Moros, R. (Eds.), 1996. Application of Genetic Algorithms and Neural Networks in Analysis of Multicomponent Mixtures by NMR-Spectroscopy, Springer, Frankfurt.

Meusinger, R., Moros, R., 2001. Determination of octane numbers of gasoline compounds from their chemical structure by C-13 NMR spectroscopy and neural networks. Fuel 80, 613–621.

Meusinger, R., Fischer, G., Moros, R., 1999. The calculation of sensitometric properties of 1,2,4-triazolo[1,5-a] pyrimidines by use of a neural network. J. Prak. Chem.-Chem. Ztg. 341, 449–454.

Meyer, B., Hansen, T., Nute, D., Albersheim, P., Darvill, A., York, W., Sellers, J., 1991. Identification of the H-1-NMR spectra of complex oligosaccharides with artificial neural networks. Science 251, 542–544.

Michon, L., Hanquet, B., Diawara, B., Martin, D., Planche, J.P., 1997. Asphalt study by neuronal networks. Correlation between chemical and rheological properties. Energy Fuels 11, 1188–1193.

Mihalic, Z., Nikolic, S., Trinajstic, N., 1992. Comparative-study of molecular descriptors derived from the distance matrix. J. Chem. Inf. Comp. Sci. 32, 28–37.

Mountford, C.E., Somorjai, R.L., Malycha, P., Gluch, L., Lean, C., Russell, P., Barraclough, B., Gillett, D., Himmelreich, U., Dolenko, B., Nikulin, A.E., Smith, I.C.P., 2001. Diagnosis and prognosis of breast cancer by magnetic resonance spectroscopy of fine-needle aspirates analysed using a statistical classification strategy. Br. J. Surg. 88, 1234–1240.

Muehlenbein, H., Schlierkamp-Voosen, D., 1993. Predictive models for the breeder genetic algorithm I. Evol. Comput. 1, 25–49.

Munk, M.E., Madison, M.S., Robb, E.W., 1996. The neural network as a tool for multispectral interpretation. J. Chem. Inf. Comp. Sci. 36, 231–238.

Negendank, W., 1992. Studies of human tumors by MRS: a review. NMR Biomed. 5, 303–324.

Nicholson, J.K., Lindon, J.C., Holmes, E., 1999. 'Metabo-nomics': understanding the metabolic responses of living systems to pathophysiological stimuli via multivariate statistical analysis of biological NMR spectroscopic data. Xenobiotica 11, 1181–1189.

Nikulin, A.E., Dolenka, B., Bezabeh, T., Somorjai, C.L., 1998. Near-optimal region selection for feature space reduction: novel preprocessing methods for classifying MR spectra. NMR Biomed. 11, 209–217.

Ott, K.H., Aranibar, N., Singh, B.K., Stockton, G.W., 2003. Metabonomics classifies pathways affected by bioactive compounds. Artificial neural network classification of NMR spectra of plant extracts. Phytochemistry 62, 971–985.

Panaye, A., Doucet, J.P., Fan, B.T., Feuilleaubois, E., Elazzouzi, S.R., 1994. Artificial neural-network simulation of C-13 NMR shifts for methyl-substituted cyclohexanes. Chemom. Intell. Lab. 24, 129–135.

Pearlman, D.A., 1996. FINGAR: a new genetic algorithm-based method for fitting NMR data. J. Biomol. NMR 8, 49–66.

Pearlman, D.A., 1999. Automated detection of problem restraints in NMR data sets using the FINGAR genetic algorithm method. J. Biomol. NMR 13, 325–335.

Pons, J.L., Delsuc, M.A., 1999. RESCUE: an artificial neural network tool for the NMR spectral assignment of proteins. J. Biomol. NMR 15, 15–26.

Poptani, H., 1999. Diagnostic assessment of brain tumors and non-neoplastic brain disorders in vivo using proton nuclear magnetic resonance spectroscopy and artificial neural networks. J. Cancer Res. Clin. Oncol. 125, 343–349.

Preul, M.C., Caramanos, Z., Collins, D.L., 1996. Accurate, noninvasive diagnosis of human brain tumors by using proton magnetic resonance spectroscopy. Nat. Med. 2, 323–325.

Robien, W., 2000. In: Robien, W., (Ed.), CSEARCH-NMR, University of Vienna, Austria.

Ruan, R.R., (1998). www.bae.umn.edu/annrpt/1998/research/food11.html

Russell, P., Lean, C.L., Delbridge, L., May, G.L., Dowd, S., Mountford, C.E., 1994. Proton magnetic resonance and human thyroid neoplasia. I. Discrimination between benign and malignant neoplasms. Am. J. Med. 96, 383–388.

Rutter, A., Hugenholtz, H., Saunders, J.K., Smith, I.C.P., 1995. Classification of brain tumors by ex vivo 1 H NMR spectroscopy. J. Neurochem. 64, 1655–1661.

Schaper, K.J., Pickert, M., Frahm, A.W., 1999. Substituted xanthones as antimycobacterial agents Part 3: QSAR investigations. Arch. Pharm. 332, 91–102.

Sharma, A.K., Sheikh, S., Pelczer, I., Levy, G.C., 1994. Classification and clustering—using neural networks. J. Chem. Inf. Comp. Sci. 34, 1130–1139.

Somorjai, R.L., Nikulin, A.E., Pizzi, N., Jackson, D., Scarth, G., Dolenko, B., Gordon, H., Russell, P., Lean, C.L., Delbridge, L., Mountford, C.E., Smith, I.C.P., 1995. Computerized consensus diagnosis—a classification strategy for the robust analysis of MR Spectra. 1. Application to H-1 spectra of thyroid neoplasms. Magn. Reson. Med. 33, 257–263.

Somorjai, R.L., Dolenka, B., Nikulin, A.E., Pizzi, N., Scarth, G., Zhilkin, P., Halliday, W., Fewer, D., Hill, N., Ross, I., West, M., Smith, I.C., Donnelly, S.M., Kuesel, A.C., Briere, K.M., 1996. Classification of 1H MR spectra of human brain biopsies: the influence of preprocessing and computerized consensus diagnosis on classification accuracy. J. Magn. Reson. Imaging 6, 437–444.

Somorjai, R.L., Dolenko, B., Nikulin, A., Nickerson, P., Rush, D., Shaw, A., Glogowski, M., Rendell, J., Deslauriers, R., 2002. Distinguishing normal from rejecting renal allografts: application of a three-stage classification strategy to MR and IR spectra of urine. Vibr. Spectrosc. 28, 97–102.

Soper, R., Himmelreich, U., Painter, D., Somorjai, R.L., Lean, C.L., Dolenko, B., Mountford, C.E., Russell, P., 2002. Pathology of hepatocellular carcinoma and its precursors using proton magnetic resonance spectroscopy and a statistical classification strategy. Pathology 34, 417–422.

SpecInfo, 2000. SpecInfo Database (Chemical Concepts), Wiley-VCH, Weinheim.

Stephenson, D.S., Binsch, G., 1980. Automated analysis of high-resolution NMR spectra. I. Principles and computational strategy. J. Magn. Reson. 37, 395–407.

Svozil, D., Pospichal, J., Kvasnicka, V., 1995. Neural-network prediction of C-13 NMR chemical-shifts of alkanes. J. Chem. Inf. Comp. Sci. 35, 924–928.

Tate, A.R., Howells, S.L., 1998. Pattern recognition analysis (Editorial). NMR Biomed. 11, 147.

Thomsen, J.U., Meyer, B., 1989. Pattern-Recognition of the H-1-NMR spectra of sugar alditols using a neural network. J. Magn. Reson. 84, 212–217.

Usenius, J.P., Kauppinen, R.A., Vinio, P.A., Hernesniemi, J.A., Vapalahti, M.P., Paljarvi, L.A., Soimakallio, S., 1994. Quantitative metabolite patterns of human brain tumors: detection by 1 H NMR spectroscopy in vivo and in vitro. J. Comput. Assist. Tomogr. 18, 705–713.

Usenius, J.P., Tuohimetsa, S., Vainio, P., AlaKorpela, M., Hiltunen, Y., Kauppinen, R.A., 1996. Automated classification of human brain tumours by neural network analysis using in vivo H-1 magnetic resonance spectroscopic metabolite phenotypes. Neuroreport 7, 1597–1600.

Vaananen, T., Koskela, H., Hiltunen, Y., Ala-Korpela, M., 2002. Application of quantitative artificial neural network analysis to 2D NMR spectra of hydrocarbon mixtures. J. Chem. Inf. Comp. Sci. 42, 1343–1346.

Valafar, H., Valafar, F., 2002. Data mining and knowledge discovery in proton nuclear magnetic resonance (H-1-NMR) spectra using frequency to information transformation (FIT). Knowl-Based Syst. 15, 251–259.

Webb, S., Collins, D.J., Leach, M.O., 1992. Quantitative magnetic resonance spectroscopy by optimized numerical curve fitting. NMR Biomed. 5, 87–94.

Weber, O.M., Duc, C.O., Meier, D., Boesiger, P., 1998. Heuristic optimization algorithms applied to the quantification of spectroscopic data. Magn. Reson. Med. 39, 723–730.

West, G.M.J., 1993. Predicting phosphorus NMR shifts using neural networks. J. Chem. Inf. Comp. Sci. 33, 577–589.

Williamson, M.P., Asakura, T., 1993. Empirical comparisons of models for chemical-shift calculation in proteins. J. Magn. Reson. Ser. B. 101, 63–71.

Wu, X.L., Xu, P., Freeman, R., 1991. Delayed-focus pulses for magnetic resonance imaging: an evolutionary approach. Magn. Reson. Med. 20, 165–170.

CHAPTER 11

A QSAR model for predicting the acute toxicity of pesticides to Gammarids

James Devillers

CTIS, 3 Chemin de la Gravière, 69140 Rillieux La Pape, France

1. Introduction

The knowledge about systematic relationships between the structure and activities of the organic compounds dates back to the early days of the modern toxicology and pharmacology. Thus, Cros (1863) stressed that within a homologous series of alcohols, toxicity increased with the increasing number of carbon and hydrogen atoms and decreasing water solubility of the molecules. Rabuteau (1870) postulated that the toxicity of the $C_nH_{2n+2}O$ alcohols increased with their number of CH_2. Five years later, Dujardin-Beaumetz and Audigé (1875) indicated that the toxicity of the alcohols mathematically followed their atomic composition and Overton (1901) derived more formal relationships between oil–water partition coefficients of molecules and their narcotic activity on tadpoles. However, most of these structure–activity relationships were purely qualitative. The dramatic change resulted from the systematic use, in 1962–1964, of linear regression analyses for correlating biological activities of congeneric series of molecules with their physicochemical properties (Hansch et al., 1962; Hansch and Fujita, 1964) or some of their structural features encoded by means of Boolean descriptors (Free and Wilson, 1964). These contributions started the development of two QSAR methodologies later termed Hansch analysis (or linear free energy-related approach, extrathermodynamic approach (Kubinyi, 1993a)) and Free-Wilson analysis. It is noteworthy that because the Hansch and Free-Wilson analyses are closely related, a mixed approach was rapidly proposed for modeling larger and structurally diverse data sets (e.g. see numerous examples in Hansch and Leo, 1995).

A considerable improvement in the design of these linear models was made by the use of the Partial Least Squares (PLS) regression analysis (Geladi and Tosato, 1990; Devillers et al., 2002a) now popularized in 3D QSAR with the Comparative Molecular Field Analysis (CoMFA) approach (Kubinyi, 1993b).

Last, in the early 1990s (Aoyama et al., 1990), a new important stage was reached by the introduction of artificial neural networks (ANNs) in QSAR and especially the three-layer

Data Handling in Science and Technology, Volume 23
ISSN: 0922-3487

DOI: 10.1016/S0922-3487(03)23011-2

feedforward neural network trained by the back-propagation algorithm (Rumelhart et al., 1986) which allowed to derive complex relationships between the structure of the molecules and their activity. This supervised neural network, being fault tolerant and able to handle noisy and incomplete data, is now increasingly used in environmental QSAR (Devillers et al., 1995, 2002b; Kaiser et al., 1997; Zakarya et al., 1997; Devillers, 2001a).

Exploiting the high flexibility of this nonlinear statistical device, we have recently designed a new type of QSAR model integrating molecular descriptors and variables influencing the ecotoxicological behaviour of chemicals such as biotic factors and test conditions (Devillers, 2000, 2001b; Devillers and Flatin, 2000).

Because the ecotoxicity of pesticides to Gammarids (*Gammarus fasciatus*) is widely affected by the life stage and size of the organisms and the experimental conditions in which the tests are performed (i.e. temperature, pH, hardness, time of exposure, etc.), the aim of the present study was to compare the respective ability of PLS regression analysis and ANN to derive a QSAR model integrating both types of variables.

2. Materials and methods

2.1. Toxicity data

The acute toxicity data and the experimental conditions, in which they were obtained, were retrieved from literature (Mayer and Ellersieck, 1986). Pesticides with missing Chemical Abstracts Service Registry Number (CAS RN) or dubious names were eliminated. The LC_{50} (lethal concentration 50%) values obtained on *G. fasciatus* after 24 h and/or 96 h of exposure and under static conditions were recorded and confronted with the solubility in water of the different selected pesticides (Shiu et al., 1990; Montgomery, 1993; Tomlin, 1994) in order to discard the unrealistic toxicity results. In the same way, results with imprecise values (i.e. $<$, $>$) were eliminated. This strategy resulted in the selection of 130 LC_{50} values for 51 pesticides presenting various structures and mechanisms of toxicity (Table 1). The LC_{50} values presented by Mayer and Ellersieck (1986) were in units of μg/l or mg/l. For modeling purposes, these toxicity data were translated into log ($1/LC_{50}$ in mmol/l). For the same reasons, the database was randomly split into a training set of 112 toxicity results and a testing set of 18 LC_{50} values.

2.2. Molecular descriptors

In order to account for all the structural characteristics of the pesticides, the autocorrelation method (Moreau and Broto, 1980a,b; Broto and Devillers, 1990) was used for describing the molecules. Briefly, the autocorrelation descriptors are simple 2D molecular descriptors designed from the hydrogen-suppressed graphs of the molecules. Autocorrelation vectors can be derived for all physicochemical properties which can be calculated from atomic contributions. They consist of autocorrelation components corresponding to the different interatomic distances, which can be computed within the studied molecule. The autocorrelation vectors are then truncated to obtain strings of descriptors of same dimensionality with a reduced number of null values. In the classical

Table 1
Pesticides with their CAS RN, experimental conditions, and observed toxicity values

No.	Pesticide	St.	T (°C)	pH	H	TEh	log (1/C)obs.
	Anilazine [101-05-3][a]						
1		M	15	7.1	44	24	2.34
2		M	15	7.1	44	96	3.01
	Bensulide [741-58-2]						
3		M	15	7.1	44	24	1.72
4		M	15	7.1	44	96	2.45
	Butylate [2008-41-5]						
5		M	15	7.4	262	24	0.92
6[b]		M	15	7.4	262	96	1.30
7		M	15	7.4	44	24	1.11
8		M	15	7.4	44	96	1.26
	Carbaryl [63-25-2]						
9		M	21	7.1	44	24	3.60
10[b]		M	21	7.1	44	96	3.89
	Carbophenothion [786-19-6]						
11		M	21	7.1	40	24	3.88
12		M	21	7.1	40	96	4.82
	Chlordane [57-74-9]						
13		M	21	7.1	44	96	4.01
	Cyanazine [21725-46-2]						
14		M	15	7.4	272	96	2.08
	DDD [72-54-8]						
15[b]		M	21	7.1	44	24	4.84
16		M	21	7.1	44	96	5.73
17		M	21	7.4	272	24	5.00
18[b]		M	21	7.4	272	96	5.55
	DDT [50-29-3]						
19		M	21	7.1	44	96	5.04
20		M	21	7.4	272	24	4.93
21[b]		M	21	7.4	272	96	5.29
	DEF [78-48-8]						
22		M	21	7.1	44	24	3.14
23		M	21	7.1	44	96	3.50
	Dichlofenthion [97-17-6]						
24[b]		M	15	7.1	44	96	3.46
25		M	15	7.4	272	96	3.08
	Dioxathion [78-34-2]						
26[b]		M	15	7.1	44	24	4.28
27		M	15	7.1	44	96	4.72
28		M	15	7.4	272	24	3.91
29		M	15	7.4	272	96	4.46
	Diphenamid [957-51-7]						
30		M	15	7.4	272	96	0.38
	Disulfoton [298-04-4]						
31		M	15	7.1	44	24	3.40
32		M	15	7.1	44	96	3.72
33[b]		M	15	7.4	272	24	3.44
34		M	15	7.4	272	96	4.12

(*continued on next page*)

Table 1 (*continued*)

No.	Pesticide	St.	T (°C)	pH	H	TEh	log (1/C)obs.
	Diuron [330-54-1]						
35		M	21	7.1	44	24	2.52
36		M	21	7.1	44	96	3.16
	DNOC [534-52-1]						
37		M	21	7.1	44	24	1.92
38		M	21	7.1	44	96	2.26
	Endosulfan [115-29-7]						
39		M	21	7.1	44	24	4.61
40		M	21	7.1	44	96	4.83
	Endrin [72-20-8]						
41[b]		M	21	7.1	44	24	4.33
42		M	21	7.1	44	96	4.95
43		M	15	7.4	272	24	4.58
44		M	15	7.4	272	96	5.47
	EPN [2104-64-5]						
45		M	15	7.1	44	24	3.83
46[b]		M	15	7.1	44	96	4.68
	EPTC [759-94-4]						
47		M	15	7.4	272	96	0.46
	Ethion [563-12-2]						
48		M	21	7.1	44	24	4.84
49		M	21	7.1	44	96	5.33
	Fenitrothion [122-14-5]						
50		M	15	7.4	272	24	3.64
51		M	15	7.4	272	96	4.97
	Fenthion [55-38-9]						
52		M	17	7.1	44	24	3.10
53[b]		M	17	7.1	44	96	3.40
	Heptachlor [76-44-8]						
54		M	15	7.1	44	96	3.82
55		M	21	7.4	272	96	3.97
	Malathion [121-75-5]						
56		M	21	7.2	44	24	4.94
57		M	21	7.2	44	96	5.64
58		M	21	7.4	272	24	5.01
59[b]		M	21	7.4	272	96	5.56
	Methoxychlor [72-43-5]						
60		M	15	7.1	44	24	4.79
61		M	15	7.1	44	96	5.26
	Molinate [2212-67-1]						
62		M	21	7.4	40	24	1.28
63		M	21	7.4	40	96	1.62
	Monocrotophos [6923-22-4]						
64		M	15	7.1	44	96	2.93
	Norea [2163-79-3]						
65		M	16	7.1	44	24	1.89
66		M	16	7.1	44	96	2.21
	Parathion [56-38-2]						
67		M	21	7.1	44	24	4.96
68		M	21	7.1	44	96	5.35

Table 1 (*continued*)

No.	Pesticide	St.	T (°C)	pH	H	TEh	log (1/C)obs.
69		M	21	7.4	272	24	4.69
70		M	21	7.4	272	96	5.14
	Phorate [298-02-2]						
71		M	15	7.4	44	24	4.04
72[b]		M	15	7.4	44	96	4.81
73		M	15	7.4	272	24	4.81
74		M	15	7.4	272	96	5.64
	Phosmet [732-11-6]						
75		M	15	7.4	44	24	4.80
76		M	15	7.4	44	96	5.20
77		M	15	7.4	272	24	4.79
78		M	15	7.4	272	96	4.88
	Picloram [1918-02-1]						
79		M	21	7.1	44	24	0.68
80		M	21	7.1	44	96	0.95
	Propanil [709-98-8]						
81		M	15	7.4	272	24	0.61
82		M	15	7.4	272	96	1.13
	Propham [122-42-9]						
83		M	15	7.4	272	24	0.52
84		M	15	7.4	272	96	0.97
	Aramite [140-57-8]						
85		M	21	7.1	44	96	3.75
	Azinphos-methyl [86-50-0]						
86		M	21	7.1	44	24	5.75
87		M	21	7.1	44	96	6.33
88		M	15	7.4	272	24	5.86
89[b]		M	15	7.4	272	96	6.50
	Bufencarb [2282-34-0]						
90		M	15	7.4	44	24	4.62
91[b]		M	15	7.4	44	96	5.34
92		M	15	7.4	272	24	4.84
93		M	15	7.4	272	96	5.34
	Chlorfenvinphos [470-90-6]						
94		M	21	7.4	44	24	4.12
95		M	21	7.4	44	96	4.57
	Crufomate [299-86-5]						
96		I	15	7.4	262	24	1.72
97		I	15	7.4	262	96	1.90
98		I	15	7.4	44	24	3.08
	Diazinon [333-41-5]						
99		M	21	7.1	44	24	4.58
100		M	21	7.1	44	96	6.18
	Dicrotophos [141-66-2]						
101		M	21	7.1	44	24	1.63
102		M	21	7.1	44	96	1.96
	Fensulfothion [115-90-2]						
103		M	15	7.1	44	24	3.82
104		M	15	7.1	44	96	4.49

(*continued on next page*)

Table 1 (*continued*)

No.	Pesticide	St.	*T* (°C)	pH	*H*	TEh	log (1/*C*)obs.
	Lindane [58-89-9]						
105		I	15	7.1	44	24	3.96
106[b]		I	15	7.1	44	96	4.46
107		I	15	7.4	272	24	4.02
108		I	15	7.4	272	96	4.42
	Methyl parathion [298-00-0]						
109		M	15	7.1	44	24	4.42
110		M	15	7.1	44	96	4.84
111		M	15	7.4	272	24	4.22
112		M	15	7.4	272	96	4.79
	Methyl trithion [953-17-3]						
113		M	21	7.4	44	24	3.80
114		M	21	7.4	44	96	4.46
	Mexacarbate [315-18-4]						
115		M	21	7.1	44	24	3.44
116		M	21	7.1	44	96	3.74
	Naled [300-76-5]						
117		M	15	7.4	272	24	4.08
118		M	15	7.4	272	96	4.33
	Pebulate [1114-71-2]						
119		I	15	7.4	272	24	1.00
120		I	15	7.4	272	96	1.31
	Phosphamidon [13171-21-6]						
121		M	21	7.4	44	24	4.57
122		M	21	7.4	44	96	5.00
123		M	15	7.4	44	24	3.57
124		M	15	7.4	44	96	4.36
125[b]		M	15	7.4	272	24	3.67
126		M	15	7.4	272	96	4.08
	Vernolate [1929-77-7]						
127		I	15	7.4	44	24	0.93
128[b]		I	15	7.4	44	96	1.16
129		I	15	7.4	272	24	0.93
130		I	15	7.4	272	96	1.05

St., stage (mature (M) or immature (I)); *T* (°C), temperature in °C; *H*, water hardness in mg/l as $CaCO_3$; TEh, time of exposure in hours.

[a] CAS RN, Chemical Abstracts Service Registry Number.

[b] Testing set.

algorithm proposed by Moreau and Broto (1980a,b), the different autocorrelation components for a property are obtained by summations of products. This can be damaging when negative atomic contributions have to be used such as for encoding the lipophilicity of some functional groups. Indeed, in that case the physicochemical meaning of the autocorrelation components is not straightforward. To overcome this problem, a slightly different algorithm was used (Devillers et al., 1992). More information on the original and modified autocorrelation algorithms can be found in a recent publication (Devillers, 1999).

On the basis of our previous QSAR studies based on the same methodology and database (Devillers and Flatin, 2000; Devillers, 2001b), the 51 pesticides under study were described by means of three different autocorrelation vectors. From the fragmental constants of Rekker and Mannhold (1992), autocorrelation vectors *H* representing lipophilicity were derived. A truncation was performed in order to obtain six autocorrelation components (distances 0–5 in the molecular graphs). Autocorrelation vectors encoding the H-bonding acceptor ability (HBA) and H-bonding donor ability (HBD) of the pesticides were also calculated by means of Boolean contributions (i.e. 0/1). For these two vectors, only the first component was kept. The autocorrelation vectors were calculated by means of AUTOCOR™ 2.4 using SMILES notation as inputs.

2.3. Statistical analyses

The PLS-QSAR model was derived from STATISTICA™ by using the NIPALS (Nonlinear estimation by Iterative Partial Least Squares) algorithm with the option no-intercept and autoscale.

In our previous studies (Devillers and Flatin, 2000; Devillers, 2001b), a classical three-layer feedforward neural network trained by the back-propagation algorithm (Rumelhart et al., 1986) was used as statistical tool to find nonlinear relationships between the autocorrelation descriptors and experimental variables and the acute toxicity data. Even if

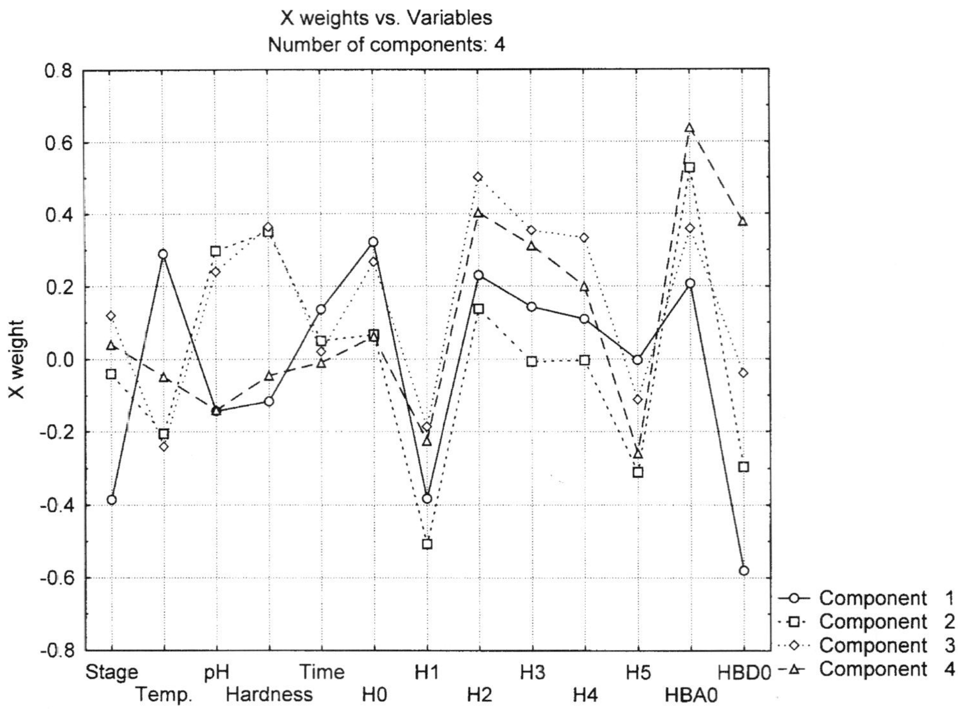

Fig. 1. Weights for the predictor variables of the PLS model at four components.

the use of the back-propagation algorithm yields good modeling results, it is well known that it suffers from some limitations (Devillers, 1996). One of them is a slow terminal convergence. To overcome this problem, in the present study, two different algorithms were simultaneously employed during the training phases. Thus, during a short period (no more than 100 cycles) the back-propagation algorithm (Rumelhart et al., 1986) was used and, in a second phase, no more than 500 cycles, the conjugate gradient descent algorithm (Wasserman, 1993) was employed. The preprocessing step of the data consisted of a classical min/max transformation. The three-layer neural network model was derived from STATISTICA™.

It is noteworthy that only the descriptor encoding the stage of the organisms (mature or immature) was treated as categorical variable in the PLS and ANN analyses performed with STATISTICA™.

3. Results and discussion

3.1. PLS model

The five experimental variables and eight autocorrelation descriptors (H_0–H_5, HBA_0, HBD_0) yielded the design of a PLS model at four significant components ($r^2 = 0.581$).

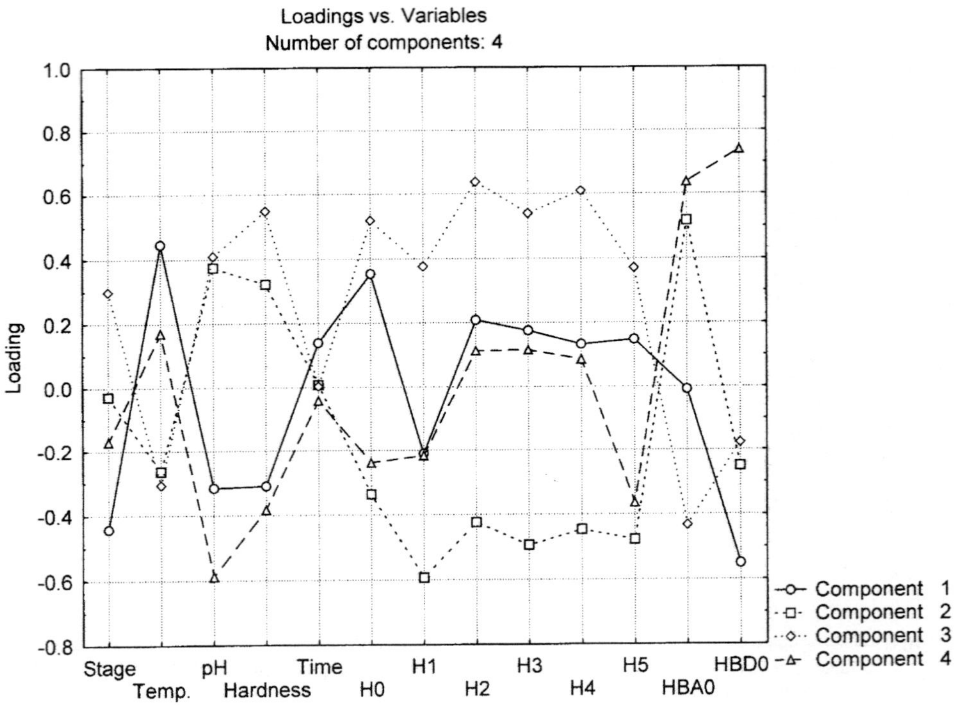

Fig. 2. Loadings for the four components of the PLS model.

The graphical analysis of the contribution of each predictor to the respective components reveals (Fig. 1) that stage, H_0, H_1, and HBD_0 are important contributors to component 1. Hardness, H_1, H_5, and HBA_0 are the most important contributors to component 2. Hardness and HBA_0 also significantly contribute to component 3 with H_2, H_3, and H_4. Last, H_2, H_3, HBA_0, and HBD_0 are the highest contributors for component 4. In the same way, Fig. 2 shows the influential variables for the selected QSAR model at four components. If we consider the unscaled regression coefficients, which have to be used to compute predicted values for new data, we can underline that all the variables present a positive sign except stage, pH, H_1, H_5, and HBD_0 (Fig. 3). Obviously, the graphical display of the scaled coefficients shows the same trend (Fig. 4) and allows to confirm that all the variables are necessary in the model. Neither the opposition between HBA_0 and HBD_0 nor the negative sign of H_5 is surprising (Figs. 3 and 4). Conversely, this between H_0 and H_1 seems to be less legitimate if we consider their significance in term of distance in the molecular graphs. It is worth noting that the removal of one of them decreased the performances of the PLS model. The calculated LC_{50} values and corresponding residuals are given in Table 2. The observed versus calculated toxicity values are also displayed in Fig. 5. The correlation between both types of data is fairly limited ($r = 0.738$). This is not surprising because inspection of Table 2 shows that in numerous situations, the model is not able to simulate the acute toxicity of pesticides to *G. fasciatus*.

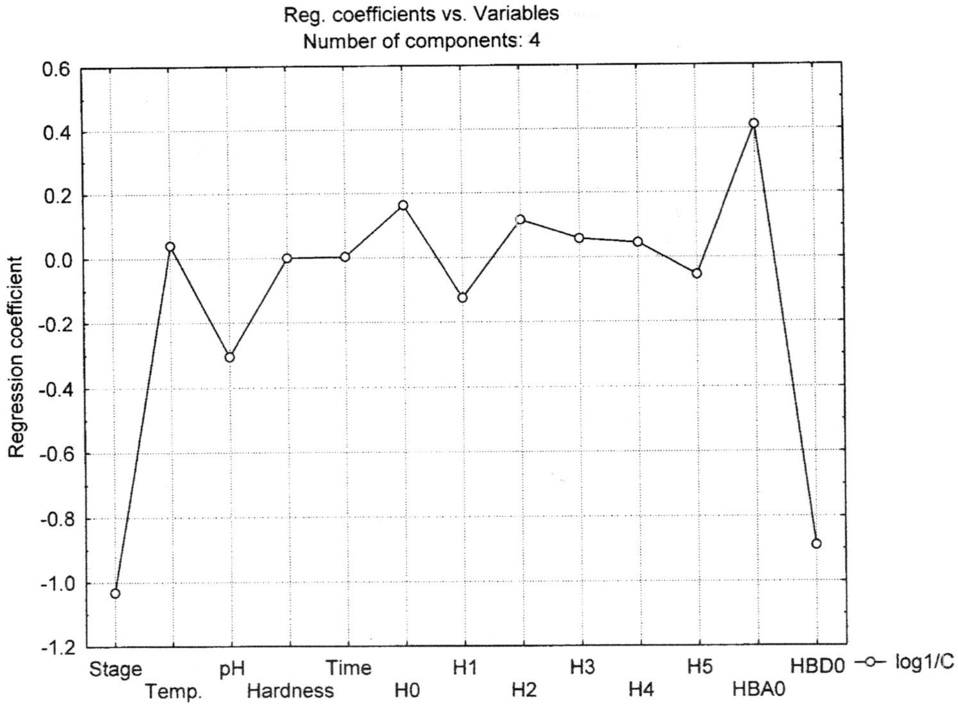

Fig. 3. PLS regression coefficients versus the different variables included in the model.

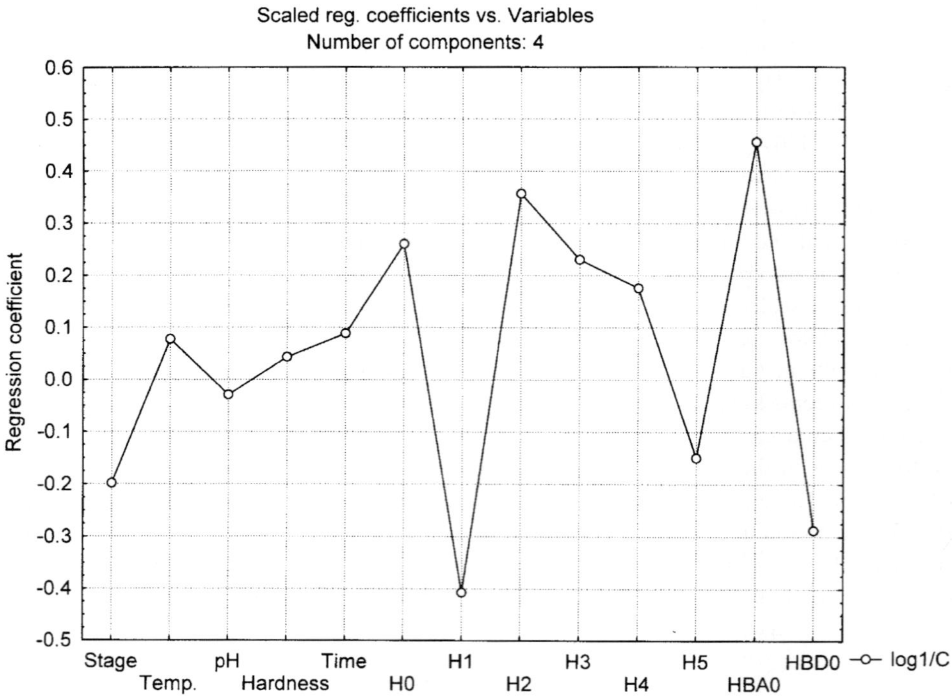

Fig. 4. Scaled regression coefficients versus the different variables included in the model.

It is important to stress that other PLS models have also been derived by changing the composition of the training and testing sets. In all cases, the former set included 112 toxicity results and a testing set of 18 LC_{50} values was used. No significant differences were found.

3.2. ANN model

Different assays were performed with the five experimental variables and eight autocorrelation descriptors (H_0–H_5, HBA_0, HBD_0) as inputs in the ANN. A 13/3/1 ANN yielded the best simulation results within 500 cycles. The correlation coefficients (r) between the observed and calculated toxicity data for both sets ranged from 0.951 to 0.967. Undoubtedly, the modeling results obtained with the ANN outperformed those previously recorded with the PLS regression analysis. However, because within this type of ANN it is always necessary to have a minimum number of connections, attempts were made to reduce the number of input neurons. Feature selection by forward and backward procedures and also with a genetic algorithm (Leardi, 1996; Putavy et al., 1996) were experienced from the STATISTICA™ package. Obviously the runs yielding the elimination of experimental variables were not considered. However, it is interesting to note that surprisingly, the time of exposure was the most proposed variable to eliminate. Among the eight autocorrelation descriptors, HBA_0 was widely

Table 2
Observed and calculated (PLS and ANN) toxicity values

No.	Obs.	PLS	Res.[a]	ANN	Res.
1	2.34	1.80	0.54	2.43	−0.09
2	3.01	2.08	0.93	2.92	0.09
3	1.72	2.41	−0.69	1.68	0.04
4	2.45	2.69	−0.24	2.15	0.30
5	0.92	2.48	−1.56	1.23	−0.31
6[b]	1.30	2.75	−1.45	1.55	−0.25
7	1.11	2.34	−1.23	1.19	−0.08
8	1.26	2.62	−1.36	1.36	−0.10
9	3.60	2.74	0.86	3.77	−0.17
10[b]	3.89	3.01	0.88	4.11	−0.22
11	3.88	4.23	−0.35	3.97	−0.09
12	4.82	4.51	0.31	4.36	0.46
13	4.01	4.98	−0.97	4.01	0.00
14	2.08	0.60	1.48	2.07	0.01
15[b]	4.84	3.90	0.94	4.77	0.07
16	5.73	4.18	1.55	4.83	0.90
17	5.00	3.95	1.05	4.77	0.23
18[b]	5.55	4.22	1.33	4.83	0.72
19	5.04	4.39	0.65	4.77	0.27
20	4.93	4.16	0.77	4.68	0.25
21[b]	5.29	4.43	0.86	4.78	0.51
22	3.14	3.45	−0.31	3.15	−0.01
23	3.50	3.73	−0.23	3.66	−0.16
24[b]	3.46	5.04	−1.58	3.35	0.11
25	3.08	5.09	−2.01	3.40	−0.32
26[b]	4.28	4.46	−0.18	3.73	0.55
27	4.72	4.74	−0.02	4.44	0.28
28	3.91	4.51	−0.60	3.77	0.14
29	4.46	4.79	−0.33	4.46	0.00
30	0.38	3.41	−3.03	1.44	−1.06
31	3.40	3.35	0.05	3.62	−0.22
32	3.72	3.62	0.10	4.13	−0.41
33[b]	3.44	3.39	0.05	3.46	−0.02
34	4.12	3.67	0.45	4.44	−0.32
35	2.52	2.32	0.20	1.58	0.94
36	3.16	2.59	0.57	1.91	1.25
37	1.92	2.56	−0.64	2.44	−0.52
38	2.26	2.83	−0.57	2.94	−0.68
39	4.61	4.93	−0.32	4.81	−0.20
40	4.83	5.20	−0.37	4.85	−0.02
41[b]	4.33	4.70	−0.39	4.57	−0.24
42	4.95	5.00	−0.05	4.71	0.24
43	4.58	4.53	0.05	4.24	0.34
44	5.47	4.80	0.67	4.51	0.96
45	3.83	4.46	−0.63	4.49	−0.66
46[b]	4.68	4.73	−0.05	4.66	0.02
47	0.46	2.17	−1.71	0.97	−0.51
48	4.84	5.41	−0.57	4.64	0.20
49	5.33	5.69	−0.36	5.56	−0.23

(continued on next page)

Table 2 (*continued*)

No.	Obs.	PLS	Res.[a]	ANN	Res.
50	3.64	4.76	−1.12	4.10	−0.46
51	4.97	5.03	−0.06	4.51	0.46
52	3.10	4.74	−1.64	3.29	−0.19
53[b]	3.40	5.02	−1.62	3.91	−0.51
54	3.82	4.51	−0.69	3.95	−0.13
55	3.97	4.80	−0.83	4.45	−0.48
56	4.94	5.09	−0.15	5.78	−0.84
57	5.64	5.36	0.28	6.06	−0.42
58	5.01	5.17	−0.16	5.39	−0.38
59[b]	5.56	5.44	0.12	5.42	0.14
60	4.79	4.39	0.40	4.29	0.50
61	5.26	4.66	0.60	4.54	0.72
62	1.28	2.06	−0.78	0.91	0.37
63	1.62	2.33	−0.71	0.94	0.68
64	2.93	2.81	0.12	3.77	−0.84
65	1.89	2.22	−0.33	1.79	0.10
66	2.21	2.49	−0.28	2.18	0.03
67	4.96	4.53	0.43	4.54	0.42
68	5.35	4.81	0.54	4.70	0.65
69	4.69	4.58	0.11	4.61	0.08
70	5.14	4.86	0.28	5.54	−0.40
71	4.04	3.26	0.78	3.94	0.10
72[b]	4.81	3.54	1.27	4.35	0.46
73	4.81	3.40	1.41	4.61	0.20
74	5.64	3.68	1.96	5.64	0.00
75	4.80	3.87	0.93	5.08	−0.28
76	5.20	4.15	1.05	5.32	−0.12
77	4.79	4.01	0.78	5.46	−0.67
78	4.88	4.28	0.60	5.46	−0.58
79	0.68	1.38	−0.70	1.23	−0.55
80	0.95	1.65	−0.70	1.41	−0.46
81	0.61	1.82	−1.21	1.16	−0.55
82	1.13	2.09	−0.96	1.31	−0.18
83	0.52	1.78	−1.26	0.88	−0.36
84	0.97	2.05	−1.08	0.89	0.08
85	3.75	3.92	−0.17	2.95	0.80
86	5.75	4.14	1.61	5.47	0.28
87	6.33	4.41	1.92	5.47	0.86
88	5.86	3.94	1.92	5.46	0.40
89[b]	6.50	4.22	2.28	5.47	1.03
90	4.62	2.57	2.05	4.89	−0.27
91[b]	5.34	2.85	2.49	4.90	0.44
92	4.84	2.71	2.13	4.08	0.76
93	5.34	2.99	2.35	4.35	0.99
94	4.12	4.02	0.10	4.67	−0.55
95	4.57	4.29	0.28	4.78	−0.21
96	1.72	3.30	−1.58	1.40	0.32
97	1.90	3.57	−1.67	2.24	−0.34
98	3.08	3.16	−0.08	3.01	0.07
99	4.58	4.16	0.42	4.69	−0.11

Table 2 (*continued*)

No.	Obs.	PLS	Res.[a]	ANN	Res.
100	6.18	4.43	1.75	5.42	0.76
101	1.63	4.00	−2.37	1.59	0.04
102	1.96	4.27	−2.31	2.22	−0.26
103	3.82	4.17	−0.35	4.09	−0.27
104	4.49	4.44	0.05	4.42	0.07
105	3.96	2.52	1.44	4.50	−0.54
106[b]	4.46	2.80	1.66	4.67	−0.21
107	4.02	2.57	1.45	3.93	0.09
108	4.42	2.84	1.58	4.29	0.13
109	4.42	4.65	−0.23	4.30	0.12
110	4.84	4.93	−0.09	4.55	0.29
111	4.22	4.70	−0.48	4.31	−0.09
112	4.79	4.98	−0.19	4.64	0.15
113	3.80	4.56	−0.76	3.93	−0.13
114	4.46	4.84	−0.38	4.43	0.03
115	3.44	3.39	0.05	4.88	−1.44
116	3.74	3.67	0.07	4.89	−1.15
117	4.08	4.25	−0.17	4.10	−0.02
118	4.33	4.53	−0.20	4.47	−0.14
119	1.00	0.95	0.05	0.88	0.12
120	1.31	1.22	0.09	0.88	0.43
121	4.57	3.77	0.80	4.37	0.20
122	5.00	4.04	0.96	5.02	−0.02
123	3.57	3.52	0.05	3.55	0.02
124	4.36	3.80	0.56	3.99	0.37
125[b]	3.67	3.66	0.01	3.03	0.64
126	4.08	3.94	0.14	3.94	0.14
127	0.93	0.96	−0.03	0.88	0.05
128[b]	1.16	1.23	−0.07	0.88	0.28
129	0.93	1.10	−0.17	0.88	0.05
130	1.05	1.37	−0.32	0.88	0.17

[a] Res. = residual = Observed − Calculated toxicity values.
[b] Testing set.

proposed to elimination, followed by H_5. These two autocorrelation descriptors were eliminated without inducing a decrease in the performances of ANNs with three neurons on the hidden layer and less than 500 cycles. The same observation was made by removing HBD_0. From a trial-and-error procedure, among the remaining autocorrelation descriptors encoding lipophilicity, it was found that the performance of the ANN models remained acceptable by also removing H_2. Thus, a neural network with stage, temperature, pH, hardness, time of exposure, H_0, H_1, H_3, and H_4 as inputs and three neurons on the hidden layer was optimized on the basis of 70 runs by changing the type of transfer functions and the values of learning rate and momentum (for the back-propagation algorithm). The best configuration was a 9/3/1 ANN obtained after 245 cycles (100 cycles for the back-propagation algorithm + 145 cycles for the conjugate gradient descent algorithm). The learning rate and momentum of the

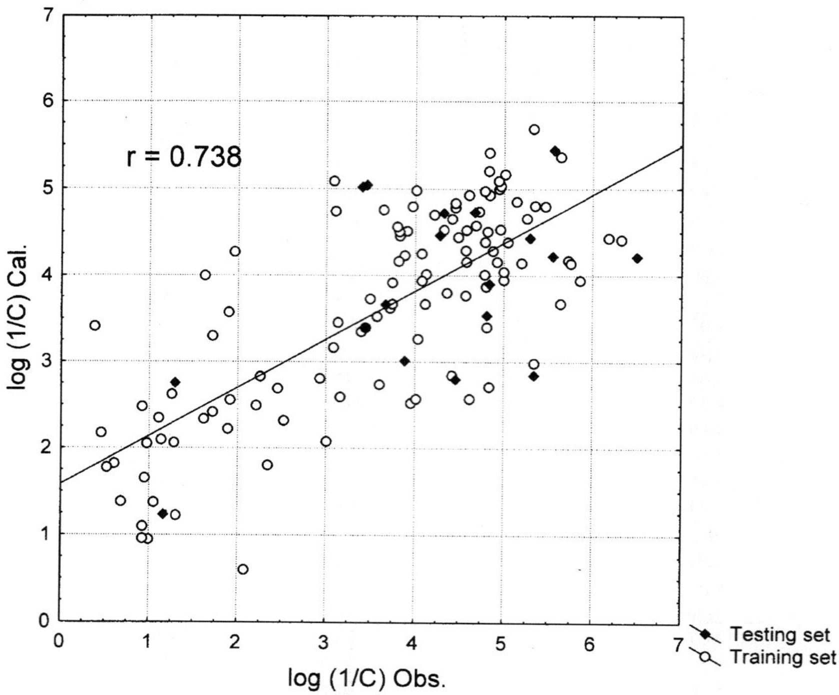

Fig. 5. Observed versus calculated LC_{50} values (PLS model).

back-propagation algorithm were 0.01 and 0.3, respectively. The synaptic functions for the three layers were all linear and the activation functions were linear, hyperbolic, and logistic from the input layer to the output layer. The learning and testing errors were equal to 0.0695 and 0.0676, respectively. The above very limited number of cycles and reduced error values guarantee that the selected ANN does not suffer from overtraining and overfitting. The calculated LC_{50} values obtained with this ANN and corresponding residuals are given in Table 2. Plot of the observed versus calculated acute toxicity values (Fig. 6) yields a correlation coefficient (r) of 0.955.

Table 3 shows the distribution of the residual values obtained with the ANN and PLS models when 11 classes are considered. Undoubtedly, the simulation performances of the former model outperform those of the latter. However, toxicity of mexacarbate (4-dimethylamino-3,5-xylyl methylcarbamate) tested at 24 and 96 h (numbers 115 and 116) is badly predicted by the ANN while the LC_{50} values calculated by the PLS model are very close to the experimental ones (Tables 1 and 2). Surprisingly, inspection of the other ANN configurations yields the same results. Acute toxicity of diuron (3-(3,4-dichlorophenyl)-1,1-dimethylurea) is underestimated by the ANN model while it is correctly simulated from the PLS model (numbers 35 and 36 in Tables 1 and 2). However, other ANN configurations provide better calculated LC_{50} values for this pesticide. Furthermore, this organophosphate pesticide was also always correctly predicted in our previous studies (Devillers and Flatin, 2000; Devillers, 2001b).

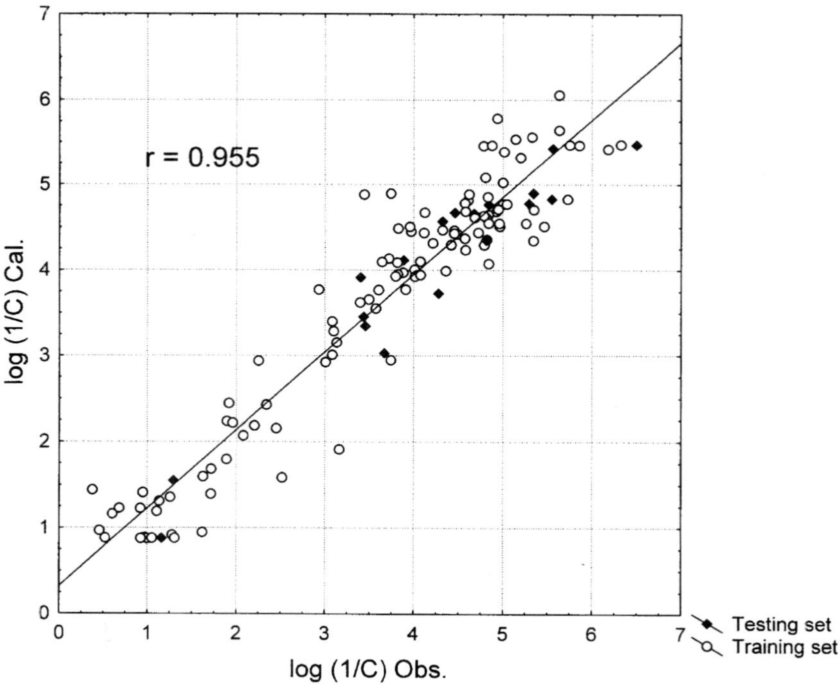

Fig. 6. Observed versus calculated LC_{50} values (ANN model).

Diphenamid (chemical number 30) is an outlier with both models with residuals equal to −3.03 and −1.06 for the PLS and ANN models, respectively (Table 2). This result is not surprising because while this molecule includes two phenyl rings (*N,N*-dimethyldiphenylacetamide), it appears not toxic to *G. fasciatus* (Table 1).

Table 3
Distribution of residuals (absolute values), differences between the observed and the calculated toxicity values computed from the PLS and ANN models

Range	PLS	ANN
<0.3	40(6)[a]	72(10)
[0.3–0.6)	23(1)	35(5)
[0.6–0.9)	22(2)	14(2)
[0.9–1.2)	10(1)	7(1)
[1.2–1.5)	11(3)	2
[1.5–1.8)	12(3)	0
[1.8–2.1)	5	0
[2.1–2.4)	5(1)	0
[2.4–2.7)	1(1)	0
[2.7–3.0)	0	0
≥3.0	1	0

[a] Including number of residuals from the external testing set.

This explains the negative sign of the residuals. However, it is interesting to note that the ANN, with a residual value of − 1.06, is able to partially encode this particularity. This confirms the high quality of the ANN model, especially when in addition, we consider that only one result was included in the database for that chemical. Toxicity of azinphos-methyl (*O*,*O*-dimethyl S-[(4-oxo-1,2,3-benzotriazin-3(4*H*)-yl)methyl]), tested on mature gammarids at 15°C with a pH of 7.4 and a hardness of 272 mg/l (as $CaCO_3$) after 96 h of exposure (number 89) is also badly predicted by both models (Table 2). However, models derived from different training and testing sets have provided better simulation results, especially for the ANN.

4. Conclusions

PLS regression analysis is a very powerful statistical tool for deriving classical QSAR models allowing to predict the toxicity of chemicals from their physicochemical properties and/or topological indices (Devillers et al., 2002a). These models are derived from toxicity results obtained in very specific experimental conditions. However, in aquatic toxicology, the correct hazard assessment of xenobiotics requires to simulate the ecotoxicological behavior of chemicals under various conditions of temperature, pH, hardness, and so on. Our study clearly shows that these experimental parameters cannot be integrated as variables in PLS models to increase their flexibility. Conversely, the ANNs do not suffer from this limitation and their introduction allows to derive very flexible environmental QSAR models.

References

Aoyama, T., Suzuki, Y., Ichikawa, H., 1990. Neural networks applied to structure–activity relationships. J. Med. Chem. 33, 905–908.

AUTOCOR (Version 2.40). CTIS, 3 Chemin de la Gravière, 69140 Rillieux La Pape, France.

Broto, P., Devillers, J., 1990. Autocorrelation of properties distributed on molecular graphs. In: Karcher, W., Devillers, J., (Eds.), Practical Applications of Quantitative Structure-Activity Relationships (QSAR) in Environmental Chemistry and Toxicology, Kluwer Academic Publishers, Dordrecht, pp. 105–127.

Cros, A.F.A., 1863. Action de l'Alcool Amylique sur l'Organisme. Thesis, Strasbourg, p. 37.

Devillers, J., 1996. Strengths and weaknesses of the backpropagation neural network in QSAR and QSPR studies. In: Devillers, J., (Ed.), Neural Networks in QSAR and Drug Design, Academic Press, London, pp. 1–46.

Devillers, J., 1999. Autocorrelation descriptors for modeling (eco)toxicological endpoints. In: Devillers, J., Balaban, A.T., (Eds.), Topological Indices and Related Descriptors in QSAR and QSPR, Gordon and Breach, The Netherlands, pp. 595–612.

Devillers, J., 2000. Prediction of toxicity of organophosphorus insecticides against the midge, *Chironomus riparius*, via a QSAR neural network model integrating environmental variables. Toxicol. Meth. 10, 69–79.

Devillers, J., 2001a. QSAR modeling of large heterogeneous sets of molecules. SAR QSAR Environ. Res. 12, 515–528.

Devillers, J., 2001b. A general QSAR model for predicting the acute toxicity of pesticides to *Lepomis macrochirus*. SAR QSAR Environ. Res. 11, 397–417.

Devillers, J., Bintein, S., Domine, D., Karcher, W., 1995. A general QSAR model for predicting the toxicity of organic chemicals to luminescent bacteria (Microtox® test). SAR QSAR Environ. Res. 4, 29–38.

Devillers, J., Chezeau, A., Thybaud, E., 2002a. PLS-QSAR of the adult and developmental toxicity of chemicals to *Hydra attenuata*. SAR QSAR Environ. Res. 13, 705–712.

Devillers, J., Domine, D., Chastrette, M., 1992. A new method of computing the octanol/water partition coefficient. Proceedings of QSAR92, July 19–23, 1992, Duluth, MN, USA, p. 12.

Devillers, J., Flatin, J., 2000. A general QSAR model for predicting the acute toxicity of pesticides to *Oncorhynchus mykiss*. SAR QSAR Environ. Res. 11, 25–43.

Devillers, J., Pham-Delègue, M.H., Decourtye, A., Budzinski, H., Cluzeau, S., Maurin, G., 2002b. Structure–toxicity modeling of pesticides to honey bees. SAR QSAR Environ. Res. 13, 641–648.

Dujardin-Beaumetz, G., Audigé, 1875. Sur les propriétés toxiques des alcools par fermentation. C. R. Acad. Sci. Paris LXXX, 192–194.

Free, S.M., Wilson, J.W., 1964. A mathematical contribution to structure–activity studies. J. Med. Chem. 1, 395–399.

Geladi, P., Tosato, M.L., 1990. Multivariate latent variable projection methods: SIMCA and PLS. In: Karcher, W., Devillers, J., (Eds.), Practical Applications of Quantitative Structure–Activity Relationships (QSAR) in Environmental Chemistry and Toxicology, Kluwer Academic Publishers, Dordrecht, The Netherlands, pp. 171–179.

Hansch, C., Fujita, T., 1964. ρ-σ-π analysis. A method for the correlation of biological activity and chemical structure. J. Am. Chem. Soc. 86, 1616–1626.

Hansch, C., Leo, A., 1995. Exploring QSAR Fundamentals and Applications in Chemistry and Biology, American Chemical Society, Washington, p. 557.

Hansch, C., Maloney, P.P., Fujita, T., Muir, R.M., 1962. Correlation of biological activity of phenoxyacetic acids with Hammett substituent constants and partition coefficients. Nature 194, 178–180.

Kaiser, K.L.E., Niculescu, S.P., Schüürmann, G., 1997. Feed forward backpropagation neural networks and their use in predicting the acute toxicity of chemicals to the fathead minnow. Water Qual. Res. J. Can. 32, 637–657.

Kubinyi, H., 1993a. QSAR: Hansch Analysis and Related Approaches, VCH, Weinheim, p. 240.

Kubinyi, H., 1993b. 3D QSAR in Drug Design, Theory, Methods and Applications, ESCOM, Leiden, The Netherlands, p. 759.

Leardi, R., 1996. Genetic algorithms in feature selection. In: Devillers, J., (Ed.), Genetic Algorithms in Molecular Modeling, Academic Press, London, pp. 67–86.

Mayer, F.L., Ellersieck, M.R., 1986. Manual of Acute Toxicity: Interpretation and Data Base for 410 Chemicals and 66 Species of Freshwater Animals, U.S. Fish Wildl. Serv., Resour. Publ. 160, p. 506.

Montgomery, J.H., 1993. Agrochemicals Desk Reference, Environmental Data, Lewis Publishers, Boca Raton, p. 625.

Moreau, G., Broto, P., 1980a. The autocorrelation of a topological structure: A new molecular descriptor. Nouv. J. Chim. 4, 359–360.

Moreau, G., Broto, P., 1980b. Autocorrelation of molecular structures, application to SAR studies. Nouv. J. Chim. 4, 757–764.

Overton, E., 1901. Studien über die Narkose, Gustav Fischer, Jena.

Putavy, C., Devillers, J., Domine, D., 1996. Genetic selection of aromatic substituents for designing test series. In: Devillers, J., (Ed.), Genetic Algorithms in Molecular Modeling, Academic Press, London, pp. 243–269.

Rabuteau, A., 1870. De quelques propriétés nouvelles ou peu connues de l'alcool du vin ou alcool éthylique. Déductions thérapeutiques de ces propriétés. Des effets toxiques des alcools butylique et amylique. Application à l'alcoolisation du vin improprement appelé vinage. Union Médicale 91, 165–173.

Rekker, R.F., Mannhold, R., 1992. Calculation of Drug Lipophilicity, The Hydrophobic Fragmental Constant Approach, VCH, Weinheim, p. 112.

Rumelhart, D.E., Hinton, G.E., Williams, R.J., 1986. Learning representations by back-propagating errors. Nature 323, 533–536.

Shiu, W.Y., Ma, K.C., Mackay, D., Seiber, J.N., Wauchope, R.D., 1990. Solubilities of pesticide chemicals in water. Part I: Environmental physical chemistry. Rev. Environ. Contam. Toxicol. 116, 35–221.

STATISTICA™. Version 6, StatSoft, 31 cours des Juilliottes, 94700 Maison-Alfort, France.

Tomlin, C., 1994. The Pesticide Manual, Incorporating the Agrochemicals Handbook, The British Crop Protection Council and The Royal Society of Chemistry, 10th edn, UK, p.1341.

Wasserman, P.D., 1993. Advanced Methods in Neural Computing, Van Nostrand Reinhold, New York, p. 255.

Zakarya, D., Boulaamail, A., Larfaoui, E.M., Lakhlifi, T., 1997. QSARs for toxicity of DDT-type analogs using neural network. SAR QSAR Environ. Res. 6, 183–203.

CONCLUSION

CHAPTER 12

Applying genetic algorithms and neural networks to chemometric problems

Brian T. Luke

SAIC-Frederick Inc., Advanced Biomedical Computing Center, NCI Frederick, P.O. Box B, Frederick, MD 21702, USA

1. Introduction

When Genetic Algorithms (GAs) and/or Neural Networks (NN) are applied to chemometric problems, several decisions have to be made that will affect the quality of the resulting algorithm. For GAs, various factors promote or retard the formation of schema which focuses the search onto a particular region of search space. These factors include the use of a mutation operator, procedures for selecting parent pairs, and the form of the mating operator. For the single class of feedforward backpropagation NNs, the results not only depend upon the number of input nodes, but also on the number of nodes in the hidden layer, whether or not a bias is applied, the exact form of the sigmoid function, the initial values of the weights connecting the nodes, and the learning procedure. This chapter reviews each of these factors and presents numerical examples showing their effects.

The first chapter of this book briefly showed that GAs and NNs are complimentary, not competitive, methodologies for solving chemometric problems. In addition, the introductory chapters for Parts I and II showed that they are heuristics and not particular algorithms. In order to solve a given problem, several decisions need to be made to construct specific algorithms, and many of these decisions can affect the ability of the resulting algorithm to find the optimum solution.

It is well beyond the scope of this chapter to examine all of the possible algorithms that can be constructed using each heuristic, and so only a particular form of a GA and NN will be used to examine a particular problem that is representative of several presented in this book. The problem is to generate a good Quantitative Structure–Property Relationship (QSPR) using the data generated by Breneman and Rehm (1997). In this study, the authors generated a set of molecular surface property descriptors using a transferable atom equivalent (TAE) method and used them to construct relationships that predict HPLC column capacity factors. For the purposes of this chapter, 118 TAE generated

Data Handling in Science and Technology, Volume 23
ISSN: 0922-3487

DOI: 10.1016/S0922-3487(03)23012-4

descriptors for 22 high-energy compounds will be used to predict the log(k') values for an ODS column.

In this study, as in many cases that use experimental or theoretical descriptors (Livingstone, 2000; Tounge et al., 2002), there are many more descriptors than values to be fitted. Therefore, most QSPR/QSAR studies represent *over-determined systems*. In general, constructing a good QSPR (or QSAR) is actually a four-step process. The first step is to examine the descriptors. In some cases it is necessary or advantageous to scale the data. For example, if each sample represented mixtures of compounds taken from different patients, it may be necessary to normalize each sample to a constant total mass or number of molecules or particular atoms. Conversely, each descriptor can be normalized to the same distribution so that one descriptor does not dominate simply because it has larger values. This normalization can force all descriptors to have values in the range [0,1] or to become *standardized* such that the mean is zero and the standard deviation is one. Other examinations of the data may have the goal of removing descriptors that have zero variance or are strongly correlated with each other (Luke, 2000), or to create orthogonal descriptors using either Gram-Schmidt Orthogonalization (Lucic and Trinajstic, 1999) or principle component analysis. Finally, it may be advantageous to increase the number of descriptors by using products of descriptors or functions of the descriptors since adding non-linearity between the descriptors and the target value can increase the quality of the relationship (Lucic and Trinajstic, 1999).

The second step is to determine which small set of descriptors should be used to generate a good relationship. This is the Feature Selection problem, and a GA or many of the methods outlined in Chapter 1 can be used to search for the optimal set of features.

The third step passes a putative set of descriptors to a procedure that constructs the relationship. This can be a simple least-squares fit (Luke, 1999), a NN (Agrafiotis et al., 2002), or a procedure that determines a target value using k-nearest neighbors (Tropsha and Zheng, 2001; Shen et al., 2002). Conversely, these descriptors can be used to cluster or classify the objects and give them categorical labels instead of specific target values. Various classification methods include machine learning (Wolberg et al., 1994), linear discriminate analysis, binary logistic regression (Cronin et al., 2002), NNs (Aires-de-Sousa, 2002; Izrailev and Agrafiotis, 2002), fuzzy clustering (Luke, 2003), and using a regression tree model (Izrailev and Agrafiotis, 2001).

The fourth step is to use some metric to determine the quality of the results generated in the third step using the descriptors chosen by the second step. This metric can be a standard or cross-validated correlation coefficient (So and Karplus, 1996), a standard error of estimate (Lucic and Trinajstic, 1999), a lack-of-fit value (Rogers and Hopfinger, 1994), or any other statistical or cluster quality metric.

The second through fourth steps are inter-related in that the Feature Selection method tries to choose the "best" set of descriptors. This determination depends upon the error metric which uses the target values or classifications generated in the third step. Changing either the error metric or the way the target values or classifications are done will also change the topology of the search space that the Feature Selection method scans and will usually lead to a different set of optimum features.

The methods described in this book deal with the second and third steps in this QSAR/QSPR generation process. Though a GA can be used to supply a set of features to a NN, this chapter examines these methods independently. In particular, Breneman and Rehm (1997) scaled each of the descriptors so that they covered the range [0,1]. Therefore, no further scaling, reduction, or increase in the values or number of descriptors will be done. A GA will then be used to search for the optimum set of five descriptors that will be used in a least-squares fit to the ODS $\log(k')$ values. The accuracy of each fit will be determined by calculating the root-mean-squared (RMS) error in these values.

These methods of determining the target values and generating an error metric are used because a previous study (Luke, 1999) used them to examine all possible sets (APS) of five descriptors and therefore located the globally optimum set of descriptors. In addition, this fitting method and error metric produces a search space that is *deceptive*, meaning that the search is much more likely to converge on a sub-optimal solution than find the optimal set.

The globally optimum set of descriptors for this fitting method and error metric will then be used in a multilayer feedforward backpropagation NN. The network will have its weights adjusted until the error in the 22 $\log(k')$ values is acceptably small. Variations in certain parameters will be examined to determine their effect on the rate of convergence of the network and to compare changes in the calculated values caused by relatively small changes in the inputs.

Section 2 describes the actual form of the GA that is used to search for the optimum set of five descriptors, as well as the parameters that will be varied. This is followed by Section 3 that describes the deceptive nature of this problem and examines the results of each of the searches. Section 4 describes the structure of the NN and the parameters that will be varied in different runs. Section 5 presents the results obtained from each network, and Section 6 contains concluding remarks.

2. Structure of the GA

To briefly restate the problem, the goal of the GA is to select the optimum set of five descriptors from the full set of 118 TAE generated descriptors (Breneman and Rehm, 1997) such that a linear fit to the $\log(k')$ values produces a minimum RMS error for the 22 compounds. Each putative set of five descriptors is used in a least-squares fit to the ODS $\log(k')$ values, and the *cost* of each descriptor set is determined by (Luke, 1999)

$$\text{Cost} = \text{Base}^{|k\text{ term}-n\text{ term}|}(\text{RMS-Error})\prod_i [W_i(n)]$$

The first term represents a penalty if the number of descriptors (k term) in a set differs from the required number of descriptors (n term $= 5$). For all of the tests presented here, Base is fixed at 1.3. The second term is simply the RMS error of the fitted relationship and the third is a penalty if any of the descriptors are raised to a non-unit exponent. $W_i(n)$ represents a weight if the ith descriptor is raised to the nth power. In this study, $W_i(1) = 1.0$, $W_i(2) = 2.0$, and $W_i(n) = 100.0$ if n is not 1 or 2. This means that the cost is not changed if the expression is linear in a descriptor and increased by a factor of 2.0 if it is quadratic in a descriptor. Therefore, a 6-descriptor relationship is competitive with a 5-descriptor relationship if the formers' RMS error is approximately smaller by a factor of 1.3. Similarly, if one of the descriptor values is squared, the resulting relationship is still competitive if its RMS error is approximately one-half of a relationship that is linear in all descriptors. If a putative solution contains six descriptors, the descriptor values are squared for two of them and linear for the rest and the resulting RMS error is 0.05, the cost of this solution would be

$$\text{Cost} = 1.3^{|6-5|}(0.05)(2.0)^2(1.0)^4 = 0.26$$

This cost function is more than what would be required if the search was strictly confined to 5-descriptor, linear relationships, but as stated earlier, this particular problem is deceptive and searches using only the RMS error as the cost function will probably converge onto sub-optimum solutions (see below). It may be possible that the optimum set of five descriptors will have a larger presence in good solutions if relationships with more descriptors and/or non-unit exponents are also sampled. The earlier study (Luke, 1999) used a particular evolutionary programming algorithm and was only able to find the global optimum when the cost function given above was used.

Since a putative solution has to include the possibility of having a variable number of descriptors, and the values of certain descriptors can be raised to a non-unit exponent, the genetic vector for a solution is comprised of a 118-element integer array. If the *i*th element of this vector is zero, the *i*th descriptor is not used in the relationship. Otherwise, the value of this integer represents the exponent to which the *i*th descriptor's values are raised. In addition, these tests will use a haploid GA, which means that only a single genetic vector is used to describe a putative solution.

With the forms of the genetic vector and cost function established, the next step is to choose a population size (μ). In most of these tests, μ is set to 1000 though a few tests are performed to determine the effect of increasing the population size to 3000. The initial population of μ putative solutions only allows unit exponents for the descriptors (non-unit exponents are created with a mutation operator). Therefore, the initial genetic vectors are only composed of 0s and 1s. The number of 1s in the genetic vector is controlled by the parameter f_5. This parameter determines the fraction of solutions in the initial population that have exactly five 1s. For example if $\mu = 1000$ and $f_5 = 0.7$, 700 of the 1000 initial solutions have five 1s and 113 0s in the genetic vector. The remaining 300 solutions have six 1s and 112 0s in the genetic vector. The parameter DIV controls how the initial population is created. If DIV = 0, a random number generator is used to randomly place the five or six 1s in the genetic vector and this putative solution is added to the initial population. If DIV = 6, the same procedure is followed, with the caveat that each new solution is only added to the initial population if its genetic vector differs from those already in the initial population. This means that the initial population contains μ putative solutions such that the Hamming distance for all pairs of genetic vectors is at least six. Therefore, any two initial solutions can have at most two positions that contain a 1 in both.

The next step in creating a specific algorithm is to decide how two parents will be selected for each mating. This selection procedure depends upon two parameters, FIT and PSEL. FIT controls how the *fitness* of each solution is determined since several methods described in Chapter 1 are fitness-based. If FIT = 0, no fitness is needed, such as for a rank-based selection method. If FIT = 1, the fitness is determined from the expression

$$\text{Fitness} = 1.0/\text{Cost}$$

since the cost function given above is positive everywhere (the RMS error is never zero given the small number of descriptors used in the relationship). Conversely, if FIT = 2, the following expression is used

$$\text{Fitness} = (\text{Max(Cost)} - \text{Cost}) + 0.1$$

To use this expression, the parent population is checked and the maximum cost (minimum fitness) is determined. This value is used to determine the fitness of all members, and the additive constant is used to ensure that none of the solutions have zero fitness and therefore has a finite probability of being chosen.

PSEL is then used to determine the actual selection method, and six different methods will be examined here. If PSEL = 0, one of the parents is always the dominant solution (i.e. the solution with the lowest-cost/highest-fitness). This solution is mated with all solutions in the parent population, including itself. This is the *line breeding operator* of Hollstein (1971). If PSEL = 1, an unscaled roulette wheel selection method is used for both parents. If PSEL = 2, a roulette wheel selection is used, but the fitness values (determined by the value of FIT) are scaled using a relationship of the form

$$f'(i) = af(i) + b$$

where (Goldberg, 1989)

$$f'_{\text{max}} = C_{\text{mult}} f_{\text{avg}}$$

$$f'_{\text{min}} = C_{\text{min}}$$

In these expressions, C_{mult} is set to 1.6, C_{min} is set to 0.1, and f_{avg} is the average of the unscaled fitness values for the parent population.

If PSEL = 3, a similar procedure is used, only the coefficients of the scaling equation are determined by requiring

$$f'_{\text{max}} = C_{\text{max}} f_{\text{avg}}$$

$$f'_{\text{avg}} = f_{\text{avg}}$$

For the studies presented here, C_{max} is also set to 1.6.

If PSEL = 4, a fitness-based roulette wheel selection method is also used, but the fitness of a solution is based upon its rank. If the solutions are ranked from lowest-to-highest cost (highest-to-lowest fitness), a solution with a rank number of z has a fitness of (Brown and Sumichrast, 2003)

$$\text{Fitness}(z) = 2(\mu - z + 1)/[\mu(\mu + 1)]$$

Finally if PSEL = 5, a rank-based selection procedure is used (Whitley, 1989). When the parent solutions are ranked from lowest-to-highest cost (highest-to-lowest fitness), the rank number of a selected parent is given by

$$z = \mu[\alpha - (\alpha^2 - 4(\alpha - 1)r)^{1/2}]/[2(\alpha - 1)]$$

For the results presented here, α is set to 1.3 and r is a random number in the range [0,1]. A histogram showing the approximate probability that a particular rank number is chosen is shown in Fig. 1. This figure shows that the first (dominant) parent solution is chosen most often and that the probability of being selected decreases as the rank number increases, with a significant reduction in the probability of choosing the worst member of the population.

From these descriptions, an actual fitness value (FIT = 1 or 2) is only needed if PSEL is 1, 2 or 3. Therefore, if PSEL is 0, 4 or 5, FIT can be set to 0.

Once the parents are selected, a mating operator is used to generate a complimentary pair of offspring. Using the notation of Chapter 1, $P_{mate} = 1.0$ which means that the mating operator is applied for all parent-pairs. In this study, the parameter CROSS controls the form of the mating operator. If CROSS = 1 a 1-point crossover is used, if CROSS = 2 a 2-point crossover is used, while if CROSS = 0 a uniform crossover is used. In all cases, these crossover operators create a complimentary pair of offspring. Each offspring may then have a chance to be mutated. For each offspring, a random number in the range [0,1] is compared to P_{mut}. If the random number is less than this mutation probability, a mutation occurs. Two different mutation operators can be used in this investigation. The first simply chooses a random location in the genetic vector and increases or decreases the value at this position by one. The second mutation operator chooses two positions in the offspring's genetic vector such that the value at one is zero and the value at the other is non-zero and the values at these two positions are switched. The probability of using one mutation operator or the other is controlled by the value P_{chg}. If a random number in [0,1] is less than P_{chg} the first mutation operator is used; otherwise the second one is used.

This means that up to four random numbers (r_1–r_4) in the range [0,1] may be needed for each offspring. If $r_1 < P_{mut}$, a mutation will occur. If $r_2 < P_{chg}$, the value at a random position in the genetic vector is increased or decreased by one. This position is determined by taking the integer portion of $r_3\mu$ and adding one (the position is taken to be the last of $r_3 = 1.0$). If $r_4 < 0.5$, the value at this position is decreased by one; otherwise it is

Fig. 1. Probability of selecting a particular parent using the rank selection procedure (Whitley, 1989) with $\alpha = 1.3$.

increased by 1. Conversely, if $r_2 \geq P_{chg}$; two random positions are chosen such that at position i_0 the value is zero while at i_1 it is not. These positions are determined by setting μ_0 equal to the number of zeroes in the genetic vector. i_0 is located by taking the integer part of $r_3\mu_0$ and adding one, and then searching down the genetic vector until this many zeroes are found. A similar procedure is used to find i_1, only one is added to the integer part of $r_4(\mu - \mu_0)$ to determine the number of non-zero positions that must be found.

It should be stressed that four different random numbers are generated for each offspring so that a mutation can occur in both, only one, or neither. The only exception to this is when a parent is mated with itself. In this case, a mutation is forced to occur in each offspring independent of their value of r_1. In addition, the algorithm has a user-supplied parameter called UNIQ, and if this parameter is set to 1 a mutation is forced to occur if the genetic vectors of the parents are the same, irrespective of whether the parents are the same member of the parent population.

Once the offspring are created, their cost is determined and only the offspring with the lowest cost is added to a new population. This process is continued until $\lambda(= \mu)$ offspring have been added. At this point a *generation* is completed and therefore the algorithm that will be tested here is a generational-GA. The final step is to create a parent population for the next generation. Though many procedures for choosing the next generation's parent population are described in Chapter 1, the algorithm tested here uses a deterministic $(\mu + \lambda)$ selection method. This means that the parent and offspring populations are combined, ordered from lowest-to-highest cost, and the best μ solutions become parents in the next generation. This procedure is run for 60 generations and the lowest-cost solution in the then final combined population represents the reported result.

The algorithm described above has several user-supplied parameters that are designed to strike a balance between exploitation and exploration. These parameters are DIV, FIT, PSEL, CROSS (which can be constant for the entire search or for only a smaller number of generations), P_{mut} (which can be constant or change from one generation to the next), P_{chg}, UNIQ, SEED (the seed to the random number generator), and f_5.

If DIV $= 0$ the initial population is randomly generated and there is a good probability that each element of the genetic vector will have a non-zero value in at least one member. When DIV $= 6$ the probability should be even larger since each member of the initial population can have at most two non-zero elements in common with any other member of the population.

When PSEL $= 0$ the initial generations should promote exploitation since all offspring will take approximately 50% of their genetic vectors from the dominant member. It should result in a fairly rapid generation of schema. When PSEL $= 1$, the algorithm is also exploitative. Since the fitness values are unscaled, the best few solutions will be chosen a majority of the time and the formation of schema should be more rapid. If a fitness scaling is used (PSEL $= 2$ or 3) weaker members should be chosen more often and therefore exploration should be promoted. If the fitness is based on the rank of the solution (PSEL $= 4$), exploitation should be promoted since the weakest member is chosen only 1/1000th as often as the dominant member (assuming $\mu = 1000$). A member of the best 294 solutions should be selected 50% of the time, while approximately 90% of the parents will be part of the best 685 solutions. Conversely, if the selection is made by choosing the rank number (PSEL $= 5$) with $\alpha = 1.3$, the best parent (lowest-cost) should be selected only 5.5 times more often than the worst (see Fig. 1). This means that 50% of the parents will be chosen from the best 427 members of the parent population, and 90% of the parents will be from the best 865 solutions. Therefore exploration will be promoted much more when PSEL = 5 as compared to PSEL = 4. Finally, there may be some dependence on the exploration/exploitation nature of the algorithm with different forms of the fitness function (FIT = 1 or 2) when PSEL is 1, 2 or 3.

The value of CROSS also controls whether exploration or exploitation is promoted. 1- and 2-point crossovers (CROSS = 1 or 2, respectively) are known to promote exploitation though Hasancebi and Erbatur (2000) found the 2-point crossover to promote this more in their test problems. Conversely, it is well established that a uniform crossover (CROSS = 0) promotes exploration. In addition, mutation is also known to promote exploration, so the larger the value of P_{mut} the more the exploration is promoted. This should be true for both mutation operators, though the increasing/decreasing operator will most likely change the number of non-zero elements in the genetic vector while the switching operator does not. The fact that a mutation operation is forced when a parent mates with itself, or with another parent with an identical genetic vector (if UNIQ = 1) will also aid in exploration and promote diversity in the combined parent + offspring population. The effect of increasing the population size (μ) is less clear. Though a larger population size will allow more of the search space to be covered in the initial population and will increase the probability that an initial solution is reasonably close to the global minimum, it is also possible that a large population size could "dilute" this close solution to the point where it is not able to influence the focusing of the search. This may be especially true for a deceptive problem like the one treated here.

Finally, the seed to the random number generator (SEED) will not promote either exploration or exploitation, but it could influence the overall search because it results in different initial populations and may influence the amount of mutation in the first few generations, where this exploration may be more important. The value of f_5 also affects the initial population since it controls the number of members that use five and six descriptors in the initial population. It, and the value of P_{chg}, therefore affects how much of the search is confined to the 5-descriptor, linear space and how much explores other spaces (though the crossover operators can also create offspring with the number of non-zero elements being different than five).

3. Results for the GAs

Before presenting the calculated results, it is important to examine the deceptive nature of this search space. The previous study (Luke, 1999) also used a least-square linear fit with the same cost function presented above. Before the search was attempted, all descriptors were examined in a pair-wise fashion, and descriptors were removed from the dataset if a very strong linear correlation with another descriptor was found. This reduced the number of descriptors from 118 to 111. These descriptors were used in an algorithm that determined the RMS error for APS of five descriptors in a linear relationship. The sets of descriptors that produced the lowest RMS errors in this APS run are listed in Table 1. Also included in this table is the number of times each descriptor is used in the top 100 sets.

Table 1
Review of sets of five descriptors that produce least-squares fits with a low RMS error

RMS error	Descriptors				
0.04130	DELRHONA8 (2)	SIKA2 (3)	SIGA3 (6)	PIP17 (3)	PIP19 (2)
0.04216	SIGA1 (6)	SIKA2 (3)	SIGA3 (6)	PIP17 (3)	PIP19 (2)
0.04147	SIDELGN (4)	DELRHONA2 (10)	SUMSIGMA (11)	SIGMANEW (11)	PIP18 (14)
0.04219	DELGNIA (5)	DELRHONA2 (10)	SUMSIGMA (11)	SIGMANEW (11)	PIP18 (14)
0.04263	SIDELGN (4)	DELRHONA2 (10)	PIP5 (4)	SIGMANEW (11)	PIP18 (14)
0.04276	DELKMIN (4)	DELRHONA2 (10)	SUMSIGMA (11)	SIGMANEW (11)	PIP18 (14)
0.04276	DELKMAX	DELRHONA2 (10)	SUMSIGMA (11)	SIGMANEW (11)	PIP18 (14)
0.04276	DELGNMAX	DELRHONA2 (10)	SUMSIGMA (11)	SIGMANEW (11)	PIP18 (14)
0.04276	DELGNA10	DELRHONA2 (10)	SUMSIGMA (11)	SIGMANEW (11)	PIP18 (14)
0.04418	SIEPMIN (4)	DELRHONA2 (10)	SUMSIGMA (11)	SIGMANEW (11)	PIP18 (14)
0.04437	SIDELKN (9)	DELRHONA2 (10)	SUMSIGMA (11)	SIGMANEW (11)	PIP18 (14)
0.04155	DELGNA3 (2)	EP3 (1)	EP8 (4)	PIP17 (3)	PIP18 (14)
0.04316	DELKMIN (4)	DELKIA (68)	SIGA9 (77)	SIEPA1 (74)	PIV (78)
0.04316	DELKMAX	DELKIA (68)	SIGA9 (77)	SIEPA1 (74)	PIV (78)
0.04316	DELGNMAX	DELKIA (68)	SIGA9 (77)	SIEPA1 (74)	PIV (78)
0.04316	DELGNA10	DELKIA (68)	SIGA9 (77)	SIEPA1 (74)	PIV (78)
0.04339	SUMSIGMA (11)	DELKIA (68)	SIGA9 (77)	SIEPA1 (74)	PIV (78)
0.04414	SIGMAPV (1)	DELKIA (68)	SIGA9 (77)	SIEPA1 (74)	PIV (78)
0.04414	SIEPA8 (2)	DELKIA (68)	SIGA9 (77)	SIEPA1 (74)	PIV (78)
0.04415	EP10 (1)	DELKIA (68)	SIGA9 (77)	SIEPA1 (74)	PIV (78)
0.04420	SIKA4 (1)	DELKIA (68)	SIGA9 (77)	SIEPA1 (74)	PIV (78)
0.04444	SIKA2 (3)	DELKIA (68)	SIGA9 (77)	SIEPA1 (74)	PIV (78)

The number in parentheses is the number of times this descriptor was found in the top 100 results when all possible sets of five descriptors was done using a reduced set of 111 descriptors.

The optimum set of five descriptors (DELRHONA8, SIKA2, SIGA3, PIP17 and PIP19) produce an RMS error of 0.04130. This solution will be hard to find using virtually any population-based search method because these descriptors are not present in many low-cost (high-fitness) solutions. When the 100 best sets of five descriptors obtained from the APS search over the 111 descriptors are examined, the optimum descriptors are found in only 2, 3, 6, 3, and 2 sets, respectively. Since one of the occurrences of each is the global optimum, four of the five descriptors are only found in one or two other good sets. In addition, there is only one other set of five descriptors with a low RMS error that shares four of the five descriptors. This means that a good population containing many of these descriptors will not form, and no optimum schema will be produced.

The second-best set of five descriptors (SIDELGN, DELPHONA2, SUMSIGMA, SIGMANEW, and PIP18) produce an RMS error of 0.04147. This solution should be easier to find because four of the five descriptors are used in more than 10% of the best sets. In addition, there are eight other sets of five descriptors that use four of them and produce a relationship with a low RMS error. This means that there is a larger chance that schema will form using some of these descriptors and this increases the probability of finding this second-best set.

It should be noted that there is no number in parenthesis next to DELKMAX, DELGNMAX, and DELGNA10. This is because these descriptors have a very good linear dependence with DELKMIN, and they were removed from consideration in Luke (1999). Since the search presented here uses all 118 descriptors, sets using these three descriptors can be formed. Therefore, these sets are also added to Table 1.

The third-best set of five descriptors (DELGNA3, EP3, EP8, PIP17 and PIP18) produces an RMS error of 0.04155, and this set will probably not be found using anything less than an APS search. The reason for this is that EP3 is not found in any of the other 100 best sets from Luke (1999). In addition, only PIP17 and PIP18 are found in any of the good sets displayed in Table 1, and they are never used together in any other of the best 100 sets. Therefore, any method that uses one or two good solutions to generate another solution will not be able to find this third-best set. By decreasing the value of Base, $W(2)$, and/or some other $W(n)$ in the cost function it may be possible to generate other sets that use these descriptors. Abruptly changing these values to the ones used here may allow the search to drop from a higher-dimensional search space into this one and may generate this set of descriptors, but this strategy is not guaranteed to work.

The next-best set of five descriptors not already listed produces a relationship that has an RMS error of 0.04316. Four of the five descriptors are found in a majority of the best 100 sets from Luke (1999), and the fifth descriptor is one of the degenerate set of four (DELKMIN, DELKMAX, DELGNMAX, or DELGNA10). In addition, six other sets of five descriptors have a low RMS error and differ only in this descriptor.

This means that any population-base search that is able to explore a region of search space containing low-cost solutions will probably be able to build schema for DELKIA, SIGA9, SIEPA1, and/or PIV. Future generations will then be focused on this region of search space and a solution with an RMS error of 0.04316 will be found. The search space of five descriptors therefore contains four low-cost regions. The region containing the 0.04316 solutions is the largest (largest number of good solutions that use many of the required descriptors). The region that contains the second-best set of descriptors is the next largest, but it is significantly smaller. Finally, the regions that contain the best and third-best sets of descriptors are extremely small and hard to locate. Since the largest region of good solutions produce a sub-optimal solution and the four regions shown in Table 1 are well separated from each other, there is no chance of hopping from one region to another and this search space is deceptive. Finally, it should be noted that the search space generated when five descriptors are chosen from a set of 118 contains approximately 1.75×10^8 unique sets. This means that a search procedure that is able to efficiently locate any of the low-cost sets listed in Table 1 should be considered a good method.

The first three groups of tests examined here are restricted to search the space of only five descriptors. This is done by setting f_5 to 1.0 (meaning that only 5-descriptor solutions appear in the initial population) and $P_{chg} = 0.0$ (which means that a mutation can only swap a zero and non-zero entry in the genetic vector). In addition, the population size μ is set to 1000, the search runs for 60 generations, and a mutation is only forced if a member of the parent population mates with itself (UNIQ = 0).

In the first group of tests a 1-point crossover mating operator is used, and the probability of mutation linearly decreases from 0.5 to 0.01 with each generation. The results of these tests are shown in Table 2. This table shows the final results for 54 different calculations

Table 2
Lowest *cost* obtained after 60 generations of a GA using a 1-point crossover mating operator (CROSS = 1) as a function of the method of generating the initial population (DIV), form of the *fitness* (FIT), parent selection strategy (PSEL), and seed to the random number generator (SEED)

Run	DIV	FIT	PSEL	SEED		
				371779.0	43297 187.0	900417 229.0
1a	0	0	0	0.04316 (8)	0.05521 (5)	0.04929 (5)
1b	0	1	1	0.04147 (44)	0.04316 (23)	0.04929 (20)
1c	0	1	2	0.05439 (23)	0.04929 (15)	0.04316 (15)
1d	0	1	3	0.04680 (21)	0.04680 (15)	0.04503 (16)
1e	0	0	4	0.04680 (18)	0.04316 (30)	0.04929 (6)
1f	0	0	5	0.05197 (15)	0.04316 (29)	0.04929 (18)
1g	0	2	1	0.04503 (27)	0.04316 (38)	0.04316 (45)
1h	0	2	2	0.04658 (32)	0.04316 (29)	0.04421 (18)
1i	0	2	3	0.05439 (3)	0.05521 (2)	0.04929 (2)
1j	6	0	0	0.05238 (5)	0.05028 (3)	0.04929 (2)
1k	6	1	1	0.04316 (37)	0.04721 (36)	0.04929 (14)
1l	6	1	2	0.05197 (16)	0.04998 (13)	0.04929 (10)
1m	6	1	3	0.04711 (19)	0.04695 (14)	0.04421 (14)
1n	6	0	4	0.04648 (26)	0.04316 (23)	0.04721 (20)
1o	6	0	5	0.04938 (13)	0.04316 (34)	0.04421 (12)
1p	6	2	1	0.04316 (36)	0.04316 (38)	0.04658 (49)
1q	6	2	2	0.04680 (23)	0.04316 (29)	0.04316 (35)
1r	6	2	3	0.05439 (3)	0.06053 (1)	0.04929 (2)

The first number is the RMS error between calculated and experimental $\log(k')$ values for the 22 compounds while the integer in parenthesis is the generation when the best result first appears. The other adjustable parameters are fixed as follows (see text for details): $\mu = 1000$, P_{mut} linearly decreases from 0.5 to 0.01 with each generation, $P_{chg} = 0.0$, $f_5 = 1.0$, and UNIQ = 0.

that use all possible combinations of generating the initial population (DIV), form of the fitness function (FIT), and parent selection method (PSEL) for three different seeds to the random number generator (SEED). Each entry lists the RMS error of the best set of five descriptors found in the search and, in parenthesis, the generation number when this solution first appeared. An examination of this table yields the following observations.

1. Using the line breeding operator (Hollstein, 1971) causes the population to converge very rapidly, and most of the simulations produce poor results. This is expected since approximately 50% of the genetic vectors in the offspring population come from this dominant solution. This promotes exploitation of this single solution, the formation of schema, and a rapid convergence of the population.
2. Poor results are obtained if the fitness is a linear function of the cost (FIT = 2) and then scaled based on the maximum and average fitness of the parent population (PSEL = 3), good results are obtained if these fitness values are unscaled (PSEL = 1), and intermediate results are obtained if the scaling is based on the minimum and maximum fitness of the parent population prior to the roulette wheel selection procedure.

3. Different results are obtained if the fitness is the inverse of the cost (FIT = 1). In this case, scaling based on the maximum and average fitness of the parent population (PSEL = 3) produced results that are as good as those found without scaling (PSEL = 1).
4. The final results for these runs that use a 1-point crossover with a population size of 1000 appear to be very dependent upon the seed to the random number generator. This seed affects the solutions that are placed in the initial population, the selection of parents for each mating in all cases except PSEL = 0, the location of the cut point in the mating of solutions, and the probability that a mutation occurs when the parents are not the same solution. When SEED = 43 297 187.0 a solution with an RMS error of 0.04316 is consistently found, while if SEED = 900 417 229.0 a result with an RMS error of 0.04929 is found in many of the runs. Conversely, if SEED = 371 779.0, no single result is obtained in a large number of the runs. This suggests that the distribution of solutions in the initial population controls the final result. The first two seeds probably have multiple good solutions in the vicinity of the result that is regularly found, while the third seed produces an initial population that is well spread over the search space and the actual search procedure determines what final result will be obtained.

The second group of runs explores the effect of the mutation. These runs use the same parameters as the first group, but instead of changing SEED, it is held constant at 371 779.0 and the values of P_{mut} are changed. These results are shown in Table 3. The first column of results (fifth column overall) again has the mutation probability changing in a linear fashion from 0.5 to 0.01, and so these results are the same as shown in the fifth column of Table 2. The next column turns off the mutation (as is often done in practice), and the last column keeps the mutation probability at 0.5. From these results, the following observations are made.

1. Removing the mutation operator clearly degrades the quality of the final result.
2. Keeping a high mutation rate throughout the simulation produces the best overall results, though the best result is obtained when the mutation rate is reduced in each generation. This is contrary to conventional wisdom which states that as the search proceeds, the nature of the search should switch from exploration to exploitation. This means that the effect of mutation, which is purely exploratory, should be reduced in later generations. This is probably due to the fact that a modest population size (1000) is used. Reducing the mutation rate from 0.5 to 0.01 may produce better results for larger population sizes, and very large population sizes may be able to produce good results without a mutation operator.

Table 4 repeats the runs of Tables 2 and 3, with only the crossover operator varied. The fifth column repeats the results using a 1-point crossover that is shown in the previous tables, while the sixth and seventh columns list the results when 2-point and uniform crossover operators are used, respectively. The last column lists the results when a *mixed crossover* operator is used (Hasancebi and Erbatur, 2000). In particular, a uniform crossover is used for the first 30 generations and a 1-point crossover is used for the last 30. The results in this table yield the following observations.

Table 3
Lowest *cost* obtained after 60 generations of a GA using a 1-point crossover mating operator (CROSS = 1) as a function of the method of generating the initial population (DIV), form of the *fitness* (FIT), parent selection strategy (PSEL), and the probability of mutation (P_{mut})

Run	DIV	FIT	PSEL	P_{mut}		
				0.5–0.01	0.0	0.5
2a	0	0	0	0.04316 (8)	0.05618 (2)	0.05096 (5)
2b	0	1	1	0.04147 (44)	0.04919 (11)	0.04316 (45)
2c	0	1	2	0.05439 (23)	0.05594 (14)	0.05429 (8)
2d	0	1	3	0.04680 (21)	0.05099 (13)	0.04316 (22)
2e	0	0	4	0.04680 (18)	0.05802 (23)	0.04621 (13)
2f	0	0	5	0.05197 (15)	0.05197 (30)	0.04648 (18)
2g	0	2	1	0.04503 (27)	0.05825 (9)	0.04316 (50)
2h	0	2	2	0.04658 (32)	0.04993 (9)	0.04316 (44)
2i	0	2	3	0.05439 (3)	0.05439 (3)	0.05439 (3)
2j	6	0	0	0.05238 (5)	0.06083 (52)	0.05439 (3)
2k	6	1	1	0.04316 (37)	0.06047 (13)	0.04316 (31)
2l	6	1	2	0.05197 (16)	0.04919 (45)	0.04316 (10)
2m	6	1	3	0.04711 (19)	0.05439 (10)	0.04316 (28)
2n	6	0	4	0.04648 (26)	0.05751 (5)	0.04316 (33)
2o	6	0	5	0.04938 (13)	0.05796 (40)	0.04938 (9)
2p	6	2	1	0.04316 (36)	0.05493 (20)	0.04316 (42)
2q	6	2	2	0.04680 (23)	0.05847 (8)	0.04316 (31)
2r	6	2	3	0.05439 (3)	0.05439 (3)	0.05439 (3)

The first number is the RMS error between calculated and experimental $\log(k')$ values for the 22 compounds while the integer in parenthesis is the generation when the best result first appears. The other adjustable parameters are fixed as follows (see text for details): $\mu = 1000$, $P_{chg} = 0.0$, $f_5 = 1.0$, UNIQ = 0 and SEED = 371779.0.

1. The mixed crossover operator produces the best overall set of results, and the 1-point crossover produces the worst overall results. This again may be due to the relatively small population size.
2. The best solution found overall has an RMS error of 0.04147, and this solution is found most often with the 2-point crossover operator. Therefore, this operator may do the best job of exploiting good solutions, as suggested by Hasancebi and Erbatur (2000).

The results shown in Tables 5 and 6 represent a concerted effort to find the optimum set of descriptors. In these runs, the mixed crossover operator described above is used, and P_{mut} linearly decreases from 0.5 to 0.01. In addition, these runs are designed to search a larger parameter space. In particular, f_5 is set to 0.7 which means that 70% of the initial population contains sets of five descriptors and the remaining 30% contains sets of six. In addition, P_{chg} is 0.15 which means that 15% of all mutations choose a random location in the genetic vector and increase or decrease its value by one. Since only a small number of descriptors are selected in any genetic vector, the probability is high that the randomly selected element has a value of zero. Therefore, this mutation has the effect of adding a descriptor to the set used in the fit. On the other hand, since there is an equal probability of increasing and decreasing the value in the genetic vector, and $W(-1) = 100.0$ in the cost

Table 4
Lowest *cost* obtained after 60 generations of a GA as a function of the crossover mating operator (CROSS), method of generating the initial population (DIV), form of the *fitness* (FIT), and parent selection strategy (PSEL)

Run	DIV	FIT	PSEL	CROSS			
				1	2	0	0 (30)/1 (30)
3a	0	0	0	0.04316 (8)	0.04316 (8)	0.04316 (9)	0.04316 (9)
3b	0	1	1	0.04147 (44)	0.04147 (34)	0.04316 (48)	0.04316 (51)
3c	0	1	2	0.05439 (23)	0.04711 (18)	0.04316 (29)	0.04316 (29)
3d	0	1	3	0.04680 (21)	0.04943 (21)	0.04147 (50)	0.04147 (51)
3e	0	0	4	0.04680 (18)	0.04316 (20)	0.04316 (48)	0.04316 (48)
3f	0	0	5	0.05197 (15)	0.04421 (28)	0.04316 (53)	0.04316 (48)
3g	0	2	1	0.04503 (27)	0.04721 (21)	0.04316 (54)	0.04316 (51)
3h	0	2	2	0.04658 (32)	0.04147 (37)	0.04339 (53)	0.04316 (51)
3i	0	2	3	0.05439 (3)	0.05439 (3)	0.05439 (3)	0.05439 (3)
3j	6	0	0	0.05238 (5)	0.05439 (7)	0.05439 (6)	0.05439 (6)
3k	6	1	1	0.04316 (37)	0.04147 (36)	0.05096 (25)	0.04818 (45)
3l	6	1	2	0.05197 (16)	0.04929 (2)	0.04316 (22)	0.04316 (22)
3m	6	1	3	0.04711 (19)	0.04316 (19)	0.04680 (46)	0.04680 (37)
3n	6	0	4	0.04648 (26)	0.04316 (38)	0.04316 (46)	0.04316 (40)
3o	6	0	5	0.04938 (13)	0.04316 (55)	0.04929 (23)	0.04647 (36)
3p	6	2	1	0.04316 (36)	0.04316 (29)	0.04702 (47)	0.04316 (56)
3q	6	2	2	0.04680 (23)	0.04316 (31)	0.04316 (56)	0.04721 (46)
3r	6	2	3	0.05439 (3)	0.05439 (3)	0.05439 (3)	0.05439 (3)

The first number is the RMS error between calculated and experimental $\log(k')$ values for the 22 compounds while the integer in parenthesis is the generation when the best result first appears. The other adjustable parameters are fixed as follows (see text for details): $\mu = 1000$, $P_{\text{chg}} = 0.0$, $f_5 = 1.0$, UNIQ = 0 and SEED = 371779.0.

function, at most only 7.5% of the mutations result in a useable set with an increased number of descriptors.

The only difference between Tables 5 and 6 is in the value of UNIQ. In Table 5, UNIQ = 0 which means that a mutation is only forced when a member of the parent population mates with itself. In Table 6, UNIQ = 1 which means that a mutation is forced each time the mating parents have identical genetic vectors.

Columns 5–7 display the results for three different seeds to the random number generator and the population size (μ) is 1000. The last column uses the first value of SEED, but increases μ to 3000.

The results in Table 5 produce the following observations.

1. The overall results are better than those in Tables 2–4.
2. The extra exploration and flexibility of the search causes the runs using the initial population with SEED = 900 417 229.0 to often escape the 0.04929 minimum found in Table 2 and find a better solution. In addition, one run (4j) found the globally optimum set of features.
3. Increasing the population size from 1000 to 3000 for SEED = 371 779.0 does not substantially improve the results. The same result is found nine times, better results five times and worse results are obtained in four of the runs.

Table 5
Lowest *cost* obtained after 60 generations of a GA using a uniform crossover for 30 generations and a 1-point crossover mating operator for another 30 generations as a function of the method of generating the initial population (DIV), form of the *fitness* (FIT), parent selection strategy (PSEL), and seed to the random number generator (SEED)

Run	DIV	FIT	PSEL	SEED			
				371779.0	43297187.0	900417229.0	371779.0($\mu = 3000$)
4a	0	0	0	0.05439 (3)	0.05937 (3)	0.04929 (6)	0.05439 (6)
4b	0	1	1	0.04147 (49)	0.04316 (48)	0.04929 (39)	0.04316 (51)
4c	0	1	2	0.04316 (28)	0.04680 (27)	0.04721 (36)	0.04316 (30)
4d	0	1	3	0.04721 (35)	0.04316 (38)	0.04929 (32)	0.04316 (39)
4e	0	0	4	0.04316 (42)	0.04711 (40)	0.04316 (38)	0.04316 (36)
4f	0	0	5	0.05439 (49)	0.04316 (34)	0.04316 (45)	0.04339 (52)
4g	0	2	1	0.04219 (55)	0.04316 (47)	0.04316 (43)	0.04316 (42)
4h	0	2	2	0.04316 (46)	0.04316 (49)	0.04219 (45)	0.04316 (51)
4i	0	2	3	0.05439 (3)	0.04316 (21)	0.04929 (2)	0.04147 (33)
4j	6	0	0	0.04316 (5)	0.05876 (8)	0.04130 (7)	0.05548 (7)
4k	6	1	1	0.04316 (53)	0.04316 (35)	0.04929 (24)	0.04316 (41)
4l	6	1	2	0.04316 (30)	0.04147 (28)	0.04929 (22)	0.04316 (27)
4m	6	1	3	0.04147 (42)	0.04316 (34)	0.04316 (37)	0.04316 (39)
4n	6	0	4	0.04316 (38)	0.04147 (48)	0.04316 (34)	0.04316 (16)
4o	6	0	5	0.04316 (58)	0.04316 (49)	0.04147 (49)	0.04316 (35)
4p	6	2	1	0.04316 (42)	0.04339 (53)	0.04316 (46)	0.04147 (57)
4q	6	2	2	0.04316 (48)	0.04316 (55)	0.04316 (41)	0.04316 (40)
4r	6	2	3	0.05439 (4)	0.04503 (21)	0.04316 (29)	0.04147 (30)

The first number is the RMS error between calculated and experimental $\log(k')$ values for the 22 compounds while the integer in parenthesis is the generation when the best result first appears. The other adjustable parameters are fixed as follows (see text for details): $\mu = 1000$ (except the last column, where it is 3000), P_{mut} linearly decreases from 0.5 to 0.01 with each generation, $P_{chg} = 0.15, f_5 = 0.7$, and UNIQ = 0.

When UNIQ is set to 1, exploration should be promoted since mating is forced each time two parents with identical genetic vectors are mated. The results in Table 6 produce the following observations.

1. Though the results in Tables 5 and 6 are comparable (run 5j with SEED = 900 417 229.0 again finds the globally optimum set of descriptors), there are several specific cases where the final solutions are slightly worse. This suggests that multiple copies of good solutions may be necessary to find these solutions and this is easily done if identical genetic vectors are mated without mutation.
2. There is a slight improvement in the results when the population size is increased to 3000 for SEED = 371 779.0 in that the same result is obtained nine times, a better results is obtained eight times, and the result is worse only once.
3. Run 5f ($\mu = 3000$) is very interesting in that the 0.04219 solution is found, but the final population of solutions finds the descriptors used in the first 0.04316 to be present in 2993 members. This suggests that the simulation has become stuck with one or more solutions in the 0.04147 region and a large number of solutions at 0.04316. This means

that a *niching* of solutions has occurred. By continuing the search it is possible that this 0.04147 solution would be found and eventually dominate the parent population, though this would not be possible if only one member of the population has an RMS error of 0.04219 and UNIQ = 1 will slow the rate of schema formation in this region.

One final point concerning the results in Tables 5 and 6 is that when the dominant member mates with all parents (PSEL = 0), the worst results are obtained when $\mu =$ 3000, independent of how the initial population is formed (DIV = 0 or 6), while this is the only method that finds the globally optimum set when SEED = 900 417 229.0. Therefore, the quality of a mating operator not only depends upon how the initial population is formed, but may also depend on the population size.

In order to gain insight into the general progress of various runs, some information on schema formation for the $\mu = 3000$ runs shown in the last columns of Tables 5 and 6 are listed in Table 7. The exact definition of a schema is the presence of a particular descriptor in *all* members of the parent population, but for this work the formation of a schema starts when a particular descriptor is present in more than one-half (1500) of the parents. For each run (value of DIV, FIT, PSEL and UNIQ), four different fields are listed in Table 7; *First*, *Final*, *Right*, and *Conv*.

Table 6
Lowest *cost* obtained after 60 generations of a GA using a uniform crossover for 30 generations and a 1-point crossover mating operator for another 30 generations as a function of the method of generating the initial population (DIV), form of the *fitness* (FIT), parent selection strategy (PSEL), and seed to the random number generator (SEED)

Run	DIV	FIT	PSEL	SEED			
				371 779.0	43 297 187.0	900 417 229.0	371 779.0 ($\mu = 3000$)
5a	0	0	0	0.05439 (3)	0.05521 (7)	0.04929 (6)	0.05439 (6)
5b	0	1	1	0.04316 (46)	0.04316 (41)	0.04316 (43)	0.04316 (41)
5c	0	1	2	0.04316 (28)	0.04680 (24)	0.04751 (38)	0.04316 (33)
5d	0	1	3	0.04316 (39)	0.04721 (37)	0.04316 (38)	0.04147 (41)
4e	0	0	4	0.04680 (29)	0.04316 (44)	0.04316 (38)	0.04316 (37)
5f	0	0	5	0.04316 (57)	0.04316 (40)	0.04316 (38)	0.04219 (41)
5g	0	2	1	0.04316 (52)	0.04147 (56)	0.04316 (43)	0.04316 (46)
5h	0	2	2	0.04316 (52)	0.04316 (46)	0.04316 (52)	0.04316 (49)
5i	0	2	3	0.05439 (3)	0.04316 (24)	0.04929 (2)	0.04147 (33)
5j	6	0	0	0.04316 (5)	0.04316 (19)	0.04130 (6)	0.05580 (3)
5k	6	1	1	0.04680 (23)	0.04316 (35)	0.04316 (42)	0.04316 (40)
5l	6	1	2	0.04316 (35)	0.04147 (34)	0.04929 (25)	0.04316 (27)
5m	6	1	3	0.05164 (29)	0.04316 (35)	0.04316 (39)	0.04316 (38)
5n	6	0	4	0.04316 (57)	0.04316 (39)	0.04316 (33)	0.04316 (29)
5o	6	0	5	0.04316 (52)	0.04316 (39)	0.04147 (47)	0.04316 (41)
5p	6	2	1	0.04513 (60)	0.04680 (35)	0.04316 (44)	0.04316 (51)
5q	6	2	2	0.04316 (46)	0.04658 (39)	0.04316 (51)	0.04316 (43)
5r	6	2	3	0.05439 (4)	0.04951 (21)	0.04316 (29)	0.04316 (29)

The first number is the RMS error between calculated and experimental log(k') values for the 22 compounds while the integer in parenthesis is the generation when the best result first appears. The other adjustable parameters are fixed as follows (see text for details): $\mu = 1000$ (except the last column, where it is 3000), P_{mut} linearly decreases from 0.5 to 0.01 with each generation, $P_{chg} = 0.15, f_5 = 0.7$, and UNIQ = 1.

Table 7
Schema formation for the $\mu = 3000$ runs shown in Tables 5 and 6 (last column for UNIQ = 0 and UNIQ = 1, respectively)

Run	UNIQ = 0				UNIQ = 1			
	First[a]	Final[b]	Right[c]	Conv[d]	First[a]	Final[b]	Right[c]	Conv[d]
4a/5a	2	5	5	10	2	5	5	31
4b/5b	13	5	5	59	13	5	5	53
4c/5c	11	5	5	33	11	5	5	–
4d/5d	15	4	4	–	15	4	4	–
4e/5e	12	5	5	43	12	5	5	–
4f/5f	15	5	0	–	15	5	0	–
4g/5g	16	5	5	52	16	5	5	53
4h/5h	19	5	5	59	19	5	5	57
4i/5i	11	4	4	–	11	4	4	–
4j/5j	3	5	5	12	3	5	5	27
4k/5k	14	5	5	53	14	5	5	51
4l/5l	10	5	5	33	10	5	5	–
4m/5m	17	4	4	–	17	4	4	–
4n/5n	13	5	5	39	13	5	5	–
4o/5o	14	5	5	47	14	5	5	51
4p/5p	14	5	4	–	14	5	5	60
4q/5q	14	5	5	52	14	5	5	52
4r/5r	12	4	4	–	12	4	4	–

[a] 'First' means the generation number when the first schema started to appear, which means that the same descriptor was used in more than 1500 of the best 3000 solutions from the combined parent and offspring populations.

[b] 'Final' means the number of schema (using the definition above) present after the 60th generation.

[c] 'Right' means the number of schema in (b) that represented descriptors used in the best set of descriptors found after 60 generations.

[d] 'Conv' means the generation number when the schema contained the five descriptors found in the best solution and this schema was present in all 3000 of the best solutions from the combined parent and offspring populations (i.e. the simulation had converged).

First represents the generation number when the first schema starts to form. For example, the first runs examined (4a/5a) states that by the end of the 2nd generation the best 3000 sets of descriptors taken from the parent and offspring populations contain the same descriptor in more than 1500 of them. *Final* represents the number of schemas that are present in more than 1500 of the best 3000 sets at the end of the 60th generation. Table 7 shows that for 4a and 5a five descriptors are present in at least 1500 of the best sets of descriptors. *Right* simply shows the number of these descriptors that are present in the best solution found in the run. Therefore, for 4a and 5a, the five schemas present at the end of the 60th generation are the five that are present in the reported (though inferior) set of descriptors found in the run. Finally, *Conv* is concerned with the question of whether or not the calculation has converged. A run is considered converged if all 3000 best members of the combined parent and offspring populations represent the best solution found in this run. For 4a, the 3000 members selected at the end of the 10th generation are all identical to the best solution found in the entire run, while in 5a this solution did not take over the entire parent population until the end of the 31st generation. In both cases, the parent population contained 3000 copies of this solution for the rest of the run. The results in Tables 5–7 yield the following observations.

1. The *First* column in Table 7 show that the generation when the first schema starts to form is unaffected by the value of UNIQ. Conversely, the rate at which the best solution is found (i.e. the generation number when it first appears) is affected by the value of UNIQ, but there is no indication that one value of this parameter is better than the other. In addition, the data in Tables 5–7 generally support the idea that run converges more slowly when UNIQ = 1 as opposed to UNIQ = 0, with some exceptions.
2. As expected, if the fitness values are not scaled (PSEL = 1) when a roulette wheel is used to select parents for mating convergence is generally promoted, though this convergence is rather slow.
3. The fastest convergence occurs when the fitness is the inverse of the cost (FIT = 1), it is scaled by fixing the minimum and maximum fitness values (PSEL = 2), and UNIQ = 0, independent of whether the initial population is random (DIV = 0) or has an internal Hamming distance of six (DIV = 6). Conversely, if mutation is forced any time the parents are identical (UNIQ = 1) and all other parameters are the same, the run never converges.
4. If the fitness is a linear function of the cost (FIT = 2) and this scaling is used, the run converges independent of the values of DIV and UNIQ, though again this convergence is slow.
5. If the fitness is scaled based upon the maximum and average values of the population (PSEL = 3), the run does not converge for any values of DIV, FIT and UNIQ. In addition, Tables 5 and 6 show that very good results are obtained in all eight PSEL = 3 runs.
6. If a rank-based fitness is used in the roulette wheel selection of the parents (PSEL = 4), the run converges fairly rapidly if UNIQ = 0 and does not converge if UNIQ = 1. Conversely, if the selection is based on the rank of the solution (PSEL = 5), the runs converge if the initial population has an internal Hamming distance of six (DIV = 6) for both values of UNIQ, while the runs do not if the initial population is purely random.

The results of these tests allow the following general observations to be made concerning the use of a GA for this particular problem. First, the final results are generally dependent upon the initial population. This means that each specific algorithm should be run multiple times with different seeds to the random number generator. Changing this seed often produces effects that are as large as the way the initial population is generated and the specifics of the parent selection and mating operators.

Secondly, the results in Table 3 show that for this problem, and probably for Feature Selection problems in general, the GA should always include a mutation operator with a non-zero probability of being used. Otherwise, it is very likely that the search will become trapped at a sub-optimal solution.

The results shown in Table 4 suggest that multipoint crossover operators perform better than a 1-point crossover on average. Though solutions shown in Table 1 are found more often when a uniform or mixed crossover operator is used, the 2-point crossover finds the second-best solution more often.

A final troubling result is that when the dominant solution mates with all other solutions the convergence is fast, the results are generally very poor, and this is the only parent selection procedure that is able to locate the globally optimum set of descriptors in this deceptive search.

In addition to these conclusions obtained from the results presented in Tables 2–7, two additional conclusions can be made if specifics of the runs are examined. The first is that great care must be taken if a Meta-GA is used, or a Parallel GA where a good-performing mating operator also gets transferred from one population to another. As stated in Chapter 1, the minimization of the cost (maximization of the fitness) in the early stages of the search does not necessarily correlate with the quality of the final result (Wolpert and Macready, 1995). One of probably many examples of this is to compare the 1-point crossover and uniform crossover results for the runs labeled 3e in Table 4 (the entries in the first and third columns). A plot of the RMS error of the best solution as a function of the generation is shown in Fig. 2 for these searches. Both of these runs use a randomly generated initial population with the same seed to the random number generator. Therefore the two initial populations are identical. They use the same form for the fitness function, the same-scaled parent selection procedure, and they have the same mutation probabilities at each generation.

Except for a brief crossing between generations 12 and 16, the 1-point crossover GA outperforms the algorithm that uses the uniform crossover for the first 28 generations. Therefore, if a Meta- or Parallel GA allowed each individual GA to run for 10 or 20 generations, the algorithm would determine that the 1-point crossover method does the

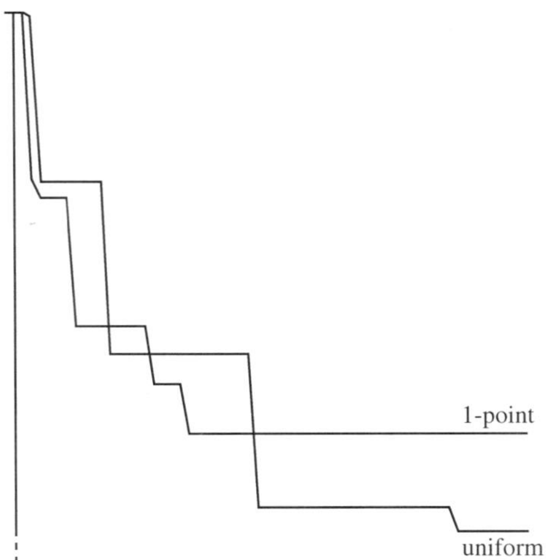

Fig. 2. RMS error of the best set of descriptors as a function of the generation number using 1-point and uniform crossovers in Run 3e.

best job and would therefore choose this procedure to use for the remainder of the simulation. Fig. 2 shows that in the 29th generation, the uniform crossover GA finds a better solution and continues to improve, while the 1-point crossover GA does not. Therefore, the quality of the final result does not necessarily correspond to the initial decrease in the cost (increase in the fitness) of the best solution in the early generations.

Another interesting aspect of these results is that, on rare occasions, the concept of a converged calculation does not hold. In Chapter 1 various convergence metrics are presented which basically use the range of fitness values in a parent population, either between the best and worst solution or between the best and average solution. An extreme case of this would be when all members in a parent population are identical. Here the ranges are zero and all offspring would also have to be identical *in the absence of any mutation*. Since the mutation probability is never zero in most runs, the above extreme condition does not guarantee convergence.

A case in point is the $\mu = 3000$ run labeled 4p in Tables 5 and 7. By the 45th generation, the simulation has converged to a solution that uses the descriptors DELGNIA, SIGA3, SUMSIGMA, SIGMANEW, and PIP18 and has an RMS error of 0.04651. This means that at the start of the 46th generation, the parent population contains 3000 copies of this solution. By this generation the mutation probability had decreased to 0.126, which means that there is a 12.6% chance that any offspring could be mutated. There was also a 1-in-3000 chance that both selected parents are the same entry and this would also force a mutation (since UNIQ = 0). This is enough of a probability that in this 46th generation a new solution is found that used the descriptors DELRHONA2, DELGNIA, SUMSIGMA, SIGMANEW, and PIP18 and it has an RMS error of only 0.04219. By the start of the 57th generations, these five descriptors are present in 1985, 2998, 3000, 3000, and 3000 of the parents, respectively, but because of the presence of two parents that do not use DELGNIA and/or a small but finite mutation probability (0.0349 plus 1-in-3000), a new solution that uses the descriptors DELRHONA2, SIDELGN, SUMSIGMA, SIGMANEW, and PIP18 is found and it has an RMS error of 0.04147. By the start of the 58th generation, the parent population contains one copy of this new solution and 2999 copies of DELRHONA2, DELGNIA, SUMSIGMA, SIGMANEW, and PIP18, and at the end of the 60th generation, the best 3000 solutions still contain 2977 copies of this suboptimal solution. This is why Table 7 shows that the final population contains five schemas, but only four of them are correct and the simulation has not converged. Therefore, a non-zero mutation probability is necessary to find a good solution (Table 3) but it also means that one cannot say that a solution is converged even if all members of the parent population are the same.

4. Structure of the NN

As stated in the Introduction to this chapter, the next two sections describe the use of a basic multilayer feedforward backpropagation NN (Wasserman, 1989) to determine the $\log(k')$ values for these 22 compounds (Breneman and Rehm, 1997) using the same five descriptors that produce the smallest RMS error in a least-squares linear fit (DELRHONA8, SIKA2, SIGA3, PIP17, and PIP19) (Luke, 1999). A schematic of this network is shown in Fig. 3.

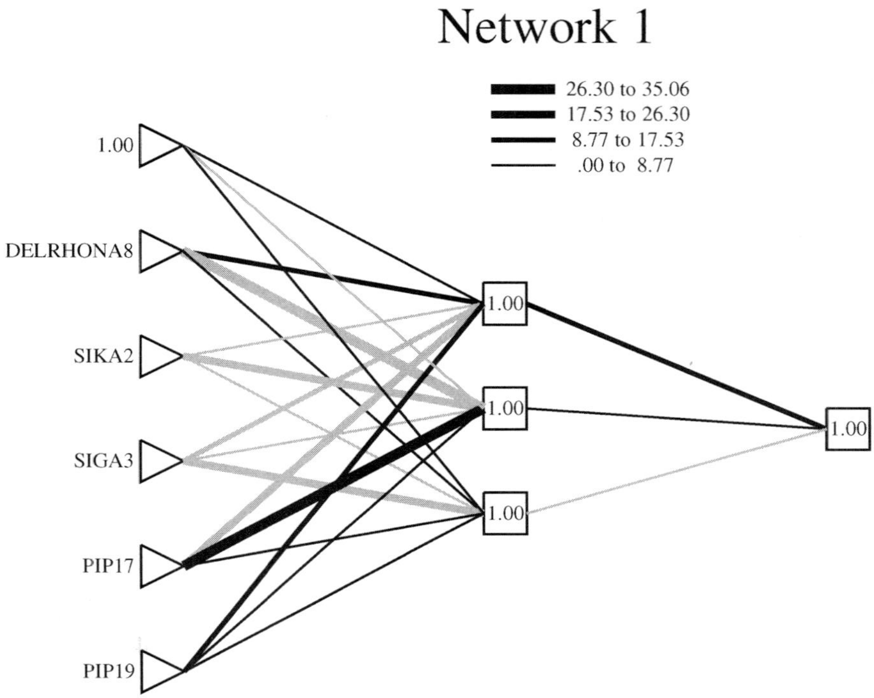

Fig. 3. Schematic of the trained NN labeled **1** in Table 8. Positive weights are black and negative weights are gray. The width of the line corresponds to the magnitude of the weight.

This network contains an input node for each descriptor and, in this general example, an input bias. This bias is simply a constant value of 1.0 for all compounds. Each input node is connected to all processing nodes in the hidden layer and each of these nodes is connected to the single processing node in the output layer. Each of these connections represents a weight that is applied to the quantity being transmitted.

To obtain a response ($\log(k')$ value) for the jth compound, the values of these five descriptors are fed into the network ($V_{i,j}$, $i = 1, 5$). The net input received by the mth node in the hidden layer, $\mathrm{NET}_{m,h}$ is simply the weighted sum of all the input values.

$$\mathrm{NET}_{m,h} = \sum_{i=0,5} (W_{i,m} V_{i,j})$$

$W_{i,m}$ is the weight of the connection between input node i and hidden node m, and V_0 is 1.0 if a bias is used and 0.0 if it is not. This net input is used to generate an output from the mth node, $\mathrm{OUT}_{m,h}$, by using a sigmoid function (which is also called a *transfer function*, an *activation function*, or a *squashing function*).

$$\mathrm{OUT}_{m,h} = f(\mathrm{NET}_{m,h}) = 1/(1 + \mathrm{e}_{m,h}^{-\kappa}\mathrm{NET}_{m,h})$$

The numbers inside of the squares representing each processing node in Fig. 3 represent the value of κ for that node. Therefore, this example network uses a constant value of 1.0

for all nodes, which is the common practice. The response of this network for the jth compound, R_j, is then determined from the following expressions

$$\mathrm{NET}_{\mathrm{out}} = \sum_m (W_{m,o}\mathrm{OUT}_{m,h})$$

$$R_j = 1/(1 + \mathrm{e}_{\mathrm{out}}^{-\kappa}\mathrm{NET}_{\mathrm{out}})$$

The goal is to adjust the weights ($W_{i,m}$ and $W_{m,o}$) such that the error between the target value, T_{j}, and the response is minimized. The procedure used here to solve this is called backpropagation and was independently developed by three different groups (Werbos, 1974; Parker, 1982; Rumelhart et al., 1986). Backpropagation uses a gradient descent method to minimize the error-squared divided by two. The change in the weights for the connections leading to the output layer is given by

$$W'_{m,o} = W_{m,o} + \Delta W_{m,o}$$

where

$$\Delta W_{m,o} = \eta\delta_{m,o}\mathrm{OUT}_{m,h}$$

$$\delta_{m,o} = \kappa R_j(1 - R_j)(T_j - R_j)$$

In this last equation, $\delta_{m,o}$ is just the negative of the derivative of the error-squared divided by two with respect to the response $[T_j - R_j]$ times the derivative of the response with respect to $\mathrm{NET}_{\mathrm{out}}$ $[\kappa R_j(1 - R_j)]$. Since the derivative of $\mathrm{NET}_{\mathrm{out}}$ with respect to $W_{m,o}$ is just $\mathrm{OUT}_{m,h}$, $\Delta W_{m,o}$ is just a constant (η) times the derivative of the error-squared divided by two with respect to $W_{m,o}$. This constant η is known as the *learning rate* and is set to a value that is usually less than or equal to 1.0.

Since the output from each node of the hidden layer has no target value, a different value has to be used for $\delta_{i,m}$. Since the network treated here only has a single node in the output layer, the derivative of the error-squared divided by two is replaced by the weight leading to the output layer $[W_{m,o}]$ times $\delta_{m,o}$. Therefore,

$$\delta_{i,m} = \kappa\,\mathrm{OUT}_{m,h}(1 - \mathrm{OUT}_{m,h})(W_{m,o}\delta_{m,o})$$

$$\Delta W_{i,m} = \eta\delta_{i,m}V_{i,j}$$

and

$$W'_{i,m} = W_{i,m} + \Delta W_{i,m}$$

In practice, the data is divided into a training set and a test set. The network is constructed and the initial values of the weights are set to small, non-zero values. Since the output of the sigmoid function is a positive number in the range [0.0,1.0], some of the weights will turn out to be negative. Therefore, small initial values are used so that the network is not initially biased with large positive and negative weights. The first member of the training set is fed into the network and its response value is calculated. Since all log(k') values in this problem are in the range (0.0,1.0), no transformation of the response from the output node needs to be made. The error in the response is then used to adjust the weights leading to the output node and then the weights between the input and hidden nodes. This same training member can be used several more times to minimize its error. At that point, the next member of the training set is used and the process continues until all members of the training set have been used. This completes a single training cycle. The training process continues with a second cycle, and the network is continually trained until either there is no further improvement in the quality of the network, or a given accuracy is achieved.

With this background, it should be clear why a NN should produce better approximations to the $\log(k')$ values than a GA that feeds the selected descriptors into a least-squares fitting routine. If five descriptors are used, a least-squares fit has six adjustable parameters that will be set to minimize one-half of the sum of the errors squared. Conversely, the network shown in Fig. 3 has 21 adjustable weights. This extra flexibility in the model allows a better fit to the data. If an extra node were added to the hidden layer, the number of adjustable parameters would increase to 28. Since there are only 22 compounds in this problem, this would produce an *over-determined* system. As stated in Chapter 7, the *number of adjustable weights should never be larger than the number of training samples* since this would produce a situation where many combinations of weights could exist that exactly fit the data.

As the number of adjustable parameters increases, it is possible that oscillations in the fitting function can occur. For example, a fit of a collection of N points in two dimensions (x_i, y_i) can be done by approximating y as a polynomial in x. As the number of terms in this polynomial increases, the resulting polynomial will come closer to the N points, but will generally begin to oscillate between these points. Therefore, such a fitting function will do a better job of reproducing the training samples, but will be less useful as an interpolating function that predicts the value of new samples. The presence or absence of oscillations will be examined in the tests performed here.

In particular, all 22 compounds will be used to train 15 different networks that differ in initial conditions, learning procedure, or structure of the network. Each network will then be used to calculate a $\log(k')$ value for 88 test compounds. This test set is constructed from the training set using the equations

$$P_{i,j} = (1 - 2R_{i,j})P_{\max}$$

$$V'_{i,j} = (1 + P_{i,j})V_{i,j}$$

In the first equation, $P_{\max}$ is a maximum allowed fractional change in a descriptor value and $R_{i,j}$ is a random number in [0,1]. Therefore each $P_{i,j}$ will be a randomly created value in the range $[-P_{\max}, P_{\max}]$. This value is added to one and multiplied by the descriptor value present in the training set. The first set of 22 test compounds have $P_{\max} = 0.02$. This means that each descriptor value in a test compound is within 2% of the value in the corresponding training compound. The next three groups of 22 test compounds have $P_{\max}$ values of 0.05, 0.10, and 0.15, which means that the descriptor values differ from the training values by at most 5, 10, and 15%, respectively.

5. Results for the NN

For a given number of descriptors, the parameters that must be set to build a particular network are the number of nodes in the hidden layer (NHID), the learning rate (η), the maximum absolute value of an initial weight MAX$(W)_i$, the seed to the random number generator (SEED), the constant in the exponent of the sigmoid functions (κ), the number of times each member of the training set will be used to adjust the weights in each cycle (NPER), and whether or not an input bias is used (BIAS).

The 15 NNs that will be examined are described in Table 8. The first NN, labeled **1**, contains a bias input node, three nodes in the hidden layer and has a learning rate of 0.7. Each of the 21 weights in the network is randomly assigned an initial value between −0.05 and 0.05, and 371 779.0 is the seed to the random number generator that is used to randomly pick these weights. All computed nodes in the hidden and output layers use a sigmoid function with an exponential constant (κ) of 1.0. Each training point is examined 10 times per cycle to optimize the weights, and a bias input is used. The second network (**2**) is identical to the first except that the input bias is removed. Therefore, this network has only 18 adjustable weights.

Networks **3** and **4** have the same initial values as network **0**; the only difference is that the learning rate (η) is changed to 0.5 and 0.9, respectively. Networks **5** and **6** simply increase the magnitude of the initial weights. Since the same seed is used as in network **1**, the initial weights are simply multiplied by three and six, respectively. Networks **7** and **8** are the same as network **1** with the exception that the seed to the random number changes. Therefore, these networks have different initial weighs, but these weights are still constrained to lie between −0.05 and 0.05.

Table 8
Parameters used to control the 15 NNs

NN	NINP	NHID	η	MAX(W)$_i$	SEED	κ[a]	NPER	BIAS	RMS	E_{max}
1	5	3	0.7	0.05	371 779.0	C (1.0)	10	Y	0.009	0.023 (18)
2	5	3	0.7	0.05	371 779.0	C (1.0)	10	N	0.016	0.034 (18)
3	5	3	0.5	0.05	371 779.0	C (1.0)	10	Y	0.025	0.082 (9)
4	5	3	0.9	0.05	371 779.0	C (1.0)	10	Y	0.011	0.025 (7)
5	5	3	0.7	0.15	371 779.0	C (1.0)	10	Y	0.011	0.029 (7)
6	5	3	0.7	0.30	371 779.0	C (1.0)	10	Y	0.018	0.046 (9)
7	5	3	0.7	0.05	43 297 187.0	C (1.0)	10	Y	0.025	0.058 (7)
8	5	3	0.7	0.05	900 417 229.0	C (1.0)	10	Y	0.013	0.032 (4)
9	5	3	0.7	0.05	371 779.0	V (0.75-1.25)	10	Y	0.028	0.068 (6)
10	5	3	0.7	0.05	371 779.0	V (0.50-2.00)	10	Y	0.011	0.033 (18)
11	5	3	0.7	0.05	371 779.0	C (1.0)	7	Y	0.023	0.072 (9)
12	5	3	0.7	0.05	371 779.0	C (1.0)	15	Y	0.019	0.051 (9)
13	4	3	0.7	0.05	371 779.0	C (1.0)	10	N	0.062	0.134 (15)
14	4	3	0.7	0.05	371 779.0	C (1.0)	10	Y	0.065	0.172 (15)
15	4	4	0.7	0.05	371 779.0	C (1.0)	10	N	0.013	0.034 (9)

They are the number of nodes in the input layer (NINP), the number of nodes in the hidden layer (NHID), the learning rate (η), the maximum magnitude of an initial weight (MAX(W)$_i$), the seed to the random number generator (SEED), the constant in the exponent of the sigmoid function (κ), the number of update cycles per training structure (NPER), and whether or not a bias is used (BIAS). The last two columns give the RMS error in the final network (RMS) and the maximum error in the 22 training compounds (E_{max}) after 400 000 training cycles. The compound number with the maximum error is listed in parenthesis.

[a] C(1.0) means that κ was constant at 1.0 for all processing nodes and V(A–B) means that it was randomly assigned a value between A and B.

Networks **9** and **10** are significantly different than network **1** in that each node in the hidden and output layers has a different constant (κ) in the exponent of the sigmoid function. In **9** κ is a randomly chosen value between 0.75 and 1.25, while in **10** it is between 0.5 and 2.0. Networks **11** and **12** start with the same initial values as network **0**, they simply change the number of times each training compound is tested in each cycle to 7 and 15, respectively.

It must be emphasized that each of these networks use the same set of descriptors that are found to be optimal for the least squares linear function (DELRHONA8, SIKA2, SIGA3, PIP17, and PIP19; see Table 1), even though these descriptors may not yield the optimal network. The actual form of this linear function is as follows.

$$\log(k') = 0.3448 - 0.7457(\text{DELRHONA8}) - 0.2859(\text{SIKA2}) + 1.523(\text{SIGA3}) - 1.070(\text{PIP17}) + 0.6502(\text{PIP19})$$

This linear relationship produces an RMS error of 0.413 over the 22 compounds and has a maximum error of 0.094 at Compound 21. To test the effects of reducing the number of nodes/weights, the last three networks only use four descriptors. For these tests, SIKA2 is excluded since it has the smallest coefficient in the equation above and all descriptors have maximum values of 1.0. Networks **13** and **14** are the same as Networks **2** and **1**, respectively, except that the number of input notes (NINP) is reduced to four. Therefore Network **13**, which does not have an input bias, has a total of 15 adjustable weights, while Network **14**, which has an input bias, has a total of 18 adjustable weights. Finally, Network **15** does not have an input bias, but increases the number of hidden nodes to four. This network then has 20 adjustable weights.

Each network trains on the set of 22 compounds for a total of 400 000 cycles. The second to last column in Table 8 lists the RMS error of the final network over these compounds, and the last column gives the maximum error. It is interesting to note that all changes to network **1** resulted in an increase in both the RMS and maximum error. In addition, the two networks with four input nodes and three hidden nodes (**13** and **14**) have significantly larger RMS and maximum errors, while moving a node from the input to the hidden layer (Network **15**) did not seriously degrade its ability to reproduce the $\log(k')$ values for the 22 training compounds. Though it is generally agreed that introducing a bias increases the accuracy of the resulting network, the results in Table 8 are contradictory. Network **2** is Network **1** without the bias and it has a larger RMS and maximum error. Conversely, Network **13** does not contain the bias and its RMS and maximum error is slightly less than Network **14**, which does have the input bias.

Each trained network is then used to predict the $\log(k')$ for the 88 test compounds. The change in values for each test compound relative to the $\log(k')$ value in the corresponding training compound is graphically displayed in Fig. 4. The change is displayed above the base line if the test compound has a larger predicted $\log(k')$ and below the base line if it is smaller. The largest absolute change in the 22 test compounds for each value of $P_{\max}$ is listed in Table 9. As a point of reference, the 22 compounds have $\log(k')$ values that range from 0.952 to 0.053 and they are ordered in decreasing $\log(k')$. Since the $\log(k')$ values are in (0.0,1.0), the result from the output node does not need to be scaled.

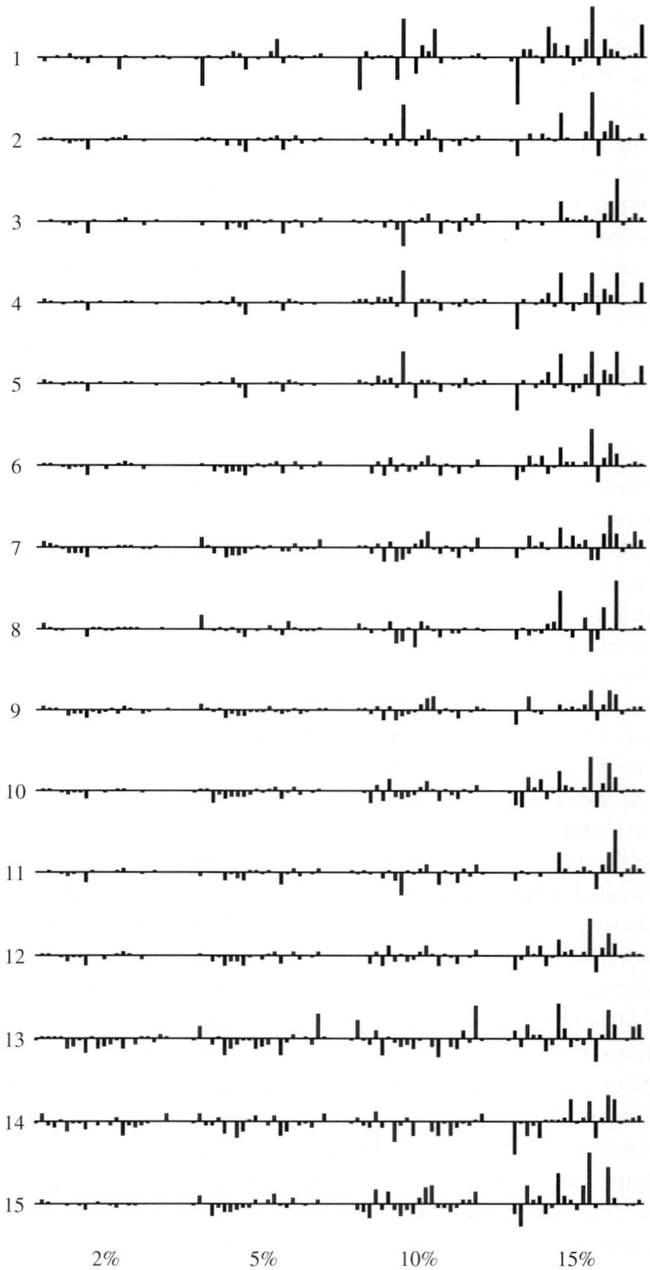

Fig. 4. Change in log(k') values in the 88 test compounds for the 15 trained NNs.

Table 9
Maximum absolute error in the log(k') values of the test compounds for each of the 15 NNs described in Table 8. The number in parenthesis is the compound in the group with this maximum error

NN	P_{max}			
	0.02	0.05	0.10	0.15
1	0.145 (14)	0.349 (2)	0.469 (9)	0.634 (14)
2	0.122 (9)	0.154 (9)	0.431 (9)	0.575 (14)
3	0.140 (9)	0.138 (15)	0.296 (9)	0.523 (14)
4	0.100 (9)	0.160 (9)	0.405 (9)	0.375 (18)
5	0.100 (9)	0.163 (9)	0.397 (9)	0.411 (18)
6	0.118 (9)	0.121 (9)	0.127 (6)	0.450 (14)
7	0.120 (9)	0.134 (2)	0.208 (13)	0.406 (17)
8	0.112 (9)	0.167 (2)	0.233 (11)	0.604 (18)
9	0.090 (9)	0.112 (6)	0.166 (14)	0.246 (14)
10	0.093 (9)	0.144 (4)	0.149 (4)	0.427 (14)
11	0.133 (9)	0.140 (15)	0.285 (9)	0.528 (18)
12	0.122 (9)	0.126 (9)	0.134 (6)	0.449 (14)
13	0.167 (9)	0.299 (21)	0.404 (21)	0.426 (9)
14	0.172 (15)	0.202 (8)	0.255 (8)	0.410 (2)
15	0.071 (9)	0.149 (4)	0.220 (14)	0.620 (14)

As expected, the overall change (Fig. 4) and maximum change in the log(k') values (Table 9) for the 88 testing samples increases as P_{max} increases. Network **1**, which has the smallest RMS and maximum errors over the training set, is found to have the largest changes in the log(k') values for all values of P_{max} relative to all networks with five input nodes. Conversely, Network **9** has the largest RMS error and the second largest maximum error over the training set, but the smallest maximum change in the log(k') values for $P_{max} = 0.02$, 0.05 and 0.15, and a relatively small maximum change in the log(k') values for $P_{max} = 0.10$. Similarly, Network **14** has the same change in the log(k') values for $P_{max} = 0.02$ as was found in the training set, and the maximum change in these values grows relatively slowly as P_{max} increases. If two samples have values of the five descriptors that are within 15% of each other, most of the networks tested produce significantly different log(k') values.

Several conclusions can be drawn from these results. The first is that there seems to be no way to a priori set the initial values of a network such that the resulting network is improved. Adding a bias increased the fit to the training data in the case of five input nodes, but with four input nodes it had a slight negative effect. Similarly, both increasing and decreasing the learning rate relative to Network **1** (Networks **3** and **4**) produced final networks that increased the RMS and maximum error in the training set.

In addition, even minor changes in the initial conditions can cause the network to train to qualitatively different networks. This is graphically shown for networks **1**, **7** and **8** in Figs. 3, 5 and 6, respectively. These three networks differ only in the seed to the random number generator and therefore start with different values of the weights (thought they all lie between -0.05 and 0.05). In these figures, a weight is black if it is positive and gray if it is negative, and the width of the connection corresponds to the magnitude of the weight.

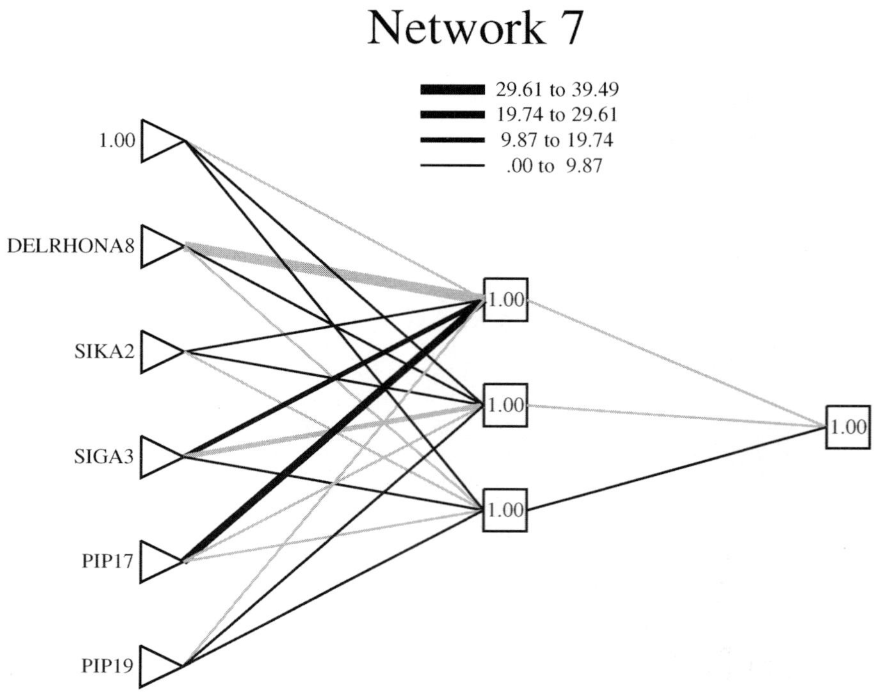

Fig. 5. Schematic of the trained NN labeled **7** in Table 8. Positive weights are black and negative weights are gray. The width of the line corresponds to the magnitude of the weight.

By simply examining the sign and magnitude of the weights leading to the output layer, it is clear that these networks are qualitatively different.

The results shown in Table 9 also suggest that, as in many over-determined systems, the response value can oscillate and change rapidly with small changes in the input values. One could argue that there are some peculiar properties about Compounds 9 and 14 since they show the largest change in $\log(k')$ most of the time. This may be partly true, but it could also be that their $\log(k')$ values (0.494 and 0.364) lie in the region where the response R varies the most with NET_{out}. Since the derivative of R with respect to NET_{out} is $[\kappa R(1-R)]$, it has a maximum value when $R = 0.5$. On the other hand, this maximum slope is only $\kappa/4$, so when $\kappa = 1$ NET_{out} must change by at least 0.4 for R to change by 0.1. Therefore, small changes in the initial values result in significantly larger changes in NET_{out}. Conversely, Compounds 8 and 10 have the maximum change in their $\log(k')$ values twice and never, respectively, even though their values are also in the range where the response has a large slope (0.516 and 0.468, respectively). The maximum change is found in Compounds 2 and 21 four and two times, respectively, even though their target values are far away from this region (0.700 and 0.134, respectively). Since 12 of the 22 compounds have the maximum change one or more times in Table 9, this suggests that oscillations as well as the sensitivity of the final response has an important effect.

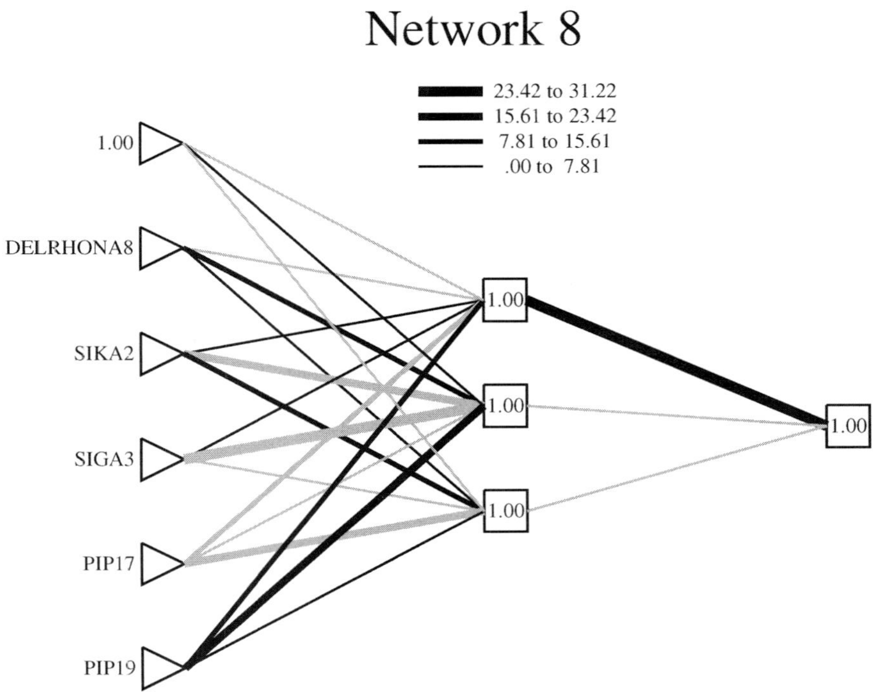

Fig. 6. Schematic of the trained NN labeled **8** in Table 8. Positive weights are black and negative weights are gray. The width of the line corresponds to the magnitude of the weight.

Though these oscillations may be caused by the large number of adjustable weights, they can also be amplified by *over fitting*. In other words, as the network continues to minimize the errors in the $\log(k')$ values of the training points, it may introduce oscillations between these points. As discussed in Chapter 7, it is advisable to monitor the fit to a set of test points, which are not used in training, during the optimization of the weights. The errors in these test points should decrease and then either maintain a relatively constant error, or begin to increase in the error. If the latter occurs, the network is trying to over fit the training points. Since there are only 22 compounds in this examination, removing a given number of them from training is not possible. Since changing the descriptor values by a maximum of 2% ($P_{max} = 0.02$) should not significantly affect the target response ($\log(k')$ values), these 22 test cases will be used to monitor the training process. Every 1000 training cycles the RMS error of the $\log(k')$ values for these compounds will be determined, and the network that produces the smallest RMS error will be used to test the compounds with $P_{max} = 0.05$, 0.10, and 0.15.

The results of these runs are shown in Table 10. The first column gives the number of the network and the second lists the training cycle number that produces the smallest RMS error in the $P_{max} = 0.02$ compounds. The next two columns list the RMS error for this network over the training and $P_{max} = 0.02$ compounds, respectively. The last five columns

Table 10
Number of training cycles (NCYC) when the $P_{max} = 0.02$ test set reached a minimum RMS error as well as the RMS and maximum absolute error in the log(k') values of the training and test compounds for each of the 15 NNs described in Table 8

NN	NCYC	RMS error		Train	P_{max}			
		Train	$P_{max} = 0.02$		0.02	0.05	0.10	0.15
1	60 000	0.017	0.033	0.035 (6)	0.114 (9)	0.188 (9)	0.495 (9)	0.630 (14)
2	14 000	0.017	0.031	0.042 (3)	0.099 (9)	0.115 (9)	0.164 (9)	0.374 (14)
3	30 000	0.020	0.034	0.053 (6)	0.101 (9)	0.166 (15)	0.213 (9)	0.460 (18)
4	473 000	0.011	0.028	0.025 (7)	0.100 (9)	0.160 (9)	0.400 (9)	0.378 (18)
5	192 000	0.011	0.028	0.026 (7)	0.099 (9)	0.160 (9)	0.403 (9)	0.388 (18)
6	52 000	0.021	0.035	0.048 (18)	0.114 (9)	0.118 (9)	0.129 (15)	0.513 (14)
7	28 000	0.025	0.037	0.047 (6)	0.090 (9)	0.091 (6)	0.147 (13)	0.291 (18)
8	32 000	0.017	0.033	0.045 (3)	0.104 (9)	0.132 (2)	0.119 (8)	0.419 (9)
9	280 000	0.028	0.040	0.068 (6)	0.090 (9)	0.112 (6)	0.158 (13)	0.252 (14)
10	740 000	0.011	0.029	0.034 (18)	0.092 (9)	0.142 (4)	0.148 (4)	0.421 (14)
11	31 000	0.020	0.034	0.053 (6)	0.100 (9)	0.165 (15)	0.208 (9)	0.461 (18)
12	53 000	0.022	0.036	0.049 (18)	0.118 (9)	0.120 (9)	0.127 (15)	0.506 (14)
13	32 000	0.061	0.068	0.146 (15)	0.140 (9)	0.206 (15)	0.258 (21)	0.382 (9)
14	605 000	0.064	0.068	0.171 (15)	0.171 (15)	0.212 (8)	0.260 (8)	0.413 (2)
15	313 000	0.013	0.026	0.034 (9)	0.072 (9)	0.149 (4)	0.221 (14)	0.619 (14)

The number in parenthesis is the compound in the group with this maximum error.

give the maximum absolute error in the training and $P_{max} = 0.02$, 0.05, 0.10, and 0.15 compounds, respectively, along with the number of the compound that has this maximum error.

The first thing to notice in Table 10 is that three of the networks required more than the 400 000 training cycles used above, while several of the networks found the best network in far fewer cycles. Comparison of the second-to-last column in Table 8 with the third column of Table 10 shows that in most cases there is a very small difference between the final RMS errors in the training compounds for each of the 15 networks. Of notable exception is Network **1** where stopping after 60 000 training cycles almost doubled the RMS error. In addition, Network **3** (and some of the others) has a smaller RM error over the training points at 30 000 training cycles than at 400 000 training cycles. This says that by continuing the training further, the fit actually became worse. This can be caused by the network oscillating around a minimum and not being at the minimum after the last training cycle.

As expected, the maximum change in the $P_{max} = 0.02$ compounds listed in Table 10 is at least as small as the comparable maximum changes listed in Table 9. Since these compounds are used to determine when the training of the network stops, this cannot be considered a test of the interpolative quality of these networks. The maximum change in the log(k') values for the $P_{max} = 0.05$ compounds in Table 10 is again the same size or smaller than the comparable changes in Table 9 with the exception of Networks **11** and **14**. Here the change is slightly larger in Table 10. As P_{max} increases, the difference in maximum changes between the networks displayed in Tables 9 and 10 decreases, and $P_{max} = 0.15$ the change in the log(k') values is still quite large. The exception to this is

Network **7**. In Table 9 only Network **9** has a maximum change of less than 0.3 when $P_{max} = 0.15$, while in Table 10 both Networks **7** and **9** have a maximum change of less than 0.3. This suggests that using a set of testing points to determine when to stop the network training yields better values in the vicinity of these points, but the effects of oscillations are still seen as the distance from these training and testing points increases, and often be relatively large for reasonably small separations.

6. Conclusions

This chapter presented "real world" applications of GAs and multilayer feedforward backpropagation NNs to two aspects of a chemometric problem. Both of these computational methodologies require several parameters to be set before an actual algorithm can be generated, and the tests have shown that different initial parameters can result in quantitatively and qualitatively different results.

The GA needs to efficiently search a very large feature space, but the results in Tables 5 and 6 (for $\mu = 3000$) show that many combinations of parameter values can yield a good solution. On the other hand, a deceptive problem like the one treated here makes finding the global optimum particularly hard. Finding the global optimum is only achieved by incorporating flexibility into the search and, with a fair degree of luck, choosing particular initial conditions. The only other conclusion is that good results are more often achieved if a mutation operator can be used with some non-zero probability in each generation. This is both good and bad news in that it allows the search to escape from a sub-optimal solution. This means that convergence cannot be guaranteed, even if all members of the parent population are identical.

The tests performed on the NNs reinforce the points made in Chapter 7. Since the network that did the best job on fitting the training points did the worst job of producing stable results for small changes in the input values for all networks that use five input nodes, simply fitting the training data is not sufficient to justify a good predictive model. As stated in Chapter 7, the optimum situation is one where there are a sufficient number of samples so that three independent sets can be constructed. The training set will be used to construct the networks with either different starting conditions and/or different topologies and a testing set should then be used to determine which of these networks does the best job of interpolating the response for new samples. Finally, a validation set should be used to determine the quality of this network. This also means that the training and testing points should sufficiently span the space of the samples so that none of the validation points is too far from a member of either set.

Therefore, the use of NNs should be limited to cases where the number of samples greatly exceeds the number of weights in the network. In this under-determined system, problems can arise during training since the backpropagation algorithm may not yield the optimum set of weights. This is why small changes in the initial conditions produce qualitatively different networks when trained on the same data. Different updating algorithms can be used (Rumelhart et al., 1986; Sejnowski and Rosenberg, 1987) or a Boltzmann (Geman and Geman, 1984) or Cauchy machine (Szu and Hartley, 1987; Wasserman, 1988) can be tried. As stated above, a GA can also be used to choose an optimal set of initial parameters, but it is important to remember that training even a simple NN like the ones examined here can take on the order of a minute of computer time. This is much longer than a nearly instantaneous least-squares fit, and a combined GA/NN strategy may be prohibitively time consuming.

Acknowledgements

This work was funded in whole or in part with federal funds from the US National Cancer Institute, National Institutes of Health, under contract no. NO1-CO-12400. The content of this publication does not necessarily reflect the views or policies of the Department of Health and Human Services, nor does the mention of trade names, commercial products or organizations imply endorsement by the US Government.

References

Agrafiotis, D.K., Cedeno, W., Lobanov, V.S., 2002. On the use of neural network ensembles in QSAR and QSPR. J. Chem. Inf. Comput. Sci. 42, 903–911.

Aires-de-Sousa, J., 2002. JATOON: Java tools for neural networks. Chemom. Intell. Lab. Syst. 61, 167–173.

Breneman, C.M., Rehm, M., 1997. QSPR analysis of HPLC column capacity factors for a set of high-energy materials using electronic van der Waals surface property descriptors computed by transferable atom equivalent method. J. Comput. Chem. 18, 182–197.

Brown, E.C., Sumichrast, R.T., 2003. Impact of the replacement heuristic in a grouping genetic algorithm. Computers Oper. Res. 30, 1575–1593.

Cronin, M.T.D., Aptula, A.O., Dearden, J.C., Duffy, J.C., Netzeva, T.I., Patel, H., Rowe, P.H., Schultz, T.W., Worth, A.P., Voutzoulidis, K., Schuurmann, G., 2002. Structure-based classification of antibacterial activity. J. Chem. Inf. Comput. Sci. 42, 869–878.

Geman, S., Geman, D., 1984. Stochastic relaxation, Gibbs distributions and Bayesian restoration of images. IEEE Trans. Pattern Anal. Mach. Intell. 6, 721–741.

Goldberg, D.E., 1989. Genetic Algorithms in Search, Optimization, and Machine Learning, Addison-Wesley, Berkeley, CA.

Hasancebi, O., Erbatur, F., 2000. Evaluation of crossover techniques in genetic algorithm based optimum structural design. Computers Struct. 78, 435–448.

Hollstein, R.B., 1971. Artificial genetic adaptation in computer control systems. Doctoral dissertation, University of Michigan, Dissertation Abstracts International, 32, 1510B (University Microfilms No. 71-23,773).

Izrailev, S., Agrafiotis, D., 2001. A novel method for building regression tree models for QSAR based on artificial ant colony systems. J. Chem. Inf. Comput. Sci. 41, 176–180.

Izrailev, S., Agrafiotis, D., 2002. Variable selection for QSAR by artificial ant colony systems. SAR QSAR Environ. Res. 3-4, 417–423.

Livingstone, D.J., 2000. The characterization of chemical structures using molecular properties. A Survey. J. Chem. Inf. Comput. Sci. 40, 195–209.

Lucic, B., Trinajstic, N., 1999. Multivariate regression outperforms several robust architectures of neural networks in QSAR modeling. J. Chem. Inf. Comput. Sci. 39, 121–132.

Luke, B.T., 1999. Comparison of three different QSAR/QSPR generation techniques. J. Mol. Struct. (Theochem) 468, 13–20.

Luke, B.T., 2000. Comparison of different data set screening methods for use in QSAR/QSPR generation studies. J. Mol. Struct. (Theochem) 507, 229–238.

Luke, B.T., 2003. Fuzzy structure-activity relationships. SAR QSAR Environ. Res. 14, 41–57.

Parker, D.B., 1982. Learning Logic, Invention Report, S81-64, File 1, Office of Technology Licensing, Stanford University.

Rogers, D., Hopfinger, A.J., 1994. Application of genetic function approximation to quantitative structure–activity relationships and quantitative structure–property relationships. J. Chem. Inf. Comput. Sci. 34, 854–866.

Rumelhart, D.E., Hilton, G.E., Williams, R.J., 1986. Learning internal representations by error propagation, parallel distributed processing, vol. 1, MIT Press, Cambridge, MA, pp. 318–362.

Sejnowski, T.J., Rosenberg, C.R., 1987. Parallel networks that learn to pronounce English text. Complex Syst. 1, 145–168.

Shen, M., LeTiran, A., Xiao, Y., Golbraikh, A., Kohn, H., Tropsha, A., 2002. Quantitative structure–activity relationship analysis of functionalized amino acid anticonvulsant agents using *k* nearest neighbor and simulated annealing PLS methods. J. Med. Chem. 45, 2811–2823.

So, S., Karplus, M., 1996. Evolutionary optimization in quantitative structure–activity relationship: an application of genetic neural networks. J. Med. Chem. 39, 1521–1530.

Szu, H., Hartley, R., 1987. Fast simulated annealing. Phys. Lett. 1222, 157–162.

Tounge, B.A., Pfahler, L.B., Reynolds, C.H., 2002. Chemical information based scaling of molecular descriptors: a universal chemical scale for library design. J. Chem. Inf. Comput. Sci. 42, 879–884.

Tropsha, A., Zheng, W., 2001. Identification of the descriptor pharmacophores using variable selection QSAR: application to database mining. Curr. Pharm. Des. 7, 599–612.

Wasserman, P.D., 1988. Combined Backpropagation/Cauchy Machine, Neural Networks: Abstracts of the First INNS Meeting, Boston, vol. 1. Pergamon Press, Elmsford, NY, p. 556.

Wasserman, P.D., 1989. Neural Computing: Theory and practice, Van Nostrand Reinhold, New York.

Werbos, P.J., (1974). Beyond regression: new tools for prediction and analysis in the behavioral sciences, Masters Thesis, Harvard University.

Whitley, D., 1989. The GENITOR algorithm and selection pressure: why rank-based allocation of reproductive trials is best. In: Schaffer, J.D., (Ed.), Proceedings of the Third International Conference on Genetic Algorithms, Morgan Kaufmann, Los Altos, CA, pp. 116–121.

Wolberg, W.H., Street, W.N., Mangasarian, O.L., 1994. Machine learning techniques to diagnose breast cancer from fine-needle aspirates. Cancer Lett. 77, 163–171.

Wolpert, D.H., Macready, W.G., 1995. No Free Lunch Theorems for Search, Technical Report TR-92-02-010, The Santa Fe Institute, Santa Fe, NM, USA.

SUBJECT INDEX

A

ACO *see* ant colony optimisation
Activation function 363
Adaptation 5
Adaptive least squares analysis (ALS) 289
Adaptive mutation 40
Adaptive parallel genetic algorithms 41–42
Adaptivity 79
All possible sets (APS) 345, 350
Alleles 5
ALS *see* adaptive least squares analysis
Amino acids 287
Amperometry 258
Analytical chemistry 56
Analytical neural networks 80–81
ANNs *see* artificial neural networks
Ant colony optimization (ACO) 4, 45
Anti-coagulant drugs 243
APS *see* all possible sets
Aquatic toxicity 239–242
Artefacts 285
Artificial neural networks (ANNs) 199–226
- applications 285–303, 308, 323
- molecular structures 231–253
- toxicity data 332–338
- voltammetric data 261–269

Aspen Technology 75
Assortative mating 31
Assortive mating 31
Asymptotic Q^2 rule 152
Autocorrelation 178–179, 301, 324

B

Back-propagation algorithm 329, 364, 373
Bias 202
Biased uniform crossover 17
Binary coding 9, 112
Binary logistic regression 344
Binding sites 307
Binning technique 283
Biochemistry 282
Biological samples 309
Biological systems 3
Biomass estimation 74
Biomedical NMR spectroscopy 309–315
Biosensors 257
Bloch equations 285, 304
Blood glucose biosensors 257
Boltzmann machine 373
Boolean descriptors 323
Bottle-neck mapping 224
Branch-and-bound algorithm 118

C

CADD *see* computer-aided drug design
Capillary probes 281
Cascade learning 288
Cataclysmic mutation 36
Cauchy machine 373
Chemical shift 283, 294, 307
Chemometric problems 333–374
Chromosomes 5, 10, 12, 141, 173–175, 269, 303, 307
Chronoamperometric techniques 258
Classification 236, 286–290, 308
Classification problems 204, 223
Clinical chemistry 282
Clinical microbiology 314–315
Clustering 236
Clustering algorithm 64
Coding schemes 8
Competitive learning 235
Computer-aided drug design (CADD) 109
Conformational searches 114, 119–124
Conformations 109

Consensus diagnosis 312
Constitutional descriptors 232
Convergence 61, 304, 330, 353
Cost function 346, 352, 361
Counter propagation neural networks 213–215, 233–237
Coupling 60
Cross-validation 178
Crossover 111, 141, 142, 145–148, 250, 271, 303, 348
Crossover operator 17, 39, 116, 304, 354
Cryogenic probes 281
Cycle crossover (CX) 20

D
Data complexity 286
Data compression 303
Data condensation 89, 101
Data mapping 224
Data partitioning 4
Data processing 304–345
DCS *see* distributed control systems
Deceptive problems 7
Delta coding 8, 43
Deoxyribose nucleic acid (DNA) 5
Descriptors 237, 253, 344, 345, 367
DG *see* distance geometry
Diabetes 257
Dihedral angles 112, 115
Dimensionality 84, 112, 267, 286
Diploids 11, 16
Direct design variable exchange crossover 37
Directed tweak 122
Discrete recombination 304
Dissortative mating 31
Distance geometry (DG) 122
Distributed control systems (DCS) 99
DKP *see* dynamic knapsack problem
DNA *see* deoxyribose nucleic acid
Docking procedures 124
Dominance rules 11, 358
DPSV *see* dual pulse staircase voltammetry
Drug design 237–251
Dual pulse staircase voltammetry (DPSV) 259
Dynamic knapsack problem (DKP) 11

E
Ecotoxicity 324
Edge recombination crossover (ERX) 22
Effective crowding operator 35
Electroanalytical chemistry 257–278
Electrodes 258
Elite chromosome 271
Elitist strategy 27, 115
Embedding frequencies 295
Emission monitoring 74, 103
Empirical descriptors 233
Encoding rules 111
Energy minimisation 130
Entropy 144, 226
Environmental emission monitoring 74
Enzyme-binding neural network model 242
EP *see* evolutionary programming
Epochs 264, 265
Error backpropagation 199, 204–206
Error metric 344
ERX *see* edge recombination crossover
ES *see* evolution strategies
Evaluation 142
Evolution 154–145
Evolution strategies (ES) 44–45
Evolutionary programming (EP) 4, 44, 69
Excited neuron 209
Expert errors 310
Exploitation 6, 28, 36, 40, 349
Exploration 6, 28, 36, 40, 349, 356
Extrapolation 104

F
Families 47
Fast Fourier transformation (FFT) 304
FB *see* fragment-based methods
Feature reduction 309
Feature selection 4, 344
Feed forward neural networks 269–278
FFLD program 129

FFT *see* fast Fourier transformation
FINGAR 306
FIT *see* frequency to information transfer
Fitness 5, 31, 39, 173, 353, 361
Fitness function 7–28, 63, 110, 225, 303, 307, 346
Flexibility 356
Flow-probes 281
Focus points 27
Focusing 6
Fogging 292
Fragment-based (FB) methods 125
Frame-shift operator 24
Frequency to information transfer (FIT) 287
FT-IR spectra 182
Functions 93
Fuzzy clustering 344
Fuzzy solutions 226

G
Gasoline 290
GASP *see* genetic algorithm similarity program
Gene-based coding 8
Generalisation 71, 79
Generation gap 111
Generation-apart elitism 35
Generational algorithms 26–27
Generations 349
Genes 269
Genetic algorithm similarity program (GASP) 123
Genetic alphabets 112
Genetic diversity 35
Genetic drift 31, 32
Genetic local search algorithm 25
Genetic neural network 300
Genetic programming (GP) 90–99, 303
Genetic vectors 5, 8, 29–30, 141, 346, 355
Genotype 5, 110
Genotypic assortative mating 31
Gibbs sampling 43
Ginseng 294
Global kernels 89
Glucose oxidase 257
GOLD software 129
GP *see* genetic programming
Gray coding 9
Greedy heuristics 56
Growing algorithms 127

H
H-bonding acceptor ability (HBA) 329
H-bonding donor ability (HBD) 329
Hamming cliff 9
Hamming distance 30, 156
Haploids 10
Hardware sensors 76
Hardy–Weinberg law 31
HBA *see* H-bonding acceptor ability
HBD *see* H-bonding donor ability
Herbicides 288
Heterozygous 11
Hidden layers 202, 365
Hidden neurons 268, 275
Hierarchically ordered spherical description of environment (HOSE) code 296
High-throughput screening 282
Homozygous 11
HOSE *see* hierarchically ordered spherical description of environment
Hybrid genetic algorithm 25, 55–66, 115
Hybridisation 176
Hydrophobic interactions 134
Hypermutation 40
Hypersurface 114

I
In vivo NMR spectroscopy 305
Inbreeding 31, 32
Incest prevention 32, 40
Industrial data 71
Inference mechanisms 72
Inferential models 102
Inferential sensors 76, 78, 105
Initial populations 13–14, 30

Input mutation probability 147
Inputs 71, 366
Integer coding 29
Intermediate recombination 18, 304
Intermolecular variables 307
Interval crossover 19
Introns 93
Inversion 25

J

Junk code 93

K

Kalman filter 72
Kernels 86, 89
Kohonen networks 206–213, 233, 311
Kurtosis 144

L

Lamarckian genetic algorithm (LGA) 130
Lambert–Beer law 169
Lattice models 132
LDA *see* linear discriminant analysis
Learning 220–223, 235, 288, 315, 364
Least-squares fit 344, 350
Levels 212
LGA *see* Lamarckian genetic algorithm
Ligand-receptor complexes 242
Line breeding operators 33, 37, 347, 353
Linear discriminant analysis (LDA) 308, 344
Linear order crossover (LOX) 23
Linear rank-based selection 304
Linear regression 72
Local optimizers 59
Look-up table 215
LOX *see* linear order crossover

M

Machine learning 344
Magnetisation transfer 283
MAGS *see* molecular abstract graph space template
Mapping 224, 296
Masking 12, 16
Mating operators 6, 16–23, 33, 36, 348
Mating pool 14, 36
Maturation operators 25–26, 34
Maximisation 7
MC *see* Monte Carlo
MD *see* molecular dynamics
Medical diagnosis 56
Memetic algorithm 25
MEP *see* molecular electrostatic potential
Messy genetic algorithms 42
Meta-genetic algorithms 42, 361
Metabolic disease 282
Metabonomics 310
Micro-probes 281
Mixed crossover 37, 354
MLR *see* multiple linear regression
MobyDigs software 141–166
Model building 73
Model distance 155–158
Modeling 225
Molecular abstract graph space template (MAGS) 296
Molecular descriptors 324
Molecular dynamics (MD) 125
Molecular electrostatic potential (MEP) 247
Molecular modelling 109–135
Molecular representation 231
Molecular structures 231–253
see also quantitative structure–activity relationship
Monte Carlo (MC) methods 125
Multi-hybrids 56
Multidimensional vectors 231
Multiple linear regression (MLR) techniques 216
Multiple solutions 57
Multipoint crossover operators 360
Multivariate calibration 170, 262
Multivariate data 226
Mutation operators 6, 10, 23–25, 34, 348, 354, 360

Mutations 111, 120, 141, 142, 145–148, 173, 250, 272, 303

N

Nano-probes 281
k -nearest neighbour 344
Negative assortive mating 31
Network design 221
Network optimisation 264–265
Network performance 264
Network weights 265
Neural networks (NNs) 3, 62, 69, 344
  chemometric problems 333–374
  electroanalytical chemistry 257–278
  NMR spectroscopy 281–315
Neurons 200, 202–203, 223, 233, 264
NHID *see* number of nodes in the hidden layer
Niches 115
Niching 32, 40, 129, 358
NIR spectra 182
NMR *see* nuclear magnetic resonance
NMR-derived distance constraints 120
NNs *see* neural networks
Node-based coding 8, 16, 24
NOE *see* Nuclear Overhauser Effect
Noise 78, 226, 304
Non-generational algorithms 26
Non-linear information 274
Non-linear problems 199
Non-linear relationships 70
Normalisation 344
Nuclear magnetic resonance (NMR) spectroscopy 124, 281–315
Nuclear Overhauser Effect (NOE) 120, 283, 306
Number of nodes in the hidden layer (NHID) 365, 367

O

Objects 223
Octane number (ON) 301
OCX *see* orthogonal crossover operator
Offspring 26–27, 34–35, 110, 146, 348
Offspring generation function (OGF) 33
ON *see* octane number
One-dimensional descriptors 232
One-point crossover 120
Optimisation 25–26, 66, 177, 225, 249, 264, 269, 276–278, 303, 315
Ordered crossover (OX) 20
Orthogonal crossover operator (OCX) 33
Oscillations 365, 370
Outliers 236
Output mutation probability 147
Outputs 71, 222, 235
Over-determined systems 344, 365
Over-training 221
Overfitting 78, 178, 311, 371
OX *see* ordered crossover
Oxidation 258

P

PAD *see* pulsed amperometric detection
Parallel genetic algorithms 41, 361
Parallel hybrids 56
Parent selection 145
Parents 142, 146, 270, 346
Partial least-square (PLS) 3, 169–195, 257, 323, 338
Partially matched crossover (PMX) 19
Particle swarm optimisation (PSO) 4, 46
Pattern recognition 286, 308
Pavilion Technologies 75
PBX *see* position based crossover
PCA *see* principal component analysis
PCs *see* principal components
Peptidomimetics 244
Pesticides 323–337
Pharmacophores 119, 121–124
Phenotype 5, 110
Phenotypic assortative mating 31
Pheromones 45
Pioneer search 26
Planes 212
PLS *see* partial least-square
PMX *see* partially matched crossover
Polyploids 11

Population analysis 164
Population diversity 129
Population junction 155
Population size 111
Population splitting 116
Population transfer 155
Populations 141, 143, 172, 349
Position based crossover (PBX) 21
Positive assortive mating 31
Potential energy 114
Pre-selection 26
Prediction 236
Pretreatment 181–182
Principal component analysis (PCA) 118, 262, 308, 309
Principal components (PCs) 264, 273–274
Prior knowledge 79
Process quality 70
Production composition 75
Proportional selection 14
Protein folding 114, 131–132
Protein-ligand docking 112, 119, 124–131
PSO *see* particle swarm optimisation
Pulsed amperometric detection (PAD) 259

Q

QPLS *see* quadratic partial least squares
QSAR *see* quantitative structure-activity relationship
QSPR *see* quantitative structure-property relationship
Quadratic partial least squares (QPLS) 63
Quality criteria 118
Quantitative structure-activity relationship (QSAR) 225, 290, 323–337
Quantitative structure-property relationship (QSPR) 290, 343
Quantum chemical descriptors 232
QUIK rule 152

R

Radial basis function (RBF) 86, 216–220
Random selection 145
Random variables 144–145
Randomisation tests 176
RBF *see* radial basis function
Real-number encoding 116
Real-valued coding 24
Receptor-dependent descriptors 233
Recombination operators 304
Reduction 258
Regression tree model 344
Relaxation time 283
Relocation 25
Replicates 174
Reproduction 173, 303, 304
Resolution 283
Restart mechanisms 40
RMSD *see* root mean squared distances
RMSE *see* root-mean-square-error
Robustness 79
Root mean squared distances (RMSD) 125
Root-mean-square-error (RMSE) 220, 264
Roulette wheel selection 14, 120, 145, 270, 304, 353
RQK fitness functions 151–154

S

SA *see* simulated annealing
Sampling errors 310
SANN *see* stacked analytical neural networks
SAR *see* structure–activity relationships
Scaling 181–182
Scoring functions 126
Search space 352
Seed 350, 360, 366
Selection pressure 145, 148–150
Selection schemes 14–16, 31–33, 36
Self-diagnosis 80
Self-organized map (SOM) 206, 210–213, 233
Sensitivity analysis 84
Sequential variable selection 170
Serial hybrids 56
Sexual reproduction 5, 36
SGA *see* simple genetic algorithm
Sharing technique 115

SICS *see* substituent induced chemical shift
Signal recognition 304
Simple genetic algorithm (SGA) 4, 48
Simulated annealing (SA) method 120, 286
Single-point crossover 304
Soft sensors 69–106
SOM *see* self-organized map
Spectral data sets 169–195
Spectral quantification 304
Spectroscopy 261
SPX *see* swap path crossover
Squashing function 363
SRM *see* structural minimisation principle
Stacked analytical neural networks (SANN) 80–82
STATISTICA 329
Structural minimisation principle (SRM) 85
Structure determination 305–307
Structure prediction 308
Structure–activity relationships (SAR) 282
Substituent induced chemical shift (SICS) 294
Support vector machines (SVM) 85–90
Survival of the fittest 250
SVM *see* support vector machines
Swap path crossover (SPX) 21
Switching 24
Systematic search 122

T

Tabu list 143–144
Tabu search 43
TAE *see* transferable atom equivalent
Targets 222
Terminals 93
Termination metrics 27–28
Testing 364
TGA *see* transformation-based genetic algorithm
Three-dimensional descriptors 232
Three-parent crossover 116
Thrombin 243
Top-map *see* self-organized map
Tournament selection 145, 304
Toxicity 238
Toxicology 237–251, 323
Training 220, 262, 364
Training epochs 265, 272
Transfer function 201, 363
Transferable atom equivalent (TAE) method 343
Transformation 37
Transformation-based genetic algorithm (TGA) 37
Translation 114
Translocation operator 24
Trial and error 265
Truncation selection 304
Tumour classification 311–314
Twins 174
Two point crossover 304
Two-dimensional descriptors 232

U

U-matrix 211
Unbiased uniform crossover 146
Underfitting 78
Uniform crossover 17, 116
Uniqueness operator 26
United-atom model 132
Univariate variable selection 170
Updating algorithms 373

V

Validation 164–165, 222, 263, 373
Van der Waals interactions 134
Variable frequency analysis 165–166
Variable selection 170–171, 249
Variables 182
Voltammetric data 257–278
Voltammetry 259

W

Wavelength selection 169–195
Weights 212

X

X-ray crystallography 124